Ramanujan's Notebooks

Part II

Springer

New York
Berlin
Heidelberg
Barcelona
Hong Kong
London
Milan
Paris
Singapore
Tokyo

Bust of Ramanujan by Paul Granlund

Bruce C. Berndt

Ramanujan's Notebooks

Part II

Springer

Bruce C. Berndt
Department of Mathematics
University of Illinois
Urbana, IL 61801
USA

The following journals have published earlier versions of
chapters in this book:
L'Enseignement Mathématique **26** (1980), 1–65.
Journal of the Indian Mathematical Society **46** (1982), 31–76.
Bulletin London Mathematical Society **15** (1983), 273–320.
Expositiones Mathematicae **2** (1984), 289–347.
Journal für die reine und angewandte Mathematik **361** (1985), 118–134.
Rocky Mountain Journal of Mathematics **15** (1985), 235–310.
Acta Arithmetica **47** (1986), 123–142.

Mathematics Subject Classification (1980): 11-03, 11P99

Library of Congress Cataloging-in-Publication Data
(Revised for volume 2)
Ramanujan Aiyangar, Srinivasa, 1887–1920.
Ramanujan's notebooks.
Includes bibliographies and index.
1. Mathematics.   I. Berndt, Bruce C., 1939–
II. Title.
QA3.R33   1985   510   84-20201

Printed on acid-free paper.

Typeset by Asco Trade Typesetting Ltd., Hong Kong.
Printed and bound by Braun-Brumfield, Inc., Ann Arbor, MI.
Printed in the United States of America

9 8 7 6 5 4 3 2 (Corrected second printing, 1999)

ISBN 0-387-96794-X Springer-Verlag New York Berlin Heidelberg
ISBN 3-540-96794-X Springer-Verlag Berlin Heidelberg New York   SPIN 10692231

Dedicated to my mother
Helen
and the memory of my father
Harvey

The relation between Hardy and Ramanujan is unparalleled in scientific history. Each had enormous respect for the abilities of the other. Mrs. Ramanujan told the author in 1984 of her husband's deep admiration for Hardy. Although Ramanujan returned from England with a terminal illness, he never regretted accepting Hardy's invitation to visit Cambridge.

Photograph reprinted with permission from *Collected Papers of G. H. Hardy*, Vol. 1, Oxford University Press, Oxford, 1969.

Preface

During the years 1903–1914, Ramanujan recorded many of his mathematical discoveries in notebooks without providing proofs. Although many of his results were already in the literature, more were not. Almost a decade after Ramanujan's death in 1920, G. N. Watson and B. M. Wilson began to edit his notebooks, but never completed the task. A photostat edition, with no editing, was published by the Tata Institute of Fundamental Research in Bombay in 1957.

This book is the second of four volumes devoted to the editing of Ramanujan's notebooks. Part I, published in 1985, contains an account of Chapters 1–9 in the second notebook as well as a description of Ramanujan's quarterly reports. In this volume, we examine Chapters 10–15 in Ramanujan's second notebook. If a result is known, we provide references in the literature where proofs may be found; if a result is not known, we attempt to prove it. Except in a few instances when Ramanujan's intent is not clear, we have been able to establish each result in these six chapters.

Chapters 10–15 are among the most interesting chapters in the notebooks. Not only are the results fascinating, but for the most part, Ramanujan's methods remain a mystery. Much work still needs to be done. We hope readers will strive to discover Ramanujan's thoughts and further develop his beautiful ideas.

Urbana, Illinois   *Bruce C. Berndt*
November 1987

Contents

Preface   ix

Introduction   1

CHAPTER 10
Hypergeometric Series, I   7

CHAPTER 11
Hypergeometric Series, II   48

CHAPTER 12
Continued Fractions   103

CHAPTER 13
Integrals and Asymptotic Expansions   185

CHAPTER 14
Infinite Series   240

CHAPTER 15
Asymptotic Expansions and Modular Forms   300

References   339

Index   355

Introduction

We take up something—we know it is finite; but as soon as we begin to analyze it, it leads us beyond our reason, and we never find an end to all its qualities, its possibilities, its powers, its relations. It has become infinite.

Vivekananda

In a certain sense, mathematics has been advanced most by those who are distinguished more for intuition than for rigorous methods of proof.

Felix Klein

For now we see through a glass, darkly; but then face to face: now I know in part; but then shall I know even as also I am known.

First Corinthians 13:12

The quoted passages of Vivekananda, Klein, and St. Paul each point to a certain facet of Ramanujan's work. First, on June 1–5, 1987, the centenary of Ramanujan's birth was celebrated at the University of Illinois with a series of 28 expository lectures and several contributed papers that traced Ramanujan's influence to many areas of current research; see the conference Proceedings edited by Andrews et al. [1]. Thus, Ramanujan's mathematics continues to generate a vast amount of research in a variety of areas. Second, in the sequel, we shall see many instances where Ramanujan made profound contributions but for which he probably did not have rigorous proofs; for example, see Entry 10 of Chapter 13. Third, although St. Paul's passage is eschatological in nature, it points to the great need to learn how Ramanujan reasoned and made his discoveries. Perhaps we can prove Ramanujan's claims, but we may not know the well from which they sprung. These three aspects of Ramanujan's work will frequently be made manifest in the pages that follow.

In this book, we examine Chapters 10–15 in Ramanujan's second notebook. In many respects, these chapters contain some of Ramanujan's most fascinating and enigmatic discoveries. Our goal has been to prove each claim made by Ramanujan. With a few possible exceptions where the meaning is obscure, we either give a proof or indicate where in the literature proofs can be found. We emphasize that many (perhaps most) of our proofs are undoubtedly different from those found by Ramanujan. In particular, we have often employed the theory of functions of a complex variable, a subject with which Ramanujan had no familiarity. In no way should our proofs, or this book, be regarded as definitive. In many instances, more transparent proofs, especially those that might give insight into Ramanujan's reasoning, should be sought.

Each of Chapters 10–13 and 15 contains 12 pages, while Chapter 14 encompasses 14 pages in Ramanujan's second notebook. The number of theorems, corollaries, and examples in each chapter is listed in the following table.

Chapter	Number of Results
10	116
11	103
12	113
13	92
14	87
15	94
Total	605

In the sequel, we have employed Ramanujan's designations of corollary, example, and so on, although the appellations may not be optimal. Generally, we have adhered to Ramanujan's notation so that the reader following our account with a copy of Ramanujan's notebooks at hand will have an easier task. At times, for clarity, we have changed notation, especially in Chapter 14 where we make heavy use of complex function theory. Except for some minor alterations, especially in Chapter 15, we have also preserved Ramanujan's order of presentation.

Many of the theorems communicated by Ramanujan in his famous letters to G. H. Hardy on January 16, 1913 and February 27, 1913 may be found in Chapters 10–15. In the table below, we list these results.

Location in Collected Papers	Location in Notebooks
p. xxvi, V, (2)	Chapter 10, Section 7, Example 15
p. xxvi, V, (3)	Chapter 10, Section 7, Example 14
p. xxvi, V, (4)	Chapter 14, Section 13, Corollary (iii)
p. xxvi, V, (5)	Chapter 14, Entry 25(ii)
p. xxvi, V, (6)	Chapter 14, Entry 25(vii)
p. xxvi, VI, (3)	Chapter 11, Section 20, Example 2
p. xxvi, VII, (2)	Chapter 12, Entry 48, Corollary of Entry 48
	Chapter 13, Entry 6
p. xxvi, VII, (3)	Chapter 13, Corollary (ii) of Entry 10
p. xxvii, VII, (7)	Chapter 15, Section 2, Example (iv)
p. xxvii, IX, (1)	Chapter 12, Section 25, Corollary 1
p. xxviii, (3)	Chapter 10, Equation (31.1)
p. xxviii, (10)	Chapter 11, Entry 29(i)
p. xxix, (14)	Chapter 12, Entry 27
p. 349, V, (7)	Chapter 14, Entry 25(xi)
p. 349, V, (8)	Chapter 14, Entry 25(xii)
p. 350, VI, (4)	Chapter 13, Corollary of Entry 21
p. 350, VI, (5)	Chapter 13, Example for Corollary of Entry 21
p. 350, IX, (2)	Chapter 12, Entry 34
p. 351, last formula in first letter	Chapter 10, Entry 29(b)
p. 352, penultimate paragraph of 3	Chapter 15, Section 2, Example (ii)
p. 352, last paragraph of 3	Chapter 15, Section 2, Example (iv)
p. 353, (16)	Chapter 12, Corollary to Entry 34

Several of Ramanujan's published papers and problems posed in the *Journal of the Indian Mathematical Society* have their origins in the notebooks. In most cases, only a small portion of the published paper is actually found in the notebooks. We list below those papers with their geneses in Chapters 10–15, together with the respective locations in the notebooks.

Paper	Location in Notebooks
On question 330 of Prof. Sanjana	Chapter 10, Section 13
Modular equations and approximations to π	Chapter 14, Section 8, Example
On the product $\displaystyle\prod_{n=0}^{n=\infty}\left[1+\left(\frac{x}{a+nd}\right)^3\right]$	Chapter 13, Section 27
Some definite integrals	Chapter 13, Entries 14, 15, 16(iii), Corollary of Entry 19, Entry 21, Corollary of Entry 21, Entry 22
	Chapter 14, Section 6
Some definite integrals connected with Gauss's sums	Chapter 14, Entry 22(ii)

Paper	Location in Notebooks
On certain arithmetical functions	Chapter 15, Sections 9, 10, 12, 13, and 14
On certain trigonometrical sums and their applications in the theory of numbers	Chapter 14, Entry 13
Asymptotic formulae in combinatory analysis (with G. H. Hardy)	Chapter 15, Section 2, Example (iv)
A class of definite integrals	Chapter 13, Sections 23–25
Question 289	Chapter 12, Section 4, Examples (i), (ii)
Question 294	Chapter 12, Section 48
	Chapter 13, Entry 6
Question 295	Chapter 13, Section 21, Example
Question 358	Chapter 14, Corollary of Entry 14
Question 387	Chapter 14, Section 8, Example
Question 769	Chapter 13, Entry 11(iii)

We now provide brief summaries for each of Chapters 10–15. More detailed descriptions may be found at the beginning of each chapter.

Of all the topics examined by Ramanujan in his notebooks, only modular equations received more attention than hypergeometric series. Chapter 10 is the first of two chapters devoted almost entirely to the latter subject. In 1923, Hardy [1], [7, pp. 505–516] published a brief overview of the corresponding chapter in the first notebook. Ramanujan rediscovered most of the classical formulas in the subject, including those attached to the names of Gauss, Kummer, Dougall, Dixon, and Saalschütz. Ramanujan possessed the uncanny ability for finding the most important examples of theorems, and Chapter 10 contains many elegant examples of infinite series summed in closed form. Ramanujan was the first to discover identities for certain partial sums of hypergeometric series, and these may be found in the latter parts of Chapter 10. Ramanujan continues his study of hypergeometric series in Chapter 11. Two topics dominate the chapter. The first concerns products of hypergeometric series, and most of these results are original with Ramanujan. Second, Ramanujan offers several beautiful asymptotic formulas for hypergeometric functions. By far, the most interesting is Corollary 2 in Section 24. Quadratic transformations of hypergeometric series are also featured in Chapter 11.

Chapter 12 is almost entirely devoted to continued fractions and is one of the most fascinating chapters in the notebooks. Ramanujan's published papers contain only one continued fraction! However, Ramanujan submitted some continued fractions as problems to the *Journal of the Indian Mathematical Society*, and his letters to Hardy contain some of his most beautiful theorems on continued fractions. Nonetheless, the great majority of the results in Chapter 12 are new. Perhaps the most exquisite theorems are the many continued fraction expansions for products and quotients of gamma functions. We have no idea how Ramanujan discovered these formulas. Especially awe inspiring is Entry 40 involving several parameters.

Equally astonishing is Chapter 13. In the first 11 sections, one finds several beautiful, deep asymptotic expansions for integrals and series. Entries 7 and 10 are perhaps highlights. Ramanujan left us no clues of how he discovered these fascinating theorems. Are these results prototypes for further yet undiscovered theorems? Although we have given proofs, we do not have a firm understanding of how these wonderful theorems fit with the rest of mathematics.

Those readers who are fascinated by elegant series evaluations and identities will take great pleasure in reading Chapter 14. Here, one can find several series identities that have a symmetry that one often associates with certain applications of the Poisson summation formula, which, however, does not seem to be applicable in most cases here. Several closed form evaluations of series involving hyperbolic functions are given. Some of the results in this chapter can be established by employing partial fraction decompositions. We have utilized two additional primary tools: contour integration and some theorems of the author on transformations of Eisenstein series. Since neither of these techniques was in Ramanujan's arsenal, we do not know how Ramanujan discovered most of the results in Chapter 14.

Chapter 15 is the most unorganized of all the chapters in the second notebook. The first seven sections are primarily devoted to interesting asymptotic expansions of several series. Entry 8 offers an elegant transformation formula for a modified theta-function.

In the sequel, equation numbers refer to equations in the same chapter, unless another chapter is indicated. Unless otherwise stated, page numbers refer to pages in Ramanujan's second notebook [15] in the pagination of the Tata Institute. Part I refers to the author's account [9] of Chapters 1–9, and Part III refers to his account [11] of Chapters 16–21.

In what follows, the principal value of the logarithm is always denoted by Log. The set of all (finite) complex numbers is denoted by $\mathscr{C}$. The residue of a function f at an isolated singularity a will be denoted by $R(a)$. (The identity of f will always be clear.)

A small portion of this book has been aided by notes left by G. N. Watson and B. M. Wilson in their efforts to edit Ramanujan's notebooks. We are grateful to the Master and Fellows of Trinity College, Cambridge, for providing a copy of these notes and for permission to use this material in this book.

We sincerely appreciate the collaboration of Robert L. Lamphere on Chapter 12 and Ronald J. Evans on Chapters 13 and 15. Because of their efforts, our accounts of these chapters are decidedly better than what we would have accomplished without their help. Most of the material in this book appeared in previously published versions of these chapters. We are grateful for the cooperation shown by each of the journals publishing our earlier accounts. A table below indicates the bibliographic data for the original publications. (Portions of Chapter 15 were published in two parts.)

Chapter	Coauthors	Publication
10		*J. Indian Math. Soc.* **46** (1982), 31–76
11		*Bull. London Math. Soc.* **15** (1983), 273–320
12	R. L. Lamphere, B. M. Wilson	*Rocky Mt. J. Math.* **15** (1985), 235–310
13	R. J. Evans	*Expos. Math.* **2** (1984), 289–347
14		*L'Enseign. Math.* **26** (1980), 1–65
15	R. J. Evans	*J. Reine Angew. Math.* **361** (1985), 118–134
15	R. J. Evans	*Acta Arith.* **47** (1986), 123–142

Although only one author is listed on the cover of this book, several mathematicians have made valuable contributions. We are very grateful to George Andrews, Richard Askey, Henri Cohen, Ronald Evans, Jerry Fields, P. Flajolet, M. L. Glasser, Mourad Ismail, Lisa Jacobsen, Robert Lamphere, David Masser, F. W. J. Olver, R. Sitaramachandrarao, and Don Zagier for the many proofs and suggestions that they have contributed. In particular, Askey, Evans, and Jacobsen have each supplied several proofs and offered many helpful comments, and we are especially indebted to them. Others, not named, have made helpful comments, and we publicly offer them our thanks as well.

The author bears the responsibility for all errors and would like to be notified of such, whether they be minor or serious.

The manuscript was typed by the three best technical typists in Champaign–Urbana—Melody Armstrong, Hilda Britt, and Dee Wrather. We thank them for the superb quality of their typing.

Lastly, we express our deep gratitude to James Vaughn and the Vaughn Foundation for the generous funding that they have given the author during summers. This book could not have been completed without the support of the Vaughn Foundation.

Hypergeometric Series, I

In 1923, Hardy published a paper [1], [7, pp. 505–516] providing an overview of the contents of Chapter 12 of the first notebook. This chapter, which corresponds to Chapter 10 of the second notebook, is concerned primarily with hypergeometric series. It should be emphasized that Hardy gave only a brief survey of Chapter 12; this chapter contains many interesting results not mentioned by Hardy, and Chapter 10 of the second notebook possesses material not found in the first. Quite remarkably, Ramanujan independently discovered a great number of the primary classical theorems in the theory of hypergeometric series. In particular, he rediscovered well-known theorems of Gauss, Kummer, Dougall, Dixon, Saalschütz, and Thomae, as well as special cases of Whipple's transformation. Unfortunately, Ramanujan left us little knowledge as to how he made his beautiful discoveries about hypergeometric series. The first notebook contains a few brief sketches of proofs, but the only sketch in the second notebook is found after Entry 8, which is Gauss's theorem. We shall present this argument of Ramanujan in the sequel.

As the reader will see, this chapter contains a wealth of beautiful evaluations of hypergeometric functions, usually at the argument $+1$ or -1. In this connection, we mention the recent work of R. Wm. Gosper, I. Gessel, and D. Stanton. By employing "splitting functions" and the computer algebra system MACSYMA, Gosper discovered many new hypergeometric function evaluations. Most of these, in the terminating cases, were ingeniously proved by Gessel and Stanton [1]. Two conjectures of Gosper were established by P. W. Karlsson [1].

Many elegant and useful binomial coefficient sums can be evaluated, usually quite simply, by employing the theorems of Gauss, Dixon, Saalschütz, Kummer, and others. See the paper by R. Roy [2] for many illustrations.

We now offer several remarks about notation. As usual, we put

$$(a)_k = \frac{\Gamma(a + k)}{\Gamma(a)},$$

where k is any complex number. The generalized hypergeometric series $_pF_q$ is defined by

$$_pF_q\left[\begin{matrix} \alpha_1, \alpha_2, \ldots, \alpha_p \\ \beta_1, \beta_2, \ldots, \beta_q \end{matrix}; x\right] = \sum_{k=0}^{\infty} \frac{(\alpha_1)_k(\alpha_2)_k \cdots (\alpha_p)_k}{(\beta_1)_k(\beta_2)_k \cdots (\beta_q)_k} \frac{x^k}{k!}, \tag{0.1}$$

where p and q are nonnegative integers and $\alpha_1, \alpha_2, \ldots, \alpha_p$ and $\beta_1, \beta_2, \ldots, \beta_q$ are complex numbers. If the number of parameters is "small," we may sometimes use the notation $_pF_q(\alpha_1, \alpha_2, \ldots, \alpha_p; \beta_1, \beta_2, \ldots, \beta_q; x)$ in place of the notation on the left side of (0.1). In this chapter, we are concerned only with the cases when $p = q + 1$. In these instances, the series defining $_pF_q$ converges when $|x| < 1$ for all choices of the parameters α_i, β_j, $1 \le i \le q + 1$, $1 \le j \le q$. However, $_{q+1}F_q$ can be continued analytically into the complex plane cut at $[1, \infty)$. If $x = 1$, the series converges for $\text{Re}(\alpha_1 + \cdots + \alpha_{q+1}) < \text{Re}(\beta_1 + \cdots + \beta_q)$; if $x = -1$, there is convergence for $\text{Re}(\alpha_1 + \cdots + \alpha_{q+1}) < \text{Re}(\beta_1 + \cdots + \beta_q) + 1$. In all the theorems and examples that follow, when $x = \pm 1$, we state the conditions for convergence, but without further comment. Moreover, as is customary, if $x = 1$, we omit the argument in the notation (0.1). It should be remarked that Ramanujan has no notation for hypergeometric series. All formulas are stated by writing out the first few terms in each series. This practice has one distinct advantage in that the elegance of formulas involving series is often more easily discerned. Frequently, a compact notation obscures the aesthetic beauty of a series relation. For brevity, we usually use a compact notation, but, at times, in particularly elegant instances, we follow Ramanujan's practice. To aid readers examining this chapter in conjunction with the second notebook, we have usually adhered to Ramanujan's notations for the parameters.

For the most part, we refer only to primary sources. For example, we give a reference to Dougall's paper wherein his famous theorem is initially proved, but we do not usually offer further references to other proofs, applications, and so on. The classical texts of Appell and Kampé de Fériet [1], Klein [1], Bailey [4], and Slater [1] contain excellent bibliographies on which it would be difficult to elaborate. In the sequel, Bailey's well-known tract [4] will be our basic reference. We also indicate which formulas have been discussed by Hardy [1] in his overview. For those readers wishing to learn more about the history of hypergeometric functions, we recommend the papers of Askey [1], Dutka [3], and Bühler [1].

In the sequel, always, $\psi(z) = \Gamma'(z)/\Gamma(z)$. Frequent use is made of the classical representation (e.g., see Luke's text [1, p. 12])

$$\psi(z + 1) = -\gamma + \sum_{k=1}^{\infty} \left(\frac{1}{k} - \frac{1}{k + z}\right), \tag{0.2}$$

where γ denotes Euler's constant. We also often employ the simple differentiation formulas

$$\frac{d}{du}(\pm u)_k \bigg|_{u=0} = \pm(k-1)!, \qquad k \geq 1. \tag{0.3}$$

Entry 1. *Suppose that at least one of the quantities $x, y, z, u,$ or $-x - y - z - u - 2n - 1$ is a positive integer. Then*

$$_7F_6\left[\begin{array}{c} n, \frac{1}{2}n + 1, -x, -y, -z, -u, x + y + z + u + 2n + 1 \\ \frac{1}{2}n, x + n + 1, y + n + 1, z + n + 1, u + n + 1, -x - y - z - u - n \end{array}\right]$$

$$= \frac{\Gamma(x+n+1)\Gamma(y+n+1)\Gamma(z+n+1)\Gamma(u+n+1)\Gamma(x+y+z+n+1)}{\Gamma(n+1)\Gamma(x+y+n+1)\Gamma(y+z+n+1)\Gamma(x+u+n+1)\Gamma(z+u+n+1)}$$

$$\times \frac{\Gamma(y+z+u+n+1)\Gamma(x+u+z+n+1)\Gamma(x+y+u+n+1)}{\Gamma(x+z+n+1)\Gamma(y+u+n+1)\Gamma(x+y+z+u+n+1)}.$$

$$\tag{1.1}$$

Ramanujan did not indicate that (1.1) holds when $-x - y - z - u - 2n - 1$ is a positive integer.

Entry 1 is originally due to Dougall [1] in 1907, which is probably less than three years before Ramanujan discovered the theorem. Hardy [1, Eq. (2.1)] has thoroughly discussed Entry 1 and gives Dougall's proof, as does Bailey [4, p. 34].

Entry 2. *If either $x, y,$ or z is a positive integer, then*

$$_3F_2\left[\begin{array}{c} -x, -y, -z \\ n + 1, -x - y - z - n \end{array}\right]$$

$$= \frac{\Gamma(n+1)\Gamma(x+y+n+1)\Gamma(y+z+n+1)\Gamma(z+x+n+1)}{\Gamma(x+n+1)\Gamma(y+n+1)\Gamma(z+n+1)\Gamma(x+y+z+n+1)}.$$

Entry 2 is known as Saalschütz's theorem [1], [2], although according to Jacobi [1], [2] and Askey [1], the result was first established by Pfaff [1] in 1797. In Hardy's paper [1], Entry 2 corresponds to Eq. (5.1) there. It should be mentioned that Hardy's formulation is incorrect. For a proof of Entry 2, see Bailey's tract [4, p. 9].

Entry 3. *If $x, y, z,$ or $-x - y - z - 2n$ is a positive integer, then*

$$_6F_5\left[\begin{array}{c} \frac{1}{2}n + 1, 1, -x, -y, -z, x + y + z + 2n \\ \frac{1}{2}n, x + n + 1, y + n + 1, z + n + 1, -x - y - z - n + 1 \end{array}\right]$$

$$= \frac{(x+n)(y+n)(z+n)(x+y+z+n)}{n(x+y+n)(y+z+n)(x+z+n)}.$$

PROOF. Set $u = -1$ in Entry 1.   □

Entry 4. *If either x, y, z, or $-x - y - z - 2n - 1$ is a positive integer, then*

$$\sum_{k=1}^{\infty} \frac{(n + 2k)(-x)_k(-y)_k(-z)_k(x + y + z + 2n + 1)_k}{k(n + k)(x + n + 1)_k(y + n + 1)_k(z + n + 1)_k(-x - y - z - n)_k}$$

$$= \psi(x + n + 1) + \psi(y + n + 1) + \psi(z + n + 1) + \psi(x + y + z + n + 1)$$

$$- \psi(n + 1) - \psi(x + y + n + 1) - \psi(y + z + n + 1) - \psi(z + x + n + 1).$$

PROOF. Logarithmically differentiate both sides of (1.1) with respect to u and then set $u = 0$. Using (0.3), we complete the proof after a little simplification.   □

Example (i). *If x is a positive integer, then*

$$1 - 3\left(\frac{x - 1}{x + 1}\right)^4 \frac{4x - 1}{4x - 3} + 5\left(\frac{(x - 1)(x - 2)}{(x + 1)(x + 2)}\right)^4 \frac{(4x - 1)(4x)}{(4x - 3)(4x - 4)} - \cdots$$

$$= \frac{\Gamma^4(x + 1)\Gamma^4(3x - 1)}{\Gamma^6(2x)\Gamma(4x - 2)}.$$

PROOF. In Entry 1, put $n = 1$, replace x by $x - 1$, and set $y = z = u = x - 1$. After some simplification, the desired equality follows.   □

Example (ii). *If x is an odd, positive integer, then*

$$\frac{(x - 1)^3(3x - 1)}{(x + 1)^3(3x - 3)} + \frac{1}{2}\left(\frac{(x - 1)(x - 3)}{(x + 1)(x + 3)}\right)^3 \frac{(3x - 1)(3x + 1)}{(3x - 3)(3x - 5)} + \cdots$$

$$= \frac{1}{2}\left\{\psi\left(\frac{3x - 1}{2}\right) + 3\psi\left(\frac{x + 1}{2}\right) - 3\psi(x) - \psi(1)\right\}.$$

PROOF. In Entry 4, put $n = 0$, replace x by $\frac{1}{2}(x - 1)$ and set $y = z = \frac{1}{2}(x - 1)$. The proposed equality now readily follows.   □

Example (iii). *If x is a positive integer, then*

$$1 + 3\left(\frac{x - 1}{x + 1}\right)^3 \frac{3x - 1}{3x - 3} + 5\left(\frac{(x - 1)(x - 2)}{(x + 1)(x + 2)}\right)^3 \frac{(3x - 1)(3x)}{(3x - 3)(3x - 4)} + \cdots$$

$$= \frac{x^3(3x - 2)}{(2x - 1)^3}.$$

PROOF. In Entry 3, set $n = 1$, replace x by $x - 1$, and let $y = z = x - 1$. The displayed equality now easily follows.   □

Example (iv). *If x is a nonnegative integer, then*

$$1 + \left(\frac{x}{1!}\right)^2 \frac{x}{3x} + \left(\frac{x(x-1)}{2!}\right)^2 \frac{x(x-1)}{3x(3x-1)} + \cdots = \frac{\Gamma^3(2x+1)}{\Gamma^3(x+1)\Gamma(3x+1)}.$$

PROOF. In Entry 2, set $n = 0$ and $x = y = z$ to achieve the desired result. □

In the notebooks (p. 118), Ramanujan has mistakenly put $\Gamma(3x + 1)$ in the numerator instead of the denominator in Example (iv).

Example (v). *If x is a positive integer, then*

$$1 + \frac{x}{1!}\frac{x-1}{x+1}\frac{x}{4x-1} + \frac{x(x-1)}{2!}\frac{(x-1)(x-2)}{(x+1)(x+2)}\frac{x(x-1)}{(4x-1)(4x-2)} + \cdots$$

$$= \frac{8\Gamma^3(3x+1)\Gamma(x+1)}{9\Gamma^3(2x+1)\Gamma(4x+1)}.$$

PROOF. In Entry 2, put $n = z = x$ and $y = x - 1$. The proposed equality readily follows. □

Entry 5. *If $\mathrm{Re}(x + y + z + n + 1) > 0$, then*

$$
{}_5F_4\left[\begin{matrix} \tfrac{1}{2}n + 1, n, -x, -y, -z \\ \tfrac{1}{2}n, x+n+1, y+n+1, z+n+1 \end{matrix}\right]
$$

$$
= \frac{\Gamma(x+n+1)\Gamma(y+n+1)\Gamma(z+n+1)\Gamma(x+y+z+n+1)}{\Gamma(n+1)\Gamma(x+y+n+1)\Gamma(y+z+n+1)\Gamma(x+z+n+1)}. \tag{5.1}
$$

Entry 5 is again due to Dougall [1]. Hardy [1] discusses Entry 5 ((3.1) in his paper) and gives a proof based on a theorem of Carlson. For another proof, see Bailey's monograph [4, p. 27]. It is interesting that a q-analogue of Entry 5 was established by L. J. Rogers [1] in 1895, twelve years before Dougall's discovery.

Wilson [1] has shown that Dougall's theorem is intimately connected with the orthogonality of certain orthogonal polynomials. Moreover [1, p. 694],

$$
\int_0^\infty \left| \frac{\Gamma(a+ix)\Gamma(b+ix)\Gamma(c+ix)\Gamma(d+ix)}{\Gamma(2ix)} \right|^2 dx
$$

is a continuous analogue of the sum in Entry 5. The special case $c = 0$, $d = \tfrac{1}{2}$ was, in fact, evaluated by Ramanujan [8], [16, p. 57]. For further related comments, see Section 22 of Chapter 13.

For brevity, let

$$
{}_{p+1}F_p\left[\begin{matrix} \alpha_1, \ldots, \alpha_{p+1} \\ \beta_1, \ldots, \beta_p \end{matrix}\,\middle|\, m\right]
$$

denote the sum of the first $m + 1$ terms of ${}_{p+1}F_p(\alpha_1, \ldots, \alpha_{p+1}; \beta_1, \ldots, \beta_p; 1)$.

Entry 6. *If* $\alpha + \beta + \gamma + 1 = n$, *then*

$$\frac{\Gamma(n + 2)\Gamma(\alpha + 1)\Gamma(\beta + 1)\Gamma(\gamma + 1)}{\Gamma(n - \alpha + 1)\Gamma(n - \beta + 1)\Gamma(n - \gamma + 1)}$$

$$\times \; {}_5F_4\left[\begin{matrix} \frac{1}{2}(n + 3), n + 1, \alpha + 1, \beta + 1, \gamma + 1 \\ \frac{1}{2}(n + 1), n - \alpha + 1, n - \beta + 1, n - \gamma + 1 \end{matrix}\middle| m \right]$$

$$\sim 2 \operatorname{Log} m - \psi(\alpha + 1) - \psi(\beta + 1) - \psi(\gamma + 1) - C,$$

as m tends to ∞, where C denotes Euler's constant.

In our originally published account of Chapter 10 (see the reference in the Introduction), we gave a proof of Entry 6 supplied to us by J. L. Fields based on his paper [1]. R. J. Evans [1] has since found a much simpler proof of a slightly stronger result. We reformulate this stronger version of Entry 6 and give Evans's proof.

Entry 6 (Second Version). *If a, b, c, and $a + b + c$ are not nonpositive integers, then as m tends to ∞,*

$$\frac{\Gamma(a + b + c)\Gamma(a)\Gamma(b)\Gamma(c)}{\Gamma(b + c)\Gamma(a + c)\Gamma(a + b)} \, {}_5F_4\left[\begin{matrix} \frac{1}{2}(a + b + c + 1), a + b + c - 1, a, b, c \\ \frac{1}{2}(a + b + c - 1), b + c, a + c, a + b \end{matrix}\middle| m \right]$$

$$= 2 \operatorname{Log} m - \gamma - \psi(a) - \psi(b) - \psi(c) + O\left(\frac{\operatorname{Log} m}{m}\right),$$

where γ denotes Euler's constant.

PROOF. Recall Whipple's transformation [1] (Bailey [4, p. 25]),

$$\begin{aligned} & {}_7F_6\left[\begin{matrix} a, 1 + \frac{1}{2}a, b, c, d, e, -m \\ \frac{1}{2}a, 1 + a - b, 1 + a - c, 1 + a - d, 1 + a - e, 1 + a + m \end{matrix}\right] \\ & = \frac{(1 + a)_m(1 + a - d - e)_m}{(1 + a - d)_m(1 + a - e)_m} \, {}_4F_3\left[\begin{matrix} 1 + a - b - c, d, e, -m \\ 1 + a - b, 1 + a - c, d + e - a - m \end{matrix}\right], \end{aligned}$$

$$\tag{6.1}$$

where m is a nonnegative integer. Replacing a, d, and e by $a + b + c - 1, a$, and $a + b + c + m + \varepsilon$, respectively, in (6.1), where $\varepsilon > 0$, we find that

$$\begin{aligned} & {}_7F_6\left[\begin{matrix} a + b + c - 1, \frac{1}{2}(a + b + c + 1), b, c, a, a + b + c + m + \varepsilon, -m \\ \frac{1}{2}(a + b + c - 1), a + c, a + b, b + c, -m - \varepsilon, a + b + c + m \end{matrix}\right] \\ & = \frac{(a + b + c)_m(-a - m - \varepsilon)_m}{(b + c)_m(-m - \varepsilon)_m} \, {}_4F_3\left[\begin{matrix} a, a, a + b + c + m + \varepsilon, -m \\ a + c, a + b, a + 1 + \varepsilon \end{matrix}\right]. \end{aligned}$$

Letting ε tend to 0, we deduce that

$$_5F_4\left[\begin{array}{c}\frac{1}{2}(a+b+c+1),\,a+b+c-1,\,a,\,b,\,c\\[4pt]\frac{1}{2}(a+b+c-1),\,a+c,\,a+b,\,b+c\end{array}\bigg|\,m\right]$$

$$=\frac{(a+b+c)_m(a+1)_m}{(b+c)_m(1)_m}\,{}_4F_3\left[\begin{array}{c}a,\,a,\,a+b+c+m,\,-m\\[4pt]a+c,\,a+b,\,a+1\end{array}\right].$$

Thus, the left side of Entry 6 is equal to

$$\frac{\Gamma(a+b+c)\Gamma(a)\Gamma(b)\Gamma(c)}{\Gamma(b+c)\Gamma(a+c)\Gamma(a+b)}\frac{(a+b+c)_m(a+1)_m}{(b+c)_m(1)_m}$$

$$\times\,{}_4F_3\left[\begin{array}{c}a,\,a,\,a+b+c+m,\,-m\\[4pt]a+c,\,a+b,\,a+1\end{array}\right]. \qquad (6.2)$$

We now apply a transformation for 1-balanced terminating $_4F_3$ series found in Bailey's tract [4, p. 56]. If $u+v+w=x+y+z-m+1$, then

$$_4F_3\left[\begin{array}{c}x,\,y,\,z,\,-m\\[4pt]u,\,v,\,w\end{array}\right]=\frac{(v-z)_m(w-z)_m}{(v)_m(w)_m}\,{}_4F_3\left[\begin{array}{c}u-x,\,u-y,\,z,\,-m\\[4pt]1-v+z-m,\,1-w+z-m,\,u\end{array}\right].$$
$$(6.3)$$

Letting $x=a+b+c+m,\,y=a,\,z=a,\,u=a+b,\,v=a+c,$ and $w=a+1,$ we find that

$$_4F_3\left[\begin{array}{c}a,\,a,\,a+b+c+m,\,-m\\[4pt]a+c,\,a+b,\,a+1\end{array}\right]$$

$$=\frac{(c)_m(1)_m}{(a+c)_m(a+1)_m}\,{}_3F_2\left[\begin{array}{c}a,\,b,\,-c-m\\[4pt]a+b,\,1-c-m\end{array}\bigg|\,m\right].$$

Using this equality in (6.2), we find that the left side in Entry 6 equals

$$\frac{\Gamma(a+b+c+m)\Gamma(c+m)}{\Gamma(b+c+m)\Gamma(a+c+m)}R_m, \qquad (6.4)$$

where

$$R_m=\frac{\Gamma(a)\Gamma(b)}{\Gamma(a+b)}\,{}_3F_2\left[\begin{array}{c}a,\,b,\,-c-m\\[4pt]a+b,\,1-c-m\end{array}\bigg|\,m\right]$$

$$=\sum_{k=0}^{m}\frac{\Gamma(a+k)\Gamma(b+k)(c+m)}{\Gamma(a+b+k)\Gamma(1+k)(c+m-k)}.$$

By Stirling's formula, the coefficient of R_m in (6.4) equals $1+O(1/m)$. Examining Entry 6, we see that it remains to show that

$$R_m=2\,\mathrm{Log}\,m-\gamma-\psi(a)-\psi(b)-\psi(c)+O\left(\frac{\mathrm{Log}\,m}{m}\right). \qquad (6.5)$$

Let

$$R_m=U_m+V_m,$$

where

$$U_m = \sum_{k=0}^{m} \frac{\Gamma(a+k)\Gamma(b+k)}{\Gamma(a+b+k)\Gamma(1+k)}$$

and

$$V_m = \sum_{k=1}^{m} \frac{\Gamma(a+k)\Gamma(b+k)}{\Gamma(a+b+k)\Gamma(k)(m+c-k)}.$$

From Luke's book [1, p. 110, Eq. (35)],

$$U_m = \text{Log } m - \gamma - \psi(a) - \psi(b) + O(1/m), \qquad (6.6)$$

as m tends to ∞. (A slightly weaker version is given in Entry 15 below. See also (24.5) of Chapter 11 for (6.6).) By Stirling's formula and (0.2),

$$V_m = \sum_{k=1}^{m} \frac{1}{m+c-k}(1 + O(1/k))$$

$$= \sum_{k=1}^{m} \frac{1}{c+k-1} + O\left(\frac{1}{m+c}\sum_{k=1}^{m}\left(\frac{1}{m+c-k} + \frac{1}{k}\right)\right)$$

$$= \psi(m+c) - \psi(c) + O(1/m)$$

$$= \text{Log } m - \psi(c) + O(1/m), \qquad (6.7)$$

by Stirling's formula for $\psi(z)$ (Luke [1, p. 33]). Combining (6.6) and (6.7), we deduce (6.5) to complete the proof. $\qquad\qquad\qquad\qquad\qquad\qquad\square$

Corollary. *Let $0 < x < 1$. Then as x tends to 0,*

$$\frac{\pi^2}{4} \, {}_5F_4\left[\begin{array}{c} \frac{5}{4}, \frac{1}{2}, \frac{1}{2}, \frac{1}{2}, \frac{1}{2} \\ \frac{1}{4}, 1, 1, 1 \end{array}; 1-x\right] \sim -\text{Log } x + 3 \text{ Log } 2.$$

PROOF. Let $n = \alpha = \beta = \gamma = -\frac{1}{2}$ in Entry 6 to obtain the formula

$$\frac{\pi^2}{2} \, {}_5F_4\left[\begin{array}{c} \frac{5}{4}, \frac{1}{2}, \frac{1}{2}, \frac{1}{2}, \frac{1}{2} \\ \frac{1}{4}, 1, 1, 1 \end{array}\middle| m\right] \sim 2 \text{ Log } m - 3\psi(\tfrac{1}{2}) - \gamma,$$

as m tends to ∞, where on the right side above γ now denotes Euler's constant. Since $\psi(\tfrac{1}{2}) = -2 \text{ Log } 2 - \gamma$ (see Luke's book [1, p. 13]), we find that

$$\frac{\pi^2}{4} \, {}_5F_4\left[\begin{array}{c} \frac{5}{4}, \frac{1}{2}, \frac{1}{2}, \frac{1}{2}, \frac{1}{2} \\ \frac{1}{4}, 1, 1, 1 \end{array}\middle| m\right] - 3 \text{ Log } 2 - \gamma \sim \text{Log } m,$$

as m tends to ∞. It follows that

$$\frac{\pi^2}{4} + \sum_{k=1}^{\infty}\left\{\frac{\Gamma(k+\frac{5}{4})\Gamma^4(k+\frac{1}{2})}{\Gamma(k+\frac{1}{4})(k!)^4} - \frac{1}{k}\right\} = 3 \text{ Log } 2.$$

Hence,

$$\lim_{x\to 1-}\left(\frac{\pi^2}{4} + \sum_{k=1}^{\infty}\left\{\frac{\Gamma(k+\frac{5}{4})\Gamma^4(k+\frac{1}{2})}{\Gamma(k+\frac{1}{4})(k!)^4} - \frac{1}{k}\right\}x^k\right) = 3 \text{ Log } 2.$$

Therefore, as x tends to $1-$,

$$\frac{\pi^2}{4} {}_5F_4\left[\begin{array}{c} \frac{5}{4}, \frac{1}{2}, \frac{1}{2}, \frac{1}{2}, \frac{1}{2} \\ \frac{1}{4}, 1, 1, 1 \end{array} ; x\right] \sim -\text{Log}(1-x) + 3\,\text{Log}\,2.$$

The corollary now follows. $\square$

For further expansions of hypergeometric functions in the neighborhoods of logarithmic singularities, see Section 15 of this chapter and Sections 24–26 in Chapter 11. B. C. Carlson [1] has established expansions about logarithmic branch points for several classes of related functions.

Entry 7. If $\text{Re}(x + y + \frac{1}{2}n + 1) > 0$, then

$$
{}_3F_2\left[\begin{array}{c} n, -x, -y \\ x+n+1, y+n+1 \end{array}\right]
$$

$$
= \frac{\Gamma(x+n+1)\Gamma(y+n+1)\Gamma(\frac{1}{2}n+1)\Gamma(x+y+\frac{1}{2}n+1)}{\Gamma(n+1)\Gamma(x+y+n+1)\Gamma(x+\frac{1}{2}n+1)\Gamma(y+\frac{1}{2}n+1)}.
$$

PROOF. Set $z = -\frac{1}{2}n$ in Entry 5. $\square$

Entry 7 is a famous theorem of Dixon [1]. In Hardy's paper [1], see (3.2). A terminating version of Dixon's theorem can be used to evaluate Selberg's integral in two dimensions (Andrews [3]). The case $n = 3$ of the Dyson–Gunson–Wilson identity can also be established from a terminating case of Dixon's theorem (Andrews [1]). Gessel and Stanton [2] have found new short proofs of both Saalschütz's theorem (Entry 2) and Dixon's theorem by computing the constant terms in certain Laurent series in two variables.

Corollary 1. If $\text{Re}(x + y + n + 1) > 0$, then

$$
\sum_{k=1}^{\infty} \left(\frac{1}{k} + \frac{1}{n+k}\right) \frac{(-x)_k(-y)_k}{(x+n+1)_k(y+n+1)_k}
$$

$$
= \psi(x+n+1) + \psi(y+n+1) - \psi(n+1) - \psi(x+y+n+1). \quad (7.1)
$$

PROOF. Logarithmically differentiate both sides of (5.1) with respect to z and then set $z = 0$. With the aid of (0.3), we obtain the identity above after a little simplification. $\square$

Corollary 2. If $\text{Re}(x + y + 1) > 0$, then

$$
{}_5F_4\left[\begin{array}{c} \frac{1}{2}n+1, n, n, -x, -y \\ \frac{1}{2}n, x+n+1, y+n+1, 1 \end{array}\right]
$$

$$
= \frac{\Gamma(x+n+1)\Gamma(y+n+1)\Gamma(x+y+1)}{\Gamma(n+1)\Gamma(x+y+n+1)\Gamma(x+1)\Gamma(y+1)}.
$$

PROOF. Set $z = -n$ in Entry 5.   □

Corollary 3. *If* $\text{Re}(x + y + n) > 0$, *then*

$$
{}_4F_3\left[\begin{array}{c} \frac{1}{2}n + 1, -x, -y, 1 \\ \frac{1}{2}n, x + n + 1, y + n + 1 \end{array}\right] = \frac{(x + n)(y + n)}{n(x + y + n)}.
$$

PROOF. Put $z = -1$ in Entry 5.   □

Corollary 4. *If* $\text{Re}(x + y + \frac{1}{2}(n + 1)) > 0$, *then*

$$
{}_4F_3\left[\begin{array}{c} \frac{1}{2}n + 1, n, -x, -y \\ \frac{1}{2}n, x + n + 1, y + n + 1 \end{array}\right]
$$

$$
= \frac{\Gamma(x + n + 1)\Gamma(y + n + 1)\Gamma(\frac{1}{2}(n + 1))\Gamma(x + y + \frac{1}{2}(n + 1))}{\Gamma(n + 1)\Gamma(x + y + n + 1)\Gamma(x + \frac{1}{2}(n + 1))\Gamma(y + \frac{1}{2}(n + 1))}.
$$

PROOF. Set $z = -\frac{1}{2}(n + 1)$ in Entry 5.   □

Corollary 5. *For* $\text{Re}(2x + 2y + n + 2) > 0$,

$$
{}_4F_3\left[\begin{array}{c} \frac{1}{2}n + 1, n, -x, -y \\ \frac{1}{2}n, x + n + 1, y + n + 1 \end{array}; -1\right] = \frac{\Gamma(x + n + 1)\Gamma(y + n + 1)}{\Gamma(n + 1)\Gamma(x + y + n + 1)}. \tag{7.2}
$$

PROOF. Corollary 5 follows from Entry 5 by letting z tend to ∞. The details are easily justified by using Stirling's formula.   □

Bailey [4, p. 28] gives a proof of Corollary 5 based on Whipple's transformation (6.1).

Corollary 6. *If* $\text{Re}(x + n + 1) > 0$, *then*

$$
\sum_{k=1}^{\infty}\left(\frac{1}{k} + \frac{1}{n + k}\right)\frac{(-x)_k(k - 1)!}{(x + n + 1)_k(n + 1)_k} = \sum_{k=1}^{\infty}\frac{1}{(k + x + n)^2} - \sum_{k=1}^{\infty}\frac{1}{(k + n)^2}. \tag{7.3}
$$

PROOF. Differentiate both sides of (7.1) with respect to y and then set $y = 0$. With the use of (0.2) and (0.3), we complete the proof.   □

On the left side of (7.3), Ramanujan (p. 119) has written $k!$ instead of $(k - 1)!$.

Corollary 7. *If* $\text{Re}(x - n + 1) > 0$, *then*

$$
{}_5F_4\left[\begin{array}{c} \frac{1}{2}n + 1, n, n, n, -x \\ \frac{1}{2}n, x + n + 1, 1, 1 \end{array}\right] = \frac{\sin(\pi n)\Gamma(x + n + 1)\Gamma(x - n + 1)}{\pi n \Gamma^2(x + 1)}. \tag{7.4}
$$

PROOF. Set $y = z = -n$ in Entry 5.   □

Corollary 8. *If* $\mathrm{Re}(x - \frac{1}{2}n + 1) > 0$, *then*

$$_3F_2\left[\begin{array}{c} n, n, -x \\ x + n + 1, 1 \end{array}\right] = \frac{\Gamma(x + n + 1)\Gamma(\frac{1}{2}n + 1)\Gamma(x - \frac{1}{2}n + 1)}{\Gamma(n + 1)\Gamma(x + 1)\Gamma(1 - \frac{1}{2}n)\Gamma(x + \frac{1}{2}n + 1)}.$$

PROOF. Put $y = -n$ in Entry 7. □

Corollary 9. *If* $\mathrm{Re}(x - \frac{1}{2}n + \frac{1}{2}) > 0$, *then*

$$_4F_3\left[\begin{array}{c} \frac{1}{2}n + 1, n, n, -x \\ \frac{1}{2}n, x + n + 1, 1 \end{array}\right] = \frac{\Gamma(x + n + 1)\Gamma(\frac{1}{2}n + \frac{1}{2})\Gamma(x - \frac{1}{2}n + \frac{1}{2})}{\Gamma(n + 1)\Gamma(x + 1)\Gamma(\frac{1}{2} - \frac{1}{2}n)\Gamma(x + \frac{1}{2}n + \frac{1}{2})}.$$

PROOF. In Entry 5, set $y = -n$ and $z = -\frac{1}{2}(n + 1)$. □

Corollary 10. *If* $\mathrm{Re}(2x - n + 2) > 0$, *then*

$$_4F_3\left[\begin{array}{c} \frac{1}{2}n + 1, n, n, -x \\ \frac{1}{2}n, x + n + 1, 1 \end{array}; -1\right] = \frac{\Gamma(x + n + 1)}{\Gamma(n + 1)\Gamma(x + 1)}.$$

PROOF. Let $y = -n$ in Corollary 5. □

Corollary 11. *If* $\mathrm{Re}\, x > -1$, *then*

$$_3F_2\left[\begin{array}{c} \frac{1}{2}n, n, -x \\ \frac{1}{2}n + 1, x + n + 1 \end{array}\right] = \frac{\Gamma(x + n + 1)\Gamma^2(\frac{1}{2}n + 1)\Gamma(x + 1)}{\Gamma(n + 1)\Gamma^2(x + \frac{1}{2}n + 1)}.$$

PROOF. Put $y = -\frac{1}{2}n$ in Entry 7. □

Corollary 12. *If* $\mathrm{Re}\, x > -\frac{1}{2}$, *then*

$$_2F_1(n, -x; x + n + 1) = \frac{\Gamma(x + n + 1)\Gamma(2x + 1)}{\Gamma(2x + n + 1)\Gamma(x + 1)}.$$

Ramanujan probably deduced Corollary 12 from Entry 7 by setting $y = -\frac{1}{2}(n + 1)$ and then using Legendre's duplication formula to simplify the resulting evaluation. However, in fact, Corollary 12 is a special case of Gauss's theorem, which is given by Ramanujan in complete generality in Entry 8 below. See Bailey's monograph [4, pp. 2, 3] for a proof.

Corollary 13. *If* $\mathrm{Re}\, x > -1$, *then*

$$_2F_1(n, -x; x + n + 1; -1) = \frac{\Gamma(x + n + 1)\Gamma(\frac{1}{2}n + 1)}{\Gamma(x + \frac{1}{2}n + 1)\Gamma(n + 1)}.$$

Corollary 13 is known as Kummer's theorem [1], [2, pp. 75–166] and is most commonly proved by using a quadratic transformation also due to Kummer. See Bailey's tract [4, pp. 9, 10] for details. Ramanujan evidently derived Corollary 13 by letting y tend to ∞ in Entry 7.

Corollary 14. *If* Re $x > -\frac{1}{2}$, *then*

$$_3F_2\left[\begin{matrix} \frac{1}{2}n + 1, n, -x \\ \frac{1}{2}n, x + n + 1 \end{matrix}; -1\right] = \frac{\Gamma(x + n + 1)\Gamma(\frac{1}{2}n + \frac{1}{2})}{\Gamma(n + 1)\Gamma(x + \frac{1}{2}n + \frac{1}{2})}.$$

PROOF. Put $y = -\frac{1}{2}(n + 1)$ in Corollary 5.   □

Corollary 15. *If* Re $n < \frac{1}{2}$, *then*

$$_5F_4\left[\begin{matrix} \frac{1}{2}n + 1, n, n, n, n \\ \frac{1}{2}n, 1, 1, 1 \end{matrix}\right] = \frac{\Gamma^2(n) \sin(\pi n) \tan(\pi n)}{\pi^2 \Gamma(2n + 1)}.$$

PROOF. Let $x = y = z = -n$ in Entry 5 and use the reflection formula to simplify the resulting evaluation.   □

Corollary 15 is Eq. (3.33) in Hardy's paper [1].

Corollary 16. *If* Re $n < \frac{1}{3}$, *then*

$$_4F_3\left[\begin{matrix} \frac{1}{2}n + 1, n, n, n \\ \frac{1}{2}n, 1, 1 \end{matrix}\right] = \frac{\sin(\pi n)\Gamma(\frac{1}{2}n + \frac{1}{2})\Gamma(\frac{1}{2} - \frac{3}{2}n)}{\pi n \Gamma^2(\frac{1}{2} - \frac{1}{2}n)}.$$

PROOF. Put $x = y = -n$ in Corollary 4.   □

Corollary 16 is (3.31) in Hardy's paper [1] and can also be found in Bailey's text [4, p. 96].

Corollary 17. *If* Re $n < \frac{2}{3}$, *then*

$$_4F_3\left[\begin{matrix} \frac{1}{2}n + 1, n, n, n \\ \frac{1}{2}n, 1, 1 \end{matrix}; -1\right] = \frac{\sin(\pi n)}{\pi n}.$$

PROOF. Set $x = y = -n$ in Corollary 5.   □

Corollary 17 is Eq. (3.32) in Hardy's paper [1] and is recorded by Bailey [4, p. 96].

Corollary 18. *If* Re $n < 1$, *then*

$$_3F_2\left[\begin{matrix} \frac{1}{2}n, n, n \\ \frac{1}{2}n + 1, 1 \end{matrix}\right] = \frac{2 \tan(\frac{1}{2}\pi n)\Gamma^4(\frac{1}{2}n + 1)}{\pi n \Gamma^2(n + 1)}.$$

PROOF. Put $x = -n$ in Corollary 11 and use the reflection principle to simplify the resulting equality.   □

Corollary 19. *If* Re $n < 2$, *then*

$$_3F_2\left[\begin{matrix} \frac{1}{2}n, \frac{1}{2}n, n \\ \frac{1}{2}n + 1, \frac{1}{2}n + 1 \end{matrix}\right] = \frac{\pi n \Gamma^2(\frac{1}{2}n + 1)}{2 \sin(\frac{1}{2}\pi n)\Gamma(n + 1)}.$$

PROOF. Put $x = -\frac{1}{2}n$ in Corollary 11.   □

For the evaluation of certain other classes of $_3F_2$ and $_4F_3$ series at the argument 1, see the papers by Lavoie [1], [2].

Corollary 20. *If* $\mathrm{Re}(2x + n + 2) > 0$, *then*

$$\sum_{k=1}^{\infty} \left(\frac{1}{k} + \frac{1}{n+k}\right) \frac{(-1)^k(-x)_k}{(x+n+1)_k} = \psi(x+n+1) - \psi(n+1).$$

PROOF. Take the logarithmic derivative of both sides of (7.2) with respect to y and then set $y = 0$. Simplifying with the aid of (0.3), we achieve the desired equality.   □

Corollary 21. *If* $\mathrm{Re}\ x > -1$, *then*

$$\sum_{k=1}^{\infty} \left(\frac{1}{k} + \frac{1}{n+k}\right) \frac{(-x)_k(n)_k}{(x+n+1)_k(1)_k} = \sum_{k=1}^{\infty} \left(\frac{1}{k+n} + \frac{1}{k+x} - \frac{1}{k+x+n} - \frac{1}{k}\right).$$

PROOF. In Entry 5, set $z = -n$, logarithmically differentiate both sides of (5.1) with respect to y, and then set $y = 0$. Using (0.2) and (0.3), we deduce the desired result.   □

Ramanujan (p. 120) neglected to record the summands $-1/k$, $1 \le k < \infty$, in Corollary 21.

Corollary 22. *If* $\mathrm{Re}\ n > 0$, *then*

$$\sum_{k=0}^{\infty} \left(\frac{1}{k+1} + \frac{1}{n+k}\right) \frac{(1)_k^2}{(n+1)_k^2} = 2n^2 \sum_{k=0}^{\infty} \frac{1}{(k+n)^3}. \tag{7.5}$$

PROOF. In (7.3), replace n by $n - 1$, differentiate both sides with respect to x, and then set $x = 0$. Use (0.3) in completing the proof.   □

Corollary 23. *If* $\mathrm{Re}\ n > -2$, *then*

$$\sum_{k=1}^{\infty} \left(\frac{1}{k} - \frac{1}{n+k}\right) \frac{(k-1)!}{(n+1)_k} = \sum_{k=1}^{\infty} \frac{1}{(k+\frac{1}{2}n)^2} - \sum_{k=1}^{\infty} \frac{1}{(k+n)^2}.$$

PROOF. In Corollary 6, set $x = -\frac{1}{2}n$. After a little simplification, the desired result follows.   □

Corollary 24. *If* $\mathrm{Re}\ n < 1$, *then*

$$\sum_{k=1}^{\infty} \left(\frac{1}{k} + \frac{1}{n+k}\right) \frac{(n)_k^2}{(k!)^2} = \sum_{k=1}^{\infty} \left(\frac{1}{k+n} + \frac{1}{k-n}\right).$$

PROOF. Differentiate both sides of (7.4) in Corollary 7 with respect to x and then set $x = 0$. Using (0.2) and (0.3) and simplifying, we complete the proof.   □

Example 1. *If* Re $x > \frac{1}{3}$, *then*

$$\sum_{k=0}^{\infty} (2k + 1) \frac{(1 - x)_k^3}{(1 + x)_k^3} = \frac{\Gamma^3(x + 1)\Gamma(3x - 1)}{\Gamma^3(2x)}.$$

PROOF. In Entry 5, let $n = 1$, replace x by $x - 1$, and set $y = z = x - 1$.   □

Example 1 has been given by both Hardy [1, Eq. (3.45)] and Bailey [4, p. 96]. The following example is also recorded by Hardy [1, Eq. (3.43)].

Example 2. *If* Re $x > \frac{1}{2}$, *then*

$$\sum_{k=0}^{\infty} (2k + 1) \frac{(1 - x)_k^2}{(1 + x)_k^2} = \frac{x^2}{2x - 1}.$$

PROOF. In Corollary 2, let $n = 1$, replace x by $x - 1$, and set $y = x - 1$.   □

Example 3. *If* Re $x > \frac{1}{4}$, *then*

$$1 + \left(\frac{x - 1}{x + 1}\right)^2 + \left(\frac{(x - 1)(x - 2)}{(x + 1)(x + 2)}\right)^2 + \cdots = \frac{2x\Gamma^4(x + 1)\Gamma(4x + 1)}{(4x - 1)\Gamma^4(2x + 1)}.$$

PROOF. In Entry 7, put $n = 1$, replace x by $x - 1$, and let $y = x - 1$. After using Legendre's duplication formula to simplify, we obtain the proposed formula.   □

Example 3 is found in Hardy's paper [1, Eq. (3.49)] and Bailey's book [4, p. 96]. The next example is equality (3.44) in Hardy's paper [1].

Example 4. *If* Re $x > \frac{1}{4}$, *then*

$$\sum_{k=0}^{\infty} (-1)^k (2k + 1) \frac{(1 - x)_k^2}{(1 + x)_k^2} = \frac{\Gamma^2(x + 1)}{\Gamma(2x)}.$$

PROOF. In Corollary 5, let $n = 1$, replace x by $x - 1$, and set $y = x - 1$.   □

Example 5. *If* Re $x > \frac{1}{2}$, *then*

$$1 + 3\frac{x - 1}{x + 1} + 5\frac{(x - 1)(x - 2)}{(x + 1)(x + 2)} + \cdots = x.$$

PROOF. Put $n = 1$ and replace x by $x - 1$ in Corollary 14. □

Example 5 is given by both Hardy [1, Eq. (3.41)] and Bailey [4, p. 96]. Example 6 is also given by Hardy [1, Eq. (3.46)].

Example 6. *If* Re $x > 0$, *then*

$$1 + \frac{x - 1}{x + 1} + \frac{(x - 1)(x - 2)}{(x + 1)(x + 2)} + \cdots = \frac{2^{2x-1}\Gamma^2(x + 1)}{\Gamma(2x + 1)}.$$

PROOF. In Corollary 13, put $n = 1$, replace x by $x - 1$, and use Legendre's duplication formula to simplify. □

Example 7. *If* Re $x > \frac{1}{2}$, *then*

$$1 - \frac{x - 1}{x + 1} + \frac{(x - 1)(x - 2)}{(x + 1)(x + 2)} - \cdots = \frac{x}{2x - 1}.$$

PROOF. In Corollary 12, set $n = 1$ and replace x by $x - 1$. □

Examples 7 and 8 are given by Hardy [1, Eqs. (3.47), (3.42)]. See also Bailey's tract [4, p. 96] for Example 8.

Example 8. *If* Re $x > 1$, *then*

$$1 - 3\frac{x - 1}{x + 1} + 5\frac{(x - 1)(x - 2)}{(x + 1)(x + 2)} - \cdots = 0.$$

PROOF. Put $n = 1$ and replace x by $x - 1$ in Corollary 9. □

Example 9. *If* Re $x > 0$, *then*

$$\sum_{k=0}^{\infty} \frac{(-1)^k (1 - x)_k}{(k + 1)(1 + x)_k} = \frac{2^{2x-2}\Gamma^2(x)}{\Gamma(2x)} + \frac{1}{2}\sum_{k=1}^{\infty}\left(\frac{1}{k} - \frac{1}{k + x - 1}\right). \tag{7.6}$$

PROOF. Replace x by $x - 1$ in Kummer's formula, Corollary 13. Then logarithmically differentiate both sides with respect to n and set $n = 0$. Using (0.2) and (0.3), we find that

$$\sum_{k=1}^{\infty} \frac{(-1)^k (1 - x)_k}{k(x)_k} = \frac{1}{2}\psi(x) - \frac{1}{2}\psi(1)$$

$$= \frac{1}{2}\sum_{k=1}^{\infty}\left(\frac{1}{k} - \frac{1}{k + x - 1}\right). \tag{7.7}$$

Now,

$$\sum_{k=1}^{\infty} \frac{(-1)^k (1-x)_k}{k(x)_k} = \frac{1}{x} \sum_{k=1}^{\infty} \frac{(-1)^k (1-x)_{k-1}(k-x)}{k(1+x)_{k-1}}$$

$$= \sum_{k=1}^{\infty} \frac{(-1)^{k-1}(1-x)_{k-1}}{k(1+x)_{k-1}} - \frac{1}{x} \sum_{k=1}^{\infty} \frac{(-1)^{k-1}(1-x)_{k-1}}{(1+x)_{k-1}}$$

$$= \sum_{k=0}^{\infty} \frac{(-1)^k (1-x)_k}{(k+1)(1+x)_k} - \frac{2^{2x-1}\Gamma^2(x+1)}{x\Gamma(2x+1)}, \qquad (7.8)$$

by Example 6. Combining (7.7) and (7.8), we deduce the desired result. $\qquad \square$

Example 10. *If* Re $x > 0$, *then*

$$\sum_{k=0}^{\infty} \frac{(1-x)_k}{(k+1)(1+x)_k} = \frac{1}{2x} + \sum_{k=1}^{\infty} \left(\frac{1}{k+x} - \frac{1}{k+2x} \right).$$

PROOF. By Entry 9 below and (0.2),

$$\sum_{k=0}^{\infty} \frac{(1-x)_k}{(k+1)(1+x)_k} = -\sum_{k=1}^{\infty} \frac{(-x)_k}{k(x)_k}$$

$$= \psi(2x) - \psi(x)$$

$$= \frac{1}{x} - \frac{1}{2x} + \sum_{k=1}^{\infty} \left(\frac{1}{k+x} - \frac{1}{k+2x} \right),$$

and the proof is complete. $\qquad \square$

The next example is in Hardy's paper [1, Eq. (3.48)] and Bailey's book [4, p. 96].

Example 11. *If* Re $x > 0$, *then*

$$1 - \frac{x-1}{3(x+1)} + \frac{(x-1)(x-2)}{5(x+1)(x+2)} - \cdots = \frac{2^{4x}\Gamma^4(x+1)}{4x\Gamma^2(2x+1)}.$$

PROOF. In Corollary 11, put $n = 1$ and replace x by $x - 1$. After using the Legendre duplication formula, we easily obtain the proposed equality. $\qquad \square$

Example 12. *If* x *is a positive integer, then*

$$\sum_{k=0}^{\infty} \frac{(1-x)_k}{(k+1)^2(1+x)_k} = \frac{1}{x} \sum_{k=1}^{x} \frac{1}{k+x} + \frac{1}{2} \sum_{k=1}^{x} \frac{1}{k^2}.$$

PROOF. Consider Dixon's formula, Entry 7, and logarithmically differentiate both sides with respect to y. Setting $y = 0$ and using (0.2) and (0.3), we find that

$$-\sum_{k=1}^{\infty} \frac{(n)_k(-x)_k}{k(x+n+1)_k(n+1)_k}$$

$$= \sum_{k=1}^{\infty} \left(\frac{1}{k+x+n} - \frac{1}{k+n} \right) + \sum_{k=1}^{\infty} \left(\frac{1}{k+\frac{1}{2}n} - \frac{1}{k+x+\frac{1}{2}n} \right). \quad (7.9)$$

Next, replace x by $x - 1$ and differentiate both sides of (7.9) with respect to n. Setting $n = 0$ and using (0.3), we deduce that

$$\sum_{k=1}^{\infty} \frac{(1-x)_k}{k^2(x)_k} = \frac{1}{2} \sum_{k=1}^{\infty} \left(\frac{1}{(k+x-1)^2} - \frac{1}{k^2} \right). \quad (7.10)$$

On the other hand, by Example 10,

$$\sum_{k=1}^{\infty} \frac{(1-x)_k}{k^2(x)_k} = \sum_{k=1}^{\infty} \frac{(1-x)_{k-1}(k-x)}{k^2(1+x)_{k-1}x}$$

$$= \frac{1}{x} \sum_{k=0}^{\infty} \frac{(1-x)_k}{(k+1)(1+x)_k} - \sum_{k=0}^{\infty} \frac{(1-x)_k}{(k+1)^2(1+x)_k}$$

$$= \frac{1}{x} \left\{ \frac{1}{2x} + \sum_{k=1}^{\infty} \left(\frac{1}{k+x} - \frac{1}{k+2x} \right) \right\}$$

$$- \sum_{k=0}^{\infty} \frac{(1-x)_k}{(k+1)^2(1+x)_k}. \quad (7.11)$$

By combining (7.10) and (7.11) and using the fact that x is a positive integer, we complete the proof.   □

Example 13. *If* $\operatorname{Re} x > \frac{3}{2}$, *then*

$$1^3 + 3^3 \frac{x-1}{x+1} + 5^3 \frac{(x-1)(x-2)}{(x+1)(x+2)} + \cdots = x(4x-3).$$

PROOF. We shall apply Entry 31 below with $n = 1$, $y = -1$, $z = u = -\frac{3}{2}$, and x replaced by $x - 1$. Accordingly, we find that

$$_6F_5 \left[\begin{matrix} 1, \frac{3}{2}, \frac{3}{2}, \frac{3}{2}, 1-x, 1 \\ \frac{1}{2}, \frac{1}{2}, \frac{1}{2}, 1+x, 1 \end{matrix} ; -1 \right] = \frac{\Gamma(1+x)}{\Gamma(x)} {}_3F_2 \left[\begin{matrix} -1, 1, -x, 1 \\ \frac{1}{2}, \frac{1}{2} \end{matrix} \right]$$

$$= x \left\{ 1 + \frac{(-1)(1-x)}{\frac{1}{4}} \right\}$$

$$= x(4x-3). \qquad □$$

Example 14

$$1 - 5\left(\frac{1}{2}\right)^3 + 9\left(\frac{1\cdot 3}{2\cdot 4}\right)^3 - \cdots = \frac{2}{\pi}.$$

PROOF. Let $n = -x = -y = \frac{1}{2}$ in Corollary 5 to obtain

$$_4F_3\left[\begin{array}{c} \frac{5}{4}, \frac{1}{2}, \frac{1}{2}, \frac{1}{2} \\ \frac{1}{4}, 1, 1 \end{array}; -1\right] = \frac{1}{\Gamma(\frac{3}{2})\Gamma(\frac{1}{2})},$$

which is equivalent to the proposed formula. $\square$

Examples 14 and 15 were communicated by Ramanujan in his first letter to Hardy [16, pp. xxvi, xxv, respectively]. Hardy [2], [7, pp. 517, 518] has observed the simple proofs that we offer here. Evidently, Example 14 was first established in 1859 by Bauer [1]. Examples 14 and 15 may also be found in Bailey's tract [4, p. 96] and Hardy's book [9, p. 7].

Example 15

$$1 + 9\left(\frac{1}{4}\right)^4 + 17\left(\frac{1\cdot 5}{4\cdot 8}\right)^4 + \cdots = \frac{2\sqrt{2}}{\sqrt{\pi}\,\Gamma^2(\frac{3}{4})}.$$

PROOF. Set $x = y = z = -n = -\frac{1}{4}$ in Entry 5, and the proposed equality follows forthwith. $\square$

Example 16

$$1 + \frac{1}{5}\left(\frac{1}{2}\right)^2 + \frac{1}{9}\left(\frac{1\cdot 3}{2\cdot 4}\right)^2 + \cdots = \frac{\pi^2}{4\Gamma^4(\frac{3}{4})}.$$

PROOF. In Dixon's theorem, Entry 7, let $x = -\frac{1}{2}$, $y = -\frac{1}{4}$, and $n = \frac{1}{2}$. $\square$

Example 17

$$1 + \frac{1}{5^2}\left(\frac{1}{2}\right) + \frac{1}{9^2}\left(\frac{1\cdot 3}{2\cdot 4}\right) + \cdots = \frac{\pi^{5/2}}{8\sqrt{2}\,\Gamma^2(\frac{3}{4})}.$$

PROOF. In Dixon's theorem, Entry 7, put $n = \frac{1}{2}$ and $x = y = -\frac{1}{4}$. $\square$

Example 18

$$1 + \left(\frac{1}{2}\right)^3 + \left(\frac{1\cdot 3}{2\cdot 4}\right)^3 + \cdots = \frac{\pi}{\Gamma^4(\frac{3}{4})}.$$

PROOF. Set $-x = -y = n = \frac{1}{2}$ in Dixon's theorem, Entry 7. $\square$

Example 19

$$1 - \left(\frac{1}{2}\right)^2 + \left(\frac{1\cdot 3}{2\cdot 4}\right)^2 - \cdots = \frac{\sqrt{\pi}}{\sqrt{2}\,\Gamma^2(\frac{3}{4})}.$$

PROOF. In Kummer's theorem, Corollary 13, set $n = -x = \frac{1}{2}$. $\square$

Example 20. *If* Re $n < \frac{2}{3}$, *then*

$$1 + \left(\frac{n}{1!}\right)^3 + \left(\frac{n(n+1)}{2!}\right)^3 + \cdots = \frac{6 \sin(\frac{1}{2}\pi n) \sin(\pi n)\Gamma^3(\frac{1}{2}n + 1)}{\pi^2 n^2(1 + 2 \cos(\pi n))\Gamma(\frac{3}{2}n + 1)}.$$

PROOF. In Dixon's theorem, Entry 7, set $x = y = -n$. After several applications of the reflection principle and some simplification, we deduce the desired formula.   □

Example 20 is due to Morley [1] in 1902. See Bailey's tract [4, p. 13] for further references.

Entry 8. *If* Re$(x + y + n + 1) > 0$, *then*

$$_2F_1(-x, -y; n + 1) = \frac{\Gamma(n + 1)\Gamma(x + y + n + 1)}{\Gamma(x + n + 1)\Gamma(y + n + 1)}. \tag{8.1}$$

As mentioned earlier, Entry 8 is Gauss's theorem [1]. Following Entry 8, Ramanujan indicates, in one sentence, how he deduced Entry 8. This is the only clue to the methods used by Ramanujan in his derivations of the several theorems in Chapter 10.

Assume that n and x are integers with $n \geq 0$ and $n + x \geq 0$. Expanding $(1 + u)^{y+n}$ and $(1 + 1/u)^x$ in their formal binomial series and taking their product, we find that, if a_n is the coefficient of u^n,

$$a_n = \sum_{k=0}^{\infty} \binom{y+n}{k+n}\binom{x}{k} = \frac{\Gamma(y + n + 1)}{\Gamma(n + 1)\Gamma(y + 1)} \sum_{k=0}^{\infty} \frac{(-x)_k(-y)_k}{(n + 1)_k(1)_k}. \tag{8.2}$$

On the other hand, expanding $(1 + u)^{x+y+n}$ in its binomial series and dividing by u^x, we find that

$$a_n = \binom{x + y + n}{x + n} = \frac{\Gamma(x + y + n + 1)}{\Gamma(x + n + 1)\Gamma(y + 1)}. \tag{8.3}$$

Comparing (8.2) and (8.3), we deduce (8.1).

Entry 9. *If* Re$(\alpha - \beta) > 0$, *then*

$$\sum_{k=1}^{\infty} \frac{(\beta)_k}{k(\alpha)_k} = \psi(\alpha) - \psi(\alpha - \beta). \tag{9.1}$$

PROOF. In Gauss's theorem, Entry 8, put $\beta = -x$ and $\alpha = n + 1$. Take the logarithmic derivative of both sides of (8.1) with respect to y and set $y = 0$. Using (0.3), we complete the proof.   □

Entry 10. *If* Re $x > -1$, *then*

$$\sum_{k=0}^{\infty} \frac{(-x)_k}{(n + k)k!} = \frac{\Gamma(n)\Gamma(x + 1)}{\Gamma(n + x + 1)}. \tag{10.1}$$

PROOF. In Gauss's theorem, Entry 8, let $y = -n$.   □

Example 1. *If* Re $n > -1$, *then*

$$\sum_{k=1}^{\infty} \frac{(k-1)!}{k(n+1)_k} = \sum_{k=1}^{\infty} \frac{1}{(k+n)^2}.$$

PROOF. Differentiate both sides of (9.1) with respect to β and set $\beta = 0$ and $\alpha = n + 1$. Using (0.3), we complete the proof.   □

Example 2. *If* Re $n < 1$, *then*

$$\frac{1}{n} + \frac{n}{(n+1)1!} + \frac{n(n+1)}{(n+2)2!} + \cdots = \frac{\pi}{\sin(\pi n)}.$$

PROOF. Set $x = -n$ in Entry 10.   □

Example 3. *If n is arbitrary, then*

$$\frac{1}{n+1} + \frac{1}{n+2}\left(\frac{1}{2}\right) + \frac{1}{n+3}\left(\frac{1\cdot3}{2\cdot4}\right) + \cdots = \frac{\sqrt{\pi}\,\Gamma(n+1)}{\Gamma(n+\frac{3}{2})}.$$

PROOF. Let $x = -\frac{1}{2}$ and replace n by $n + 1$ in Entry 10.   □

Example 4. *If* Re $n > -1$, *then*

$$1 - \frac{n}{3\cdot1!} + \frac{n(n-1)}{5\cdot2!} - \cdots = \frac{\sqrt{\pi}\,\Gamma(n+1)}{2\Gamma(n+\frac{3}{2})}.$$

PROOF. In Entry 10, replace x by n and n by $\frac{1}{2}$.   □

Example 5. *If* Re $x > -1$, *then*

$$\sum_{k=0}^{\infty} \frac{(-x)_k}{(n+k)^2 k!} = \frac{\Gamma(n)\Gamma(x+1)}{\Gamma(n+x+1)} \sum_{k=1}^{\infty}\left(\frac{1}{k+n-1} - \frac{1}{k+n+x}\right).$$

PROOF. Differentiate both sides of (10.1) with respect to n and use (0.2).   □

Example 6. *If n is arbitrary, then*

$$\frac{1}{(n+1)^2} + \frac{1}{(n+2)^2}\left(\frac{1}{2}\right) + \frac{1}{(n+3)^2}\left(\frac{1\cdot3}{2\cdot4}\right) + \cdots$$

$$= \frac{\sqrt{\pi}\,\Gamma(n+1)}{\Gamma(n+\frac{3}{2})} \sum_{k=1}^{\infty}\left(\frac{1}{k+n} - \frac{1}{k+n+\frac{1}{2}}\right).$$

PROOF. Let $x = -\frac{1}{2}$ and replace n by $n + 1$ in Example 5.   □

Example 7. *If* Re $n < 1$, *then*

$$\frac{1}{n^2} + \frac{n}{(n+1)^2 1!} + \frac{n(n+1)}{(n+2)^2 2!} + \cdots = \frac{\pi}{\sin(\pi n)} \sum_{k=1}^{\infty} \left(\frac{1}{k+n-1} - \frac{1}{k} \right).$$

PROOF. Let $x = -n$ in Example 5. □

Entry 11. *Let* $n > 0$ *and suppose that* Re$(\alpha - \beta - 1) > 0$. *Then*

$$\sum_{k=0}^{\infty} \{(\alpha + k)^n - (\beta + 1 + k)^n\} \left\{ \frac{(\beta + 1)_k}{(\alpha + 1)_k} \right\}^n = \alpha^n.$$

PROOF. Observe that, for each positive integer m,

$$\sum_{k=0}^{m} \{(\alpha + k)^n - (\beta + 1 + k)^n\} \left\{ \frac{(\beta + 1)_k}{(\alpha + 1)_k} \right\}^n = \alpha^n - \left\{ \frac{(\beta + 1)_{m+1}}{(\alpha + 1)_m} \right\}^n.$$

Thus, it suffices to show that

$$\lim_{m \to \infty} \frac{(\beta + 1)_m}{(\alpha)_m} = 0.$$

Since Re$(\beta + 1) <$ Re α, the statement above is true by Stirling's formula. □

Corollary 1. *If* Re$(\alpha - \beta - 1) > 0$, *then*

$$\sum_{k=1}^{\infty} \frac{(\beta)_k}{(\alpha)_k} = \frac{\beta}{\alpha - \beta - 1}.$$

PROOF. If $n = 1$, Entry 11 yields

$$(\alpha - \beta - 1) \sum_{k=0}^{\infty} \frac{(\beta + 1)_k}{(\alpha + 1)_k} = \alpha.$$

Multiplying both sides by $\beta / \{\alpha(\alpha - \beta - 1)\}$, we obtain the desired formula.
    Alternatively, in Entry 8, set $n + 1 = \alpha$, $x = -1$, and $y = -\beta$, and the formula of Corollary 1 readily follows. □

Corollary 2. *If* Re$(\alpha - \beta - 1) > 0$, *then*

$$\sum_{k=1}^{\infty} (\alpha + \beta + 2k - 1) \frac{(\beta)_k^2}{(\alpha)_k^2} = \frac{\beta^2}{\alpha - \beta - 1}.$$

PROOF. Apply Entry 11 with $n = 2$. Since

$$(\alpha + k)^2 - (\beta + 1 + k)^2 = (\alpha - \beta - 1)(\alpha + \beta + 2k + 1),$$

we find that

$$(\alpha - \beta - 1) \sum_{k=0}^{\infty} (\alpha + \beta + 2k + 1) \frac{(\beta + 1)_k^2}{(\alpha + 1)_k^2} = \alpha^2.$$

Multiplying both sides by $\beta^2/\{\alpha^2(\alpha - \beta - 1)\}$, we complete the proof. $\square$

An alternate proof can be obtained by letting $x = y = -\beta$ and $n = \alpha + \beta - 1$ in Corollary 3 of Section 7.

Entry 12(a). *Suppose that* $f(x) = \sum_{k=1}^{\infty} (A_k x^k/k)$ *in some neighborhood of the origin. Define* P_k, $0 \le k < \infty$, *by*

$$e^{f(x)} = \sum_{k=0}^{\infty} P_k x^k. \tag{12.1}$$

Then $P_0 = 1$ *and, for* $n \ge 1$,

$$nP_n = \sum_{k=1}^{n} A_k P_{n-k}.$$

PROOF. It is clear that $P_0 = 1$. Differentiating both sides of (12.1) with respect to x, we find that

$$\sum_{j=0}^{\infty} P_j x^j \sum_{k=1}^{\infty} A_k x^{k-1} = \sum_{n=1}^{\infty} P_n n x^{n-1}.$$

Equating coefficients of x^{n-1} on both sides, we deduce the required recursion formula. $\square$

Entry 12(b) is an instance of the inclusion–exclusion principle, but Ramanujan cleverly deduces Entry 12(b) from Entry 12(a). According to Macmahon [1, p. 6], Entry 12(b) is due to Newton.

Entry 12(b). *For positive integers* n *and* r, *define*

$$S_r = S_r(n) = \sum_{k=1}^{n} a_k^r$$

and

$$\mathscr{P}_r = \mathscr{P}_r(n) = \sum_{\substack{1 \le k_i \le n \\ k_1 < k_2 < \cdots < k_r}} a_{k_1} a_{k_2} \cdots a_{k_r}, \qquad r \le n,$$

where $a_1, a_2, \ldots, a_n$ *are arbitrary nonzero complex·numbers. Then, if* $r \ge 1$,

$$r\mathscr{P}_r = \sum_{k=1}^{r} (-1)^{k+1} S_k \mathscr{P}_{r-k}, \tag{12.2}$$

where $\mathscr{P}_0 = 1$.

PROOF. In Entry 12(a), let

$$A_j = (-1)^{j+1} S_j, \qquad j \ge 1.$$

Let $\alpha = \max_{1 \le k \le n} |a_k|$. Then if $|x| < 1/\alpha$,

$$\exp\left(\sum_{j=1}^{\infty} \frac{A_j x^j}{j}\right) = \exp\left(\sum_{k=1}^{n} \sum_{j=1}^{\infty} \frac{(-1)^{j+1}(a_k x)^j}{j}\right)$$

$$= \exp\left(\sum_{j=1}^{n} \mathrm{Log}(1 + a_k x)\right)$$

$$= \prod_{k=1}^{n} (1 + a_k x)$$

$$= \sum_{r=0}^{n} \mathscr{P}_r x^r.$$

Hence, in the notation of Entry 12(a), $P_r = \mathscr{P}_r$, and (12.2) follows immediately from the conclusion of Entry 12(a).   □

In preparation for Entry 13, we need to make two definitions and prove one lemma. For each positive integer r, define

$$S_r = S_r(n, x) = \sum_{k=0}^{\infty} \left(\frac{1}{(k+n)^r} - \frac{1}{(k+n+x+1)^r}\right). \tag{13.1}$$

Let $\varphi(0) = 1$, and define $\varphi(n, x, r) = \varphi(r)$, $r \ge 1$, recursively by

$$r\varphi(r) = \sum_{k=1}^{r} S_k \varphi(r - k). \tag{13.2}$$

Lemma. *If r is a positive integer, then*

$$\frac{d}{dn} \varphi(r) = -\sum_{k=1}^{r} S_{k+1} \varphi(r - k). \tag{13.3}$$

PROOF. We proceed by induction on r. If $r = 1$, equality (13.3) implies that

$$\frac{d}{dn} \varphi(1) = \frac{d}{dn} S_1 = -S_2,$$

which is easily verified from the definition (13.1).
    Now assume that

$$\frac{d}{dn} \varphi(j) = -\sum_{k=1}^{j} S_{k+1} \varphi(j - k), \qquad 1 \le j \le r - 1. \tag{13.4}$$

Hence, by (13.2), (13.1), and (13.4),

$$\frac{d}{dn} \varphi(r) = \frac{1}{r} \frac{d}{dn} \sum_{k=1}^{r} S_k \varphi(r - k)$$

$$= \frac{1}{r}\left(-\sum_{k=1}^{r} k S_{k+1} \varphi(r - k) - \sum_{k=1}^{r} S_k \sum_{j=1}^{r-k} S_{j+1} \varphi(r - k - j)\right)$$

$$= -\frac{1}{r}\left(\sum_{k=1}^{r} kS_{k+1}\varphi(r-k) + \sum_{j=1}^{r} S_{j+1}\sum_{k=1}^{r-j} S_k\varphi(r-k-j)\right)$$

$$= -\frac{1}{r}\left(\sum_{k=1}^{r} kS_{k+1}\varphi(r-k) + \sum_{j=1}^{r} S_{j+1}(r-j)\varphi(r-j)\right)$$

$$= -\sum_{j=1}^{r} S_{j+1}\varphi(r-j),$$

which completes the proof. $\square$

Entry 13. *If* Re $x > -1$ *and* r *is any positive integer, then*

$$\sum_{k=0}^{\infty} \frac{(-x)_k}{(n+k)^{r+1}k!} = \frac{\Gamma(n)\Gamma(x+1)}{\Gamma(n+x+1)}\varphi(r). \tag{13.5}$$

PROOF. Now by Example 5 in Section 10,

$$\sum_{k=0}^{\infty} \frac{(-x)_k}{(n+k)^2 k!} = \frac{\Gamma(n)\Gamma(x+1)}{\Gamma(n+x+1)}S_1 = \frac{\Gamma(n)\Gamma(x+1)}{\Gamma(n+x+1)}\varphi(1).$$

Thus, (13.5) is valid for $r = 1$.

   Proceeding by induction, we assume that (13.5) holds for any fixed positive integer r and show that (13.5) is true with r replaced by $r + 1$. Differentiating both sides of (13.5) with respect to n and using the foregoing lemma, we find that

$$-(r+1)\sum_{k=0}^{\infty} \frac{(-x)_k}{(n+k)^{r+2}k!}$$

$$= \frac{\Gamma(n)\Gamma(x+1)}{\Gamma(n+x+1)}\left(\{\psi(n) - \psi(n+x+1)\}\varphi(r) + \frac{d}{dn}\varphi(r)\right)$$

$$= \frac{\Gamma(n)\Gamma(x+1)}{\Gamma(n+x+1)}\left(-S_1\varphi(r) - \sum_{k=1}^{r} S_{k+1}\varphi(r-k)\right)$$

$$= -\frac{\Gamma(n)\Gamma(x+1)}{\Gamma(n+x+1)}\sum_{k=0}^{r} S_{k+1}\varphi(r-k)$$

$$= -\frac{\Gamma(n)\Gamma(x+1)}{\Gamma(n+x+1)}(r+1)\varphi(r+1),$$

from which (13.5), with r replaced by $r + 1$, follows. $\square$

Corollary 1. *Let* $S_r(n, x)$ *and* $\varphi(n, x, r)$ *be defined by* (13.1) *and* (13.2), *respectively. If* $n = \frac{1}{2}$ *and* $x = -\frac{1}{2}$, *then* $S_1 = 2 \text{ Log } 2$, $S_r = (2^r - 2)\zeta(r), r \geq 2$, *and*

$$1 + \frac{1}{3^{r+1}}\left(\frac{1}{2}\right) + \frac{1}{5^{r+1}}\left(\frac{1\cdot 3}{2\cdot 4}\right) + \cdots = \frac{\pi}{2^{r+1}}\varphi(r), \qquad r \geq 1. \tag{13.6}$$

PROOF. The proposed formulas for $S_r, r \geq 1$, are easily determined from (13.1) after brief calculations. Setting $n = \frac{1}{2}$ and $x = -\frac{1}{2}$ in Entry 13 yields

$$\frac{1}{(\frac{1}{2})^{r+1}} + \frac{1}{(\frac{3}{2})^{r+1}}\left(\frac{1}{2}\right) + \frac{1}{(\frac{5}{2})^{r+1}}\left(\frac{1\cdot 3}{2\cdot 4}\right) + \cdots = \pi\varphi(r),$$

from which (13.6) trivially follows.   □

In the notebooks (p. 124), Ramanujan redefines S_r for Corollary 1. We emphasize that his formulation of Corollary 1 is correct, however. Likewise, in Corollary 2, Ramanujan has redefined S_r in the notebooks. In fact, Ramanujan has proved Corollary 1 in his second published paper [1], [16, pp. 15–17] by another method. Entry 13 and Example 1 below are also given in [1].

Corollary 2. *Let S_r and $\varphi(r)$ be defined by (13.1) and (13.2), respectively, with $n = 1$ and $x = -\frac{1}{2}$. Then $S_1 = 2 - 2\,\text{Log}\,2$, $S_r = (2 - 2^r)\zeta(r) + 2^r, r \geq 2$, and*

$$1 + \frac{1}{2^{r+1}}\left(\frac{1}{2}\right) + \frac{1}{3^{r+1}}\left(\frac{1\cdot 3}{2\cdot 4}\right) + \cdots = 2\varphi(r), \qquad r \geq 1.$$

PROOF. Let $n = 1$ and $x = -\frac{1}{2}$ in Entry 13. The proof is completely analogous to that of Corollary 1.   □

Example 1

$$1 + \frac{1}{3^3}\left(\frac{1}{2}\right) + \frac{1}{5^3}\left(\frac{1\cdot 3}{2\cdot 4}\right) + \cdots = \frac{\pi}{4}\,\text{Log}^2\,2 + \frac{\pi^3}{48}.$$

PROOF. Letting S denote the infinite series above, we find from Corollary 1 and (13.2) that

$$S = \frac{\pi}{8}\varphi(2) = \frac{\pi}{16}\{S_1\varphi(1) + S_2\varphi(0)\}$$

$$= \frac{\pi}{16}\{S_1^2 + S_2\}$$

$$= \frac{\pi}{16}\left\{4\,\text{Log}^2\,2 + \frac{\pi^2}{3}\right\}.$$ □

Example 2

$$\int_0^{\pi/2} \theta \cot\theta \,\text{Log}(\sin\theta)\,d\theta = -\frac{\pi}{4}\,\text{Log}^2\,2 - \frac{\pi^3}{48}.$$

PROOF. Letting $u = \sin\theta$ and integrating by parts, we first find that

$$\int_0^{\pi/2} \theta \cot\theta \,\text{Log}(\sin\theta)\,d\theta = -\frac{1}{2}\int_0^1 \frac{\text{Log}^2\,u}{\sqrt{1 - u^2}}\,du. \tag{13.7}$$

Next, for each nonnegative integer k, an elementary calculation shows that

$$\frac{1}{2} \int_0^1 u^k \operatorname{Log}^2 u \, du = \frac{1}{(k+1)^3}. \tag{13.8}$$

Lastly, recall that

$$(1 - u^2)^{-1/2} = 1 + \tfrac{1}{2}u^2 + \frac{1 \cdot 3}{2 \cdot 4} u^4 + \cdots, \qquad |u| < 1. \tag{13.9}$$

Now substitute (13.9) into (13.7) and integrate termwise with the help of (13.8) to obtain

$$-\int_0^{\pi/2} \theta \cot \theta \operatorname{Log}(\sin \theta) \, d\theta = 1 + \frac{1}{3^3}\left(\frac{1}{2}\right) + \frac{1}{5^3}\left(\frac{1 \cdot 3}{2 \cdot 4}\right) + \cdots.$$

Using Example 1, we complete the proof.   □

In preparation for Entry 14, define

$$C_{m,n}(x) = \sum_{k=0}^{\infty} \frac{1}{(k+x)^m} \sum_{j=0}^{k-1} \frac{1}{(j+x)^n}$$

and

$$S_m(x) = \sum_{k=0}^{\infty} \frac{1}{(k+x)^m},$$

where m and n are positive integers with $m \geq 2$.

Entry 14. *Let n be an integer with $n \geq 2$. Then*

$$2 \sum_{k=1}^{\infty} \frac{1}{(k+x)^n} \sum_{j=1}^{k} \frac{1}{j} = n S_{n+1}(x) - \sum_{r=1}^{n-2} S_{r+1}(x) S_{n-r}(x)$$

$$- 2 \sum_{k=1}^{\infty} \frac{1}{(k+x)^n} \sum_{j=0}^{k} \frac{1}{j+x}$$

$$+ 2 \sum_{k=1}^{\infty} \sum_{j=0}^{k-1} \frac{1}{(j+x)^n} \left(\frac{1}{k-j} - \frac{1}{k+x} \right).$$

PROOF. Consider the decomposition from Nielsen's book [1, p. 48]

$$\frac{1}{(k+x)^n(k-j)} = -\sum_{r=0}^{n-1} \frac{1}{(k+x)^{n-r}(j+x)^{r+1}} + \frac{1}{(j+x)^n(k-j)}.$$

Summing on j, $0 \leq j \leq k - 1$, we find that

$$\frac{1}{(k+x)^n} \sum_{j=1}^{k} \frac{1}{j} = -\sum_{r=0}^{n-1} \frac{1}{(k+x)^{n-r}} \sum_{j=0}^{k-1} \frac{1}{(j+x)^{r+1}} + \sum_{j=0}^{k-1} \frac{1}{(j+x)^n(k-j)}.$$

Next, sum on k, $1 \leq k < \infty$, to obtain

$$\sum_{k=1}^{\infty} \frac{1}{(k+x)^n} \sum_{j=1}^{k} \frac{1}{j} = -\sum_{r=1}^{n-2} C_{n-r,r+1} - C_{n,1}(x)$$

$$+ \sum_{k=1}^{\infty} \sum_{j=0}^{k-1} \frac{1}{(j+x)^n} \left(\frac{1}{k-j} - \frac{1}{k+x} \right). \tag{14.1}$$

Observe that

$$S_m(x)S_n(x) = S_{m+n}(x) + C_{m,n}(x) + C_{n,m}(x), \qquad m, n \geq 2.$$

Thus, (14.1) may be written in the form

$$2\sum_{k=1}^{\infty} \frac{1}{(k+x)^n} \sum_{j=1}^{k} \frac{1}{j} = (n-2)S_{n+1}(x) - \sum_{r=1}^{n-2} S_{r+1}(x)S_{n-r}(x)$$

$$- 2\sum_{k=1}^{\infty} \frac{1}{(k+x)^n} \sum_{j=0}^{k} \frac{1}{j+x} + 2S_{n+1}(x)$$

$$+ 2\sum_{k=1}^{\infty} \sum_{j=0}^{k-1} \frac{1}{(j+x)^n} \left(\frac{1}{k-j} - \frac{1}{k+x} \right).$$

This completes the proof. □

Ramanujan's formulation of Entry 14 (p. 124) is somewhat imprecise. For several other results of this type, see Chapter 9 and the relevant references mentioned in Part I [9].

Entry 15. *If α and β are arbitrary complex numbers, then*

$$\sum_{k=0}^{n-1} \frac{\Gamma(\alpha+k+1)\Gamma(\beta+k+1)}{\Gamma(\alpha+\beta+k+2)k!} \sim \text{Log } n - \psi(\alpha+1) - \psi(\beta+1) - \gamma,$$

as n tends to ∞.

PROOF. From a theorem in Luke's book [1, p. 110, Eq. (35)],

$$\frac{\Gamma(a)\Gamma(b)}{\Gamma(a+b)} \sum_{k=0}^{n-1} \frac{(a)_k(b)_k}{(a+b)_k k!} \sim \text{Log } n - \psi(a) - \psi(b) - \gamma,$$

as n tends to ∞. Putting $a = \alpha + 1$ and $b = \beta + 1$, we deduce Entry 15. □

Corollary. *Let $0 < x < 1$. Then as x tends to 0,*

$$\pi \, {}_2F_1(\tfrac{1}{2}, \tfrac{1}{2}; 1; 1-x) \sim \text{Log } x + 4\,\text{Log } 2.$$

PROOF. From a general theorem in Luke's text [1, p. 87, Eq. (11)],

$${}_2F_1(a, b; a+b; 1-x) \sim -\frac{\Gamma(a+b)}{\Gamma(a)\Gamma(b)}(\text{Log } x + \psi(a) + \psi(b) + 2\gamma), \tag{15.1}$$

as x tends to 0, $0 < x < 1$. The corollary now follows by putting $a = b = \tfrac{1}{2}$ and using the fact that $\psi(\tfrac{1}{2}) = -\gamma - 2\,\text{Log } 2$ (Luke [1, p. 13]). □

It follows from Entry 15 that

$$\sum_{k=0}^{n-1} \frac{\Gamma(\alpha + k + 1)\Gamma(\beta + k + 1)}{\Gamma(\alpha + \beta + k + 2)k!} \sim \text{Log } n,$$

as n tends to ∞. This weaker result is due to Hill [1], [2]. See also Copson's book [2, p. 266]. According to Copson [2, p. 267], Gauss showed that

$$\lim_{x \to 1-} \frac{{}_2F_1(a, b; a + b; x)}{\text{Log}\{1/(1-x)\}} = \frac{\Gamma(a+b)}{\Gamma(a)\Gamma(b)},$$

which is a consequence of (15.1). See also Whittaker and Watson's text [1, p. 299].

Entry 16. *If $A_0, A_1, \ldots, A_r$ are any complex numbers and*

$$P_r = \sum_{k=0}^{r} A_k(-1)^k \binom{r}{k}, \qquad r \geq 0,$$

then

$$A_r = \sum_{k=0}^{r} P_k(-1)^k \binom{r}{k}, \qquad r \geq 0.$$

A proof of this well-known inversion formula can be found in Riordan's book [1, pp. 43, 44].

Entry 17. *Suppose that*

$$f(x) = f(r, x) = \sum_{k=0}^{\infty} \frac{(r)_k A_k}{k! x^{r+k}} \tag{17.1}$$

is analytic for $|x| > R$. For $|x| > \sup(R, |h|)$, write

$$f(x) = \sum_{k=0}^{\infty} \frac{(r)_k B_k}{k!(x+h)^{r+k}}. \tag{17.2}$$

Then

$$B_k = \sum_{j=0}^{k} A_j h^{k-j} \binom{k}{j}, \qquad k \geq 0.$$

PROOF. For $|x| > R, |h|$,

$$f(x) = \sum_{k=0}^{\infty} \frac{(r)_k B_k}{k! x^{r+k}(1 + h/x)^{r+k}}$$

$$= \sum_{k=0}^{\infty} \frac{(r)_k B_k}{k! x^{r+k}} \sum_{j=0}^{\infty} \binom{-r-k}{j}\left(\frac{h}{x}\right)^j. \tag{17.3}$$

Now equate coefficients of x^{-r-n} in (17.1) and (17.3) to deduce that

$$\frac{(r)_n A_n}{n!} = \sum_{k=0}^{n} \frac{(r)_k B_k}{k!} \binom{-r-k}{n-k} h^{n-k}, \qquad n \geq 0.$$

After a straightforward calculation, the foregoing equality yields, for $h \neq 0$,

$$\frac{A_n(-1)^n}{h^n} = \sum_{k=0}^{n} B_k(-h)^k \binom{n}{k}, \qquad n \geq 0.$$

Applying the inversion formula of Entry 16 and simplifying, we conclude that

$$B_n h^{-n} = \sum_{k=0}^{n} A_k h^{-k} \binom{n}{k}, \qquad n \geq 0,$$

which implies the desired conclusion. $\qquad\qquad\qquad\qquad\qquad\qquad\quad\square$

Entry 18(i). *Suppose that (17.1) holds. Assume also that*

$$f(x) = \sum_{k=0}^{\infty} \frac{(-1)^k (r)_k A_k}{k!(x-1)^{r+k}} \tag{18.1}$$

for $|x| > \sup(R, 1)$. Furthermore, assume that $\sum_{k=0}^{\infty}(A_k x^k/k!)$ is analytic for $|x| < R^*$. Then for $|x| < R^*$,

$$e^x \sum_{k=0}^{\infty} \frac{A_k(-x)^k}{k!} = \sum_{k=0}^{\infty} \frac{A_k x^k}{k!}. \tag{18.2}$$

PROOF. Apply Entry 17 with $h = -1$. Comparing (17.2) and (18.1), we find that

$$A_k = \sum_{j=0}^{k} A_j(-1)^j \binom{k}{j}, \qquad k \geq 0. \tag{18.3}$$

On the other hand, for $|x| < R^*$, by the Cauchy multiplication of power series,

$$e^x \sum_{j=0}^{\infty} \frac{A_j(-x)^j}{j!} = \sum_{k=0}^{\infty} \frac{C_k x^k}{k!}, \tag{18.4}$$

where

$$C_k = \sum_{j=0}^{k} (-1)^j A_j \binom{k}{j}, \qquad k \geq 0. \tag{18.5}$$

By (18.3) and (18.5), $A_k = C_k$, $k \geq 0$. Thus, (18.4) becomes the equality that we sought to prove. $\qquad\qquad\qquad\qquad\qquad\qquad\qquad\qquad\qquad\qquad\quad\square$

In Entry 18(ii), Ramanujan claims that if (17.1) and (18.1) hold, then

$$\frac{1}{\varphi^r(x)} \sum_{k=0}^{\infty} \frac{(r)_k A_k}{k!} \left\{ \frac{\varphi(x) - \varphi(-x)}{\varphi(x)} \right\}^k$$

is always an even function of x. This is clearly false. For example, letting $\varphi(x) = x$ and $r = 1$ provides a counterexample.

Entry 18(iii). *Suppose that* (17.1) *and* (18.1) *hold. Then, if n is an even integer,*

$$\tfrac{1}{2}nA_{n-1} = \sum_{k=1}^{n/2} \binom{n}{2k}(2^{2k} - 1)B_{2k}A_{n-2k}, \qquad n \geq 2, \qquad (18.6)$$

where B_j denotes the jth Bernoulli number.

PROOF. From the generating function (Abramowitz and Stegun [1, p. 804]),

$$\frac{x}{e^x - 1} = \sum_{n=0}^{\infty} \frac{B_n x^n}{n!}, \qquad |x| < 2\pi,$$

we find that, for $|x| < \pi$,

$$\frac{x}{e^x + 1} = \frac{x}{e^x - 1} - \frac{2x}{e^{2x} - 1} = \sum_{k=1}^{\infty} \frac{B_k(1 - 2^k)x^k}{k!}. \qquad (18.7)$$

We now use the representation for e^x given by (18.2) on the left side of (18.7). After some manipulation and simplification, we deduce that, for $|x| < \min(\pi, R^*)$,

$$x \sum_{j=0}^{\infty} \frac{A_j(-x)^j}{j!} = 2 \sum_{k=1}^{\infty} \frac{B_k(1 - 2^k)x^k}{k!} \sum_{j=0}^{\infty} \frac{A_{2j}x^{2j}}{(2j)!}.$$

If we equate coefficients of x^n, with n even, on both sides above, we readily deduce (18.6). ☐

Entry 19. *Suppose that $|x|, |x - 1| > 1$. Then*

$$x^{-r} \, {}_2F_1(r, m; n; 1/x) = (x - 1)^{-r} \, {}_2F_1(r, n - m; n; -1/(x - 1)).$$

This transformation is well known (Bailey [4, p. 10]) and is generally attributed to Gauss or Kummer. However, Askey [1] has indicated that it was originally discovered by Pfaff [1]. We shall give what was evidently Ramanujan's argument.

PROOF. Apply Entry 17 with $A_k = (m)_k/(n)_k$ and $h = 1$. We then see that it suffices to show that

$$\frac{(n - m)_k}{(n)_k} = \sum_{j=0}^{k} (-1)^j \frac{(m)_j}{(n)_j} \binom{k}{j}, \qquad k \geq 0. \qquad (19.1)$$

But this is simply Vandermonde's theorem (Bailey [4, p. 3]), which is a special case of Gauss's theorem, Entry 8. ☐

Entry 20. *Let*

$$\varphi(x) = \sum_{r=0}^{\infty} \frac{\varphi^{(r)}(1)}{r!}(x - 1)^r \qquad (20.1)$$

be analytic for $|x - 1| < R$, where $R > 1$. Suppose that m and n are complex *parameters such that the order of summation in*

$$\sum_{k=0}^{\infty} \frac{(m)_k}{(n)_k k!} \sum_{r=k}^{\infty} \frac{\varphi^{(r)}(1)(-1)^r(-r)_k}{r!}$$

may be inverted. Then

$$\sum_{k=0}^{\infty} \frac{(m)_k \varphi^{(k)}(0)}{(n)_k k!} = \sum_{k=0}^{\infty} \frac{(-1)^k (n-m)_k \varphi^{(k)}(1)}{(n)_k k!}. \tag{20.2}$$

PROOF. Using (20.1) to calculate $\varphi^{(k)}(0)$, $0 \leq k < \infty$, and inverting the order of summation by hypothesis, we find that

$$\sum_{k=0}^{\infty} \frac{(m)_k \varphi^{(k)}(0)}{(n)_k k!} = \sum_{k=0}^{\infty} \frac{(m)_k}{(n)_k k!} \sum_{r=k}^{\infty} \frac{\varphi^{(r)}(1)(-1)^r(-r)_k}{r!}$$

$$= \sum_{r=0}^{\infty} \frac{(-1)^r \varphi^{(r)}(1)}{r!} \sum_{k=0}^{r} \frac{(m)_k(-r)_k}{(n)_k k!}$$

$$= \sum_{r=0}^{\infty} \frac{(-1)^r \varphi^{(r)}(1)(n-m)_r}{r!(n)_r},$$

by Vandermonde's theorem (19.1).   □

Note that if $\varphi(x) = (x-1)^r$, where r is a nonnegative integer, then (20.2) yields

$$\sum_{k=0}^{r} \frac{(m)_k(-r)_k}{(n)_k k!} = \frac{(n-m)_r}{(n)_r}.$$

Hence, in this case, (20.2) reduces to Vandermonde's theorem.

Entry 21. *For any complex numbers m, n, and x,*

$$e^x \sum_{k=0}^{\infty} \frac{(-1)^k (n-m)_k x^k}{(n)_k k!} = \sum_{r=0}^{\infty} \frac{(m)_r x^r}{(n)_r r!}.$$

PROOF. Now,

$$e^x \sum_{k=0}^{\infty} \frac{(-1)^k (n-m)_k x^k}{(n)_k k!} = \sum_{j=0}^{\infty} \frac{x^j}{j!} \sum_{k=0}^{\infty} \frac{(-1)^k (n-m)_k x^k}{(n)_k k!}.$$

The coefficient of x^r on the right side is

$$\sum_{k=0}^{r} \frac{(-1)^k (n-m)_k}{(n)_k k!(r-k)!} = \frac{(m)_r}{(n)_r r!},$$

by Vandermonde's theorem (19.1). This completes the proof.   □

Entry 21 is due to Kummer [1]. An alternate proof can be obtained from Entry 19 by replacing x by r/x and letting r tend to ∞.

Entry 22. *Suppose that* $|x|, |x + 1| > 1$. *Then*

$$(x + 1)^{-r} \, {}_2F_1(r, m; 2m; 1/(x + 1)) = x^{-r} \, {}_2F_1(r, m; 2m; -1/x).$$

PROOF. Set $n = 2m$ and replace x by $x + 1$ in Entry 19. $\square$

Entry 23. *Let m and x be any complex numbers. Then*

$$e^x \sum_{k=0}^{\infty} \frac{(-1)^k (m)_k x^k}{(2m)_k k!} = \sum_{r=0}^{\infty} \frac{(m)_r x^r}{(2m)_r r!}.$$

PROOF. Set $n = 2m$ in Entry 21. $\square$

Corollary 1. *If x is any complex number, then*

$$e^x \left(1 - \left(\frac{1}{2}\right)\frac{x}{1!} + \left(\frac{1 \cdot 3}{2 \cdot 4}\right)\frac{x^2}{2!} - \cdots\right) = 1 + \left(\frac{1}{2}\right)\frac{x}{1!} + \left(\frac{1 \cdot 3}{2 \cdot 4}\right)\frac{x^2}{2!} + \cdots.$$

PROOF. Let $m = \frac{1}{2}$ in Entry 23. $\square$

Corollary 2. *If $|x| < 1$ and* Re $x < \frac{1}{2}$, *then*

$$\, {}_2F_1\left(\frac{1}{2}, \frac{1}{2}; 1; \frac{-x}{1 - x}\right) = \sqrt{1 - x} \, {}_2F_1(\tfrac{1}{2}, \tfrac{1}{2}; 1; x).$$

PROOF. In Entry 22, let $r = m = \frac{1}{2}$ and replace x by $-1/x$. $\square$

The function $\, {}_2F_1(\frac{1}{2}, \frac{1}{2}; 1; x)$ is a constant multiple of the complete elliptic integral of the first kind and is central to the theory of elliptic functions. See Part III of our account [11] of Ramanujan's notebooks.

T. Matala-Aho and K. Väänänen [1] have studied the arithmetic properties of $\, {}_2F_1(\frac{1}{2}, \frac{1}{2}; 1; \theta)$ when θ is algebraic.

Entry 24. *Let $|x|, |x - 1| > 1$ and suppose that m is arbitrary and that* Re $n > 0$. *Then*

$$\sum_{k=0}^{\infty} \frac{(m)_k}{(n + k)k! x^{n+k}} = \sum_{k=0}^{\infty} \frac{(-1)^k (n + 1 - m)_k}{(n + k)k!(x - 1)^{n+k}}.$$

PROOF. In Entry 19, replace n by $n + 1$ and set $r = n + 1$ to obtain

$$\sum_{k=0}^{\infty} \frac{(m)_k}{k! x^{n+k+1}} = \sum_{k=0}^{\infty} \frac{(-1)^k (n + 1 - m)_k}{k!(x - 1)^{n+k+1}}.$$

Integrate both sides over $[x, \infty)$ to achieve the desired result. $\square$

Entry 25. *Let $|x|, |x - 1| > 1$ and suppose that n is arbitrary. Then*

$$\sum_{k=0}^{\infty} \frac{k!}{(n)_{k+1} x^{k+1}} = \sum_{k=0}^{\infty} \frac{(-1)^k}{(n + k)(x - 1)^{k+1}}.$$

PROOF. Put $r = m = 1$ and replace n by $n + 1$ in Entry 19, and multiply both sides by $1/n$.   □

Entry 26. *If* $|x| < 1$ *and* α, β, *and* γ *are arbitrary, then*

$$(1 - x)^{\alpha + \beta} \, _2F_1(\alpha, \beta; \gamma; x) = (1 - x)^{\gamma} \, _2F_1(\gamma - \alpha, \gamma - \beta; \gamma; x).$$

Entry 26 is elementary and well known; see Bailey's tract [4, p. 2].

Entry 27. *If* $\operatorname{Re}(n + 1) > -\operatorname{Re}(x + y)$, $-\operatorname{Re}(p + q)$, *then*

$$\frac{\Gamma(x + y + n + 1)}{\Gamma(x + n + 1)\Gamma(y + n + 1)} \, _3F_2 \left[\begin{array}{c} -p, -q, x + y + n + 1 \\ x + n + 1, y + n + 1 \end{array} \right]$$

$$= \frac{\Gamma(p + q + n + 1)}{\Gamma(p + n + 1)\Gamma(q + n + 1)} \, _3F_2 \left[\begin{array}{c} -x, -y, p + q + n + 1 \\ p + n + 1, q + n + 1 \end{array} \right].$$

Entry 27 is a famous theorem of Thomae [1] and can be derived from Entry 26. Hardy [1, p. 499], [7, p. 512] has extensively discussed Entry 27 and has given references to other proofs. In Bailey's book [4, p. 14], Entry 27 is equivalent to formula (1).

Entry 28. *If* $\operatorname{Re}(n + 1) > -\operatorname{Re}(x + y)$, $-\operatorname{Re}(p - 1)$, *then*

$$_3F_2 \left[\begin{array}{c} -x, -y, p + n \\ n, p + n + 1 \end{array} \right] = \frac{(p + n)\Gamma(n)\Gamma(x + y + n + 1)}{\Gamma(x + n + 1)\Gamma(y + n + 1)}$$

$$\times \, _3F_2 \left[\begin{array}{c} -p, 1, x + y + n + 1 \\ x + n + 1, y + n + 1 \end{array} \right].$$

PROOF. Set $q = -1$ in Entry 27.   □

Entry 29(a). *If* $\operatorname{Re} n > -1$, *then*

$$_3F_2 \left[\begin{array}{c} \frac{1}{2}, \frac{1}{2}, n + 1 \\ 1, n + 2 \end{array} \right] = \frac{4(n + 1)}{\pi} \, _3F_2 \left[\begin{array}{c} -n, 1, 1 \\ \frac{3}{2}, \frac{3}{2} \end{array} \right].$$

PROOF. In Entry 28, put $x = y = -\frac{1}{2}$, $n = 1$, and $p = n$.   □

Entry 29(b). *If* n *is a nonnegative integer, then*

$$\frac{1}{n + 1} \, _3F_2 \left[\begin{array}{c} \frac{1}{2}, \frac{1}{2}, n + 1 \\ 1, n + 2 \end{array} \right] = \frac{\Gamma^2(n + 1)}{\Gamma^2(n + \frac{3}{2})} \sum_{k=0}^{n} \frac{(\frac{1}{2})_k^2}{(k!)^2}. \qquad (29.1)$$

This extremely interesting result was communicated in Ramanujan's [16, p. 351] first letter to Hardy and was first established in print by Watson [4] in 1929. A flurry of papers was written on this formula and certain generalizations in the years 1929–1931. References may be found in Bailey's book [4, pp. 92–95]. Related results are given in Entry 32 and Section 35 below. A more recent proof of (29.1) has been given by Dutka [1]. Further identities for partial sums of hypergeometric series have been established by Lamm and Szabo [1], [2] in their work on Coulomb approximations. The finite sum on the right side of (29.1) arises in the theory of functions of one complex variable and is called Landau's constant. For details of this connection, see Watson's paper [5].

Entry 29(c). *If n is any complex number, then*

$$_3F_2\left[\begin{matrix} \frac{1}{2}, \frac{1}{2}, n+1 \\ 1, n+2 \end{matrix}\right] = \frac{2}{\pi} \, _3F_2\left[\begin{matrix} \frac{1}{2}, 1, n+\frac{3}{2} \\ \frac{3}{2}, n+2 \end{matrix}\right].$$

PROOF. In Entry 27, let $p = -n-1, q = -\frac{1}{2}, x = -n-\frac{3}{2}$, and $y = -\frac{1}{2}$, and replace n by $n+\frac{3}{2}$.   □

Entry 29(d). *If Re $n > -\frac{3}{2}$, then*

$$_3F_2\left[\begin{matrix} \frac{1}{2}, 1, n+\frac{3}{2} \\ \frac{3}{2}, n+2 \end{matrix}\right] = \frac{\sqrt{\pi}\,\Gamma(n+2)}{\Gamma(n+\frac{3}{2})} \, _3F_2\left[\begin{matrix} \frac{1}{2}, \frac{1}{2}, -n \\ 1, \frac{3}{2} \end{matrix}\right].$$

PROOF. In Entry 27, put $p = -\frac{1}{2}, q = -1, x = n, y = -\frac{1}{2}$, and $n = 1$.   □

Corollary 1. *If G denotes Catalan's constant, that is,*

$$G = \sum_{k=0}^{\infty} \frac{(-1)^k}{(2k+1)^2},\tag{29.2}$$

then

$$\tfrac{\pi}{4} \, _3F_2(\tfrac{1}{2}, \tfrac{1}{2}, \tfrac{1}{2}; 1, \tfrac{3}{2}) = G.\tag{29.3}$$

PROOF. Putting $n = -\frac{1}{2}$ in Entry 29(a), we find that

$$\tfrac{\pi}{2} \, _3F_2(\tfrac{1}{2}, \tfrac{1}{2}, \tfrac{1}{2}; 1, \tfrac{3}{2}) = \, _3F_2(\tfrac{1}{2}, 1, 1; \tfrac{3}{2}, \tfrac{3}{2}).$$

On the other hand, from Example (i) in Section 32 of Chapter 9 (see Part I [9]),

$$_3F_2(\tfrac{1}{2}, 1, 1; \tfrac{3}{2}, \tfrac{3}{2}) = 2G.$$

Combining these two equalities, we deduce (29.3).   □

Corollary 2. *As n tends to ∞,*

$$\pi \, _3F_2(\tfrac{1}{2}, \tfrac{1}{2}, n; 1, n+1) \sim \text{Log } n + 4 \text{ Log } 2 + \gamma.$$

Watson [5] has established an asymptotic formula for the finite sum on the right side of (29.1) as n tends to ∞. Thus, Corollary 2 follows from Entry 29(b), Watson's theorem, and Stirling's formula. We shall not relate any more details, because Entry 35(i) below gives a very closely related, fuller asymptotic expansion. R. J. Evans [1, Theorem 21] has generalized Corollary 2 by showing that

$$\frac{\Gamma(a)\Gamma(b)}{\Gamma(a+b)} {}_3F_2\left[\begin{array}{c} a, b, c \\ a+b, c+1 \end{array}\right] = \text{Log } c - \gamma - \psi(a) - \psi(b) + O\left(\frac{\text{Log } c}{c}\right),$$

as real c tends to ∞.

Entry 30. *If* Re $n > -$Re x, $-$Re y, *then*

$${}_3F_2\left[\begin{array}{c} -x, 1, y+n \\ n, y+n+1 \end{array}\right] = \frac{y+n}{x+n} {}_3F_2\left[\begin{array}{c} -y, 1, x+n \\ n, x+n+1 \end{array}\right].$$

PROOF. Let $y = -1$ and $p = y$ in Entry 28. □

Entry 31. *If* Re$(x+y+n+1) > 0$ *and* Re$(2x+2y+2z+2u+3n+4) > 0$, *then*

$${}_6F_5\left[\begin{array}{c} \tfrac{1}{2}n+1, n, -x, -y, -z, -u \\ \tfrac{1}{2}n, x+n+1, y+n+1, z+n+1, u+n+1 \end{array}; -1\right]$$

$$= \frac{\Gamma(x+n+1)\Gamma(y+n+1)}{\Gamma(n+1)\Gamma(x+y+n+1)} {}_3F_2\left[\begin{array}{c} -x, -y, z+u+n+1 \\ z+n+1, u+n+1 \end{array}\right].$$

Entry 31 is an immediate consequence of Whipple's transformation (6.1). See Bailey's tract [4, p. 28] for details.

It is interesting to note that although Ramanujan did not discover Whipple's transformation, he did find this important special case approximately 20 years before Whipple's proof [1] in 1926. An enlightening discussion of Whipple's theorem can be found in Askey's paper [3].

Suppose that we set $-n = x = y = z = u = -\frac{1}{2}$ in Entry 31. Then

$$1 - 5\left(\frac{1}{2}\right)^5 + 9\left(\frac{1\cdot 3}{2\cdot 4}\right)^5 - 13\left(\frac{1\cdot 3\cdot 5}{2\cdot 4\cdot 6}\right)^5 + \cdots$$

$$= \frac{1}{\Gamma(\tfrac{3}{2})\Gamma(\tfrac{1}{2})} {}_3F_2\left[\begin{array}{c} \tfrac{1}{2}, \tfrac{1}{2}, \tfrac{1}{2} \\ 1, 1 \end{array}\right] = \frac{2}{\Gamma^4(\tfrac{3}{4})}, \tag{31.1}$$

by Example 18 in Section 7. This result may be found in Ramanujan's [16, p. xxviii] first letter to Hardy as well as in Hardy's book [9, p. 7, Eq. (1.4)]. Equality (31.1) was established by Watson [6], who gave the same proof that we have given. Another proof was given by Hardy [2], [7, pp. 517, 518].

Entry 32. *If* $x + y + z = 0$ *and* x *is a positive integer, then*

$${}_3F_2\left[\begin{array}{c} n, -x, -y \\ n+1, z \end{array}\right] = \frac{\Gamma(n+1)\Gamma(x+1)}{\Gamma(n+x+1)} \sum_{k=0}^{x} \frac{(n)_k(y+z)_k}{(z)_k k!}. \tag{32.1}$$

PROOF. Consider the following result

$$\sum_{k=0}^{n-1} \frac{(a)_k(b)_k}{(f)_k k!} = \frac{\Gamma(a+n)\Gamma(b+n)}{\Gamma(n)\Gamma(a+b+n)} \; {}_3F_2\left[\begin{array}{c} a, b, f+n-1 \\ f, a+b+n \end{array}\right], \qquad (32.2)$$

due to Bailey [2], [4, p. 93]. Set $a = n$, $b = y + z$, $f = z$, and $n = x + 1$ in (32.2) and use the fact that $x + y + z = 0$ to complete the proof. $\qquad\square$

In fact, Entry 29(b) is not a special case of Entry 32. However, (32.2) does generalize Entry 29(b). The hypothesis $x + y + z = 0$ is not mentioned by Ramanujan. If $x + y + z \neq 0$, (32.1) is false in general. For example, if $x = 2$ and $y = z = -\frac{1}{2}$, then (32.1) is erroneous, as can be seen by a comparison with the correct formula (32.2) with the proper parameters. For Entry 33 below, Ramanujan does provide the hypothesis $x + y + z = 0$.

Entry 33. *If $x + y + z = 0$ and $x + y + n$ is a positive integer, then*

$$\; {}_3F_2\left[\begin{array}{c} n, -x, -y \\ n+1, z \end{array}\right] = \frac{\Gamma(n+1)\Gamma(x+y+n+1)}{\Gamma(x+n+1)\Gamma(y+n+1)} \sum_{k=0}^{x+y+n} \frac{(-x)_k(-y)_k}{(z)_k k!}.$$

PROOF. In (32.2), set $a = -x$, $b = -y$, and $f = z$, and replace n by $x + y + n + 1$. $\qquad\square$

Entry 34. *If x and y are arbitrary, then*

$$\; {}_2F_1(x, y; \tfrac{1}{2}(x + y + 1); \tfrac{1}{2}) = \frac{\sqrt{\pi}\,\Gamma(\tfrac{1}{2}x + \tfrac{1}{2}y + \tfrac{1}{2})}{\Gamma(\tfrac{1}{2}x + \tfrac{1}{2})\Gamma(\tfrac{1}{2}y + \tfrac{1}{2})}.$$

Entry 34 is due to Gauss [1]. In Bailey's text [4, p. 11], Entry 34 is Eq. (2). The following result is due to Kummer [1] and can be found in Bailey's monograph [4, p. 11, Eq. (3)].

Corollary. *If x and n are arbitrary, then*

$$\; {}_2F_1(\tfrac{1}{2} - \tfrac{1}{2}x, \tfrac{1}{2} + \tfrac{1}{2}x; \tfrac{1}{2}n + \tfrac{1}{2}; \tfrac{1}{2}) = \frac{\sqrt{\pi}\,2^{(1-n)/2}\Gamma(\tfrac{1}{2}n + \tfrac{1}{2})}{\Gamma(\tfrac{1}{4}\{n - x + 2\})\Gamma(\tfrac{1}{4}\{n + x + 2\})}.$$

We refrain from explicitly stating Examples 1 and 2 which are merely the special cases $x = 0$ and $x = \frac{1}{2}$, respectively, of the previous corollary.

In Entry 35(i), Ramanujan defines

$$\varphi(n) = \sum_{k=0}^{n-1} \frac{(\tfrac{1}{2})_k^2}{(k!)^2}$$

and then states an asymptotic formula for $\varphi(\{n + 1\}/4)$ as n tends to ∞. More properly, $\varphi(n)$ should be defined by (29.1). Thus, for all complex n, define

$$\varphi(n) = \frac{\Gamma^2(n + \tfrac{1}{2})}{\Gamma(n)\Gamma(n + 1)} \; {}_3F_2\left[\begin{array}{c} \tfrac{1}{2}, \tfrac{1}{2}, n \\ 1, n+1 \end{array}\right]. \qquad (35.1)$$

Entry 35(i) is thus an extension of Corollary 2 in Section 29. Watson [5] and Dutka [1] have each derived asymptotic expansions for $\varphi(n)$. However, the expansions of Watson, Dutka, and Ramanujan are all of different forms. We shall employ Dutka's asymptotic series to establish Ramanujan's formulation.

Entry 35(i). *Let* $\varphi(n)$ *be defined by (35.1). Then as n tends to* ∞,

$$\pi\varphi\left(\frac{n+1}{4}\right) - \psi\left(\frac{n+1}{2}\right) \sim 3 \text{ Log } 2 + \gamma + \frac{3}{4n^2} - \frac{99}{32n^4} + \frac{999}{32n^6} + \cdots.$$

PROOF. According to Dutka [1], as n tends to ∞,

$$\pi\varphi\left(\frac{n+1}{4}\right) \sim \psi\left(\frac{n+3}{4}\right) + 4 \text{ Log } 2 + \gamma - U_n,$$

where

$$U_n = \frac{1}{2(n/2 + \frac{3}{2})} + \left(\frac{1\cdot 3}{2\cdot 4}\right)\frac{1\cdot 3}{2(n/2 + \frac{3}{2})(n/2 + \frac{7}{2})}$$

$$+ \left(\frac{1\cdot 3\cdot 5}{2\cdot 4\cdot 6}\right)\frac{1\cdot 3\cdot 5}{3(n/2 + \frac{3}{2})(n/2 + \frac{7}{2})(n/2 + \frac{11}{2})} + \cdots. \qquad (35.2)$$

From Legendre's duplication formula, it is easy to show that

$$2\psi\left(\frac{n+1}{2}\right) = \psi\left(\frac{n+1}{4}\right) + \psi\left(\frac{n+3}{4}\right) + 2 \text{ Log } 2.$$

Thus, as n tends to ∞,

$$\pi\varphi\left(\frac{n+1}{4}\right) \sim \psi\left(\frac{n+1}{2}\right) + 3 \text{ Log } 2 + \gamma + \psi\left(\frac{n+1}{2}\right) - \psi\left(\frac{n+1}{4}\right)$$

$$- \text{ Log } 2 - U_n.$$

Using Stirling's formula for Log $\psi(x)$ (Luke [1, p. 33]),

$$\psi(x) \sim \text{Log } x - \frac{1}{2x} - \sum_{k=1}^{\infty} \frac{B_{2k}x^{-2k}}{2k},$$

where x tends to ∞ and B_n, $0 \le n < \infty$, denotes the nth Bernoulli number, we find that, as n tends to ∞,

$$\pi\varphi\left(\frac{n+1}{4}\right) \sim \psi\left(\frac{n+1}{2}\right) + 3 \text{ Log } 2 + \gamma + \frac{1}{n+1}$$

$$+ \sum_{k=1}^{\infty} \frac{B_{2k}2^{2k}(2^{2k} - 1)}{(2k)(n+1)^{2k}} - U_n.$$

Recalling that U_n is defined by (35.2), we now express the terms of $1/(n + 1) -$

U_n in terms of quotients of gamma functions. Thus, as n tends to ∞,

$$\pi\varphi\left(\frac{n+1}{4}\right) \sim \psi\left(\frac{n+1}{2}\right) + 3 \operatorname{Log} 2 + \gamma$$

$$+ \sum_{k=1}^{\infty} \frac{B_{2k}2^{2k}(2^{2k}-1)}{(2k)n^{2k}}\left\{1 - \frac{2k}{n} + \frac{2k(2k+1)}{2n^2}\right.$$

$$\left. - \frac{2k(2k+1)(2k+2)}{6n^3} + \frac{2k(2k+1)(2k+2)(2k+3)}{24n^4} - \cdots\right\}$$

$$+ \frac{\Gamma(\frac{1}{2}n+\frac{1}{2})}{2\Gamma(\frac{1}{2}n+\frac{5}{2})} - \frac{1^2\cdot3^2}{2^42!2}\frac{\Gamma\left(\dfrac{n+3}{4}\right)}{\Gamma\left(\dfrac{n+11}{4}\right)} - \frac{1^2\cdot3^2\cdot5^2}{2^63!3}\frac{\Gamma\left(\dfrac{n+3}{4}\right)}{\Gamma\left(\dfrac{n+15}{4}\right)}$$

$$- \frac{1^2\cdot3^2\cdot5^2\cdot7^2}{2^84!4}\frac{\Gamma\left(\dfrac{n+3}{4}\right)}{\Gamma\left(\dfrac{n+19}{4}\right)} - \frac{1^2\cdot3^2\cdot5^2\cdot7^2\cdot9^2}{2^{10}5!5}\frac{\Gamma\left(\dfrac{n+3}{4}\right)}{\Gamma\left(\dfrac{n+23}{4}\right)}$$

$$- \frac{1^2\cdot3^2\cdot5^2\cdot7^2\cdot9^2\cdot11^2}{2^{12}6!6}\frac{\Gamma\left(\dfrac{n+3}{4}\right)}{\Gamma\left(\dfrac{n+27}{4}\right)} - \cdots. \tag{35.3}$$

For each quotient of gamma functions displayed above, we use a general asymptotic formula for $\Gamma(x+a)/\Gamma(x+b)$ due to Tricomi and Erdélyi [1] and reproduced in Luke's book [1, p. 33]. Omitting the numerical calculations, we find that, as n tends to ∞,

$$\frac{\Gamma\left(\dfrac{n+1}{2}\right)}{\Gamma\left(\dfrac{n+5}{2}\right)} = \frac{2^2}{n^2}\left\{1 - \frac{4}{n} + \frac{13}{n^2} - \frac{40}{n^3} + \frac{121}{n^4}\right\} + O(n^{-7}), \tag{35.4}$$

$$\frac{\Gamma\left(\dfrac{n+3}{4}\right)}{\Gamma\left(\dfrac{n+11}{4}\right)} = \frac{2^4}{n^2}\left\{1 - \frac{10}{n} + \frac{79}{n^2} - \frac{580}{n^3} + \frac{4141}{n^4}\right\} + O(n^{-7}), \tag{35.5}$$

$$\frac{\Gamma\left(\dfrac{n+3}{4}\right)}{\Gamma\left(\dfrac{n+15}{4}\right)} = \frac{2^6}{n^3}\left\{1 - \frac{21}{n} + \frac{310}{n^2} - \frac{3990}{n^3}\right\} + O(n^{-7}), \tag{35.6}$$

$$\frac{\Gamma\left(\dfrac{n+3}{4}\right)}{\Gamma\left(\dfrac{n+19}{4}\right)} = \frac{2^8}{n^4}\left\{1 - \frac{36}{n} + \frac{850}{n^2}\right\} + O(n^{-7}), \tag{35.7}$$

$$\frac{\Gamma\left(\dfrac{n+3}{4}\right)}{\Gamma\left(\dfrac{n+23}{4}\right)} = \frac{2^{10}}{n^5}\left\{1 - \frac{55}{n}\right\} + O(n^{-7}), \tag{35.8}$$

and

$$\frac{\Gamma\left(\dfrac{n+3}{4}\right)}{\Gamma\left(\dfrac{n+27}{4}\right)} = \frac{2^{12}}{n^6} + O(n^{-7}). \tag{35.9}$$

Substituting (35.4)–(35.9) into (35.3), we now calculate the coefficients of n^{-k}, $2 \le k \le 6$. After some lengthy calculations, we find that all the coefficients agree with what Ramanujan has claimed in Entry 35(i). □

Entry 35(ii). *Let $\varphi(n)$ be defined by (35.1). Then for each nonnegative integer n,*

$$\frac{\pi^2}{4}\,\varphi(n + \tfrac{1}{2}) = \sum_{k=0}^{n-1} \frac{(k!)^2}{(\frac{3}{2})_k^2} + 2G, \tag{35.10}$$

where G is defined by (29.2).

Entry 35(iii). *Let $\varphi(n)$ be defined by (35.1). Then for each nonnegative integer n,*

$$2\varphi(n + \tfrac{1}{4}) = 1 + \frac{16\Gamma^4(\frac{3}{4})}{\pi^3} \sum_{k=0}^{n-1} \frac{(\frac{3}{4})_k^2}{(\frac{5}{4})_k^2}. \tag{35.11}$$

Entry 35(iv). *If $\varphi(n)$ is defined by (35.1), then*

$$\varphi(\tfrac{1}{2}) = \frac{8}{\pi^2}\,G \quad and \quad \varphi(\tfrac{1}{4}) = \tfrac{1}{2}.$$

We shall first prove Entry 35(iv) and then prove Entries 35(ii) and 35(iii) by induction.

PROOF OF ENTRY 35(iv). By (35.1) and Corollary 1 in Section 29,

$$\varphi(\tfrac{1}{2}) = \frac{2}{\pi}\,_3F_2(\tfrac{1}{2}, \tfrac{1}{2}, \tfrac{1}{2}; 1, \tfrac{3}{2}) = \frac{8}{\pi^2}\,G.$$

By (35.1) and Dixon's theorem, Entry 7, with $n = \frac{1}{2}$, $x = -\frac{1}{2}$, and $y = -\frac{1}{4}$, we find that

$$\varphi(\tfrac{1}{4}) = \frac{\Gamma^2(\tfrac{3}{4})}{\Gamma(\tfrac{1}{4})\Gamma(\tfrac{5}{4})} \,_3F_2(\tfrac{1}{2}, \tfrac{1}{2}, \tfrac{1}{4}; 1, \tfrac{5}{4}) = \tfrac{1}{2}. \qquad \square$$

PROOF OF ENTRY 35(ii). We proceed by induction on n. For $n = 0$, formula (35.10) is valid by Entry 35(iv). Assume now that (35.10) holds for any fixed nonnegative integer n. Thus, it remains to prove (35.10) with n replaced by $n + 1$.

We first establish the recursion formula

$$\varphi(n + 1) = \varphi(n) + \frac{\Gamma^2(n + \tfrac{1}{2})}{\pi\Gamma^2(n + 1)}, \tag{35.12}$$

where n is any complex number, or, by (35.1),

$$(2n + 1)^2 \left(\frac{1}{n + 1} + \left(\frac{1}{2}\right)^2 \frac{1}{n + 2} + \left(\frac{1 \cdot 3}{2 \cdot 4}\right)^2 \frac{1}{n + 3} + \cdots \right)$$

$$= 4n^2 \left(\frac{1}{n} + \left(\frac{1}{2}\right)^2 \frac{1}{n + 1} + \left(\frac{1 \cdot 3}{2 \cdot 4}\right)^2 \frac{1}{n + 2} + \cdots \right) + \frac{4}{\pi}. \tag{35.13}$$

In the course of proving Entry 29(b), Darling [1, p. 9, line 1] proved precisely the formula (35.13).

Hence, by (35.12) and (35.10),

$$\frac{\pi^2}{4} \varphi(n + \tfrac{3}{2}) = \frac{\pi^2}{4} \varphi(n + \tfrac{1}{2}) + \frac{\pi\Gamma^2(n + 1)}{4\Gamma^2(n + \tfrac{3}{2})}$$

$$= \sum_{k=0}^{n-1} \frac{(k!)^2}{(\tfrac{3}{2})_k^2} + 2G + \frac{(n!)^2}{(\tfrac{3}{2})_n^2}$$

$$= \sum_{k=0}^{n} \frac{(k!)^2}{(\tfrac{3}{2})_k^2} + 2G,$$

which completes the proof. $\qquad \square$

PROOF OF ENTRY 35(iii). We induct on n. If $n = 0$, then (35.11) holds by Entry 35(iv). Assume now that (35.11) holds for any fixed nonnegative integer n, and so it suffices to prove (35.11) with n replaced by $n + 1$.

By (35.12) and (35.11),

$$2\varphi(n + \tfrac{5}{4}) = 2\varphi(n + \tfrac{1}{4}) + \frac{2\Gamma^2(n + \tfrac{3}{4})}{\pi\Gamma^2(n + \tfrac{5}{4})}$$

$$= 1 + \frac{16\Gamma^4(\tfrac{3}{4})}{\pi^3} \sum_{k=0}^{n-1} \frac{(\tfrac{3}{4})_k^2}{(\tfrac{5}{4})_k^2} + \frac{2\Gamma^2(\tfrac{3}{4})(\tfrac{3}{4})_n^2}{\pi\Gamma^2(\tfrac{5}{4})(\tfrac{5}{4})_n^2}$$

$$= 1 + \frac{16\Gamma^4(\frac{3}{4})}{\pi^3} \sum_{k=0}^{n-1} \frac{(\frac{3}{4})_k^2}{(\frac{5}{4})_k^2} + \frac{16\Gamma^4(\frac{3}{4})(\frac{3}{4})_n^2}{\pi^3(\frac{5}{4})_n^2},$$

and the desired result follows. $\square$

In the first notebook (p. 239), Entry 35(iv) is listed before (35.10) and (35.11). Furthermore, prior to the latter two formulas, Ramanujan states the recursion formula (35.12). Thus, it seems clear that Ramanujan also used induction to establish (35.10) and (35.11).

At the beginning of Darling's paper [1], in conjunction with Entry 29(b), he remarks, "His (Watson's) own proof is by transformation of series, and it seems probable that Ramanujan obtained the theorem in a similar manner; but the following two proofs by induction, which will perhaps appeal more to the average analyst, may be of interest." It appears that Darling's speculation is incorrect, and that he, in fact, had likely found Ramanujan's proof.

Dutka [1] has found a different proof of Entry 35(ii).

Hypergeometric Series, II

Much of Chapter 11 is contained in Chapters 13 and 15 of the first notebook, while some formulas from Chapter 11 may be found scattered among the "working pages" of the first notebook.

In Chapter 11, Ramanujan gives many results on quadratic transformations of hypergeometric series. Several of these results can be traced back to Kummer [1], [2]. Ramanujan also offers many theorems on products of hypergeometric series. Although some of these results were established in the 19th century, most are originally due to Ramanujan. Entry 34(iii) is a particularly elegant formula which combines a product formula and a quadratic transformation. Much of Bailey's work in the 1930s on products of hypergeometric series was motivated by Ramanujan's discoveries.

Corollary 2 in Section 24 offers a certain asymptotic formula for zero-balanced $_3F_2$ series. Such formulas in the literature have previously been established only for zero-balanced $_2F_1$ series. It is interesting that this elegant formula had been overlooked for 60 years after Ramanujan's death. We provide here an elegant proof of this asymptotic formula by R. J. Evans and D. Stanton [1]. However, their proof depends on knowing the formula in advance. It would be interesting to have a more direct proof that might shed some light on Ramanujan's approach.

There are two additional formulas in Chapter 11 which are amazing indeed. The first is Entry 22, which involves a remarkable recursively defined sequence A_n and which leads to two intriguing binomial coefficient identities (22.22) and (22.23). The second is Entry 31(ii), which we were only able to prove by using the theory of second-order inhomogeneous linear differential equations and equating coefficients in *15* power series. Unfortunately, we have no idea how Ramanujan discovered these two extraordinary formulas (as well as most of the results in this chapter). Our proofs of these two theorems are certainly not those found by Ramanujan; he must have derived these formulas more naturally. Although differential equations have traditionally played a strong role in the theory of hypergeometric series, there is no evidence that Ramanujan significantly utilized this connection. The hypergeometric differential equation does appear in somewhat disguised form in Entry 31(i). The formulas in Sections 30 and 31 of Chapter 11 are the only ones in Chapters 10 and 11 with links to differential equations.

A few formulas in Chapter 11 are apparently without meaning. Entry 24 is such an example; we have not been able to find any functions for which the proposed formula is valid.

We use the notation that was set forth in the introduction to Chapter 10. In that chapter, we considered the case $p = q + 1$. Since in this chapter, we establish theorems for $p \neq q + 1$, we offer further remarks about convergence. If $p < q + 1$, then $_pF_q$ converges for all finite values of x; if $p > q + 1$, then $_pF_q$ converges for only $x = 0$ unless the series terminates. For most of the theorems and examples in the sequel, we shall not state the region of validity because it can readily be ascertained from the general remarks we have made about convergence.

In the sequel, we shall frequently appeal to the treatises of Erdélyi [1] and Bailey [4].

Entry 1. *Let φ be any function. Then, provided the series converges,*

$$\frac{1}{\varphi^r(x)} \sum_{k=0}^{\infty} \frac{(r)_k(m)_k}{(2m)_k k!} \left\{ 1 - \frac{\varphi(-x)}{\varphi(x)} \right\}^k$$

is an even function of x.

PROOF. Consider the quadratic transformation found in Erdélyi's work [1, p. 112, formula (26)] and due to Kummer [1, p. 78], [2, p. 114],

$$_2F_1(r, m; 2m; z) = (1 - z)^{-r/2} \, _2F_1\left(\tfrac{1}{2}r, m - \tfrac{1}{2}r; m + \tfrac{1}{2}; \frac{z^2}{4(z - 1)}\right).$$

Setting $z = 1 - \varphi(-x)/\varphi(x)$, we find after some simplification that

$$\frac{1}{\varphi^r(x)} \, _2F_1\left(r, m; 2m; 1 - \frac{\varphi(-x)}{\varphi(x)}\right)$$

$$= \frac{1}{\{\varphi(x)\varphi(-x)\}^{r/2}} \, _2F_1\left(\tfrac{1}{2}r, m - \tfrac{1}{2}r; m + \tfrac{1}{2}; - \frac{\{\varphi(x) - \varphi(-x)\}^2}{4\varphi(x)\varphi(-x)}\right),$$

which clearly is an even function of x. $\qquad\square$

Entry 2

$$_2F_1\left(r, m; 2m; \frac{2x}{1 + x}\right) = (1 + x)^r \, _2F_1(\tfrac{1}{2}r, \tfrac{1}{2}(r + 1); \tfrac{1}{2}(2m + 1); x^2).$$

Entry 2 is a well-known quadratic transformation (see Erdélyi's book [1, p. 111, Eq. (4)]) that is due to Kummer [1, p. 78], [2, p. 114].

Entry 3

$$_2F_1\left(r, m; 2m; \frac{4x}{(1 + x)^2}\right) = (1 + x)^{2r} \,_2F_1(r, r - m + \tfrac{1}{2}; m + \tfrac{1}{2}; x^2).$$

Entry 3 is precisely Eq. (5) of Erdélyi's treatise [1, p. 111] and is due to Gauss [1]. This formula is also mentioned by Hardy [1, p. 502], [7, p. 515].

Entry 4

$$_2F_1\left(\tfrac{1}{2}r, \tfrac{1}{2}(r + 1); \tfrac{1}{2}(2m + 1); \frac{4x}{(1 + x)^2}\right) = (1 + x)^r \,_2F_1(r, r - m + \tfrac{1}{2}; m + \tfrac{1}{2}; x).$$

PROOF. In Entry 2, replace x by $2\sqrt{x}/(1 + x)$ to find that

$$_2F_1\left(\tfrac{1}{2}r, \tfrac{1}{2}(r + 1); \tfrac{1}{2}(2m + 1); \frac{4x}{(1 + x)^2}\right)$$

$$= \frac{(1 + x)^r}{(1 + \sqrt{x})^{2r}} \,_2F_1\left(r, m; 2m; \frac{4\sqrt{x}}{(1 + \sqrt{x})^2}\right)$$

$$= (x + 1)^r \,_2F_1(r, r - m + \tfrac{1}{2}; m + \tfrac{1}{2}; x),$$

by Entry 3.   □

Entry 5

$$_2F_1\left(r, \tfrac{1}{2}; 1; \frac{4x}{(1 + x)^2}\right) = (1 + x)^{2r} \,_2F_1(r, r; 1; x^2).$$

PROOF. Put $m = \tfrac{1}{2}$ in Entry 3.   □

Entry 6

$$_2F_1\left(\tfrac{1}{2}r, \tfrac{1}{2}(r + 1); 1; \frac{4x}{(1 + x)^2}\right) = (1 + x)^r \,_2F_1(r, r; 1; x).$$

PROOF. Put $m = \tfrac{1}{2}$ in Entry 4.   □

Entry 7

$$_1F_1(m; 2m; 2x) = e^x \,_0F_1(m + \tfrac{1}{2}; x^2/4). \tag{7.1}$$

Entry 7 is due to Kummer [1, p. 140], [2, p. 134] and was recorded by Hardy [1, p. 502], [7, p. 515]. Entry 7 follows from Entry 2 by replacing x by x/r there and then letting r tend to ∞.

Corollary. $_1F_1(\tfrac{1}{2}; 1; x) = e^{x/2}\,_0F_1(1; (x/4)^2)$.

PROOF. In Entry 7, put $m = \tfrac{1}{2}$ and replace x by $x/2$.   □

Entry 8. *Let $\varphi(x)$ be analytic for $|x - 1| < R$, where $R > 1$. Suppose that m and φ are such that the order of summation in*

$$\sum_{k=0}^{\infty} \frac{2^k(m)_k}{(2m)_k k!} \sum_{n=k}^{\infty} \frac{\varphi^{(n)}(1)(-1)^n(-n)_k}{n!}$$

may be inverted. Then

$$\sum_{k=0}^{\infty} \frac{\varphi^{(k)}(0)2^k(m)_k}{(2m)_k k!} = \sum_{k=0}^{\infty} \frac{\varphi^{(2k)}(1)}{2^{2k}(m + \tfrac{1}{2})_k k!}.$$

PROOF. Since φ is analytic for $|x - 1| < R$, $R > 1$, we readily find that

$$\varphi^{(k)}(0) = \sum_{n=k}^{\infty} \frac{\varphi^{(n)}(1)(-1)^n(-n)_k}{n!}, \qquad k \geq 0.$$

Hence, inverting the order of summation, by hypothesis, we find that

$$\sum_{k=0}^{\infty} \frac{\varphi^{(k)}(0)2^k(m)_k}{(2m)_k k!} = \sum_{k=0}^{\infty} \frac{2^k(m)_k}{(2m)_k k!} \sum_{n=k}^{\infty} \frac{\varphi^{(n)}(1)(-1)^n(-n)_k}{n!}$$

$$= \sum_{n=0}^{\infty} \frac{\varphi^{(n)}(1)(-1)^n}{n!} \sum_{k=0}^{n} \frac{2^k(m)_k(-n)_k}{(2m)_k k!}. \tag{8.1}$$

Now multiply both sides of (7.1) by e^{-x} and then equate coefficients of x^n, $n \geq 0$, on both sides to obtain the evaluation

$$\frac{(-1)^n}{n!} \sum_{k=0}^{n} \frac{(-n)_k(m)_k 2^k}{(2m)_k k!} = \begin{cases} \dfrac{1}{2^n(m + \tfrac{1}{2})_{n/2}(\tfrac{1}{2}n)!}, & \text{if } n \text{ is even,} \\[2mm] 0, & \text{if } n \text{ is odd.} \end{cases} \tag{8.2}$$

If we substitute (8.2) into (8.1), we complete the proof.   □

Entry 9. *If n is an integer, then*

$$_0F_1(n + \tfrac{1}{2}; (\tfrac{1}{2}x)^2) = \frac{2^{n-1}\Gamma(n + \tfrac{1}{2})}{\sqrt{\pi}\, x^n} \left\{ e^x\,_2F_0\!\left(n, 1 - n; \frac{1}{2x}\right) \right.$$

$$\left. + \cos(n\pi)e^{-x}\,_2F_0\!\left(n, 1 - n; -\frac{1}{2x}\right) \right\}.$$

Observe that

$$_0F_1(n + \tfrac{1}{2}; (\tfrac{1}{2}x)^2) = \Gamma(n + \tfrac{1}{2})(2/x)^{n-1/2} I_{n-1/2}(x), \tag{9.1}$$

where I_ν is the Bessel function of imaginary argument usually so denoted (see Watson's treatise [9, p. 77]). Thus, Entry 9 is a well-known result in the theory of Bessel functions (ibid. [9, p. 80, formulas (10), (11)]).

Corollary. *As x tends to* ∞,

$$_0F_1(1; (\tfrac{1}{2}x)^2) \sim \frac{e^x}{\sqrt{2\pi x}}\left(1 + \frac{1^2}{2^2(2x)} + \frac{1^2 \cdot 3^2}{2^4 2!(2x)^2} + \frac{1^2 \cdot 3^2 \cdot 5^2}{2^6 3!(2x)^3} + \cdots\right). \quad (9.2)$$

PROOF. Undoubtedly, Ramanujan formally deduced this formula from Entry 9 by setting $n = \tfrac{1}{2}$ there.

However, by (9.1), which holds for any complex number n,

$$_0F_1(1; (\tfrac{1}{2}x)^2) = I_0(x).$$

Remembering that x is positive, we observe that (9.2) is precisely the asymptotic expansion of $I_0(x)$ given by Watson [9, p. 203].   □

It is possible that Ramanujan did not restrict n to be an integer in Entry 9. In such a case, the right side of Entry 9 is an asymptotic expansion for the left side as $|x|$ tends to ∞ when $|\arg x| < \tfrac{3}{2}\pi$ (Watson [9, p. 203]), provided that $\cos(n\pi)$ is replaced by $\exp(in\pi)$.

Entry 10. *If n is an integer, then*

$$_0F_1(n + \tfrac{1}{2}; -(\tfrac{1}{2}x)^2) = \frac{2^n\Gamma(n + \tfrac{1}{2})}{\sqrt{\pi}\,x^n}\left\{\cos(\tfrac{1}{2}n\pi - x) \sum_{k=0}^{\infty} \frac{(-1)^k(n)_{2k}(1 - n)_{2k}}{(2k)!(2x)^{2k}}\right.$$

$$\left. - \sin(\tfrac{1}{2}n\pi - x) \sum_{k=0}^{\infty} \frac{(-1)^k(n)_{2k+1}(1 - n)_{2k+1}}{(2k + 1)!(2x)^{2k+1}}\right\}.$$

PROOF. Replace x by ix in Entry 9 and equate real parts on both sides. After some simplification, we achieve the desired equality.   □

Corollary. *Suppose that n is an integer. Let x_0 be a root of*

$$_0F_1(n + \tfrac{1}{2}; -(\tfrac{1}{2}x)^2) = 0.$$

Let μ be an odd integer chosen so that $|x_0 - \tfrac{1}{2}\pi(\mu + n)|$ is minimal. Then if x_0 is "large,"

$$x_0 \approx \frac{\pi(\mu + n)}{2} + \frac{n(1 - n)}{\pi(\mu + n)} + \frac{n(1 - n)\{7n(1 - n) - 6\}}{3\pi^3(\mu + n)^3} + \cdots. \quad (10.1)$$

PROOF. By Entry 10, we want to approximate large roots of

$$\cos(\tfrac{1}{2}n\pi - x) \sum_{k=0}^{\infty} \frac{(-1)^k(n)_{2k}(1 - n)_{2k}}{(2k)!(2x)^{2k}}$$

$$- \sin(\tfrac{1}{2}n\pi - x) \sum_{k=0}^{\infty} \frac{(-1)^k(n)_{2k+1}(1 - n)_{2k+1}}{(2k + 1)!(2x)^{2k+1}} = 0. \quad (10.2)$$

We shall use a method of successive approximations.

From (10.2), it is clear that we should take as a first approximation

$$x = \tfrac{1}{2}\pi(\mu + n).$$

For our second approximation, consider $\tfrac{1}{2}\pi(\mu + n) + y$, where, by (10.2), y should satisfy the equation

$$\cos(\tfrac{1}{2}n\pi - \{\tfrac{1}{2}\pi(\mu + n) + y\}) - \sin(\tfrac{1}{2}n\pi - \{\tfrac{1}{2}\pi(\mu + n) + y\})\frac{n(1 - n)}{\pi(\mu + n)} = 0.$$

After a short calculation, we find that

$$\tan y = \frac{n(1 - n)}{\pi(\mu + n)}.$$

Hence, as our second approximation, we shall take

$$x = \frac{\pi(\mu + n)}{2} + \frac{n(1 - n)}{\pi(\mu + n)}.$$

For our third approximation, consider

$$\frac{\pi(\mu + n)}{2} + \frac{n(1 - n)}{\pi(\mu + n)} + z,$$

where, by (10.2), z is to satisfy the equation

$$
\begin{aligned}
&- \sin\left(\frac{n(1 - n)}{\pi(\mu + n)} + z\right)\left\{1 - \frac{n(n + 1)(1 - n)(2 - n)}{2\pi^2(\mu + n)^2}\right\} \\[2mm]
&+ \cos\left(\frac{n(1 - n)}{\pi(\mu + n)} + z\right)\left\{\frac{n(1 - n)}{\pi(\mu + n)\left(1 + \dfrac{2n(1 - n)}{\pi^2(\mu + n)^2}\right)}\right. \\[2mm]
&\left. - \frac{n(n + 1)(n + 2)(1 - n)(2 - n)(3 - n)}{6\pi^3(\mu + n)^3}\right\} = 0.
\end{aligned}
$$

Hence,

$$
\begin{aligned}
\tan\left(\frac{n(1 - n)}{\pi(\mu + n)} + z\right) &\approx \left\{\frac{n(1 - n)}{\pi(\mu + n)}\left(1 - \frac{2n(1 - n)}{\pi^2(\mu + n)^2}\right)\right. \\[2mm]
&\left. - \frac{n(n + 1)(n + 2)(1 - n)(2 - n)(3 - n)}{6\pi^3(\mu + n)^3}\right\}\left\{1 + \frac{n(n + 1)(1 - n)(2 - n)}{2\pi^2(\mu + n)^2}\right\} \\[2mm]
&\approx \frac{n(1 - n)}{\pi(\mu + n)} + \frac{n(1 - n)}{\pi^3(\mu + n)^3}\left\{-2n(1 - n) - \frac{(n + 1)(n + 2)(2 - n)(3 - n)}{6}\right. \\[2mm]
&\left. + \frac{n(n + 1)(1 - n)(2 - n)}{2}\right\}.
\end{aligned}
\tag{10.3}
$$

Now,

$$\tan\left(\frac{n(1-n)}{\pi(\mu+n)}+z\right) \approx \frac{n(1-n)}{\pi(\mu+n)}+z+\frac{n^3(1-n)^3}{3\pi^3(\mu+n)^3}. \tag{10.4}$$

Thus, from (10.3) and (10.4),

$$z \approx \frac{n(1-n)}{\pi^3(\mu+n)^3}\left\{-2n(1-n)-\frac{(n+1)(n+2)(2-n)(3-n)}{6}\right.$$
$$\left.+\frac{n(n+1)(1-n)(2-n)}{2}-\frac{n^2(1-n)^2}{3}\right\}$$

$$=\frac{n(1-n)}{\pi^3(\mu+n)^3}\left\{\frac{7}{3}n^2-\frac{7}{3}n-2\right\}.$$

Hence, our third-order approximation is precisely that claimed by Ramanujan in (10.1).   □

Entry 11. *If*

$$\int_0^x \frac{\sin u}{u}\,du = \frac{\pi}{2}-r\cos(x-\theta) \tag{11.1}$$

and

$$\int_0^x \frac{1-\cos u}{u}\,du = \gamma+\text{Log }x-r\sin(x-\theta), \tag{11.2}$$

where γ denotes Euler's constant, then

$$r\cos\theta \sim \sum_{k=0}^\infty \frac{(-1)^k(2k)!}{x^{2k+1}}, \tag{11.3}$$

$$r\sin\theta \sim \sum_{k=1}^\infty \frac{(-1)^{k+1}(2k-1)!}{x^{2k}}, \tag{11.4}$$

and

$$r^2 \sim \sum_{k=1}^\infty \frac{(-1)^{k+1}(2k-1)!}{kx^{2k}}, \tag{11.5}$$

as x tends to ∞.

PROOF. By (11.1),

$$\int_0^x \frac{\sin u}{u}\,du = \left(\int_0^\infty-\int_x^\infty\right)\frac{\sin u}{u}\,du$$

$$=\frac{\pi}{2}-r\cos x\cos\theta-r\sin x\sin\theta. \tag{11.6}$$

By successively integrating by parts, we easily find that

$$\int_x^\infty \frac{\sin u}{u} \, du \sim \cos x \sum_{k=0}^\infty \frac{(-1)^k (2k)!}{x^{2k+1}} + \sin x \sum_{k=1}^\infty \frac{(-1)^{k+1}(2k-1)!}{x^{2k}}, \quad (11.7)$$

as x tends to ∞. Thus, (11.3) and (11.4) follow from (11.6) and (11.7).

We next show that (11.2) is consistent with (11.3) and (11.4). From (11.2) and the tables of Gradshteyn and Ryzhik [1, p. 928],

$$\int_0^x \frac{1 - \cos u}{u} \, du = \gamma + \text{Log } x + \int_x^\infty \frac{\cos u}{u} \, du$$

$$= \gamma + \text{Log } x - r \sin x \cos \theta + r \cos x \sin \theta. \quad (11.8)$$

On the other hand, by successively integrating by parts,

$$\int_x^\infty \frac{\cos u}{u} \, du \sim -\sin x \sum_{k=0}^\infty \frac{(-1)^k (2k)!}{x^{2k+1}} + \cos x \sum_{k=1}^\infty \frac{(-1)^{k+1}(2k-1)!}{x^{2k}},$$

$$(11.9)$$

as x tends to ∞. Using (11.8) and (11.9), we again deduce (11.3) and (11.4).

From (11.3) and (11.4),

$$r^2 \sim \left\{ \sum_{k=0}^\infty \frac{(-1)^k (2k)!}{x^{2k+1}} \right\}^2 + \left\{ \sum_{k=1}^\infty \frac{(-1)^{k+1}(2k-1)!}{x^{2k}} \right\}^2,$$

as x tends to ∞. The coefficient of x^{-2n}, $n \geq 1$, above is equal to

$$(-1)^{n+1} \sum_{k=0}^{n-1} (2k)!(2n - 2 - 2k)! + (-1)^n \sum_{k=1}^{n-1} (2k-1)!(2n - 2k - 1)!$$

$$= (-1)^n \sum_{k=0}^{2n-2} (-1)^{k+1} k!(2n - 2 - k)!. \quad (11.10)$$

Comparing (11.5) and (11.10) and replacing n by $n + 1$, we see that it suffices to show that

$$\sum_{k=0}^{2n} (-1)^k k!(2n - k)! = \frac{(2n + 1)!}{n + 1}, \quad n \geq 0. \quad (11.11)$$

Let S_n denote the left side of (11.11). Using (32.2) in Chapter 10 with n replaced by $2n + 1$, $a = b = 1$, and $f = -2n - \varepsilon$, where $\varepsilon > 0$, we find that

$$S_n = (2n)! \sum_{k=0}^{2n} \frac{(1)_k}{(-2n)_k}$$

$$= (2n)! \lim_{\varepsilon \to 0} \sum_{k=0}^{2n} \frac{(1)_k (1)_k}{(-2n - \varepsilon)_k k!}$$

$$= (2n)! \frac{\Gamma^2(2n + 2)}{\Gamma(2n + 1)\Gamma(2n + 3)} \lim_{\varepsilon \to 0} {}_3F_2 \left[\begin{array}{c} 1, 1, -\varepsilon \\ -2n - \varepsilon, 2n + 3 \end{array} \right]$$

$$= \frac{\Gamma^2(2n + 2)}{\Gamma(2n + 3)} \left(1 + \lim_{\varepsilon \to 0} \sum_{k=2n+1}^\infty \frac{(1)_k (-\varepsilon)_k}{(-2n - \varepsilon)_k (2n + 3)_k} \right)$$

$$= \frac{(2n+1)!}{2n+2}\left(1 + \lim_{\varepsilon \to 0} \frac{(1)_{2n+1}(-\varepsilon)_{2n+1}}{(-2n-\varepsilon)_{2n+1}(2n+3)_{2n+1}} \sum_{k=0}^{\infty} \frac{(2n+2)_k(2n+1-\varepsilon)_k}{(1-\varepsilon)_k(4n+4)_k}\right)$$

$$= \frac{(2n+1)!}{2n+2}\left(1 + \frac{(2n+1)!(2n+2)!}{(4n+3)!} \, {}_2F_1\left[\begin{matrix}2n+1, 2n+2 \\ 4n+4\end{matrix}; 1\right]\right)$$

$$= \frac{(2n+1)!}{2n+2}\left(1 + \frac{(2n+1)!(2n+2)!}{(4n+3)!} \frac{\Gamma(4n+4)\Gamma(1)}{\Gamma(2n+3)\Gamma(2n+2)}\right)$$

$$= \frac{(2n+1)!}{2n+2}(1+1) = \frac{(2n+1)!}{n+1},$$

where we have employed Gauss's theorem, which is Entry 8 of Chapter 10. This completes the proof of (11.11). □

For results similar to Entry 11, see the author's [9] account of Chapter 4 of Ramanujan's second notebook.

Example 1. $\int_0^{\pi/2} \cos(\pi \sin^2 \theta) \, d\theta = 0$.

PROOF. Letting

$$\sin^2 \theta = \tfrac{1}{2}(1 - \cos 2\theta) \tag{11.12}$$

and replacing θ by $\pi/2 - \theta$, we find that

$$\int_0^{\pi/2} \cos(\pi \sin^2 \theta) \, d\theta = \int_0^{\pi/2} \sin(\tfrac{1}{2}\pi \cos 2\theta) \, d\theta$$

$$= -\int_0^{\pi/2} \sin(\tfrac{1}{2}\pi \cos 2\theta) \, d\theta,$$

from which the desired result follows. □

Example 2. $\int_0^{\pi/2} \cos(2\pi \sin^2 \theta) \, d\theta = -\int_0^{\pi/2} \cos(\pi \sin \theta) \, d\theta$.

PROOF. As above, the proof is quite elementary. First, use the identity (11.12) and then replace 2θ by $\pi/2 - \theta$. After simplifying, we obtain the desired equality. □

Example 3. $\displaystyle\int_0^{\pi/2} \cos\left(\frac{2\pi}{3} \sin^2 \theta\right) d\theta = \tfrac{1}{2}\int_0^{\pi/2} \cos\left(\frac{\pi}{3} \sin \theta\right) d\theta$.

PROOF. The steps are exactly the same as in the previous proof. □

Entry 12. *If* $x + y + z = \tfrac{1}{2}$, *then*

$${}_2F_1(-x, -y; z; p) = {}_2F_1(-2x, -2y; z; \tfrac{1}{2}(1 - \sqrt{1-p})).$$

With obvious changes in the parameters, Entry 12 is the same as equality (10) of Erdélyi's book [1, p. 111]. A formula equivalent to Entry 12 was given by Hardy [1, p. 502], [7, p. 515] in his overview. Entry 12 is due to Gauss [1].

Corollary

$$1 + \frac{1^2 + n}{4^2}x + \frac{(1^2 + n)(5^2 + n)}{4^2 \cdot 8^2}x^2 + \frac{(1^2 + n)(5^2 + n)(9^2 + n)}{4^2 \cdot 8^2 \cdot 12^2}x^3 + \cdots$$

$$= 1 + \frac{1^2 + n}{2^2}\left(\frac{1 - \sqrt{1 - x}}{2}\right) + \frac{(1^2 + n)(3^2 + n)}{2^2 \cdot 4^2}\left(\frac{1 - \sqrt{1 - x}}{2}\right)^2$$

$$+ \cdots.$$

PROOF. In Entry 12, set $x = (-1 + i\sqrt{n})/4$, $y = (-1 - i\sqrt{n})/4$, $z = 1$, and $p = x$.   □

Example 1

$$1 + \frac{x}{2^2} + \sum_{k=2}^{\infty} \frac{(1)_k}{(\frac{1}{2})_k}\left(1 - \frac{1}{2^3}\right)\left(1 - \frac{1}{4^3}\right)\cdots\left(1 - \frac{1}{(2k - 2)^3}\right)\frac{x^k}{(2k)^2}$$

$$= 1 + \sum_{k=1}^{\infty} \frac{1}{k + 1}\left(1 + \frac{1}{1^3}\right)\left(1 + \frac{1}{2^3}\right)\cdots\left(1 + \frac{1}{k^3}\right)\left(\frac{1 - \sqrt{1 - x}}{2}\right)^k.$$

PROOF. In the corollary above set $n = 3$. For $k \geq 2$, we are led to examine

$$\frac{(1^2 + 3)(5^2 + 3)(9^2 + 3)\cdots((4k - 3)^2 + 3)}{4^2 \cdot 8^2 \cdots (4k - 4)^2 (4k)^2}$$

$$= \frac{(1^2 + 3)(5^2 + 3)(9^2 + 3)\cdots((4k - 3)^2 + 3)}{2^2 \cdot 4^2 \cdots (2k - 2)^2 (2k)^2 4^k}$$

$$= \frac{(2^2 + 2 + 1)}{2^2}\frac{(4^2 + 4 + 1)}{4^2}\cdots\frac{((2k - 2)^2 + (2k - 2) + 1)}{(2k - 2)^2}\frac{1}{(2k)^2}$$

$$= \frac{2 \cdot 4 \cdots (2k - 2)}{1 \cdot 3 \cdots (2k - 3)}\frac{1(2^2 + 2 + 1)}{2^3}\frac{3(4^2 + 4 + 1)}{4^3}\cdots$$

$$\times \frac{(2k - 3)((2k - 2)^2 + (2k - 2) + 1)}{(2k - 2)^3}\frac{1}{(2k)^2}$$

$$= \frac{(1)_k}{(\frac{1}{2})_k}\left(1 - \frac{1}{2^3}\right)\left(1 - \frac{1}{4^3}\right)\cdots\left(1 - \frac{1}{(2k - 2)^3}\right)\frac{1}{(2k)^2}, \qquad (12.1)$$

where in the middle expression above we used the equality $4((2n)^2 + 2n + 1) = (4n + 1)^2 + 3$.

We are also led to examine, for $k \geq 1$,

$$\frac{(1^2 + 3)(3^2 + 3)\cdots((2k-1)^2 + 3)}{2^2 \cdot 4^2 \cdots (2k)^2}$$

$$= \frac{(2^2 - 2 + 1)\cdots(k^2 - k + 1)}{1^2 \cdot 2^2 \cdots k^2}$$

$$= \frac{2(1^2 - 1 + 1)}{1^3} \frac{3(2^2 - 2 + 1)}{2^3} \cdots \frac{(k+1)(k^2 - k + 1)}{k^3} \frac{1}{k+1}$$

$$= \frac{1}{k+1}\left(1 + \frac{1}{1^3}\right)\left(1 + \frac{1}{2^3}\right)\cdots\left(1 + \frac{1}{k^3}\right), \tag{12.2}$$

where in the second expression above we used the equality $4(n^2 - n + 1) = (2n-1)^2 + 3$.

Using (12.1) and (12.2) in the previous corollary, we obtain the desired result. $\qquad\square$

Example 2. *If $\alpha + \beta = 1$, then*

$$\left(\frac{1 + \sqrt{1-x}}{2}\right)^\gamma {}_2F_1(\tfrac{1}{2}(\alpha + \gamma), \tfrac{1}{2}(\beta + \gamma); \gamma + 1; x)$$

$$= {}_2F_1(\alpha, \beta; \gamma + 1; \tfrac{1}{2}(1 - \sqrt{1-x})).$$

Example 2 is well-known (e.g., see Erdélyi's compendium [1, p. 112, formula (22)]).

Entry 13. *If $\alpha + \beta + \gamma = 0$, then*

$${}_2F_1^2(-\alpha, -\beta; \gamma + \tfrac{1}{2}; x) = {}_3F_2(-2\alpha, -2\beta, \gamma; \gamma + \tfrac{1}{2}, 2\gamma; x).$$

Entry 13 is a famous result of Clausen [1]. Other results on products of hypergeometric series are given in the sequel. See also Bailey's tract [4, Chapter 10]. Entry 13 was mentioned by Hardy [1, p. 503], [7, p. 516].

Corollary 1

$$\left\{1 + \frac{1^2 + n}{4^2}x + \frac{(1^2 + n)(5^2 + n)}{4^2 \cdot 8^2}x^2 + \cdots\right\}^2$$

$$= 1 + \frac{1}{2}\frac{1^2 + n}{2^2}x + \frac{1 \cdot 3}{2 \cdot 4}\frac{(1^2 + n)(3^2 + n)}{2^2 \cdot 4^2}x^2 + \cdots.$$

PROOF. Put $\alpha = (-1 + i\sqrt{n})/4$, $\beta = (-1 - i\sqrt{n})/4$, and $\gamma = \tfrac{1}{2}$ in Entry 13. $\qquad\square$

Corollary 2. *If J_0 denotes the ordinary Bessel function of order 0, then for all x,*

$$J_0^2(i\sqrt{x}) = {}_1F_2(\tfrac{1}{2}; 1, 1; x).$$

PROOF. In Watson's text [9, formula (6), p. 147], set $v = 0$ and $z = i\sqrt{x}$ to find that

$$J_0^2(i\sqrt{x}) = \sum_{k=0}^{\infty} \frac{(2k)! x^k}{(k!)^4 2^{2k}}.$$

Since $(2k)!/(k! 2^{2k}) = (\tfrac{1}{2})_k$, the desired result follows.   □

According to Watson [9, p. 145], Corollary 2 is originally due to Schläfli [1].

Entry 14. *If $\alpha + \beta + 1 = \gamma + \delta$, then*

$${}_2F_1(\alpha, \beta; \gamma; \tfrac{1}{2}(1 - \sqrt{1 - x})) \, {}_2F_1(\alpha, \beta; \delta; \tfrac{1}{2}(1 - \sqrt{1 - x}))$$

$$= {}_4F_3\left[\begin{matrix} \alpha, \beta, \tfrac{1}{2}(\alpha + \beta), \tfrac{1}{2}(\gamma + \delta) \\ \gamma, \delta, \alpha + \beta \end{matrix}; x\right].$$

This equality can be found in Bailey's monograph [4, p. 88, formula (3)], where $x = 4z(1 - z)$. The first published proof of Entry 14 is due to Bailey [3] in 1935.

Entry 15. *For any x,*

$${}_0F_1(\gamma; x) \, {}_0F_1(\delta; x) = \sum_{k=0}^{\infty} \frac{(\gamma + \delta + k - 1)_k x^k}{(\gamma)_k (\delta)_k k!}.$$

A short calculation shows that

$$\frac{(\tfrac{1}{2}(\gamma + \delta))_k (\tfrac{1}{2}(\gamma + \delta - 1))_k 2^{2k}}{(\gamma + \delta - 1)_k} = (\gamma + \delta + k - 1)_k.$$

Thus, Entry 15 can be written in terms of hypergeometric series,

$${}_0F_1(\gamma; x) \, {}_0F_1(\delta; x) = {}_2F_3(\tfrac{1}{2}(\gamma + \delta), \tfrac{1}{2}(\gamma + \delta - 1); \gamma, \delta, \gamma + \delta - 1; 4x).$$

In fact, Entry 15 gives a formula for $J_{\gamma-1}(2i\sqrt{x})J_{\delta-1}(2i\sqrt{x})$, where J_v denotes the ordinary Bessel function of order v. This result is due to Schläfli [1] and thus represents a generalization of Corollary 2 in the previous section. Entry 15 is also given by Watson [9, p. 147, formula (5)], Hardy [1, p. 503], [7, p. 516], and Erdélyi [1, p. 185, formula (2)]. Bailey [1] has also established Entry 15 as well as generalizations.

Entry 16. *If x is arbitrary, then*

$${}_0F_2(m + 1, n + 1; x) \, {}_0F_2(m + 1, n + 1; -x)$$

$$= \sum_{k=0}^{\infty} \frac{(-1)^k (m + n + 2k + 1)_k x^{2k}}{(m + 1)_k (n + 1)_k (m + 1)_{2k} (n + 1)_{2k} k!}. \tag{16.1}$$

A brief calculation shows that

$$\frac{(\frac{1}{3}(m+n+1))_k(\frac{1}{3}(m+n+2))_k(\frac{1}{3}(m+n+3))_k 3^{3k}}{(\frac{1}{2}(m+n+1))_k(\frac{1}{2}(m+n+2))_k(\frac{1}{2}(m+1))_k(\frac{1}{2}(m+2))_k(\frac{1}{2}(n+1))_k(\frac{1}{2}(n+2))_k 2^{6k}}$$

$$= \frac{(m+n+2k+1)_k}{(m+1)_{2k}(n+1)_{2k}}.$$

Thus, Entry 16 may be written in the form

$${}_0F_2(m+1, n+1; x) \, {}_0F_2(m+1, n+1; -x)$$

$$= {}_3F_8\left[\begin{array}{c} \frac{1}{3}(m+n+1), \frac{1}{3}(m+n+2), \frac{1}{3}(m+n+3) \\ \frac{1}{2}(m+n+1), \frac{1}{2}(m+n+2), m+1, n+1, \frac{1}{2}(m+1), \frac{1}{2}(m+2), \frac{1}{2}(n+1), \frac{1}{2}(n+2) \end{array}; -\frac{27}{64}x^2\right].$$

The first published proof of Entry 16 is evidently due to Hardy [1, p. 503], [7, p. 516] who stated Entry 16 in the latter form. See also Erdélyi's treatise [1, p. 186, formula (7)].

Entry 17

$${}_0F_2(m+n+1, n+1; x) \, {}_0F_2(m+1, 1-n; -x)$$

$$= 1 + \sum_{k=1}^{\infty} \frac{\alpha_k(\frac{1}{2}(2m+n+k+2))_k(2x)^k}{(m+n+1)_k(m+1)_k k!}, \tag{17.1}$$

where, for $k \geq 1$,

$$\alpha_k = \begin{cases} \dfrac{n}{(n^2 - 1^2)(n^2 - 3^2)\cdots(n^2 - k^2)}, & \text{if } k \text{ is odd,} \\[4mm] \dfrac{1}{(n^2 - 2^2)(n^2 - 4^2)\cdots(n^2 - k^2)}, & \text{if } k \text{ is even.} \end{cases} \tag{17.2}$$

PROOF. For $r \geq 0$, the coefficient of x^r on the left side of (17.1) is equal to

$$c_r := \sum_{k=0}^{r} \frac{(-1)^k}{(m+n+1)_{r-k}(n+1)_{r-k}(r-k)!(m+1)_k(1-n)_k k!}$$

$$= \frac{1}{(m+n+1)_r(n+1)_r r!} \sum_{k=0}^{r} \frac{(-m-n-r)_k(-n-r)_k(-r)_k}{(m+1)_k(1-n)_k k!},$$

where we have used the elementary relation

$$(a)_{r-k} = \frac{(-1)^k(a)_r}{(-a-r+1)_k} \tag{17.3}$$

with $a = m+n+1, n+1,$ and 1.

We now apply Dixon's theorem, Entry 7 of Chapter 10, with $x = m+n+r$, $y = r$, and n replaced by $-n-r$. Accordingly, we find that

$$c_r = \frac{\Gamma(\frac{1}{2}(2 - n - r))\Gamma(m + 1)\Gamma(1 - n)\Gamma(\frac{1}{2}(2 + 2m + n + 3r))}{(m + n + 1)_r(n + 1)_r r!\Gamma(1 - n - r)\Gamma(\frac{1}{2}(2 + 2m + n + r))\Gamma(\frac{1}{2}(2 - n + r))\Gamma(1 + m + r)}$$

$$= \frac{(\frac{1}{2}(2m + n + r + 2))_r \Gamma(n + 1)\Gamma(1 - n)\Gamma(\frac{1}{2}(2 - n - r))}{(m + n + 1)_r(m + 1)_r r!\Gamma(1 + n + r)\Gamma(1 - n - r)\Gamma(\frac{1}{2}(2 - n + r))},$$

(17.4)

after a considerable amount of simplification. Comparing (17.4) with (17.1), we find that it suffices to show that

$$\frac{\Gamma(n + 1)\Gamma(1 - n)\Gamma(\frac{1}{2}(2 - n - r))}{\Gamma(1 + n + r)\Gamma(1 - n - r)\Gamma(\frac{1}{2}(2 - n + r))} = 2^r\alpha_r, \qquad r \geq 0. \quad (17.5)$$

After using the functional equation of the gamma function, we readily establish (17.5), and therefore Entry 17 is proved.   □

Entry 18

$$_1F_1(-\beta; \gamma; -x) \,_1F_1(-\beta; \gamma; x) = \,_2F_3(-\beta, \beta + \gamma; \gamma, \tfrac{1}{2}\gamma, \tfrac{1}{2}(\gamma + 1); x^2/4).$$

Evidently, the first published proof of Entry 18 was given by Hardy [1, p. 503], [7, p. 516]. (There is a misprint in Hardy's formulation; read $x^2/4$ instead of $-x^2/4$.) See also Erdélyi's book [1, p. 186, formula (5)]. For extensions and q-analogues of Entries 16 and 18, see the paper by Srivastava [1].

Ramanujan (p. 133) has an extra factor of $(\gamma + 4)$ in the denominator of the coefficient of x^4 on the right side above.

If we replace x by $-x/\beta$ in Entry 18 and let β tend to ∞, we find that

$$_0F_1(\gamma; -x) \,_0F_1(\gamma; x) = \,_0F_3(\gamma, \tfrac{1}{2}\gamma, \tfrac{1}{2}(\gamma + 1); -x^2/4). \quad (18.1)$$

Entry 19. *If α or β is a nonnegative integer,*

$$_2F_0(-\alpha, -\beta; x) \,_2F_0(-\alpha, -\beta; -x)$$

$$= \sum_{k=0}^{\infty} \frac{(-\alpha)_k(-\beta)_k(-\alpha - \beta + k)_k x^{2k}}{k!}$$

$$= \,_4F_1(-\alpha, -\beta, -\tfrac{1}{2}(\alpha + \beta), -\tfrac{1}{2}(\alpha + \beta - 1); -\alpha - \beta; 4x^2).$$

Entry 19 may be proved by multiplying termwise the two series on the left side and applying Dixon's theorem. Entry 19 may be found in Erdélyi's treatise [1, p. 186, formula (4)].

Entry 20. *If x is arbitrary and α_k, $k \geq 1$, is defined by (17.2), then*

$$_1F_1(-m; n + 1; -x) \,_1F_1(-m - n; 1 - n; x)$$

$$= 1 + \sum_{k=1}^{\infty} \frac{\alpha_k(\frac{1}{2}(-2m - n - k))_k(-2x)^k}{k!}. \quad (20.1)$$

PROOF. Using (17.3), we find that the coefficient of x^r, $r \geq 1$, on the left side of (20.1) is equal to

$$d_r := \sum_{k=0}^{r} \frac{(-m)_{r-k}(-1)^{r-k}(-m-n)_k}{(n+1)_{r-k}(r-k)!(1-n)_k k!}$$

$$= \frac{(-1)^r(-m)_r}{(n+1)_r r!} \sum_{k=0}^{r} \frac{(-r)_k(-m-n)_k(-n-r)_k}{(m-r+1)_k(1-n)_k k!}.$$

Apply Dixon's theorem (17.4) with $a = -n - r$, $b = -r$, and $c = -m - n$ and use (17.6) to get

$$d_r = \frac{\Gamma(1-n)\Gamma(\tfrac{1}{2}(2-n-r))}{(n+1)_r \Gamma(1-n-r)\Gamma(\tfrac{1}{2}(2-n+r))} \frac{(-1)^r \Gamma(\tfrac{1}{2}(2+2m+n+r))}{\Gamma(\tfrac{1}{2}(2+2m+n-r))}$$

$$\times \frac{(-m)_r \Gamma(m+1-r)}{r! \Gamma(m+1)}$$

$$= 2^r \alpha_r(\tfrac{1}{2}(-2m-n-r))_r \frac{(-1)^r}{r!}.$$

This completes the proof.   □

Example 1

$$\sum_{k=0}^{\infty} \frac{x^{3k}}{(3k)!} \sum_{k=0}^{\infty} \frac{(-x)^{3k}}{(3k)!} = \frac{1}{3} + \frac{2}{3} \sum_{k=0}^{\infty} \frac{(-3x^2)^{3k}}{(6k)!}.$$

PROOF. In Entry 16, let $m = -\tfrac{1}{3}$ and $n = -\tfrac{2}{3}$ and replace x by $(x/3)^3$. Then (16.1) becomes

$$\sum_{k=0}^{\infty} \frac{x^{3k}}{(3k)!} \sum_{k=0}^{\infty} \frac{(-x)^{3k}}{(3k)!}$$

$$= \sum_{k=0}^{\infty} \frac{(-1)^k(2k)_k x^{6k}}{(3k)!(\tfrac{2}{3})_{2k}(\tfrac{1}{3})_{2k} 3^{3k}}$$

$$= 1 + \sum_{k=1}^{\infty} \frac{(-1)^k 3^k x^{6k}}{(2k-1)!(3k)2\cdot5\cdots(6k-1)1\cdot4\cdots(6k-2)}$$

$$= 1 + \frac{2}{3} \sum_{k=1}^{\infty} \frac{(-1)^k 3^{3k} x^{6k}}{(6k)!},$$

from which the desired result follows.   □

Example 2

$$\sum_{k=0}^{\infty} \frac{x^k}{(k!)^3} \sum_{k=0}^{\infty} \frac{(-x)^k}{(k!)^3} = \sum_{k=0}^{\infty} \frac{(3k)!(-x^2)^k}{(k!)^3((2k)!)^3}.$$

PROOF. Putting $m = n = 0$ in Entry 16 yields

$$_0F_2(1, 1; x) \, _0F_2(1, 1; -x) = \sum_{k=0}^{\infty} \frac{(-1)^k(2k + 1)_k x^{2k}}{(k!)^3((2k)!)^2}.$$

The desired equality readily follows.   □

Example 2 is mentioned by Hardy in his book [9, p. 7] and is found in Ramanujan's letters [16, p. xxvi] to Hardy.

Example 3

$$\sum_{k=0}^{\infty} \frac{x^{3k+1}}{(3k + 1)!} \sum_{k=0}^{\infty} \frac{(-1)^k x^{3k+1}}{(3k + 1)!} = \frac{2}{3} \sum_{k=0}^{\infty} \frac{(-1)^k(3x^2)^{3k+1}}{(6k + 2)!}.$$

PROOF. In Entry 16, set $m = \frac{1}{3}$ and $n = -\frac{1}{3}$ and replace x by $(x/3)^3$. We then find that

$$\sum_{k=0}^{\infty} \frac{x^{3k}}{(3k + 1)!} \sum_{k=0}^{\infty} \frac{(-1)^k x^{3k}}{(3k + 1)!} = \sum_{k=0}^{\infty} \frac{(-1)^k(2k + 1)_k 3^k x^{6k}}{(3k + 1)!4 \cdot 7 \cdots (6k + 1)2 \cdot 5 \cdots (6k - 1)}$$

$$= \sum_{k=0}^{\infty} \frac{(-1)^k 3^k x^{6k}}{(3k + 1)(2k)!4 \cdot 7 \cdots (6k + 1)2 \cdot 5 \cdots (6k - 1)}$$

$$= \frac{2}{3} \sum_{k=0}^{\infty} \frac{(-1)^k 3^{3k+1} x^{6k}}{(6k + 2)!}.$$

On multiplying both sides above by x^2 we complete the proof.   □

Example 4. $\cos x \cosh x = \sum_{k=0}^{\infty} \frac{(-1)^k(2x^2)^{2k}}{(4k)!}.$

PROOF. In (18.1), set $\gamma = \frac{1}{2}$ and replace x by $x^2/4$. After some simplification, the desired result follows.   □

Example 5. $\sin x \sinh x = \sum_{k=0}^{\infty} \frac{(-1)^k(2x^2)^{2k+1}}{(4k + 2)!}.$

PROOF. In (18.1), let $\gamma = \frac{3}{2}$ and replace x by $x^2/4$.   □

Example 6. $\sum_{k=0}^{\infty} \frac{x^k}{(k!)^2} \sum_{k=0}^{\infty} \frac{(-x)^k}{(k!)^2} = \sum_{k=0}^{\infty} \frac{(-x^2)^k}{(k!)^2(2k)!}.$

PROOF. Set $\gamma = 1$ in (18.1).   □

Example 7

$$_1F_1(\tfrac{1}{2}; 1; x) \, _1F_1(\tfrac{1}{2}; 1; -x) = \, _1F_2(\tfrac{1}{2}; 1, 1; x^2/4).$$

PROOF. Set $\beta = -\frac{1}{2}$ and $\gamma = 1$ in Entry 18.   □

Example 8

$$_1F_1(1; \tfrac{3}{2}; x/2) \, _1F_1(1; \tfrac{3}{2}; -x/2) = \sum_{k=0}^{\infty} \frac{(2k)!(2x)^{2k}}{(2k+1)(4k+1)!}.$$

PROOF. Apply Entry 18 with $\beta = -1$, $\gamma = \frac{3}{2}$, and x replaced by $x/2$ to obtain

$$_1F_1(1; \tfrac{3}{2}; x/2) \, _1F_1(1; \tfrac{3}{2}; -x/2) = \, _2F_3(1, \tfrac{1}{2}; \tfrac{3}{2}, \tfrac{3}{4}, \tfrac{5}{4}; x^2/16).$$

An elementary calculation shows that

$$\frac{1}{(\tfrac{3}{4})_k (\tfrac{5}{4})_k} = \frac{2^{6k}(2k)!}{(4k+1)!}.$$

The proposed equality now follows.   □

Example 9

$$_1F_1(1; n+1; x) \, _1F_1(1; n+1; -x) = \sum_{k=0}^{\infty} \frac{nx^{2k}}{(n+k)(n+1)_{2k}}$$

$$= \, _2F_3(1, n; n+1, \tfrac{1}{2}(n+1), \tfrac{1}{2}(n+2); x^2/4).$$

PROOF. Set $\beta = -1$ and $\gamma = n+1$ in Entry 18.   □

Example 10. *If n is a nonnegative integer, then*

$$_2F_0(-n, 1; x) \, _2F_0(-n, 1; -x) = \sum_{k=0}^{\infty} (-n)_k(-n+1+k)_k x^{2k}.$$

PROOF. In Entry 19, let $\alpha = n$ and $\beta = -1$. The proposed equality easily follows.   □

Entry 21

$$_2F_1(m, n; \tfrac{1}{2}(m+n+1); \tfrac{1}{2}(1+x))$$

$$= \frac{\sqrt{\pi}\,\Gamma(\tfrac{1}{2}(m+n+1))}{\Gamma(\tfrac{1}{2}(m+1))\Gamma(\tfrac{1}{2}(n+1))} \, _2F_1(\tfrac{1}{2}m, \tfrac{1}{2}n; \tfrac{1}{2}; x^2)$$

$$+ \frac{2\sqrt{\pi}\,\Gamma(\tfrac{1}{2}(m+n+1))x}{\Gamma(\tfrac{1}{2}m)\Gamma(\tfrac{1}{2}n)} \, _2F_1(\tfrac{1}{2}(m+1), \tfrac{1}{2}(n+1); \tfrac{3}{2}; x^2).$$

Entry 21 is originally due to Kummer [1, p. 82], [2, p. 118]. See also Erdélyi's compendium [1, p. 111, formula (3)].

Entry 22. *Let m be a nonpositive integer and put*

$$p = \tfrac{1}{2}m(m-1). \tag{22.1}$$

For each nonnegative integer k, let

$$A_k = p^k - \frac{k(k-1)}{3!}p^{k-1} + \frac{k(k-1)(k-2)(3k-1)}{5!}p^{k-2}$$

$$+ \cdots + \frac{2(k-1)!(2^{2k}-1)B_{2k}}{1\cdot3\cdot5\cdots(2k-1)}p, \qquad (22.2)$$

where B_j, $0 \le j < \infty$, denotes the jth Bernoulli number. Then, if $|x| < \pi$,

$$e^{-mx}\sum_{k=0}^{-m}\frac{(m)_k(\frac{1}{2})_k}{(k!)^2}(1 - e^{-2x})^k = 1 + \sum_{k=1}^{\infty}\frac{A_k x^{2k}}{2^k(k!)^2}. \qquad (22.3)$$

Before commencing the proof of Entry 22, we make one comment. We have stated Entry 22 exactly as Ramanujan gives it. Note that, by (22.2), A_k is not well defined because there apparently is no general formula for the coefficient of p^j, $1 \le j \le k$. However, A_k is well defined by a recursion formula given by (22.13) below.

PROOF. Replacing m by $-n$ and x by ix in (22.3), we rewrite (22.3) in the form

$$f_n(x) := e^{inx}\sum_{k=0}^{n}\frac{(-n)_k(\frac{1}{2})_k}{(k!)^2}(1 - e^{-2ix})^k = 1 + \sum_{k=1}^{\infty}\frac{(-1)^k A_k x^{2k}}{2^k(k!)^2}. \qquad (22.4)$$

We show first that $f_n(x) = P_n(\cos x)$, $n \ge 0$, where P_n denotes the nth Legendre polynomial. (This fact was first kindly pointed out to us by R. J. Evans.) By Bailey's book [4, p. 4],

$$f_n(x) = e^{inx}\,{}_2F_1(-n, \tfrac{1}{2}; 1; 1 - e^{-2ix})$$

$$= \frac{\Gamma(n+\frac{1}{2})}{\Gamma(n+1)\sqrt{\pi}}e^{inx}\,{}_2F_1(-n, \tfrac{1}{2}; -n + \tfrac{1}{2}; e^{-2ix}). \qquad (22.5)$$

By using (17.3), we may easily show that

$$\frac{(-n)_{n-k}(\frac{1}{2})_{n-k}}{(-n+\frac{1}{2})_{n-k}(n-k)!} = \frac{(-n)_k(\frac{1}{2})_k}{(-n+\frac{1}{2})_k k!}.$$

Hence, from (22.5),

$$f_n(x) = 2\frac{\Gamma(n+\frac{1}{2})}{\Gamma(n+1)\sqrt{\pi}}\sum_{k=0}^{[n/2]}{}'\frac{(-n)_k(\frac{1}{2})_k}{(-n+\frac{1}{2})_k k!}\cos(n-2k)x, \qquad (22.6)$$

where the prime on the summation sign indicates that if $k = n/2$, this summand is to be multiplied by $\frac{1}{2}$. From a representation for $P_n(\cos x)$ in Whittaker and Watson's text [1, p. 303], it follows that $f_n(x) = P_n(\cos x)$. Hence, it remains to show that

$$P_n(\cos x) = 1 + \sum_{k=1}^{\infty}\frac{(-1)^k A_k x^{2k}}{2^k(k!)^2}, \qquad |x| < \pi. \qquad (22.7)$$

It is well known (e.g., see Copson's text [2, p. 273]) that $P_n(\cos x)$ is a solution of Legendre's differential equation

$$y'' + (\cot x)y' + n(n + 1)y = 0. \tag{22.8}$$

Since $P_n(1) = 1$ (Whittaker and Watson [1, p. 302]) and $P_n(\cos x)$ is an even function of x, $P_n(\cos x)$ has a power series expansion of the form

$$P_n(\cos x) = \sum_{k=0}^{\infty} a_{2k}x^{2k}, \qquad a_0 = 1. \tag{22.9}$$

Our procedure will be as follows. We shall actually assume that (22.7) holds; that is, we assume that

$$a_{2k} = \frac{(-1)^k A_k}{2^k(k!)^2}, \qquad k \geq 1, \tag{22.10}$$

and then we show that A_k has the properties evinced by the formula (22.2). Recall that

$$\cot x = \sum_{k=0}^{\infty} \frac{(-1)^k B_{2k} 2^{2k} x^{2k-1}}{(2k)!}, \qquad |x| < \pi. \tag{22.11}$$

Substituting (22.9) and (22.11) into (22.8), we find that

$$\sum_{k=1}^{\infty} a_{2k} 2k(2k - 1)x^{2k-2} + \sum_{k=0}^{\infty} \frac{(-1)^k B_{2k} 2^{2k} x^{2k-1}}{(2k)!} \sum_{k=1}^{\infty} a_{2k} 2k x^{2k-1}$$

$$+ n(n + 1) \sum_{k=0}^{\infty} a_{2k} x^{2k} = 0.$$

Equating coefficients of x^{2r-2}, $r \geq 1$, on both sides, we find that

$$(2r)^2 a_{2r} + \sum_{k=1}^{r-1} \frac{(-1)^k B_{2k} 2^{2k}(2r - 2k)a_{2r-2k}}{(2k)!} + n(n + 1)a_{2r-2} = 0. \tag{22.12}$$

Noting, from (22.1), that $p = \frac{1}{2}n(n + 1)$ and using (22.10), we find, after some simplification, that the recursion relation (22.12) takes the form

$$A_r + \sum_{k=1}^{r-1} \frac{2^{3k-1}\{(r - 1)!\}^2 B_{2k} A_{r-k}}{(r - k)!(r - k - 1)!(2k)!} - pA_{r-1} = 0, \tag{22.13}$$

where $r \geq 1$ and $A_0 = 1$.

First, letting $r = 1$ in (22.13), we find that

$$A_1 = p. \tag{22.14}$$

Second, letting $r = 2$, using (22.14), and recalling that $B_2 = \frac{1}{6}$, we find that

$$A_2 = p^2 - \tfrac{1}{3}p. \tag{22.15}$$

Third, letting $r = 3$, using (22.14) and (22.15), and recalling that $B_4 = -\frac{1}{30}$, we find that

$$A_3 = p^3 - p^2 + \tfrac{2}{3}p. \tag{22.16}$$

Observe that the formulas for A_1, A_2, and A_3 given by (22.14)–(22.16), respectively, are in complete agreement with the formula for A_k given by (22.2). In particular, the coefficient of p in (22.2) is in corroboration with (22.14)–(22.16) for $k = 1, 2, 3$.

We now proceed by induction and assume that, for $k = 1, 2, \ldots, r$, the leading three coefficients and the last coefficient of A_k are in agreement with those prescribed in the formula (22.2). Thus, from (22.13) and the inductive hypothesis,

$$A_{r+1} = pA_r - \sum_{k=1}^{r} \frac{2^{3k-1}(r!)^2 B_{2k} A_{r+1-k}}{(r+1-k)!(r-k)!(2k)!}$$

$$= pA_r - \tfrac{1}{3}rA_r + \tfrac{2}{45}r^2(r-1)A_{r-1}$$

$$+ \cdots - \frac{2^{3r-1}(r!)^2 B_{2r}}{(2r)!} p$$

$$= (p - \tfrac{1}{3}r)\left\{ p^r - \frac{r(r-1)}{6}p^{r-1} + \frac{r(r-1)(r-2)(3r-1)}{5!}p^{r-2} \right.$$

$$\left. + \cdots + \frac{2(r-1)!(2^{2r}-1)B_{2r}}{1 \cdot 3 \cdots (2r-1)}p \right\}$$

$$+ \frac{2}{45}r^2(r-1)\left\{ p^{r-1} + \cdots + \frac{2(r-2)!(2^{2r-2}-1)B_{2r-2}}{1 \cdot 3 \cdots (2r-3)}p \right\}$$

$$+ \cdots - \frac{2^{3r-1}(r!)^2 B_{2r}}{(2r)!}p.$$

The coefficient of p^{r+1} above is equal to 1 in agreement with (22.2). The coefficient of p^r above is equal to

$$-\frac{r}{3} - \frac{r(r-1)}{6} = -\frac{(r+1)r}{3!},$$

which also agrees with (22.2). The coefficient of p^{r-1} above is found to be

$$\frac{r^2(r-1)}{18} + \frac{r(r-1)(r-2)(3r-1)}{5!} + \frac{2r^2(r-1)}{45} = \frac{(r+1)r(r-1)(3r+2)}{5!},$$

which again is what we desire by (22.2).

Lastly, the coefficient of p above is equal to

$$c_r := -\sum_{k=1}^{r} \frac{2^{3k-1}(r!)^2 B_{2k}(r-k)! 2^{r+2-k}(r+1-k)!(2^{2r+2-2k}-1)B_{2r+2-2k}}{(r+1-k)!(r-k)!(2k)!(2r+2-2k)!}$$

$$= -2(r!)^2 \sum_{k=1}^{r} \frac{2^{r+2k}(2^{2r+2-2k}-1)B_{2k}B_{2r+2-2k}}{(2k)!(2r+2-2k)!}. \tag{22.17}$$

Recalling the Laurent expansions for $\coth(2x)$ and $\tanh x$, we have, for $|x| < \pi/2$,

$$\coth(2x)\tanh x = \sum_{k=0}^{\infty} \frac{2^{4k-1}B_{2k}x^{2k-1}}{(2k)!} \sum_{k=1}^{\infty} \frac{2^{2k}(2^{2k}-1)B_{2k}x^{2k-1}}{(2k)!}.$$

The coefficient of x^{2r}, $r \geq 0$, on the right side is equal to

$$d_r := 2^{r+1} \sum_{k=0}^{r} \frac{2^{r+2k}(2^{2r+2-2k}-1)B_{2k}B_{2r+2-2k}}{(2k)!(2r+2-2k)!}$$

$$= 2^{r+1}\left\{ -\frac{c_r}{2(r!)^2} + \frac{2^r(2^{2r+2}-1)B_{2r+2}}{(2r+2)!} \right\}, \tag{22.18}$$

by (22.17). On the other hand, for $|x| < \pi/2$,

$$\coth(2x)\tanh x = 1 - \tfrac{1}{2}\operatorname{sech}^2 x$$

$$= 1 - \frac{1}{2}\frac{d}{dx}\tanh x$$

$$= 1 - \sum_{k=1}^{\infty} \frac{2^{2k-1}(2^{2k}-1)(2k-1)B_{2k}x^{2k-2}}{(2k)!}.$$

Hence, we have also found that, for $r \geq 1$,

$$d_r = -\frac{2^{2r+1}(2^{2r+2}-1)(2r+1)B_{2r+2}}{(2r+2)!}. \tag{22.19}$$

Equating (22.18) and (22.19) and solving for c_r, we find that

$$c_r = \frac{2^{r+2}r!(r+1)!(2^{2r+2}-1)B_{2r+2}}{(2r+2)!}, \qquad r \geq 1.$$

Examining the coefficient of p in (22.2) when $k = r + 1$, we find that this coefficient is indeed equal to c_r. This completes the inductive proof. $\square$

Corollary. *If $p = 1$, then*

$$A_k = A_k(1) = \frac{2^k(k!)^2}{(2k)!}, \qquad k \geq 1; \tag{22.20}$$

if $p - 3$, then

$$A_k = A_k(3) = 3 \cdot 2^{2k-2}A_k(1), \qquad k \geq 1. \tag{22.21}$$

PROOF. In Entry 22, let $m = -1$. Then by (22.1), $p = 1$. From (22.6), $P_1(\cos x) = \cos x$. Thus, the coefficient of x^{2k}, $k \geq 1$, in $P_1(\cos x)$ is equal to $(-1)^k/(2k)!$. But from (22.7), the coefficient of x^{2k} is also equal to $(-1)^k A_k/\{(2^k(k!)^2\}$, $k \geq 1$. Equating these two coefficients, we deduce (22.20).

Second, let $m = -2$ in Entry 22. Then $p = 3$. From (22.6),

$$P_2(\cos x) = \tfrac{3}{4}\cos(2x) + \tfrac{1}{4}.$$

Thus, the coefficient of x^{2k}, $k \geq 1$, in $P_2(\cos x)$ is equal to $3(-1)^k 2^{2k-2}/(2k)!$. Equating this with the coefficient of x^{2k} given by (22.7), we deduce (22.21). $\square$

If we expand $\cos(n - 2k)x$, $0 \leq k \leq [n/2]$, in its Maclaurin series in (22.6) and, for $P_n(\cos x)$, equate the coefficient of x^{2j}, $j \geq 1$, with the coefficient of x^{2j} in (22.7), we obtain an elegant identity involving binomial coefficients. We shall further separate this identity into two cases. Replacing n by $2n$ and then n by $2n + 1$, we find, respectively, that

$$\sum_{k=1}^{n} \binom{2n + 2k}{n + k} \binom{2n - 2k}{n - k} k^{2j} = 2^{4n-3j-1} \binom{2j}{j} A_j(p), \qquad (22.22)$$

where $n, j \geq 1$ and $p = n(2n + 1)$, and that

$$\sum_{k=0}^{n} \binom{2n + 2k + 2}{n + k + 1} \binom{2n - 2k}{n - k}(2k + 1)^{2j} = 2^{4n-j+1} \binom{2j}{j} A_j(p), \qquad (22.23)$$

where $n \geq 0$, $j \geq 1$, and $p = (n + 1)(2n + 1)$. These identities are apparently new and cannot be found in the tables of Gould [1] or Hansen [1], for example.

Entry 23 is apparently meaningless. Ramanujan claims that if

$$\varphi(x) = c_1 + \sqrt{1} = c_2 + \sqrt{2} = \cdots = c_n + \sqrt{n}$$

and "if $c_1, c_2, c_3, \ldots, c_n$ appear to be similar," then they are all identically equal to c. He then concludes that

$$\varphi(x) = c + \sqrt{1} + \sqrt{2} + \cdots + \sqrt{n}.$$

The intent of this entry shall perhaps always remain a mystery.

Entry 24. *Let*

$$\varphi(r) = \frac{1}{\Gamma(n + 1)} \sum_{k=0}^{\infty} \binom{n}{k} P_k r^{n-k}$$

and

$$Q_r = \sum_{k=0}^{\infty} \frac{(r + 1)_k}{k!} \varphi(r + k).$$

Then

$$\sum_{k=0}^{\infty} \varphi(k)(1 - x)^k = \sum_{k=0}^{\infty} Q_k(-x)^k$$

$$+ \frac{1}{\left(\text{Log} \dfrac{1}{1 - x}\right)^{n+1}} \sum_{k=0}^{\infty} P_k \left(\text{Log} \frac{1}{1 - x}\right)^k. \qquad (24.1)$$

Corollary 1. *Let $\varphi(r)$ be defined as above and let*

$$Q'_r = \frac{\Gamma(m + 1)}{\Gamma(m + 1 - r)\Gamma(r + 1)} \sum_{k=0}^{\infty} \frac{(m + 1)_k}{(m + 1 - r)_k} \varphi(m + k).$$

Then

$$\sum_{k=0}^{\infty} \varphi(m + k)(1 - x)^{m+k} = \sum_{k=0}^{\infty} Q'_k(-x)^k$$

$$+ \frac{1}{\left(\text{Log}\,\dfrac{1}{1 - x}\right)^{n+1}} \sum_{k=0}^{\infty} P_k \left(\text{Log}\,\frac{1}{1 - x}\right)^k.$$

Entry 24 and Corollary 1 are enigmatic. It seems likely that there are no functions φ for which either of the proposed identities holds. For most choices of φ, the series for Q_r and Q'_r diverge. Employing the definition of Q_j, we formally find that

$$\sum_{j=0}^{\infty} Q_j(-x)^j = \sum_{j=0}^{\infty} \sum_{k=0}^{\infty} \varphi(j + k)\frac{(j + 1)_k}{k!}(-x)^j$$

$$= \sum_{n=0}^{\infty} \varphi(n) \sum_{j=0}^{n} \binom{n}{j}(-x)^j$$

$$= \sum_{n=0}^{\infty} \varphi(n)(1 - x)^n.$$

Comparing the formula above with (24.1), we find that the logarithmic series does not appear!

Corollary 2. *Let* $\alpha + \beta + \gamma + 1 = \delta + \varepsilon$ *with* $\gamma > -1$. *Then as* x *tends to* $0+$,

$$\frac{\Gamma(\alpha + 1)\Gamma(\beta + 1)\Gamma(\gamma + 1)}{\Gamma(\delta + 1)\Gamma(\varepsilon + 1)} \,_3F_2\left[\begin{array}{c} \alpha + 1, \beta + 1, \gamma + 1 \\ \delta + 1, \varepsilon + 1 \end{array}; 1 - x\right]$$

$$\sim -\text{Log}\,x - \psi(\alpha + 1) - \psi(\beta + 1) - 2C + \sum_{k=1}^{\infty} \frac{(\delta - \gamma)_k(\varepsilon - \gamma)_k}{(\alpha + 1)_k(\beta + 1)_k k},$$

where $\psi(z) = \Gamma'(z)/\Gamma(z)$ *and* C *denotes Euler's constant.*

We cannot see how Corollary 2 would follow from Entry 24. Corollary 2 should be compared with the more precise formula for $_2F_1$ in Entry 26 below. Corollary 2 is a very beautiful and significant formula, for it is the only asymptotic formula for zero-balanced series besides that which can be obtained from Entry 26. R. J. Evans and D. Stanton [1] have recently found an elegant proof of Corollary 2 as well as of a q-analogue. They provide a complete proof of the q-analogue and sketch a proof of Corollary 2. In fact, they establish a slightly stronger version of Corollary 2. We follow Evans and Stanton in our development below. It will be convenient to trivially alter the notation of Corollary 2 above.

Theorem 1. *If* $a + b + c = d + e$ *and* $\text{Re}\,c > 0$, *then*

$$\sum_{k=0}^{\infty} \left\{\frac{\Gamma(a + k)\Gamma(b + k)\Gamma(c + k)}{\Gamma(d + k)\Gamma(e + k)\Gamma(1 + k)} - \frac{1}{k + 1}\right\} = L, \tag{24.2}$$

where

$$L = -2\gamma - \psi(a) - \psi(b) + \sum_{k=1}^{\infty} \frac{(d-c)_k(e-c)_k}{(a)_k(b)_k k}, \qquad (24.3)$$

where γ denotes Euler's constant. Furthermore, as m tends to ∞,

$$\sum_{k=0}^{m-1} \frac{(a)_k(b)_k(c)_k}{(d)_k(e)_k k!} = \frac{\Gamma(d)\Gamma(e)}{\Gamma(a)\Gamma(b)\Gamma(c)}\{\text{Log } m + L + \gamma\} + O\left(\frac{1}{m}\right), \qquad (22.4)$$

where the implied constant depends on a, b, c, d, and e but not on m.

If $c = e$, then (24.4) reduces to the following asymptotic expansion for a partial sum of a zero-balance $_2F_1$ series (e.g., see Luke's book [1, p. 109, Eq. (34)]):

$$\sum_{k=0}^{m-1} \frac{(a)_k(b)_k}{(d)_k k!} = \frac{\Gamma(d)}{\Gamma(a)\Gamma(b)}\{\text{Log } m - \gamma - \psi(a) - \psi(b)\} + O\left(\frac{1}{m}\right), \qquad (24.5)$$

as m tends to ∞. A slightly less precise version of (24.5) is given by Ramanujan in Entry 15 of Chapter 10. It would be interesting if there existed a theorem for zero-balanced $_{q+1}F_q$ series that included (24.4) and (24.5) as special cases.

Theorem 2 below is a slightly more precise theorem than Ramanujan's Corollary 2 given above.

Theorem 2. *If $a + b + c = d + e$ and Re $c > 0$, then as x tends to 1 with $0 < x < 1$,*

$$\frac{\Gamma(a)\Gamma(b)\Gamma(c)}{\Gamma(d)\Gamma(e)} \, _3F_2\left[\begin{array}{c} a, b, c \\ d, e \end{array}; x\right] = -\text{Log}(1-x) + L + O((1-x)\,\text{Log}(1-x)),$$

$$(24.6)$$

where L is defined by (24.3).

In the sequal, we shall deduce Theorem 2 from Theorem 1.
In order to establish Theorems 1 and 2, we shall need four lemmas.

Lemma 1. *If Re $C > 0$, $S = D + E - A - B - C$, and Re $S > 0$, then*

$$_3F_2\left[\begin{array}{c} A, B, C \\ D, E \end{array}\right] = \frac{\Gamma(D)\Gamma(E)\Gamma(S)}{\Gamma(C)\Gamma(A+S)\Gamma(B+S)} \, _3F_2\left[\begin{array}{c} D-C, E-C, S \\ A+S, B+S \end{array}\right].$$

Lemma 1 is a reformulation of Entry 27 in Chapter 10.

Lemma 2. *If a and d are bounded, then as z tends to ∞ with Re $z > 0$,*

$$\frac{\Gamma(a+z)}{\Gamma(d+z)} = z^{a-d}(1 + O(1/z)).$$

Lemma 2, of course, is an easy consequence of Stirling's formula for the gamma function, which can be found in Entry 23 of Chapter 7.

Lemma 3. *Let $\varepsilon > 0$ be fixed and let a complex number E be fixed. Let $\mathrm{Re}\, z \geq \varepsilon$ and suppose that k is any positive integer. Then there exists a constant $N > 0$ such that*

$$\left(1 + \frac{z}{k}\right)^E - 1 = O\left(\frac{z^N}{k}\right), \tag{24.7}$$

where the implied constant is independent of z and k.

PROOF. Let $F = \mathrm{Re}\, E$. If $F \geq 0$, then, since $\mathrm{Re}\, z \geq \varepsilon$,

$$\left(1 + \frac{z}{k}\right)^{-E} - 1 = -\left(1 + \frac{z}{k}\right)^{-E}\left(\left(1 + \frac{z}{k}\right)^E - 1\right) = O\left(\left(1 + \frac{z}{k}\right)^E - 1\right).$$

Hence, it suffices to consider the case $F \geq 0$. Let $N = F + 1$. First, suppose that $k \leq |z|$. Then

$$\left|1 + \frac{z}{k}\right|^F \leq (1 + |z|)^F = O(z^F) = O\left(\frac{z^N}{k}\right).$$

Thus, (24.7) easily follows. Finally, suppose that $k > |z|$. Then

$$\left|\left(1 + \frac{z}{k}\right)^E - 1\right| \leq \sum_{m=1}^{\infty} \left|\binom{E}{m}\right|\left|\frac{z}{k}\right|^m \leq \left|\frac{z}{k}\right| \sum_{m=1}^{\infty} \left|\binom{E}{m}\right|$$

$$= O\left(\frac{z}{k}\right) = O\left(\frac{z^N}{k}\right),$$

where the last series does indeed converge because $F \geq 0$. This completes the proof. $\qquad\square$

Lemma 4. *Let $\mathrm{Re}\, D$ be fixed, where D is not a nonpositive integer. Let k be any positive integer, and suppose that $\mathrm{Re}\, z \geq 0$. Then*

$$\frac{(D - z)_k}{(D)_k} = O(e^{2\pi|z|/3}),$$

where the implied constant is independent of z and k.

PROOF. For some constant $N > 0$ that is independent of z and k,

$$\left|\frac{(D - z)_k}{(D)_k}\right| = \prod_{j=0}^{k-1} \left|\frac{D + j - z}{D + j}\right|$$

$$= \prod_{j=0}^{k-1} \left|1 - \frac{z}{D + j}\right|$$

$$\ll (1 + |z|)^N \prod_{\substack{j=0 \\ D+j \geq 1}}^{k-1} \left|1 - \frac{z}{D + j}\right|$$

$$= (1 + |z|)^N \prod_{\substack{j=0 \\ D+j \geq 1}}^{k-1} \left(1 - 2\,\mathrm{Re}\left(\frac{z}{D + j}\right) + \left|\frac{z}{D + j}\right|^2\right)^{1/2}$$

$$\ll (1 + |z|)^N \prod_{\substack{j=0 \\ D+j \geq 1}}^{k-1} \left(1 + \left|\frac{z}{D+j}\right|^2\right)^{1/2}$$

$$\ll (1 + |z|)^N \prod_{m=1}^{\infty} \left(1 + \frac{|z|^2}{m^2}\right)^{1/2}$$

$$= (1 + |z|)^N \left(\frac{e^{\pi|z|} - e^{-\pi|z|}}{2\pi|z|}\right)^{1/2}$$

$$\ll (1 + |z|)^N e^{\pi|z|/2} \ll e^{2\pi|z|/3}.$$

PROOF OF THEOREM 1. By Lemma 2,

$$\sum_{k=m}^{\infty} \left\{\frac{\Gamma(a+k)\Gamma(b+k)\Gamma(c+k)}{\Gamma(d+k)\Gamma(e+k)\Gamma(1+k)} - \frac{1}{k+1}\right\} = O\left(\frac{1}{m}\right),$$

as m tends to ∞. Also, from Ayoub's text [1, p. 43],

$$\sum_{k=0}^{m-1} \frac{1}{k+1} = \text{Log } m + \gamma + O\left(\frac{1}{m}\right),$$

as m tends to ∞. Thus, it is readily seen that (24.4) follows from (24.2). It remains to prove (24.2).

We first prove (24.2) for $c = 1$. Then, inducting on c, we prove (24.2) for each positive integer c. Lastly, we establish (24.2) for all c with Re $c > 0$.

For each $\varepsilon > 0$, write

$$\sum_{k=0}^{m-1} \frac{(a)_k(b)_k(1)_k}{(d)_k(e+\varepsilon)_k k!} = H_1 - H_2, \tag{24.8}$$

where

$$H_1 = {}_3F_2\left[\begin{matrix} a, b, 1 \\ d, e+\varepsilon \end{matrix}\right]$$

and

$$H_2 = \frac{(a)_m(b)_m}{(d)_m(e+\varepsilon)_m} {}_3F_2\left[\begin{matrix} 1, b+m, a+m \\ d+m, e+\varepsilon+m \end{matrix}\right],$$

upon a change of index of summation. By Lemma 1,

$$H_1 = \frac{\Gamma(d)\Gamma(e+\varepsilon)\Gamma(\varepsilon)}{\Gamma(a+\varepsilon)\Gamma(b+\varepsilon)} {}_3F_2\left[\begin{matrix} d-1, e+\varepsilon-1, \varepsilon \\ a+\varepsilon, b+\varepsilon \end{matrix}\right]$$

and

$$H_2 = \frac{\Gamma(d)\Gamma(e+\varepsilon)\Gamma(\varepsilon)\Gamma(b+m)}{\Gamma(a)\Gamma(b)\Gamma(1+\varepsilon)\Gamma(b+m+\varepsilon)} {}_3F_2\left[\begin{matrix} d-a, e-a+\varepsilon, \varepsilon \\ 1+\varepsilon, b+m+\varepsilon \end{matrix}\right].$$

Thus, we may write

$$H_1 - H_2 = G_1 + G_2 + G_3, \tag{24.9}$$

where

$$G_1 = \lim_{\varepsilon \to 0} \left(\frac{\Gamma(d)\Gamma(e + \varepsilon)\Gamma(\varepsilon)}{\Gamma(a + \varepsilon)\Gamma(b + \varepsilon)} - \frac{\Gamma(d)\Gamma(e + \varepsilon)\Gamma(\varepsilon)\Gamma(b + m)}{\Gamma(a)\Gamma(b)\Gamma(1 + \varepsilon)\Gamma(b + m + \varepsilon)} \right)$$

$$= \Gamma(d)\Gamma(e) \lim_{\varepsilon \to 0} \left(\left\{ \frac{1}{\varepsilon} + \Gamma'(1) + \cdots \right\} \left\{ \frac{1}{\Gamma(a)} - \frac{\Gamma'(a)}{\Gamma^2(a)} \varepsilon + \cdots \right\} \right.$$

$$\left. \times \left\{ \frac{1}{\Gamma(b)} - \frac{\Gamma'(b)}{\Gamma^2(b)} \varepsilon + \cdots \right\} - \frac{1}{\Gamma(a)\Gamma(b)\varepsilon} \left\{ 1 - \frac{\Gamma'(b + m)}{\Gamma(b + m)} \varepsilon + \cdots \right\} \right)$$

$$= \frac{\Gamma(d)\Gamma(e)}{\Gamma(a)\Gamma(b)} \left\{ -\gamma - \frac{\Gamma'(a)}{\Gamma(a)} - \frac{\Gamma'(b)}{\Gamma(b)} + \frac{\Gamma'(b + m)}{\Gamma(b + m)} \right\}, \qquad (24.10)$$

$$G_2 = \lim_{\varepsilon \to 0} \frac{\Gamma(d)\Gamma(e + \varepsilon)\Gamma(\varepsilon)}{\Gamma(a + \varepsilon)\Gamma(b + \varepsilon)} \sum_{k=1}^{\infty} \frac{(d - 1)_k(e + \varepsilon - 1)_k(\varepsilon)_k}{(a + \varepsilon)_k(b + \varepsilon)_k k!}$$

$$= \frac{\Gamma(d)\Gamma(e)}{\Gamma(a)\Gamma(b)} \sum_{k=1}^{\infty} \frac{(d - 1)_k(e - 1)_k}{(a)_k(b)_k k}, \qquad (24.11)$$

and

$$G_3 = \lim_{\varepsilon \to 0} \frac{\Gamma(d)\Gamma(e + \varepsilon)\Gamma(\varepsilon)\Gamma(b + m)}{\Gamma(a)\Gamma(b)\Gamma(1 + \varepsilon)\Gamma(b + m + \varepsilon)} \sum_{k=1}^{\infty} \frac{(d - a)_k(e - a + \varepsilon)_k(\varepsilon)_k}{(1 + \varepsilon)_k(b + m + \varepsilon)_k k!}$$

$$= \frac{\Gamma(d)\Gamma(e)}{\Gamma(a)\Gamma(b)} \sum_{k=1}^{\infty} \frac{(d - a)_k(e - a)_k}{(1)_k(b + m)_k k}. \qquad (24.12)$$

Since (Luke [1, p. 33, Eq. (8)]),

$$\frac{\Gamma'(b + m)}{\Gamma(b + m)} = \text{Log } m + O\left(\frac{1}{m}\right),$$

as m tends to ∞, we find from (24.10) and (24.12) that, respectively,

$$G_1 = \frac{\Gamma(d)\Gamma(e)}{\Gamma(a)\Gamma(b)} \{ -\gamma - \psi(a) - \psi(b) + \text{Log } m \} + O\left(\frac{1}{m}\right) \qquad (24.13)$$

and

$$G_3 = O\left(\frac{1}{m}\right), \qquad (24.14)$$

as m tends to ∞.

Putting (24.11), (24.13), and (24.14) in (24.9) and then (24.9) into (24.8), we conclude that we have established (24.4) for $c = 1$.

Assuming that (24.4) holds with c replaced by $1, 2, \ldots, c - 1$, we examine

$$\sum_{k=0}^{m-1} \frac{(a)_k(b)_k(c)_k}{(d)_k(e)_k k!}$$

$$= \frac{(d - 1)(e - 1)}{(b - 1)(c - 1)} \sum_{k=0}^{m-1} \frac{(a)_k(b - 1)_{k+1}(c - 1)_{k+1}}{(d - 1)_{k+1}(e - 1)_{k+1} k!}$$

$$= \frac{(d-1)(e-1)}{(b-1)(c-1)} \sum_{k=1}^{m} \frac{(b-1)_k(c-1)_k}{(d-1)_k(e-1)_k(k-1)!} \left\{ \frac{(a)_k}{k} - \frac{(a-1)_k}{k} \right\}$$

$$= \frac{(d-1)(e-1)}{(b-1)(c-1)} \sum_{k=0}^{m} \frac{(a)_k(b-1)_k(c-1)_k}{(d-1)_k(e-1)_k k!}$$

$$- \frac{(d-1)(e-1)}{(b-1)(c-1)} \sum_{k=0}^{m} \frac{(a-1)_k(b-1)_k(c-1)_k}{(d-1)_k(e-1)_k k!}$$

$$= \frac{\Gamma(d)\Gamma(e)}{\Gamma(a)\Gamma(b)\Gamma(c)} \left\{ \text{Log } m - \gamma - \psi(a) - \psi(b-1) + \sum_{k=1}^{\infty} \frac{(d-c)_k(e-c)_k}{(a)_k(b-1)_k k} \right\}$$

$$- \frac{(d-1)(e-1)}{(b-1)(c-1)} \sum_{k=0}^{\infty} \frac{(a-1)_k(b-1)_k(c-1)_k}{(d-1)_k(e-1)_k k!} + O\left(\frac{1}{m}\right),$$

as m tends to ∞. Using again Lemma 1, we deduce that

$$\sum_{k=0}^{m-1} \frac{(a)_k(b)_k(c)_k}{(d)_k(e)_k k!}$$

$$= \frac{\Gamma(d)\Gamma(e)}{\Gamma(a)\Gamma(b)\Gamma(c)} \left\{ \text{Log } m - \gamma - \psi(a) - \psi(b) + \frac{1}{b-1} \right.$$

$$\left. + \sum_{k=1}^{\infty} \frac{(d-c)_k(e-c)_k}{(a)_k(b-1)_k k} - \frac{1}{b-1} \sum_{k=0}^{\infty} \frac{(d-c)_k(e-c)_k}{(a)_k(b)_k} \right\}$$

$$= \frac{\Gamma(d)\Gamma(e)}{\Gamma(a)\Gamma(b)\Gamma(c)} \left\{ \text{Log } m - \gamma - \psi(a) - \psi(b) \right.$$

$$\left. + \sum_{k=1}^{\infty} \frac{(d-c)_k(e-c)_k}{(a)_k} \left(\frac{1}{(b-1)_k k} - \frac{1}{(b-1)(b)_k} \right) \right\}$$

$$= \frac{\Gamma(d)\Gamma(e)}{\Gamma(a)\Gamma(b)\Gamma(c)} \left\{ \text{Log } m - \gamma - \psi(a) - \psi(b) + \sum_{k=1}^{\infty} \frac{(d-c)_k(e-c)_k}{(a)_k(b)_k k} \right\}.$$

Thus, (24.4) has been established for each positive integer c. Letting m tend to ∞ in (24.4) and recalling the opening paragraph of this proof, we conclude that (24.2) holds for each positive integer c.

To prove that (24.2) is valid for all c with Re $c > 0$, it suffices by Carlson's theorem (Bailey [4, p. 39]) to prove that, for a, b, d, and $\varepsilon > 0$ fixed, both sides of (24.2) are analytic in c and equal to $O(e^{2\pi|c|/3})$ for Re $c \geq \varepsilon$.

Let $D = \text{Re}(d - \varepsilon)$, with d adjusted, if necessary, so that D is not a non-positive integer. Let $z = c + D - d$. Thus,

$$S := \sum_{k=1}^{\infty} \frac{(d-c)_k(e-c)_k}{(a)_k(b)_k k} = \sum_{k=1}^{\infty} A_k \frac{(D-z)_k}{(D)_k},$$

where

$$A_k = \frac{(a+b-d)_k(D)_k}{(a)_k(b)_k k}, \qquad k \geq 1.$$

By Lemma 2, $A_k = O(k^{-1-\varepsilon})$, while by Lemma 4, $(D-z)_k/(D)_k = O(e^{2\pi|z|/3})$. Thus, S is analytic in z and equals $O(e^{2\pi|z|/3})$ for Re $z \geq 0$. It follows that S is analytic in c and equal to $O(e^{2\pi|c|/3})$ for Re $c \geq \varepsilon$.

It remains to prove that

$$T := \sum_{k=1}^{\infty} \left\{ \frac{\Gamma(a+k)\Gamma(b+k)\Gamma(c+k)}{\Gamma(1+k)\Gamma(d+k)\Gamma(a+b-d+c+k)} - \frac{1}{k+1} \right\}$$

is analytic in c and equal to $O(e^{2\pi|c|/3})$ for Re $c \geq \varepsilon$. Let $E = d - a - b$. By Lemma 2, since Re $c \geq \varepsilon$,

$$T = \sum_{k=1}^{\infty} \left\{ k^{-E-1}(c+k)^E (1 + k^{-1}O(1)) - \frac{1}{k+1} \right\}$$

$$= \sum_{k=1}^{\infty} k^{-1} \left\{ \left(1 + \frac{c}{k}\right)^E - 1 \right\} \{1 + k^{-1}O(1)\} + O(1),$$

where the expressions $O(1)$ are bounded analytic functions of c for Re $c \geq \varepsilon$. By Lemma 3, $(1 + c/k)^E - 1 = O(c^N/k)$ for some positive constant N. Thus, T is analytic in c and equals $O(c^N)$ for Re $c \geq \varepsilon$. This then completes the proof of Theorem 1. $\qquad\qquad\square$

PROOF OF THEOREM 2. Define

$$f(k) = \frac{\Gamma(a+k)\Gamma(b+k)\Gamma(c+k)}{\Gamma(d+k)\Gamma(e+k)\Gamma(1+k)}$$

and

$$V(x) = \sum_{k=0}^{\infty} f(k)x^k + \text{Log}(1-x) - L,$$

where $0 < x < 1$ and L is defined by (24.3). We must show that

$$V(x) = O((1-x)\,\text{Log}(1-x)), \tag{24.15}$$

as x tends to 1. By (24.2),

$$V(x) = \sum_{k=0}^{\infty} \left(f(k) - \frac{1}{k+1} \right)(x^k - 1) + \sum_{k=0}^{\infty} \frac{x^k - x^{k-1}}{k+1}$$

$$= \sum_{k=0}^{\infty} \left(f(k) - \frac{1}{k+1} \right)(x^k - 1) + \frac{x-1}{x}\,\text{Log}\,(1-x). \tag{24.16}$$

Now, by Lemma 2,

$$\sum_{k=1}^{\infty} \left| \left(f(k) - \frac{1}{k+1} \right)(x^k - 1) \right| \ll \sum_{k=1}^{\infty} \frac{1 - x^k}{k^2}$$

$$= (1-x) \sum_{k=1}^{\infty} k^{-2} \sum_{n=0}^{k-1} x^n$$

$$= (1 - x) \sum_{n=0}^{\infty} x^n \sum_{k=n+1}^{\infty} k^{-2}$$

$$< (1 - x) \left\{ \frac{\pi^2}{6} + \sum_{n=1}^{\infty} \frac{x^n}{n} \right\}$$

$$\ll (1 - x) \operatorname{Log}(1 - x).$$

Using this in (24.16), we complete the proof of (24.15) and so also that of Theorem 2.   □

The special case $c = e$ of Theorem 2 gives an asymptotic expansion of a zero-balanced $_2F_1$ as x tends to $1-$. This special case is also an easy consequence of Entry 26 below. Moreover, it is equivalent to (24.5).

For further remarks on Theorems 1 and 2 as well as q-analogues, consult the paper of Evans and Stanton [1]. A generalization of Theorem 2 has recently been established by Bühring [1] who uses the differential equation satisfied by $_3F_2$. His proof has the advantage that the form of the asymptotic formula does not have to be known in advance. Because Ramanujan showed little interest in differential equations, he likely had yet a different proof.

Entry 25. *Suppose that n is not an integer. Then*

$$_2F_1 \left[\begin{matrix} a + n + 1, b + n + 1 \\ a + b + n + 2 \end{matrix} ; 1 - x \right]$$

$$= \frac{\Gamma(a + b + n + 2)\Gamma(-n)}{\Gamma(a + 1)\Gamma(b + 1)} \, _2F_1 \left[\begin{matrix} a + n + 1, b + n + 1 \\ n + 1 \end{matrix} ; x \right]$$

$$+ \frac{\Gamma(a + b + n + 2)\Gamma(n)x^{-n}}{\Gamma(a + n + 1)\Gamma(b + n + 1)} \, _2F_1 \left[\begin{matrix} a + 1, b + 1 \\ -n + 1 \end{matrix} ; x \right].$$

Entry 25 is a basic formula for the analytic continuation of hypergeometric series and can be found in the treatises of Bailey [4, p. 4] and Erdélyi [1, p. 108, formula (1)].

Corollary 1. *If n is a nonnegative integer, then*

$$_2F_1 \left[\begin{matrix} a + n + 1, b + n + 1 \\ a + b + n + 2 \end{matrix} ; 1 - x \right]$$

$$= \frac{\Gamma(a + b + n + 2)\Gamma(n)x^{-n}}{\Gamma(a + n + 1)\Gamma(b + n + 1)} \sum_{k=0}^{n-1} \frac{(a + 1)_k(b + 1)_k x^k}{(-n + 1)_k k!}$$

$$- \frac{(-1)^n \Gamma(a + b + n + 2)}{\Gamma(a + 1)\Gamma(b + 1)\Gamma(n + 1)} \sum_{k=0}^{\infty} \frac{(a + n + 1)_k(b + n + 1)_k}{(n + 1)_k k!}$$

$$\times \{ \psi(a + n + k + 1) + \psi(b + n + k + 1)$$

$$- \psi(n + k + 1) - \psi(k + 1) + \operatorname{Log} x \} x^k,$$

where $\psi(z) = \Gamma'(z)/\Gamma(z)$. If $n = 0$, the first expression on the right side above is understood to be equal to 0.

Corollary 1 can be found in Erdélyi's synopsis [1, p. 110, formula (14)].

Corollary 2. *If n is a nonpositive integer, then*

$$
{}_2F_1\left[\begin{matrix} a + n + 1, b + n + 1 \\ a + b + n + 2 \end{matrix} ; 1 - x\right]
$$

$$
= \frac{\Gamma(a + b + n + 2)\Gamma(-n)}{\Gamma(a + 1)\Gamma(b + 1)} \sum_{k=0}^{-n-1} \frac{(a + n + 1)_k (b + n + 1)_k x^k}{(n + 1)_k k!}
$$

$$
- \frac{\Gamma(a + b + n + 2)(-x)^{-n}}{\Gamma(a + n + 1)\Gamma(b + n + 1)\Gamma(1 - n)} \sum_{k=0}^{\infty} \frac{(a + 1)_k (b + 1)_k}{(1 - n)_k k!}
$$

$$
\times \{\psi(a + k + 1) + \psi(b + k + 1)
$$

$$
- \psi(k - n + 1) - \psi(k + 1) + \mathrm{Log}\, x\} x^k.
$$

If $n = 0$, we employ the same convention as in Corollary 1.

Corollary 2 is a reformulation of another formula in Erdélyi's treatise [1, p. 110, formula (12)].

Entry 26. *We have*

$$
\frac{\Gamma(a + 1)\Gamma(b + 1)}{\Gamma(a + b + 2)} \, {}_2F_1(a + 1, b + 1; a + b + 2; 1 - x)
$$

$$
+ \mathrm{Log}\, x \, {}_2F_1(a + 1, b + 1; 1; x)
$$

$$
+ \sum_{k=0}^{\infty} \frac{(a + 1)_k (b + 1)_k}{(k!)^2} \{\psi(a + k + 1) + \psi(b + k + 1)
$$

$$
- 2\psi(k + 1)\} x^k = 0.
$$

Entry 26 is simply the case $n = 0$ of either Corollary 1 or Corollary 2 above. Ramanujan has given a less precise version of Entry 26 in Chapter 10 (Section 15).

Corollary

$$
\pi \, {}_2F_1(\tfrac{1}{2}, \tfrac{1}{2}; 1; 1 - x) = \mathrm{Log}\left(\frac{16}{x}\right) {}_2F_1(\tfrac{1}{2}, \tfrac{1}{2}; 1; x)
$$

$$
- 4 \sum_{k=1}^{\infty} \frac{(\tfrac{1}{2})_k^2}{(k!)^2} \sum_{j=1}^{k} \frac{1}{(2j - 1)(2j)} x^k.
$$

PROOF. Putting $a = b = -\tfrac{1}{2}$ in Entry 26 and using familiar formulas for $\psi(k + 1)$ and $\psi(k + \tfrac{1}{2})$ (Gradshteyn and Ryzhik [1, p. 945]), we find that

$$\pi\,{}_2F_1(\tfrac{1}{2},\tfrac{1}{2};1;1-x)$$

$$= -\operatorname{Log} x\,{}_2F_1(\tfrac{1}{2},\tfrac{1}{2};1;x) - 2\sum_{k=0}^{\infty}\frac{(\tfrac{1}{2})_k^2}{(k!)^2}\{\psi(k+\tfrac{1}{2}) - \psi(k+1)\}x^k$$

$$= -\operatorname{Log} x\,{}_2F_1(\tfrac{1}{2},\tfrac{1}{2};1;x) - 2\sum_{k=0}^{\infty}\frac{(\tfrac{1}{2})_k^2}{(k!)^2}$$

$$\times\left\{2\sum_{j=1}^{k}\frac{1}{2j-1} - 2\operatorname{Log} 2 - \sum_{j=1}^{k}\frac{1}{j}\right\}x^k$$

$$= \operatorname{Log}\left(\frac{16}{x}\right){}_2F_1(\tfrac{1}{2},\tfrac{1}{2};1;x) - 2\sum_{k=1}^{\infty}\frac{(\tfrac{1}{2})_k^2}{(k!)^2}\sum_{j=1}^{k}\frac{1}{j(2j-1)}x^k, \tag{26.1}$$

which completes the proof. $\qquad\qquad\qquad\qquad\qquad\qquad\square$

Example. *If* $0 < x < 1$, *then*

$$\int_0^{\pi/2}\int_0^{\pi/2}\frac{\tan(\varphi/2)\,d\theta\,d\varphi}{\sqrt{1 - x\cos^2\theta\cos^2\varphi}} = \frac{\pi}{4}\int_0^{\pi/2}\frac{d\varphi}{\sqrt{1-(1-x)\sin^2\varphi}}$$

$$+ \tfrac{1}{4}\operatorname{Log} x\int_0^{\pi/2}\frac{d\varphi}{\sqrt{1-x\sin^2\varphi}}. \tag{26.2}$$

PROOF. First, for $|x| < 1$,

$$\int_0^{\pi/2}\frac{d\varphi}{\sqrt{1-x\sin^2\varphi}} = \int_0^{\pi/2}\sum_{k=0}^{\infty}\frac{(\tfrac{1}{2})_k}{k!}x^k\sin^{2k}\varphi\,d\varphi$$

$$= \sum_{k=0}^{\infty}\frac{(\tfrac{1}{2})_k}{k!}x^k\frac{\Gamma(k+\tfrac{1}{2})\Gamma(\tfrac{1}{2})}{2\Gamma(k+1)}$$

$$= \frac{\pi}{2}\sum_{k=0}^{\infty}\frac{(\tfrac{1}{2})_k^2}{(k!)^2}x^k = \frac{\pi}{2}\,{}_2F_1(\tfrac{1}{2},\tfrac{1}{2};1;x). \tag{26.3}$$

Second, for $|1-x| < 1$,

$$\int_0^{\pi}\frac{d\varphi}{\sqrt{1-(1-x)\sin^2\varphi}} = \frac{\pi}{2}\,{}_2F_1(\tfrac{1}{2},\tfrac{1}{2};1;1-x). \tag{26.4}$$

Third, using an integral evaluation in Gradshteyn and Ryzhik's tables [1, p. 376] and the calculation (26.1), we find that, for $|x| < 1$,

$$\int_0^{\pi/2}\int_0^{\pi/2}\frac{\tan(\varphi/2)\,d\theta\,d\varphi}{\sqrt{1-x\cos^2\theta\cos^2\varphi}}$$

$$= \sum_{k=0}^{\infty}\frac{(\tfrac{1}{2})_k}{k!}x^k\int_0^{\pi/2}\tan(\varphi/2)\cos^{2k}\varphi\,d\varphi\int_0^{\pi/2}\cos^{2k}\theta\,d\theta$$

$$= \frac{\pi}{4} \sum_{k=0}^{\infty} \frac{(\frac{1}{2})_k^2}{(k!)^2} x^k \{\psi(k+1) - \psi(k+\tfrac{1}{2})\}$$

$$= -\frac{\pi}{2} \sum_{k=0}^{\infty} \frac{(\frac{1}{2})_k^2}{(k!)^2} \sum_{j=1}^{k} \frac{1}{(2j-1)(2j)} x^k + \frac{\pi}{2}(\text{Log } 2) \, {}_2F_1(\tfrac{1}{2}, \tfrac{1}{2}; 1; x). \quad (26.5)$$

Using (26.3)–(26.5), we find that (26.2) is equivalent to the identity

$$-\frac{\pi}{2} \sum_{k=0}^{\infty} \frac{(\frac{1}{2})_k^2}{(k!)^2} \sum_{j=1}^{k} \frac{1}{(2j-1)(2j)} x^k + \frac{\pi}{2}(\text{Log } 2) \, {}_2F_1(\tfrac{1}{2}, \tfrac{1}{2}; 1; x)$$

$$= \frac{\pi^2}{8} \, {}_2F_1(\tfrac{1}{2}, \tfrac{1}{2}; 1; 1-x) + \frac{\pi}{8}(\text{Log } x) \, {}_2F_1(\tfrac{1}{2}, \tfrac{1}{2}; 1; x),$$

where $0 < x < 1$. This last identity follows from the foregoing corollary, and so the proof is complete. $\square$

The integral in (26.3) is the complete elliptic integral of the first kind, and the formula (26.3) is a basic, well-known result in the theory of elliptic functions. For further ramifications, see Section 6 of Chapter 17 in Part III [11].

Entry 27. *For* $|x| < 1$,

$$\sum_{k=1}^{\infty} \frac{(\frac{1}{2})_k^2}{(k!)^2} \sum_{j=1}^{k} \frac{1}{2j-1} x^k = -\tfrac{1}{4} \, {}_2F_1(\tfrac{1}{2}, \tfrac{1}{2}; 1; x) \, \text{Log}(1-x). \quad (27.1)$$

PROOF. For $n \geq 1$, the coefficient of x^n on the right side of (27.1) is equal to

$$\frac{1}{4} \sum_{k=1}^{n} \frac{(\frac{1}{2})_{n-k}^2}{\{(n-k)!\}^2 k} = \frac{(\frac{1}{2})_n^2}{4(n!)^2} \sum_{k=1}^{n} \frac{(-n)_k^2}{(\frac{1}{2}-n)_k^2 k},$$

where we have employed (17.3). It thus suffices to show that

$$\sum_{k=1}^{n} \frac{(-n)_k^2}{(\frac{1}{2}-n)_k^2 k} = 4 \sum_{j=1}^{n} \frac{1}{2j-1}, \qquad n \geq 1. \quad (27.2)$$

Let S_n denote the left side of (27.2) and rewrite S_n in the form

$$S_n = \sum_{k=0}^{n-1} \frac{(-n)_{k+1}^2}{(k+1)(\frac{1}{2}-n)_{k+1}^2} = \frac{n^2}{(\frac{1}{2}-n)^2} \sum_{k=0}^{n-1} \frac{(1-n)_k^2 (1)_k^2}{(\frac{3}{2}-n)_k^2 (2)_k k!}. \quad (27.3)$$

The right side of the equality above is a balanced ${}_4F_3$ and so can be transformed by (6.3) in Chapter 10. Let $y = z = 1$, $x = -n$, $u = v = \tfrac{1}{2} - n$, $w = 2$, and $m = n$. Then

$${}_4F_3 \left[\begin{array}{c} 1, 1, -n, -n \\ \tfrac{1}{2} - n, \tfrac{1}{2} - n, 2 \end{array} \right] = \frac{(-\tfrac{1}{2} - n)_n (1)_n}{(\frac{1}{2} - n)_n (2)_n} \, {}_4F_3 \left[\begin{array}{c} 1, \tfrac{1}{2}, -n - \tfrac{1}{2}, -n \\ \tfrac{1}{2} - n, \tfrac{3}{2}, -n \end{array} \right]$$

$$= \frac{2n+1}{n+1} \sum_{k=0}^{n} \frac{(\frac{1}{2})_k(-n-\frac{1}{2})_k}{(\frac{3}{2})_k(-n+\frac{1}{2})_k}$$

$$= \frac{(2n+1)^2}{n+1} \sum_{k=0}^{n} \frac{1}{(2k+1)(2n+1-2k)}$$

$$= \frac{(2n+1)^2}{2(n+1)^2} \sum_{k=0}^{n} \left(\frac{1}{2k+1} + \frac{1}{2n+1-2k} \right)$$

$$= \frac{(2n+1)^2}{(n+1)^2} \sum_{k=0}^{n} \frac{1}{2k+1}.$$

Replacing n by $n-1$ above and using the result in (27.3), we complete the proof of (27.2).   □

The expression on the left side below is fundamental in the theory of elliptic functions. See Section 6 of Chapter 17 in Part III [11].

Example 1

$$\exp \left(-\pi \frac{{}_2F_1(\tfrac{1}{2}, \tfrac{1}{2}; 1; 1-x)}{{}_2F_1(\tfrac{1}{2}, \tfrac{1}{2}; 1; x)} \right) = \frac{x}{16} \left(1 + \frac{1}{2}x + \frac{21}{64}x^2 + \cdots \right).$$

PROOF. By the corollary in Section 26,

$$\exp \left(-\pi \frac{{}_2F_1(\tfrac{1}{2}, \tfrac{1}{2}; 1; 1-x)}{{}_2F_1(\tfrac{1}{2}, \tfrac{1}{2}; 1; x)} \right)$$

$$= \exp \left(-\mathrm{Log} \left(\frac{16}{x} \right) + 4 \sum_{k=1}^{\infty} \frac{(\tfrac{1}{2})_k^2}{(k!)^2} \sum_{j=1}^{k} \frac{1}{(2j-1)(2j)} x^k \Big/ {}_2F_1(\tfrac{1}{2}, \tfrac{1}{2}; 1; x) \right)$$

$$= \frac{x}{16} \exp \left\{ \left(\frac{1}{2}x + \frac{21}{64}x^2 + \cdots \right) \Big/ \left(1 + \frac{1}{4}x + \frac{9}{64}x^2 + \cdots \right) \right\}$$

$$= \frac{x}{16} \exp \left(\frac{1}{2}x + \frac{13}{64}x^2 + \cdots \right)$$

$$= \frac{x}{16} \left\{ 1 + \frac{1}{2}x + \frac{13}{64}x^2 + \cdots + \frac{1}{2} \left(\frac{1}{2}x + \frac{13}{64}x^2 + \cdots \right)^2 + \cdots \right\},$$

from which the sought result follows.   □

Example 2

$$\exp \left(-\frac{2\pi}{\sqrt{3}} \frac{{}_2F_1(\tfrac{1}{3}, \tfrac{2}{3}; 1; 1-x)}{{}_2F_1(\tfrac{1}{3}, \tfrac{2}{3}; 1; x)} \right) = \frac{x}{27} \left(1 + \frac{5}{9}x + \cdots \right).$$

PROOF. Putting $a = -\tfrac{1}{3}$ and $b = -\tfrac{2}{3}$ in Entry 26, we find that

$$-\frac{2\pi}{\sqrt{3}} {}_2F_1(\tfrac{1}{3}, \tfrac{2}{3}; 1; 1-x)$$

$$= \mathrm{Log}\, x\, {}_2F_1(\tfrac{1}{3}, \tfrac{2}{3}; 1; x) + \sum_{k=0}^{\infty} \frac{(\tfrac{1}{3})_k(\tfrac{2}{3})_k}{(k!)^2} \{ \psi(k+\tfrac{2}{3}) + \psi(k+\tfrac{1}{3}) - 2\psi(k+1) \} x^k$$

$$= \text{Log}\left(\frac{x}{27}\right) {}_2F_1(\tfrac{1}{3}, \tfrac{2}{3}; 1; x) + \sum_{k=1}^{\infty} \frac{(\tfrac{1}{3})_k (\tfrac{2}{3})_k}{(k!)^2} \{3\psi(3k) - \psi(k) - 2\psi(k+1)\} x^k$$

$$= \text{Log}\left(\frac{x}{27}\right) {}_2F_1(\tfrac{1}{3}, \tfrac{2}{3}; 1; x) + \frac{5}{9} x + \cdots,$$

where we have used the facts (Gradshteyn and Ryzhik [1, p. 945]), $\psi(\tfrac{2}{3}) + \psi(\tfrac{1}{3}) - 2\psi(1) = -3 \text{ Log } 3$ and $\psi(k+\tfrac{2}{3}) + \psi(k+\tfrac{1}{3}) = 3\psi(3k) - \psi(k) - 3 \text{ Log } 3$, for $k \geq 1$. Hence,

$$\exp\left(-\frac{2\pi}{\sqrt{3}} \frac{{}_2F_1(\tfrac{1}{3}, \tfrac{2}{3}; 1; 1-x)}{{}_2F_1(\tfrac{1}{3}, \tfrac{2}{3}; 1; x)}\right)$$

$$= \exp\left(\text{Log}\left(\frac{x}{27}\right) + \frac{\tfrac{5}{9}x + \cdots}{1 + \cdots}\right) = \frac{x}{27}\left(1 + \frac{5}{9}x + \cdots\right). \qquad \square$$

Example 3

$$\exp\left(-\sqrt{2}\,2\pi \frac{{}_2F_1(\tfrac{1}{4}, \tfrac{3}{4}; 1; 1-x)}{{}_2F_1(\tfrac{1}{4}, \tfrac{3}{4}; 1; x)}\right) = \frac{x}{64}\left(1 + \frac{5}{8}x + \cdots\right).$$

PROOF. Putting $a = -\tfrac{1}{4}$ and $b = -\tfrac{3}{4}$ in Entry 26, we find that

$$-\sqrt{2}\pi\, {}_2F_1(\tfrac{1}{4}, \tfrac{3}{4}; 1; 1-x)$$

$$= \text{Log } x\, {}_2F_1(\tfrac{1}{4}, \tfrac{3}{4}; 1; x)$$

$$+ \sum_{k=0}^{\infty} \frac{(\tfrac{1}{4})_k (\tfrac{3}{4})_k}{(k!)^2} \{\psi(k+\tfrac{1}{4}) + \psi(k+\tfrac{3}{4}) - 2\psi(k+1)\} x^k$$

$$= \text{Log}\left(\frac{x}{64}\right) {}_2F_1(\tfrac{1}{4}, \tfrac{3}{4}; 1; x) + \sum_{k=1}^{\infty} \frac{(\tfrac{1}{4})_k (\tfrac{3}{4})_k}{(k!)^2}$$

$$\times \{4\psi(4k) - \psi(k) - \psi(k+\tfrac{1}{2}) - 2 \text{ Log } 2 - 2\psi(k+1)\} x^k$$

$$= \text{Log}\left(\frac{x}{64}\right) {}_2F_1(\tfrac{1}{4}, \tfrac{3}{4}; 1; x) + \frac{5}{8} x + \cdots,$$

where we have used the facts (Gradshteyn and Ryzhik [1, p. 945]), $\psi(\tfrac{1}{4}) + \psi(\tfrac{3}{4}) - 2\psi(1) = -6 \text{ Log } 2$ and $\psi(k+\tfrac{1}{4}) + \psi(k+\tfrac{3}{4}) = 4\psi(4k) - \psi(k) - \psi(k+\tfrac{1}{2}) - 8 \text{ Log } 2$. The proposed formula now easily follows. $\qquad \square$

Example 4

$$\exp\left(-2\pi \frac{{}_2F_1(\tfrac{1}{6}, \tfrac{5}{6}; 1; 1-x)}{{}_2F_1(\tfrac{1}{6}, \tfrac{5}{6}; 1; x)}\right) = \frac{x}{432}\left(1 + \frac{13}{18}x + \cdots\right).$$

PROOF. In Entry 26, put $a = -\tfrac{1}{6}$ and $b = -\tfrac{5}{6}$ to find that

$$-2\pi \, {}_2F_1(\tfrac{1}{6}, \tfrac{5}{6}; 1; 1 - x)$$

$$= \text{Log } x \, {}_2F_1(\tfrac{1}{6}, \tfrac{5}{6}; 1; x)$$

$$+ \sum_{k=0}^{\infty} \frac{(\tfrac{1}{6})_k (\tfrac{5}{6})_k}{(k!)^2} \{\psi(k + \tfrac{1}{6}) + \psi(k + \tfrac{5}{6}) - 2\psi(k + 1)\} x^k$$

$$= \text{Log}\left(\frac{x}{432}\right) {}_2F_1(\tfrac{1}{6}, \tfrac{5}{6}; 1; x)$$

$$+ \sum_{k=1}^{\infty} \frac{(\tfrac{1}{6})_k (\tfrac{5}{6})_k}{(k!)^2} \left\{6\psi(6k) - 3\psi(3k) - 2\sum_{j=1}^{k} \frac{1}{2j - 1} + \gamma - 2\psi(k + 1)\right\} x^k$$

$$= \text{Log}\left(\frac{x}{432}\right) {}_2F_1(\tfrac{1}{6}, \tfrac{5}{6}; 1; x) + \frac{13}{18} x + \cdots.$$

As in Examples 1–3, we have employed familiar properties of $\psi(z)$ (Gradshteyn and Ryzhik [1, p. 945]). We also have used the fact that $\psi(\tfrac{1}{6}) + \psi(\tfrac{5}{6}) - 2\psi(1) = -4 \text{ Log } 2 - 3 \text{ Log } 3$, which can be deduced from results in Chapter 8 of the second notebook. (See the author's book [9, Chap. 8, Eq. (5.2) and Corollary 3 in Sec. 6]. See also Gradshteyn and Ryzhik's tables [1, p. 944, formula (7)].) The desired formula now readily follows.   □

We do not know Ramanujan's intention in giving Examples 1–4.

Entry 28. Let φ denote a polynomial of degree m. Suppose that n is not an integer and that $\text{Re}(a + b + m + n + 1) < 0$. Then

$$\Gamma(a + 1)\Gamma(b + 1)\Gamma(n) \sum_{k=0}^{\infty} \frac{(a + 1)_k(b + 1)_k \varphi(k)}{(1 - n)_k k!}$$

$$+ \Gamma(a + n + 1)\Gamma(b + n + 1)\Gamma(-n) \sum_{k=0}^{\infty} \frac{(a + n + 1)_k(b + n + 1)_k \varphi(n + k)}{(n + 1)_k k!}$$

$$= \frac{\Gamma(a + n + 1)\Gamma(b + n + 1)\Gamma(a + 1)\Gamma(b + 1)}{\Gamma(a + b + n + 2)} \sum_{k=0}^{\infty} \frac{(a + 1)_k(b + 1)_k \Delta^k \varphi(0)}{(a + b + n + 2)_k k!}.$$

PROOF. Since $1, x, x(x - 1), \ldots, x(x - 1)\cdots(x - m + 1)$ form a basis for the set of all polynomials of degree m over the field of complex numbers, it suffices to prove the result for $\varphi(x) = \varphi_m(x) := x(x - 1)\cdots(x - m + 1)$. We first observe that

$$\varphi_m(k) = \begin{cases} 0, & k < m, \\ m!, & k = m, \\ (-1)^m(-k)_m, & k > m. \end{cases}$$

Next, since

$$\sum_{j=0}^{k} (-1)^j \binom{k}{j} j^r = 0, \qquad 0 \le r < k,$$

where r is an integer, we find that

$$\Delta^k \varphi_m(0) = \sum_{j=0}^{k} (-1)^j \binom{k}{j} \varphi_m(j) = \begin{cases} 0, & k < m, \\ (-1)^m m!, & k = m, \\ 0, & k > m. \end{cases}$$

Thus, for $\varphi(x) = \varphi_m(x)$, the proposed identity may be written as

$$\Gamma(a+1)\Gamma(b+1)\Gamma(n) \sum_{k=m}^{\infty} \frac{(a+1)_k(b+1)_k(-1)^m(-k)_m}{(1-n)_k k!}$$

$$+ \Gamma(a+n+1)\Gamma(b+n+1)\Gamma(-n)$$

$$\times \sum_{k=0}^{\infty} \frac{(a+n+1)_k(b+n+1)_k(-1)^m(-n-k)_m}{(n+1)_k k!}$$

$$= \frac{\Gamma(a+n+1)\Gamma(b+n+1)\Gamma(a+1)\Gamma(b+1)(a+1)_m(b+1)_m(-1)^m m!}{\Gamma(a+b+n+2)(a+b+n+2)_m m!}.$$
(28.1)

Let S_1 denote the first sum on the left side of (28.1). Replacing k by $k + m$, employing Gauss's theorem, Entry 8 of Chapter 10, and simplifying, we find that

$$S_1 = \Gamma(a+1)\Gamma(b+1)\Gamma(n) \sum_{k=0}^{\infty} \frac{(a+1)_{k+m}(b+1)_{k+m}(-1)^m(-k-m)_m}{(1-n)_{k+m}(k+m)!}$$

$$= \frac{\Gamma(a+1)\Gamma(b+1)\Gamma(n)(a+1)_m(b+1)_m}{(1-n)_m} \sum_{k=0}^{\infty} \frac{(a+m+1)_k(b+m+1)_k}{(1-n+m)_k k!}$$

$$= \frac{\Gamma(n)\Gamma(a+m+1)\Gamma(b+m+1)\Gamma(m-n+1)\Gamma(-a-b-m-n-1)}{(1-n)_m \Gamma(-a-n)\Gamma(-b-n)}$$

$$= -\frac{\Gamma(a+m+1)\Gamma(b+m+1)\Gamma(a+n+1)\Gamma(b+n+1) \sin \pi(a+n) \sin \pi(b+n)}{\Gamma(a+b+m+n+2) \sin(\pi n) \sin \pi(a+b+m+n+1)}.$$
(28.2)

Let S_3 denote the expression on the right side of (28.1). Then

$$S_3 = \frac{(-1)^m \Gamma(a+n+1)\Gamma(b+n+1)\Gamma(a+m+1)\Gamma(b+m+1)}{\Gamma(a+b+m+n+2)}. \quad (28.3)$$

If S_2 denotes the second series on the left side of (28.1), then, by (28.2) and (28.3), we must show that

$$S_2 = \frac{\Gamma(a+n+1)\Gamma(b+n+1)\Gamma(a+m+1)\Gamma(b+m+1)}{\Gamma(a+b+m+n+2)}$$

$$\times \left\{ (-1)^m + \frac{\sin \pi(n+a) \sin \pi(n+b)}{\sin(\pi n) \sin \pi(a+b+m+n+1)} \right\}. \quad (28.4)$$

We shall prove (28.4) by inducting on m. For $m = 0$, (28.4) is valid by Entry

25, since Entry 28 reduces to Entry 25 for $x = 1$ when $\varphi(x) \equiv 1$. Assume then that (28.4) holds with m replaced by $0, 1, 2, \ldots, m - 1$. Observe that

$$\varphi_m(n + k) = (n + k)\varphi_{m-1}(n - 1 + k).$$

Thus, we may write

$$S_2 = \Gamma(a + n + 1)\Gamma(b + n + 1)\Gamma(-n)$$

$$\times \sum_{k=0}^{\infty} \frac{(a + n + 1)_k(b + n + 1)_k\varphi_{m-1}(n - 1 + k)}{(n + 1)_{k-1}k!}$$

$$= -\Gamma(a + 1 + (n - 1) + 1)\Gamma(b + 1 + (n - 1) + 1)\Gamma(-(n - 1))$$

$$\times \sum_{k=0}^{\infty} \frac{(a + 1 + (n - 1) + 1)_k(b + 1 + (n - 1) + 1)_k\varphi_{m-1}(n - 1 + k)}{(n)_k k!}.$$

We now apply the induction hypothesis, but with a, b, and n replaced by $a + 1$, $b + 1$, and $n - 1$, respectively. Hence,

$$S_2 = -\frac{\Gamma(a + n + 1)\Gamma(b + n + 1)\Gamma(a + m + 1)\Gamma(b + m + 1)}{\Gamma(a + b + m + n + 2)}$$

$$\times \left\{ (-1)^{m-1} + \frac{\sin \pi(n + a) \sin \pi(n + b)}{\sin \pi(n - 1) \sin \pi(a + b + m + n + 1)} \right\},$$

from which (28.4) follows. This completes the proof. $\square$

Corollary. *Assume the hypotheses of Entry 28. Then*

$$\frac{\Gamma(a + 1)\Gamma(b + 1)}{\Gamma(a + b + 2)} \sum_{k=0}^{\infty} \frac{(a + 1)_k(b + 1)_k\Delta^k\varphi(0)}{(a + b + 2)_k k!}$$

$$+ \sum_{k=0}^{\infty} \frac{(a + 1)_k(b + 1)_k\varphi'(k)}{(k!)^2} + \sum_{k=0}^{\infty} \frac{(a + 1)_k(b + 1)_k\varphi(k)}{(k!)^2}$$

$$\times \{\psi(a + 1 + k) + \psi(b + 1 + k) - 2\psi(k + 1)\} = 0.$$

PROOF. After some manipulation, we write Entry 28 in the form

$$\sum_{k=0}^{\infty} \frac{\Gamma(a + 1 + k)\Gamma(b + 1 + k)\varphi(k)}{\Gamma(1 - n + k)k!}$$

$$- \sum_{k=0}^{\infty} \frac{\Gamma(a + n + 1 + k)\Gamma(b + n + 1 + k)\varphi(n + k)}{\Gamma(n + 1 + k)k!}$$

$$- \frac{\sin(\pi n)}{n}\Gamma(a + n + 1)\Gamma(b + n + 1)$$

$$\times \sum_{k=0}^{\infty} \frac{\Gamma(a + 1 + k)\Gamma(b + 1 + k)\Delta^k\varphi(0)}{\Gamma(a + b + n + 2 + k)k!} = 0.$$

Differentiating both sides with respect to n and then setting $n = 0$, we find that

$$\sum_{k=0}^{\infty} \frac{\Gamma(a + 1 + k)\Gamma(b + 1 + k)\psi(k + 1)\varphi(k)}{(k!)^2}$$

$$- \sum_{k=0}^{\infty} \frac{\Gamma'(a + 1 + k)\Gamma(b + 1 + k)\varphi(k)}{(k!)^2}$$

$$- \sum_{k=0}^{\infty} \frac{\Gamma(a + 1 + k)\Gamma'(b + 1 + k)\varphi(k)}{(k!)^2}$$

$$- \sum_{k=0}^{\infty} \frac{\Gamma(a + 1 + k)\Gamma(b + 1 + k)\varphi'(k)}{(k!)^2}$$

$$+ \sum_{k=0}^{\infty} \frac{\Gamma(a + 1 + k)\Gamma(b + 1 + k)\psi(k + 1)\varphi(k)}{(k!)^2}$$

$$- \Gamma(a + 1)\Gamma(b + 1) \sum_{k=0}^{\infty} \frac{\Gamma(a + 1 + k)\Gamma(b + 1 + k)\Delta^k\varphi(0)}{\Gamma(a + b + 2 + k)k!} = 0.$$

After some manipulation and simplification, the formula above reduces to the proposed formula.   □

Entry 29(i). *If* $\mathrm{Re}(\alpha + \beta + \gamma - \delta - \varepsilon)$, $\mathrm{Re}(\delta - \gamma - 1) < 0$, *then*

$${}_3F_2\left[\begin{matrix} \alpha, \beta, \gamma \\ \delta, \varepsilon \end{matrix}\right] = \frac{\Gamma(\delta)\Gamma(\delta - \alpha - \beta)}{\Gamma(\delta - \alpha)\Gamma(\delta - \beta)} {}_3F_2\left[\begin{matrix} \alpha, \beta, \varepsilon - \gamma \\ \alpha + \beta - \delta + 1, \varepsilon \end{matrix}\right]$$

$$+ \frac{\Gamma(\delta)\Gamma(\varepsilon)\Gamma(\alpha + \beta - \gamma)\Gamma(\delta + \varepsilon - \alpha - \beta - \gamma)}{\Gamma(\alpha)\Gamma(\beta)\Gamma(\varepsilon - \gamma)\Gamma(\delta + \varepsilon - \alpha - \beta)}$$

$$\times {}_3F_2\left[\begin{matrix} \delta - \alpha, \delta - \beta, \delta + \varepsilon - \alpha - \beta - \gamma \\ \delta - \alpha - \beta + 1, \delta + \varepsilon - \alpha - \beta \end{matrix}\right].$$

Entry 29(i) was communicated by Ramanujan in his second letter to Hardy [16, p. xxviii]. For a proof of Entry 29(i) and an illuminating discussion of this formula, see Hardy's paper [1, pp. 498, 499], [7, pp. 511, 512]. Another proof can be found in Bailey's tract [4, p. 21].

Entry 29(ii). *If* α, β, *or* γ *is a nonnegative integer,*

$${}_3F_2\left[\begin{matrix} -2\alpha, -2\beta, -\gamma \\ -\alpha - \beta + \frac{1}{2}, \delta \end{matrix}\right] = {}_4F_3\left[\begin{matrix} -\alpha, -\beta, -\gamma, \gamma + \delta \\ -\alpha - \beta + \frac{1}{2}, \frac{1}{2}\delta, \frac{1}{2}(\delta + 1) \end{matrix}\right]. \qquad (29.1)$$

PROOF. R. Askey and J. Wilson [1] have recently given a short proof of Entry 29(ii) when either α or β is a nonnegative integer. Now suppose that γ is a nonnegative integer. If we multiply both sides of (29.1) by $(-\alpha - \beta + \frac{1}{2})_\gamma$, then on each side we obtain a polynomial in α of degree γ. These two polynomials agree for each nonnegative integer α. Hence, they must be identically equal, and this completes the proof.   □

If n is a nonnegative integer, define

$$P_{2n}(x) = P_{2n}(x; \alpha, \gamma) = (-1)^n \, {}_4F_3 \left[\begin{array}{c} -n, n + \alpha + \gamma - \frac{1}{2}, \gamma + ix, \gamma - ix \\ \alpha + \gamma, \gamma, \gamma + \frac{1}{2} \end{array} \right].$$

These polynomials in x arise from the right side of (29.1) by a renaming of the parameters. Askey and Wilson [1] have shown that $\{P_{2n}(x)\}$, $0 \le n < \infty$, is an orthogonal set on $(-\infty, \infty)$ with respect to the weight function $|\Gamma(\alpha + ix)\Gamma(\gamma + ix)|^2$. As we pointed out in Chapter 10, the integral over $(-\infty, \infty)$ of this weight function was first evaluated by Ramanujan [8], [16, p. 57]. There also exists a set of similarly defined polynomials $P_{2n+1}(x)$ of odd degree $2n + 1$ so that $\{P_n(x)\}$, $0 \le n < \infty$, forms a complete orthogonal set on $(-\infty, \infty)$ with respect to the aforementioned measure [1].

Entry 30. *Let* $\alpha + \beta + 1 = \gamma + \delta$, $c = \Gamma(\alpha)\Gamma(\beta)/\{\Gamma(\gamma)\Gamma(\delta)\}$, *and*

$$y = \frac{c \, {}_2F_1(\alpha, \beta; \delta; 1 - x)}{{}_2F_1(\alpha, \beta; \gamma; x)}.$$

Then

$$y' = -\frac{x^{-\gamma}(1 - x)^{-\delta}}{{}_2F_1^2(\alpha, \beta; \gamma; x)}.$$

PROOF. From Entry 25,

$$y = \frac{cA_1 \, {}_2F_1(\alpha, \beta; \gamma; x) + cA_2 x^{1-\gamma} \, {}_2F_1(\delta - \alpha, \delta - \beta; 2 - \gamma; x)}{{}_2F_1(\alpha, \beta; \gamma; x)},$$

where A_1 and A_2 are constants with $cA_2 = 1/(\gamma - 1)$. Thus,

$$y' = \frac{1}{\gamma - 1} \frac{d}{dx} \left(\frac{x^{1-\gamma} \, {}_2F_1(\delta - \alpha, \delta - \beta; 2 - \gamma; x)}{{}_2F_1(\alpha, \beta; \gamma; x)} \right)$$

$$= \frac{1}{(\gamma - 1) \, {}_2F_1^2(\alpha, \beta; \gamma; x)} \left\{ {}_2F_1(\alpha, \beta; \gamma; x) \frac{d}{dx} (x^{1-\gamma} \, {}_2F_1(\delta - \alpha, \delta - \beta; 2 - \gamma; x)) \right.$$

$$\left. - x^{1-\gamma} \, {}_2F_1(\delta - \alpha, \delta - \beta; 2 - \gamma; x) \frac{d}{dx} \, {}_2F_1(\alpha, \beta; \gamma; x) \right\}$$

$$= \frac{1}{(\gamma - 1) \, {}_2F_1^2(\alpha, \beta; \gamma; x)} W({}_2F_1(\alpha, \beta; \gamma; x), x^{1-\gamma} \, {}_2F_1(\delta - \alpha, \delta - \beta; 2 - \gamma; x))$$

$$=: \frac{1}{(\gamma - 1) \, {}_2F_1^2(\alpha, \beta; \gamma; x)} W(x), \tag{30.1}$$

where $W(f, g) = W(x)$ denotes the Wronskian of $f(x)$ and $g(x)$. Now these two functions are linearly independent solutions of the hypergeometric differential equation (Bailey [4, p. 1])

$$x(1 - x)y'' + \{\gamma - (\alpha + \beta + 1)x\}y' - \alpha\beta y = 0. \tag{30.2}$$

By Abel's formula (e.g., see the text of Coddington [1, p. 113]),

$$W(x) = C \exp\left(-\int \frac{\gamma - (\alpha + \beta + 1)x}{x(1-x)} \, dx\right)$$

$$= C \exp\left(\int \left(-\frac{\gamma}{x} + \frac{\delta}{1-x}\right) dx\right) = Cx^{-\gamma}(1-x)^{-\delta},$$

where C is a particular constant. Suppose that we write $W(x) = x^{-\gamma}F(x)$. Then $C = F(0)$. If we perform the differentiation in (30.1), we readily find that $C = 1 - \gamma$. Thus,

$$W(x) = -(\gamma - 1)x^{-\gamma}(1-x)^{-\delta}, \tag{30.3}$$

and, by (30.1), the proposed formula for y' follows. □

Corollary. *Let*

$$y = \frac{\pi}{\sin(\pi n)} \frac{{}_2F_1(n, 1-n; 1; 1-x)}{{}_2F_1(n, 1-n; 1; x)}.$$

Then

$$y' = -\frac{1}{x(1-x) \, {}_2F_1^2(n, 1-n; 1; x)}.$$

PROOF. Apply Entry 30 with $\alpha = n$, $\beta = 1 - n$, and $\gamma = \delta = 1$. □

Entry 31(i). *Let* $y = {}_2F_1(\alpha, \beta; \gamma; x)$. *Then*

$$(\alpha - 1)(\beta - 1) \int_0^x y \, dx - x(1-x)y' = (\gamma - 1)(y - 1) - (\alpha + \beta - 1)xy.$$

PROOF. Upon differentiation, it is found that the proposed formula is equivalent to the formula

$$(\alpha - 1)(\beta - 1)y - (1-x)y' + xy' - x(1-x)y''$$

$$= (\gamma - 1)y' - (\alpha + \beta - 1)y - (\alpha + \beta - 1)xy'.$$

Upon simplification, this formula reduces to (30.2), the hypergeometric differential equation satisfied by ${}_2F_1(\alpha, \beta; \gamma; x)$. □

Entry 31(ii). *Let* $\alpha + \beta + 1 = \gamma + \delta$. *Assume that* $n > 1$ *and that* $n > \mathrm{Re}\,\gamma$. *If* $y = y(x) = {}_2F_1(\alpha, \beta; \gamma; x)$, *then*

$$y(x) \int_0^x \left\{\int_0^u t^{n-2} y(t) \, dt\right\} \frac{du}{u^{\gamma}(1-u)^{\delta}y^2(u)}$$

$$= \frac{x^{n-\gamma}(1-x)^{1-\delta}}{(n-\gamma)(n-1)} \, {}_3F_2\left[\begin{matrix} n-\alpha, n-\beta, 1 \\ n, n-\gamma+1 \end{matrix}; x\right]. \tag{31.1}$$

The conditions that we have imposed on n are needed only for the convergence of the integrals on the left side of (31.1).

Entry 31(ii) is somewhat imprecisely stated by Ramanujan.

Our method of proof will be as follows. We first show that the left side of (31.1) is a solution of the inhomogeneous hypergeometric differential equation

$$x(1 - x)z'' + \{\gamma - (\alpha + \beta + 1)x\}z' - \alpha\beta z = x^{n-\gamma-1}(1 - x)^{1-\delta}.$$

Then, with considerably more difficulty, we show that the right side of (31.1) is a solution of the same differential equation. The difference of these two solutions is, of course, a solution of the associated homogeneous hypergeometric differential equation (30.2). Now $y_1 := y = {}_2F_1(\alpha, \beta; \gamma; x)$ and $y_2 := x^{1-\gamma} {}_2F_1(\delta - \alpha, \delta - \beta; 2 - \gamma; x)$ are a pair of linearly independent solutions of (30.2). By examining the power series expansions of both sides of (31.1), we easily see that the difference of these two functions cannot possibly involve y_1 or y_2; that is, their difference is identically equal to zero. This then completes the proof.

PROOF. Letting $w = w(x)$ denote the left side of (31.1), we find trivially that

$$\frac{d}{dx}\left(\frac{w}{y}\right) = \frac{1}{x^\gamma(1 - x)^\delta y^2(x)} \int_0^x t^{n-2} y(t)\, dt$$

and

$$\frac{d}{dx}\left(x^\gamma(1 - x)^\delta y^2(x)\frac{d}{dx}\left(\frac{w}{y}\right)\right) = x^{n-2} y(x). \tag{31.2}$$

On the other hand, since $\gamma + \delta = \alpha + \beta + 1$,

$$\frac{d}{dx}\left(x^\gamma(1 - x)^\delta y^2(x)\frac{d}{dx}\left(\frac{w}{y}\right)\right)$$

$$= \frac{d}{dx}\left(x^\gamma(1 - x)^\delta y\frac{dw}{dx} - x^\gamma(1 - x)^\delta w\frac{dy}{dx}\right)$$

$$= x^{\gamma-1}(1 - x)^{\delta-1} y(x(1 - x)w'' + \{\gamma - (\alpha + \beta + 1)x\}w')$$

$$\quad - x^{\gamma-1}(1 - x)^{\delta-1} w(x(1 - x)y'' + \{\gamma - (\alpha + \beta + 1)x\}y')$$

$$= x^{\gamma-1}(1 - x)^{\delta-1} y(x(1 - x)w'' + \{\gamma - (\alpha + \beta + 1)x\}w' - \alpha\beta w), \tag{31.3}$$

where we have used the fact that y is a solution of the hypergeometric equation (30.2). Combining (31.2) and (31.3), we deduce that

$$x(1 - x)w'' + \{\gamma - (\alpha + \beta + 1)x\}w' - \alpha\beta w = x^{n-\gamma-1}(1 - x)^{1-\delta}. \tag{31.4}$$

It remains to show that the right side of (31.1) satisfies the differential equation (31.4).

Let

$$Y(x) = (1 - x)^{1-\delta} \sum_{k=0}^{\infty} \frac{(n - \alpha)_k (n - \beta)_k x^{n+k-\gamma}}{(n)_k (n - \gamma + 1)_k}.$$

Then, by (31.4), we must show that

$$x(1 - x)Y'' + \{\gamma - (\alpha + \beta + 1)x\} Y' - \alpha\beta Y$$

$$= \delta(\delta - 1)(1 - x)^{-\delta} \sum_{k=0}^{\infty} \frac{(n - \alpha)_k (n - \beta)_k x^{n+k-\gamma+1}}{(n)_k (n - \gamma + 1)_k}$$

$$+ 2(\delta - 1)(1 - x)^{1-\delta} \sum_{k=0}^{\infty} \frac{(n - \alpha)_k (n - \beta)_k (n + k - \gamma) x^{n+k-\gamma}}{(n)_k (n - \gamma + 1)_k}$$

$$+ (1 - x)^{2-\delta} \sum_{k=0}^{\infty} \frac{(n - \alpha)_k (n - \beta)_k (n + k - \gamma)(n + k - \gamma - 1) x^{n+k-\gamma-1}}{(n)_k (n - \gamma + 1)_k}$$

$$+ \gamma(\delta - 1)(1 - x)^{-\delta} \sum_{k=0}^{\infty} \frac{(n - \alpha)_k (n - \beta)_k x^{n+k-\gamma}}{(n)_k (n - \gamma + 1)_k}$$

$$+ \gamma(1 - x)^{1-\delta} \sum_{k=0}^{\infty} \frac{(n - \alpha)_k (n - \beta)_k (n + k - \gamma) x^{n+k-\gamma-1}}{(n)_k (n - \gamma + 1)_k}$$

$$+ (\alpha + \beta + 1)(1 - \delta)(1 - x)^{-\delta} \sum_{k=0}^{\infty} \frac{(n - \alpha)_k (n - \beta)_k x^{n+k-\gamma+1}}{(n)_k (n - \gamma + 1)_k}$$

$$- (\alpha + \beta + 1)(1 - x)^{1-\delta} \sum_{k=0}^{\infty} \frac{(n - \alpha)_k (n - \beta)_k (n + k - \gamma) x^{n+k-\gamma}}{(n)_k (n - \gamma + 1)_k}$$

$$- \alpha\beta(1 - x)^{1-\delta} \sum_{k=0}^{\infty} \frac{(n - \alpha)_k (n - \beta)_k x^{n+k-\gamma}}{(n)_k (n - \gamma + 1)_k}$$

$$= (n - \gamma)(n - 1)x^{n-\gamma-1}(1 - x)^{1-\delta}. \tag{31.5}$$

We cancel the factor of $(1 - x)^{-\delta}$ in the last equality and show that the coefficients of like powers of x on both sides in (31.5) are equal.

We first examine the coefficients of $x^{n-\gamma-1}$. On the left side of (31.5), this coefficient is equal to

$$(n - \gamma)(n - \gamma - 1) + \gamma(n - \gamma) = (n - \gamma)(n - 1),$$

which is in agreement with the right side of (31.5).

Next, the coefficient of $x^{n-\gamma}$ on the left side of (31.5) is equal to

$$2(\delta - 1)(n - \gamma) - 2(n - \gamma)(n - \gamma - 1) + \frac{1}{n}(n - \alpha)(n - \beta)(n - \gamma)$$

$$+ \gamma(\delta - 1) - \gamma(n - \gamma) + \frac{1}{n}\gamma(n - \alpha)(n - \beta) - (\alpha + \beta + 1)(n - \gamma) - \alpha\beta.$$

$$\tag{31.6}$$

Now it is easy to see that (31.6) may be written in the form

$$-(n - n_1)(n - n_2),$$

where n_1 and n_2 are the two roots of the quadratic polynomial (31.6). By a direct verification, it can readily be shown that 1 and γ are the roots of (31.6), although the case $n = 1$ is moderately tedious. In both computations, the hypothesis $\alpha + \beta + 1 = \gamma + \delta$ is used. Thus, the coefficients of $x^{n-\gamma}$ on both sides of (31.5) agree.

Lastly, we must show that the coefficient of $x^{n+k-\gamma}$, $k \geq 1$, on the left side of (31.5) is equal to 0. This coefficient is equal to

$$\delta(\delta - 1)\frac{(n - \alpha)_{k-1}(n - \beta)_{k-1}}{(n)_{k-1}(n - \gamma + 1)_{k-1}} + 2(\delta - 1)\frac{(n - \alpha)_k(n - \beta)_k(n + k - \gamma)}{(n)_k(n - \gamma + 1)_k}$$

$$+ 2(1 - \delta)\frac{(n - \alpha)_{k-1}(n - \beta)_{k-1}(n + k - 1 - \gamma)}{(n)_{k-1}(n - \gamma + 1)_{k-1}}$$

$$+ \frac{(n - \alpha)_{k+1}(n - \beta)_{k+1}(n + k + 1 - \gamma)(n + k - \gamma)}{(n)_{k+1}(n - \gamma + 1)_{k+1}}$$

$$- 2\frac{(n - \alpha)_k(n - \beta)_k(n + k - \gamma)(n + k - 1 - \gamma)}{(n)_k(n - \gamma + 1)_k}$$

$$+ \frac{(n - \alpha)_{k-1}(n - \beta)_{k-1}(n + k - 1 - \gamma)(n + k - 2 - \gamma)}{(n)_{k-1}(n - \gamma + 1)_{k-1}}$$

$$+ \gamma(\delta - 1)\frac{(n - \alpha)_k(n - \beta)_k}{(n)_k(n - \gamma + 1)_k} + \gamma\frac{(n - \alpha)_{k+1}(n - \beta)_{k+1}(n + k + 1 - \gamma)}{(n)_{k+1}(n - \gamma + 1)_{k+1}}$$

$$- \gamma\frac{(n - \alpha)_k(n - \beta)_k(n + k - \gamma)}{(n)_k(n - \gamma + 1)_k} + (\alpha + \beta + 1)(1 - \delta)\frac{(n - \alpha)_{k-1}(n - \beta)_{k-1}}{(n)_{k-1}(n - \gamma + 1)_{k-1}}$$

$$- (\alpha + \beta + 1)\frac{(n - \alpha)_k(n - \beta)_k(n + k - \gamma)}{(n)_k(n - \gamma + 1)_k}$$

$$+ (\alpha + \beta + 1)\frac{(n - \alpha)_{k-1}(n - \beta)_{k-1}(n + k - 1 - \gamma)}{(n)_{k-1}(n - \gamma + 1)_{k-1}}$$

$$- \alpha\beta\frac{(n - \alpha)_k(n - \beta)_k}{(n)_k(n - \gamma + 1)_k} + \alpha\beta\frac{(n - \alpha)_{k-1}(n - \beta)_{k-1}}{(n)_{k-1}(n - \gamma + 1)_{k-1}}.$$

Next, remove the factor

$$\frac{(n - \alpha)_{k-1}(n - \beta)_{k-1}}{(n)_{k-1}(n - \gamma + 1)_{k-1}}$$

from each of the 14 expressions above. Then let $u = u(k) = n + k - 1$. Therefore, it suffices to show that

$$\delta(\delta - 1) + \frac{2(\delta - 1)(u - \alpha)(u - \beta)}{u} + 2(1 - \delta)(u - \gamma)$$

$$+ \frac{(u + 1 - \alpha)(u - \alpha)(u + 1 - \beta)(u - \beta)}{(u + 1)u} - \frac{2(u - \alpha)(u - \beta)(u - \gamma)}{u}$$

$$+ (u - \gamma)(u - 1 - \gamma) + \frac{\gamma(\delta - 1)(u - \alpha)(u - \beta)}{u(u + 1 - \gamma)}$$

$$+ \frac{\gamma(u + 1 - \alpha)(u - \alpha)(u + 1 - \beta)(u - \beta)}{(u + 1)u(u + 1 - \gamma)} - \frac{\gamma(u - \alpha)(u - \beta)}{u}$$

$$+ (\alpha + \beta + 1)(1 - \delta) - \frac{(\alpha + \beta + 1)(u - \alpha)(u - \beta)}{u} + (\alpha + \beta + 1)(u - \gamma)$$

$$- \frac{\alpha\beta(u - \alpha)(u - \beta)}{u(u + 1 - \gamma)} + \alpha\beta = 0. \tag{31.7}$$

If we multiply both sides of (31.7) by $u(u + 1)(u + 1 - \gamma)$, the left side becomes a polynomial of degree 5. In order to show that this quintic polynomial is identically equal to 0, we shall show that the coefficient of u^5 is equal to 0 and that the polynomial vanishes at five distinct points. It is easy to check that the coefficient of u^5 is equal to 0. One can verify that this polynomial vanishes at $u = 0$, -1, $\gamma - 1$, α, and β. We sympathetically suppress the details. This completes the proof. □

Corollary. *If n is arbitrary and $y = {}_2F_1(n, 1 - n; 1; x)$, then*

$$x(x - 1)y' = n(n - 1) \int_0^x y \, dx.$$

PROOF. In Entry 31(i), let $\alpha = n$, $\beta = 1 - n$, and $\gamma = 1$. □

Entry 32(i). *If φ is any function, then*

$$\frac{1}{\sqrt{\varphi(x)}} \sum_{k=0}^{\infty} \frac{(\frac{1}{2})_k^2}{(k!)^2} \left\{ 1 - \frac{\varphi(-x)}{\varphi(x)} \right\}^k$$

is always an even function of x, provided the series converges.

PROOF. Set $r = m = \frac{1}{2}$ in Entry 1. □

Entry 32(ii). *If $\frac{1}{2} < x < 2$, then*

$${}_2F_1(\tfrac{1}{2}, \tfrac{1}{2}; 1; 1 - 1/x) = \sqrt{x} \; {}_2F_1(\tfrac{1}{2}, \tfrac{1}{2}; 1; 1 - x).$$

This result is a special case of a transformation

$${}_2F_1(a, b; c; z) = (1 - z)^{-a} \; {}_2F_1(a, c - b; c; z/(z - 1)) \tag{32.1}$$

that is generally attributed to Gauss [1] or Kummer [1], [2] but is due to

Pfaff [1]. Equality (32.1) is also found in Chapter 10 (Entry 19). For a proof see Chapter 10 or Bailey's tract [4, p. 10, formula (1)].

Entry 32(iii)

$$_2F_1\left(\tfrac{1}{2}, \tfrac{1}{2}; 1; 1 - \left(\frac{1-x}{1+x}\right)^2\right) = (1+x) \, _2F_1(\tfrac{1}{2}, \tfrac{1}{2}; 1; x^2).$$

PROOF. Replace x by $((1-x)/(1+x))^2$ in Entry 32(ii) and then apply Entry 5 with $r = \tfrac{1}{2}$ and x replaced by $-x$. This yields

$$_2F_1\left(\tfrac{1}{2}, \tfrac{1}{2}; 1; 1 - \left(\frac{1-x}{1+x}\right)^2\right) = \frac{1+x}{1-x} \, _2F_1\left(\tfrac{1}{2}, \tfrac{1}{2}; 1; \frac{-4x}{(1-x)^2}\right)$$

$$= (1+x) \, _2F_1(\tfrac{1}{2}, \tfrac{1}{2}; 1; x^2). \qquad \square$$

Entry 32(iv)

$$_2F_1\left(\tfrac{1}{2}, \tfrac{1}{2}; 1; 1 - \left(\frac{1-x}{1+x}\right)^4\right) = (1+x)^2 \, _2F_1(\tfrac{1}{2}, \tfrac{1}{2}; 1; x^4).$$

PROOF. From the work of Kummer [1, p. 148, Eq. (46)], [2, p. 142],

$$_2F_1(\tfrac{1}{2}, \tfrac{1}{2}; 1; c^2) = \left(\frac{2}{1+\sqrt{b}}\right)^2 \, _2F_1\left(\tfrac{1}{2}, \tfrac{1}{2}; 1; \left(\frac{1-\sqrt{b}}{1+\sqrt{b}}\right)^4\right),$$

where $c^2 = 1 - b^2$. If we put $b = ((1-x)/(1+x))^2$, Ramanujan's proposed formula easily follows. $\qquad \square$

The reader should compare Entries 32(iii) and (iv).

Entry 32(v)

$$(1+n^2)^{1/4} \, _2F_1(\tfrac{1}{2}, \tfrac{1}{2}; 1; \tfrac{1}{2}(1+in))$$

$$= \tfrac{1}{2}(1+i) \, _2F_1\left(\tfrac{1}{2}, \tfrac{1}{2}; 1; \tfrac{1}{2}\left(1 + \frac{n}{\sqrt{1+n^2}}\right)\right)$$

$$+ \tfrac{1}{2}(1-i) \, _2F_1\left(\tfrac{1}{2}, \tfrac{1}{2}; 1; \tfrac{1}{2}\left(1 - \frac{n}{\sqrt{1+n^2}}\right)\right).$$

PROOF. In Erdélyi's treatise [1, p. 111, formula (8)], let $a = b = \tfrac{1}{4}$ and $z = n^2$ to get

$$\frac{2\sqrt{\pi}}{\Gamma^2(\tfrac{3}{4})}(1+n^2)^{1/4} \, _2F_1(\tfrac{1}{4}, \tfrac{1}{4}; \tfrac{1}{2}; -n^2)$$

$$= \, _2F_1\left(\tfrac{1}{2}, \tfrac{1}{2}; 1; \tfrac{1}{2}\left(1 + \frac{n}{\sqrt{1+n^2}}\right)\right) + \, _2F_1\left(\tfrac{1}{2}, \tfrac{1}{2}; 1; \tfrac{1}{2}\left(1 - \frac{n}{\sqrt{1+n^2}}\right)\right).$$

$$(32.2)$$

Next, in the same compendium [1, p. 111, formula (9)], let $a = b = \frac{3}{4}$ and $z = n^2/(1 + n^2)$ and obtain

$$-\frac{4\sqrt{\pi}}{\Gamma^2(\frac{1}{4})} \frac{n}{\sqrt{1 + n^2}} \, _2F_1\left(\tfrac{3}{4}, \tfrac{3}{4}, \tfrac{3}{2}; \frac{n^2}{1 + n^2}\right)$$

$$= \, _2F_1\left(\tfrac{1}{2}, \tfrac{1}{2}; 1; \tfrac{1}{2}\left(1 - \frac{n}{\sqrt{1 + n^2}}\right)\right) - \, _2F_1\left(\tfrac{1}{2}, \tfrac{1}{2}; 1; \tfrac{1}{2}\left(1 + \frac{n}{\sqrt{1 + n^2}}\right)\right).$$

$$(32.3)$$

From (32.2), (32.3), and (32.1),

$$\tfrac{1}{2}(1 + i) \, _2F_1\left(\tfrac{1}{2}, \tfrac{1}{2}; 1; \tfrac{1}{2}\left(1 + \frac{n}{\sqrt{1 + n^2}}\right)\right)$$

$$+ \tfrac{1}{2}(1 - i) \, _2F_1\left(\tfrac{1}{2}, \tfrac{1}{2}; 1; \tfrac{1}{2}\left(1 - \frac{n}{\sqrt{1 + n^2}}\right)\right)$$

$$= \frac{\sqrt{\pi}}{\Gamma^2(\frac{3}{4})}(1 + n^2)^{1/4} \, _2F_1(\tfrac{1}{4}, \tfrac{1}{4}; \tfrac{1}{2}; -n^2)$$

$$+ \frac{2\sqrt{\pi}i}{\Gamma^2(\frac{1}{4})} \frac{n}{\sqrt{1 + n^2}} \, _2F_1\left(\tfrac{3}{4}, \tfrac{3}{4}, \tfrac{3}{2}; \frac{n^2}{1 + n^2}\right)$$

$$= \frac{\sqrt{\pi}}{\Gamma^2(\frac{3}{4})}(1 + n^2)^{1/4} \, _2F_1(\tfrac{1}{4}, \tfrac{1}{4}; \tfrac{1}{2}; -n^2)$$

$$+ \frac{2\sqrt{\pi}i}{\Gamma^2(\frac{1}{4})} n(1 + n^2)^{1/4} \, _2F_1(\tfrac{3}{4}, \tfrac{3}{4}; \tfrac{3}{2}; -n^2)$$

$$= (1 + n^2)^{1/4} \, _2F_1(\tfrac{1}{2}, \tfrac{1}{2}; 1; \tfrac{1}{2}(1 + in)),$$

where in the last equality we employed Entry 21 with $m = n = \frac{1}{2}$ and x replaced by in.   □

Entry 33(i)

$$_2F_1\left(\tfrac{1}{2}, \tfrac{1}{2}; 1; \frac{2x}{1 + x}\right) = \sqrt{1 + x} \, _2F_1(\tfrac{1}{4}, \tfrac{3}{4}; 1; x^2).$$

PROOF. Set $r = m = \frac{1}{2}$ in Entry 2.   □

Entry 33(ii)

$$_2F_1(\tfrac{1}{2}, \tfrac{1}{2}; 1; \tfrac{1}{2}(1 - \sqrt{1 - x})) = \, _2F_1(\tfrac{1}{4}, \tfrac{1}{4}; 1; x).$$

PROOF. In Entry 12, set $x = y = -\frac{1}{4}$ and $z = 1$ and replace p by x.   □

Entry 33(iii)

$$_3F_2(\tfrac{1}{2}, \tfrac{1}{2}, \tfrac{1}{2}; 1, 1; x) = {}_2F_1^2(\tfrac{1}{4}, \tfrac{1}{4}; 1; x).$$

PROOF. Put $\alpha = \beta = -\tfrac{1}{4}$ and $\gamma = \tfrac{1}{2}$ in Entry 13. □

Entry 33(iv)

$$_2F_1\left(\tfrac{1}{4}, \tfrac{3}{4}; 1; \frac{4x}{(1 + x)^2}\right) = \sqrt{1 + x}\, {}_2F_1(\tfrac{1}{2}, \tfrac{1}{2}; 1; x).$$

PROOF. Set $r = m = \tfrac{1}{2}$ in Entry 4. □

Entry 33(v)

$$_2F_1(\tfrac{1}{4}, \tfrac{1}{4}; 1; x) = \sqrt{1 - x}\, {}_2F_1(\tfrac{3}{4}, \tfrac{3}{4}; 1; x).$$

PROOF. Set $a = b = \tfrac{1}{4}$ and $c = 1$ in Erdélyi's book [1, p. 105, formula (1)]. □

Example (i)

$$_2F_1\left(\tfrac{1}{4}, \tfrac{3}{4}; 1; \frac{-4x}{(1 - x)^2}\right) = \sqrt{\frac{1 - x}{1 + x}}\, {}_2F_1\left(\tfrac{1}{2}, \tfrac{1}{2}; 1; \frac{x}{1 + x}\right).$$

PROOF. Replacing x by $-x$ in Entry 33(iv) and then using (32.1), we readily find the proposed formula. □

Example (ii)

$$_2F_1\left(\tfrac{1}{4}, \tfrac{1}{4}; 1; \frac{-4x}{(1 - x)^2}\right) = \sqrt{1 - x}\, {}_2F_1(\tfrac{1}{2}, \tfrac{1}{2}; 1; x).$$

PROOF. Apply Entry 33(iv), Example (i), and lastly Entry 33(ii) with $\tfrac{1}{2}(1 - \sqrt{1 - x})$ replaced by $x/(x - 1)$ to find that

$$\sqrt{1 - x}\, {}_2F_1(\tfrac{1}{2}, \tfrac{1}{2}; 1; x) = \sqrt{\frac{1 - x}{1 + x}}\, {}_2F_1\left(\tfrac{1}{4}, \tfrac{3}{4}; 1; \frac{4x}{(1 + x)^2}\right)$$

$$= {}_2F_1\left(\tfrac{1}{2}, \tfrac{1}{2}; 1; \frac{x}{x - 1}\right)$$

$$= {}_2F_1\left(\tfrac{1}{4}, \tfrac{1}{4}; 1; \frac{-4x}{(1 - x)^2}\right). \quad \square$$

Example (iii)

$$_2F_1\left(\tfrac{3}{4}, \tfrac{3}{4}; 1; \frac{-4x}{(1 - x)^2}\right) = \frac{(1 - x)^{3/2}}{1 + x}\, {}_2F_1(\tfrac{1}{2}, \tfrac{1}{2}; 1; x).$$

PROOF. Apply Entry 33(v) with x replaced by $-4x/(1-x)^2$ and then use Example (ii). □

Entry 34

(a) $\dfrac{\pi^{1/4}}{\Gamma(\frac{3}{4})} = 1.08643481121330801457531612.$

(b) $\dfrac{\Gamma^2(\frac{1}{4})}{4\sqrt{2\pi}} = 1.311028777146060.$

(c) $\dfrac{\sqrt{\pi}}{\Gamma^2(\frac{3}{4})} = 1.180340599016092.$

(d) $\dfrac{\Gamma^2(\frac{3}{4})}{\pi^{3/2}} = 0.269676300594191.$

(e) $\dfrac{\pi^{3/2}}{\Gamma^2(\frac{3}{4})} = 3.708149354602731.$

Both parts (a) and (b) are correct. The last recorded digits in (c) and (d) should be 6 and 0, respectively, and the last two digits in part (e) should read 44. Numerical values for the relevant powers of π may be found in the tables of Fletcher et al. [1, Chapter 5]. A numerical value for $\Gamma(\frac{3}{4})$ was taken from Fransén and Wrigge's tables [1].

For brevity, set

$$\mu = \frac{\sqrt{\pi}}{\Gamma^2(\frac{3}{4})} \quad \text{and} \quad \eta = \frac{\Gamma^2(\frac{3}{4})}{\pi^{3/2}}. \tag{34.1}$$

Entry 34(i). *If* $|x| < 1$, *then*

$$_2F_1(\tfrac{1}{2}, \tfrac{1}{2}; 1; \tfrac{1}{2}(1+x)) = \mu \,_2F_1(\tfrac{1}{4}, \tfrac{1}{4}; \tfrac{1}{2}; x^2) + \eta x \,_2F_1(\tfrac{3}{4}, \tfrac{3}{4}; \tfrac{3}{2}; x^2).$$

PROOF. Evaluating $_2F_1(j + \tfrac{1}{2}, j + \tfrac{1}{2}; j + 1; \tfrac{1}{2})$ below by a formula of Kummer that can be found in Bailey's tract [4, p. 11, formula (2)], we find that

$$_2F_1(\tfrac{1}{2}, \tfrac{1}{2}; 1; \tfrac{1}{2}(1+x)) = \sum_{k=0}^{\infty} \frac{(\tfrac{1}{2})_k^2}{(k!)^2 2^k} \sum_{j=0}^{k} \binom{k}{j} x^j$$

$$= \sum_{j=0}^{\infty} \frac{x^j}{j!} \sum_{k=j}^{\infty} \frac{(\tfrac{1}{2})_k^2}{k!(k-j)! 2^k}$$

$$= \sum_{j=0}^{\infty} \frac{x^j}{j!} \sum_{\mu=0}^{\infty} \frac{(\tfrac{1}{2})_{\mu+j}^2}{(\mu+j)! \mu! 2^{\mu+j}}$$

$$= \sum_{j=0}^{\infty} \frac{(\tfrac{1}{2})_j^2 x^j}{(j!)^2 2^j} \,_2F_1(j + \tfrac{1}{2}, j + \tfrac{1}{2}; j + 1; \tfrac{1}{2})$$

$$= \sqrt{\pi} \sum_{j=0}^{\infty} \frac{(\frac{1}{2})_j^2 x^j}{j! \Gamma^2(\frac{1}{2}j + \frac{3}{4}) 2^j}$$

$$= \mu \sum_{k=0}^{\infty} \frac{(\frac{1}{2})_{2k}^2 x^{2k}}{(2k)! (\frac{3}{4})_k^2 2^{2k}} + \eta \sum_{k=0}^{\infty} \frac{(\frac{3}{2})_{2k}^2 x^{2k+1}}{(2k+1)! (\frac{5}{4})_k^2 2^{2k}}$$

$$= \mu \, {}_2F_1(\tfrac{1}{4}, \tfrac{1}{4}; \tfrac{1}{2}; x^2) + \eta x \, {}_2F_1(\tfrac{3}{4}, \tfrac{3}{4}; \tfrac{3}{2}; x^2), \qquad (34.2)$$

after some simplification. $\square$

Entry 34(ii)

$${}_2F_1\left(\tfrac{1}{2}, \tfrac{1}{2}; 1; \tfrac{1}{2} + \frac{x}{1 + x^2}\right) = \mu\sqrt{1 + x^2} \, {}_2F_1(\tfrac{1}{4}, \tfrac{1}{4}; \tfrac{3}{4}; x^4)$$
$$+ \eta x (1 + x^2)^{3/2} \, {}_2F_1(\tfrac{3}{4}, \tfrac{3}{4}; \tfrac{5}{4}; x^4).$$

PROOF. First apply Entry 21 with $m = n = \frac{1}{2}$ and x replaced by $2x/(1 + x^2)$. Then make two applications of Entry 3 with x replaced by x^2 and $r = m = \frac{1}{4}$ and $r = m = \frac{3}{4}$, respectively. Thus,

$${}_2F_1\left(\tfrac{1}{2}, \tfrac{1}{2}; 1; \tfrac{1}{2} + \frac{x}{1 + x^2}\right)$$

$$= \mu \, {}_2F_1\left(\tfrac{1}{4}, \tfrac{1}{4}; \tfrac{1}{2}; \frac{4x^2}{(1 + x^2)^2}\right) + \eta x \, {}_2F_1\left(\tfrac{3}{4}, \tfrac{3}{4}; \tfrac{3}{2}; \frac{4x^2}{(1 + x^2)^2}\right)$$

$$= \mu\sqrt{1 + x^2} \, {}_2F_1(\tfrac{1}{4}, \tfrac{1}{4}; \tfrac{3}{4}; x^4) + \eta x (1 + x^2)^{3/2} \, {}_2F_1(\tfrac{3}{4}, \tfrac{3}{4}; \tfrac{5}{4}; x^4). \quad (34.3) \quad \square$$

Ramanujan (p. 141) has mistakenly written $(1 + x^2)^{1/2}$ instead of $(1 + x^2)^{3/2}$ on the right side of Entry 34(ii).

Entry 34(iii)

$$\frac{\pi}{4} \, {}_2F_1^2(\tfrac{1}{2}, \tfrac{1}{2}; 1; \tfrac{1}{2}(1 + x)) - \frac{\pi}{4} \, {}_2F_1^2(\tfrac{1}{2}, \tfrac{1}{2}; 1; \tfrac{1}{2}(1 - x))$$

$$= x \sum_{n=0}^{\infty} \frac{n! x^{2n}}{(\frac{3}{2})_n} \, {}_3F_2\left[\begin{matrix} \tfrac{1}{2}, \tfrac{1}{2}, -n \\ 1, 1 \end{matrix}\right]$$

$$= x + \frac{x^3}{2} + \frac{41x^5}{120} + \frac{21x^7}{80} + \cdots$$

$$= \frac{x}{1 - x^2} - \frac{x^3}{2(1 - x^2)^2} + \frac{41x^5}{120(1 - x^2)^3} + \cdots. \qquad (34.4)$$

PROOF. We first establish the latter two equalities. The four displayed coefficients on the right side of the second equality are simply numerical calculations of the first four coefficients of the left side. Apparently, Ramanujan does not possess a simple formula for these rational coefficients. Expanding

$(1 - x^2)^{-k}$, $k = 1, 2, 3$, in binomial series on the far right side of (34.4) and collecting coefficients of x, x^3, and x^5, we establish the last equality. Evidently, Ramanujan is not claiming to have found a general formula for the coefficient of $x^{2k-1}/(1 - x^2)^k$, $k \geq 1$.

We now prove the first equality of (34.4). By Entry 21 and (34.1),

$$\frac{\pi}{4} \, {}_2F_1^2(\tfrac{1}{2}, \tfrac{1}{2}; 1; \tfrac{1}{2}(1 + x)) - \frac{\pi}{4} \, {}_2F_1^2(\tfrac{1}{2}, \tfrac{1}{2}; 1; \tfrac{1}{2}(1 - x))$$

$$= \frac{\pi}{4}\{\mu \, {}_2F_1(\tfrac{1}{4}, \tfrac{1}{4}; \tfrac{1}{2}; x^2) + \eta x \, {}_2F_1(\tfrac{3}{4}, \tfrac{3}{4}; \tfrac{3}{2}; x^2)\}^2$$

$$- \frac{\pi}{4}\{\mu \, {}_2F_1(\tfrac{1}{4}, \tfrac{1}{4}; \tfrac{1}{2}; x^2) - \eta x \, {}_2F_1(\tfrac{3}{4}, \tfrac{3}{4}; \tfrac{3}{2}; x^2)\}^2$$

$$= x \, {}_2F_1(\tfrac{1}{4}, \tfrac{1}{4}; \tfrac{1}{2}; x^2) \, {}_2F_1(\tfrac{3}{4}, \tfrac{3}{4}; \tfrac{3}{2}; x^2)$$

$$= x \sum_{n=0}^{\infty} \frac{(\tfrac{3}{4})_n^2 x^{2n}}{(\tfrac{3}{2})_n n!} \, {}_4F_3\left[\begin{matrix} \tfrac{1}{4}, \tfrac{1}{4}, -\tfrac{1}{2} - n, n \\ \tfrac{1}{2}, \tfrac{1}{4} - n, \tfrac{1}{4} - n \end{matrix}\right], \tag{34.5}$$

where we have employed Erdélyi's work [1, p. 187, formula (14)]. In comparing (34.4) and (34.5), we find that we must show that

$$\frac{(\tfrac{3}{4})_n^2}{(n!)^2} \, {}_4F_3\left[\begin{matrix} \tfrac{1}{4}, \tfrac{1}{4}, -\tfrac{1}{2} - n, -n \\ \tfrac{1}{2}, \tfrac{1}{4} - n, \tfrac{1}{4} - n \end{matrix}\right] = {}_3F_2\left[\begin{matrix} \tfrac{1}{2}, \tfrac{1}{2}, -n \\ 1, 1 \end{matrix}\right], \qquad n \geq 0.$$

Now from Erdélyi's book [1, p. 85, formula (2)] we find that

$${}_3F_2\left[\begin{matrix} \tfrac{1}{2}, \tfrac{1}{2}, -n \\ 1, \tfrac{1}{2} - n \end{matrix}\right] = \frac{(\tfrac{3}{4})_n^2}{n!(\tfrac{1}{2})_n} \, {}_4F_3\left[\begin{matrix} \tfrac{1}{4}, \tfrac{1}{4}, -\tfrac{1}{2} - n, -n \\ \tfrac{1}{2}, \tfrac{1}{4} - n, \tfrac{1}{4} - n \end{matrix}\right], \qquad n \geq 0.$$

Thus, it remains to show that

$${}_3F_2\left[\begin{matrix} \tfrac{1}{2}, \tfrac{1}{2}, -n \\ 1, 1 \end{matrix}\right] = \frac{(\tfrac{1}{2})_n}{n!} \, {}_3F_2\left[\begin{matrix} \tfrac{1}{2}, \tfrac{1}{2}, -n \\ 1, \tfrac{1}{2} - n \end{matrix}\right], \qquad n \geq 0.$$

However, this last formula is a special case of Entry 29(i). Thus, the proof is complete.   □

Example (i)

$${}_2F_1(\tfrac{1}{2}, \tfrac{1}{2}; 1; \tfrac{1}{2}(1 + x)) = \mu(1 - x^2)^{-1/4} \, {}_2F_1\left(\tfrac{1}{4}, \tfrac{1}{4}; \tfrac{1}{2}; \frac{x^2}{x^2 - 1}\right)$$

$$+ \eta x(1 - x^2)^{-3/4} \, {}_2F_1\left(\tfrac{3}{4}, \tfrac{3}{4}; \tfrac{3}{2}; \frac{x^2}{x^2 - 1}\right).$$

PROOF. Employing Entry 34(i) and then (32.1), we easily achieve the proposed formula.   □

In Ramanujan's formulation of Example (i) (p. 142), he has written $(1 - x^2)^{-5/4}$ instead of $(1 - x^2)^{-3/4}$ on the right side.

Example (ii)

$$\sqrt{1 - x^2} \; {}_2F_1\left(\tfrac{1}{2}, \tfrac{1}{2}; 1; \tfrac{1}{2} + \frac{x}{1 + x^2}\right)$$

$$= \mu \; {}_2F_1\left(\tfrac{1}{2}, \tfrac{1}{2}; \tfrac{3}{4}; \frac{x^4}{x^4 - 1}\right) + \eta x(1 + x^2) \; {}_2F_1\left(\tfrac{1}{2}, \tfrac{1}{2}; \tfrac{5}{4}; \frac{x^4}{x^4 - 1}\right).$$

PROOF. To each of the functions on the right side of Entry 34(ii), apply (32.1). The desired result easily follows. □

On the right side of Example (ii), Ramanujan (p. 142) has written 2 instead of $1 + x^2$.

Entry 35(i). *If n is arbitrary, then*

$$\cos(2n \sin^{-1} x) = {}_2F_1(n, -n; \tfrac{1}{2}; x^2).$$

PROOF. In Erdélyi's treatise [1, p. 101, formula (11)], let $a = 2n$ and $z = \sin^{-1} x$. □

Entry 35(ii). *If n is arbitrary, then*

$$\sin(2n \sin^{-1} x) = 2nx \; {}_2F_1(\tfrac{1}{2} + n, \tfrac{1}{2} - n; \tfrac{3}{2}; x^2).$$

PROOF. In Erdélyi's book [1, p. 101, formula (12)], put $a = 2n$ and $z = \sin^{-1} x$. □

Entries 35(i) and (ii) are closely related to the Tschebyscheff polynomials.

Entry 35(iii). *If n is arbitrary, then*

$$(1 - x^2)^{-1/2} \cos(2n \sin^{-1} x) = {}_2F_1(\tfrac{1}{2} + n, \tfrac{1}{2} - n; \tfrac{1}{2}; x^2).$$

PROOF. By Entry 35(i),

$$(1 - x^2)^{-1/2} \cos(2n \sin^{-1} x) = \sum_{j=0}^{\infty} \frac{(\tfrac{1}{2})_j}{j!} x^{2j} \sum_{k=0}^{\infty} \frac{(n)_k(-n)_k}{(\tfrac{1}{2})_k k!} x^{2k}.$$

Using (17.3), we find that the coefficient of $x^{2r}, r \geq 0$, on the right side is equal to

$$\sum_{k=0}^{r} \frac{(n)_k(-n)_k(\tfrac{1}{2})_{r-k}}{(\tfrac{1}{2})_k k!(r - k)!} = \frac{(\tfrac{1}{2})_r}{r!} \sum_{k=0}^{r} \frac{(n)_k(-n)_k(-r)_k}{(\tfrac{1}{2})_k(\tfrac{1}{2} - r)_k k!}$$

$$= \frac{(\tfrac{1}{2} + n)_r(\tfrac{1}{2} - n)_r}{r!(\tfrac{1}{2})_r},$$

where we have utilized Saalschütz's theorem, Entry 2 of Chapter 10. The proposed formula immediately follows. □

Entries 36(i), (ii). *For n real and $k \geq 2$, let*

$$b_k(n) = \begin{cases} n^2(n^2 - 2^2)(n^2 - 4^2)\cdots(n^2 - (k-2)^2), & \text{if } k \text{ is even,} \\ n(n^2 - 1^2)(n^2 - 3^2)\cdots(n^2 - (k-2)^2), & \text{if } k \text{ is odd.} \end{cases}$$

If $2 - 2\sqrt{2} \leq x \leq 2 + 2\sqrt{2}$, then

$$(x + 1)^{n/2} - (x + 1)^{-n/2} = \frac{nx}{\sqrt{1 + x}} + 2 \sum_{k=1}^{\infty} \frac{b_{2k+1}(n)}{(2k + 1)!} \left(\frac{x}{2\sqrt{1 + x}}\right)^{2k+1}$$

and $\hspace{8cm}$ (36.1)

$$(x + 1)^{n/2} + (x + 1)^{-n/2} = 2 + 2 \sum_{k=1}^{\infty} \frac{b_{2k}(n)}{(2k)!} \left(\frac{x}{2\sqrt{1 + x}}\right)^{2k}. \qquad (36.2)$$

We have stated Entries 36(i), (ii) in somewhat different forms than did Ramanujan.

PROOF. From Corollary 2, Section 14 of our description of Chapter 3 [9],

$$(a + \sqrt{1 + a^2})^n = 1 + na + \sum_{k=2}^{\infty} \frac{b_k(n)a^k}{k!},$$

where $|a| \leq 1$ and n is any real number. Let $a = x/\{2\sqrt{1 + x}\}$. Then an elementary calculation shows that $a + \sqrt{1 + a^2} = \sqrt{1 + x}$. Thus,

$$(x + 1)^{n/2} = 1 + \frac{nx}{2\sqrt{1 + x}} + \sum_{k=2}^{\infty} \frac{b_k(n)}{k!} \left(\frac{x}{2\sqrt{1 + x}}\right)^k. \qquad (36.3)$$

Replacing n by $-n$ and using the definition of $b_k(n)$, we find that

$$(x + 1)^{-n/2} = 1 - \frac{nx}{2\sqrt{1 + x}} + \sum_{k=2}^{\infty} \frac{(-1)^k b_k(n)}{k!} \left(\frac{x}{2\sqrt{1 + x}}\right)^k. \qquad (36.4)$$

Subtracting (36.4) from (36.3), we deduce (36.1); adding (36.3) and (36.4), we deduce (36.2). Finally, an elementary computation shows that $|a| \leq 1$ if and only if $2 - 2\sqrt{2} \leq x \leq 2 + 2\sqrt{2}$. $\qquad\square$

Entries 36(iii), (iv). *Let n be real and suppose that*

$$\left| \frac{x}{(1 + x)^{3/2}} \right| \leq \frac{2}{3^{3/2}}. \qquad (36.5)$$

Then

$$\left(\frac{1 + \sqrt{1 + 4x}}{2} \right)^n$$

$$= 1 + nx(1 + x)^{(n-3)/2} + \frac{n(n - 5)(n - 7)}{4 \cdot 3!} x^3 (1 + x)^{(n-9)/2}$$

$$+ \frac{n(n - 7)(n - 9)(n - 11)(n - 13)}{4^2 \cdot 5!} x^5 (1 + x)^{(n-15)/2} + \cdots$$

$$\hspace{11cm} (36.6)$$

and

$$\frac{1}{2} + \frac{1}{2}\left(\frac{1 + \sqrt{1 + 4x}}{2}\right)^n = (1 + x)^{n/2} + \frac{n(n - 4)}{4 \cdot 2!}x^2(1 + x)^{(n-6)/2}$$

$$+ \frac{n(n - 6)(n - 8)(n - 10)}{4^2 \cdot 4!}x^4(1 + x)^{(n-12)/2} + \cdots.$$

$$(36.7)$$

We have presented Ramanujan's formulations of Entries 36(iii), (iv). As we shall observe below, a general formula for the coefficient of $(x/(1 + x)^{3/2})^k$, $k \geq 0$, can be given in terms of gamma functions.

PROOF. We shall apply the Lagrange inversion formula (Whittaker and Watson [1, p. 133]). Accordingly, we let

$$\varphi(x) = \frac{1}{\sqrt{1 + x}} \quad \text{and} \quad f(x) = (1 + x)^{n/2}$$

and define y by $y = x + x\varphi(y)$. If we solve this equation for y, we find that

$$y = \tfrac{1}{2}(2x - 1 + \sqrt{1 + 4x}).$$

It follows that

$$f(\tfrac{1}{2}(2x - 1 + \sqrt{1 + 4x})) = \left(\frac{1 + \sqrt{1 + 4x}}{2}\right)^n.$$

Also note that

$$\varphi^k(x)f'(x) = \tfrac{1}{2}n(1 + x)^{(n-k)/2 - 1}, \qquad k \geq 1.$$

Hence, by the Lagrange inversion formula,

$$f(y) = f(x) + \sum_{k=1}^{\infty} \frac{x^k}{k!} \frac{d^{k-1}}{dx^{k-1}}\{\varphi^k(x)f'(x)\}, \qquad (36.8)$$

we find that

$$\left(\frac{1 + \sqrt{1 + 4x}}{2}\right)^n = (1 + x)^{n/2} + \tfrac{1}{2}nx(1 + x)^{(n-3)/2} + \frac{n(n - 4)}{2^2 \cdot 2!}x^2(1 + x)^{(n-6)/2}$$

$$+ \frac{n(n - 5)(n - 7)}{2^3 \cdot 3!}x^3(1 + x)^{(n-9)/2} + \cdots$$

$$= (1 + x)^{n/2} \sum_{k=0}^{\infty} \frac{(-1)^{k+1}\Gamma(\tfrac{1}{2}(-n + 3k))}{\Gamma(\tfrac{1}{2}(-n + 3k) - k + 1)k!}\left(\frac{x}{(1 + x)^{3/2}}\right)^k.$$

$$(36.9)$$

By Stirling's formula, the series above converges for those values of x given by (36.5). (The radius of convergence of a more general class of power series has been calculated in our book [9, Sec. 14 of Chap. 3].)

We now make a second application of the Lagrange inversion formula. Set

$$\varphi(x) = -\frac{1}{\sqrt{1 + x}} \quad \text{and} \quad f(x) = (1 + x)^{n/2}$$

and define y by $y = x - x\varphi(y)$. It follows that $y = 0$. Hence, $f(y) = f(0) = 1$.
Hence, by an application of the Lagrange inversion formula (36.8) like that above,

$$1 = (1 + x)^{n/2} - \tfrac{1}{2}nx(1 + x)^{(n-3)/2}$$

$$+ \frac{n(n - 4)}{2^2 \cdot 2!}x^2(1 + x)^{(n-6)/2} - \frac{n(n - 5)(n - 7)}{2^3 \cdot 3!}x^3(1 + x)^{(n-9)/2} + \cdots .$$

$$(36.10)$$

Again, by Stirling's formula, the series (36.10) converges for those values of x given by (36.5). Subtracting (36.10) from (36.9), we deduce (36.6); adding (36.9) and (36.10), we deduce (36.7).   □

Ramanujan had an affinity for the Lagrange inversion formula or, perhaps more precisely, for the beautiful expansions that can be derived from it. Ramanujan undoubtedly learned the Lagrange inversion formula from Carr's Synopsis [1]. The Lagrange inversion formula is also found in the calculus books of Edwards [1, pp. 450–457] and Williamson [1, pp. 151–153], both of which were known to Ramanujan. In Chapter 3 of his second notebook and in his quarterly reports, Ramanujan offers many applications of the Lagrange inversion formula. Although perhaps Ramanujan first discovered some of these expansions via the Lagrange inversion formula, his primary method for deriving these results arose from one of his favorite discoveries, a type of interpolation formula in the theory of integral transforms. This theorem has been thoroughly discussed by Hardy [9, Chapter 11] and by the author [9] in his account of Ramanujan's quarterly reports.

An excellent survey on the q-Lagrange inversion formula has been given by Stanton [1].

CHAPTER 12

Continued Fractions

In assessing the content of Ramanujan's first letter, dated January 16, 1913, to him, Hardy [9, p. 9] remarked: "but (1.10)–(1.12) defeated me completely; I had never seen anything in the least like them before. A single look at them is enough to show that they could only be written down by a mathematician of the highest class. They must be true because, if they were not true, no one would have had the imagination to invent them." These comments were directed at three continued fraction representations. Indeed, Ramanujan's contributions to the continued fraction expansions of analytic functions are one of his most spectacular achievements. The three formulas that challenged Hardy's acumen are not found in Chapter 12, but this chapter, which is almost entirely devoted to the study of continued fractions, contains many other beautiful and penetrating formulas. Unfortunately, Ramanujan left us no clues as to how he discovered these elegant continued fraction formulas. Especially enigmatic are the several representations for products and quotients of gamma functions. Three of the principal formulas involving gamma functions are Entries 34, 39, and 40. Entries 20 and 22, giving Gauss's and Euler's continued fractions, respectively, for a quotient of two hypergeometric functions, also play prominent roles. Several other formulas are dependent on these five entries, and it may be helpful to schematically indicate these connections among entries.

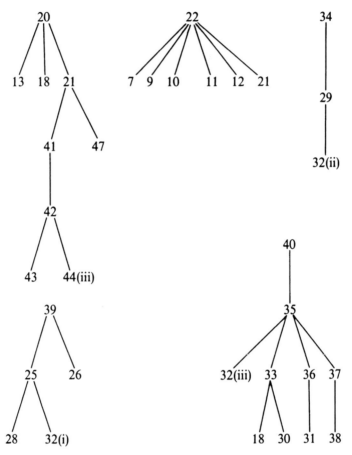

We shall use the notation for hypergeometric functions that we introduced at the beginning of Chapter 10.

In the sequel, $\psi(z)$ always denotes $\Gamma'(z)/\Gamma(z)$. We shall employ the representation (Olver [1, p. 39])

$$\psi(z) = -\gamma + \sum_{k=0}^{\infty} \left(\frac{1}{k+1} - \frac{1}{k+z} \right) \qquad (0.1)$$

several times in this chapter, usually without comment. Here γ denotes Euler's constant.

We shall usually adopt the notation

$$\frac{a_1}{b_1} + \frac{a_2}{b_2} + \frac{a_3}{b_3} + \cdots \qquad (0.2)$$

for the continued fraction

$$\cfrac{a_1}{b_1 + \cfrac{a_2}{b_2 + \cfrac{a_3}{b_3 + \cdots}}}.$$

The notation (0.2) appears to be the most convenient and widely used notation for continued fractions. For brevity, it will occasionally be convenient to employ the notation $K(a_n/b_n)$ instead of (0.2). We shall refer frequently to the well-known texts of Perron [3], Wall [1], Khovanskii [1], and Jones and Thron [1]. Because Perron's book contains several formulas that we shall employ and that are not found in the other texts, we shall make many references to this classic work.

In our initial published account of Ramanujan's work on continued fractions (see the Introduction for a complete reference), the domains of convergence were often more restrictive than necessary, and, in a few cases, they were incorrect. The account that follows has been considerably improved because of the comments and work of L. Jacobsen. In particular, she [3] has employed analytic continuation and the uniform parabola theorem to extend the domains of convergence of many of Ramanujan's continued fractions. The work of Jacobsen [1] and Waadeland [1] on tails of continued fractions has yielded a simpler, more uniform approach to several of Ramanujan's formulas.

Entry 1. *Let* $a_1, a_2, \ldots, a_r$ *and* $b_1, b_2, \ldots, b_r$ *be complex numbers such that* $a_n \neq 0$ *for each positive integer n. Define* $N_{-1} = 0$, $N_0 = 1$, $D_{-1} = 0$, $D_0 = 1$,

$$N_{k-1} = b_k N_{k-2} + a_k N_{k-3}, \qquad k \geq 2, \tag{1.1}$$

and

$$D_k = b_k D_{k-1} + a_k D_{k-2}, \qquad k \geq 1.$$

Then, for $r \geq 1$,

$$\frac{a_1}{b_1} + \frac{a_2}{b_2} + \cdots + \frac{a_r}{b_r} = a_1 \frac{N_{r-1}}{D_r} = \sum_{k=1}^{r} \frac{(-1)^{k+1} a_1 \cdots a_k}{D_{k-1} D_k}. \tag{1.2}$$

PROOF. The first equality in (1.2) is a somewhat unusual formulation of a basic elementary formula in the theory of continued fractions (Wall [1, p. 15]). For future reference, we restate the first equality of (1.2) in a more familiar fashion. Let $A_{-1} = 1$, $A_0 = 0$, $B_{-1} = 0$, $B_0 = 1$,

$$A_k = b_k A_{k-1} + a_k A_{k-2}, \qquad k \geq 1, \tag{1.3}$$

and

$$B_k = b_k B_{k-1} + a_k B_{k-2}, \qquad k \geq 1. \tag{1.4}$$

Then, for $r \geq 1$,

$$\frac{a_1}{b_1} + \frac{a_2}{b_2} + \cdots + \frac{a_r}{b_r} = \frac{A_r}{B_r}. \tag{1.5}$$

Thus, $a_1 N_k = A_{k+1}$ and $D_k = B_k$, $k \geq -1$. Note that if we define $N_{-2} = 1$, then (1.1) is valid for $k = 1$ as well. Recall that A_k and B_k are the kth numerator and denominator of the continued fraction (0.2).

The second equality in (1.2) is essentially another version of a well-known fact (Wall [1, p. 18]) due to Euler [1]. The relations (1.3) and (1.4) were first established by Wallis [1] and first studied seriously by Euler [5]. □

Corollary. *Let* $a_1, a_2, \ldots, a_r$ *be nonzero complex numbers such that* $a_j + a_{j+1} \neq 0, j \geq 1,$ *and* $r \geq 3.$ *Then*

$$\sum_{k=1}^{r} a_k = \frac{a_1}{1} - \frac{a_2}{a_1 + a_2} - \frac{a_1 a_3}{a_2 + a_3} - \frac{a_2 a_4}{a_3 + a_4} - \cdots - \frac{a_{r-2} a_r}{a_{r-1} + a_r}.$$

This corollary is due to Euler [1], and a proof may be found in Perron's book [3, p. 17].

Entry 2. *Let* $x, a_1, a_2, \ldots$ *denote nonzero complex numbers and define, for each nonnegative integer* $n,$

$$f_n(x) = \sum_{k=0}^{n} \frac{(-x)^k}{a_1 a_2 \cdots a_{k+1}}.$$

If

$$\lim_{n \to \infty} f_n(x) = \infty, \tag{2.1}$$

then

$$x = x - a_1 + \frac{a_1 x}{x - a_2} + \frac{a_2 x}{x - a_3} + \cdots. \tag{2.2}$$

PROOF. For each nonnegative integer n (Chrystal [1, p. 516, Eq. (14)]),

$$\sum_{k=1}^{n+1} \frac{a_1 a_2 \cdots a_k x^k}{b_1 b_2 \cdots b_k} = \frac{a_1 x}{b_1} - \frac{b_1 a_2 x}{b_2 + a_2 x} - \frac{b_2 a_3 x}{b_3 + a_3 x} - \cdots - \frac{b_n a_{n+1} x}{b_{n+1} + a_{n+1} x}.$$

If we set $a_j = 1$ and replace b_j by $-a_j, j \geq 1,$ we find that

$$f_n(x) = \frac{1}{a_1 - A},$$

where

$$A = \frac{a_1 x}{x - a_2} + \frac{a_2 x}{x - a_3} + \cdots + \frac{a_n x}{x - a_{n+1}}.$$

Letting n tend to ∞ and using (2.1), we deduce that $a_1 - A = 0,$ which is equivalent to (2.2). □

Of course, we could impose several sets of conditions on $x, a_1, a_2, \ldots$ in order to ensure that (2.1) holds. For example, if $\lim_{n \to \infty} |a_n| = \rho,$ then, by the ratio test, (2.1) is valid if $|x| > \rho.$

Quite possibly, Ramanujan attempted to prove (2.2) by the following nonrigorous argument. Trivially,

$$a_k = \frac{a_k x}{x - a_{k+1} + a_{k+1}}, \qquad k \geq 1. \tag{2.3}$$

If we successively employ (2.3) for $k = 1, 2, \ldots$, we find that

$$a_1 = \frac{a_1 x}{x - a_2 + a_2} = \frac{a_1 x}{x - a_2 + } \frac{a_2 x}{x - a_3 + a_3}$$

$$= \cdots = \frac{a_1 x}{x - a_2 + } \frac{a_2 x}{x - a_3 + } \frac{a_3 x}{x - a_4 + } \cdots,$$

which is equivalent to (2.2). This type of argument is valid under certain conditions which will be set forth in the next theorem.

Entry 2 and some entries in the sequel are consequences of the following result which is due to H. Waadeland [1] and L. Jacobsen [1].

Theorem. *Let* $K(a_n/b_n)$ *have a sequence* $g^{(n)}$, $0 \leq n < \infty$, *of tails; that is,*

$$g^{(n-1)}(b_n + g^{(n)}) = a_n, \qquad n \geq 1,$$

such that $g^{(n)} \neq \infty$, $0 \leq n < \infty$. *(Thus,* $g^{(n)} \neq 0$, $-b_n$, *if* $a_n \neq 0$, $1 \leq n < \infty$.) *Then* $K(a_n/b_n)$ *converges if and only if*

$$\sum_{k=0}^{\infty} \prod_{n=1}^{k} \left(-\frac{b_n + g^{(n)}}{g^{(n)}} \right) \tag{2.4}$$

converges in $\overline{\mathscr{C}} = \mathscr{C} \cup \{\infty\}$. *In particular, if* (2.4) *has the sum* ∞, *then* $K(a_n/b_n)$ *converges to* $g^{(0)}$. *More generally, if* (2.4) *has the sum* $L \in \overline{\mathscr{C}}$, *then* $K(a_n/b_n)$ *converges to* $g^{(0)}(L - 1)/L$.

Note that a continued fraction $K(a_n/b_n)$ has infinitely many sequences of tails; define $g^{(0)} \in \overline{\mathscr{C}}$ arbitrarily, and define $g^{(n)}$, $n \geq 1$, by

$$g^{(n)} = \frac{a_n}{g^{(n-1)}} - b_n.$$

We now show that Entry 2 follows readily from Waadeland and Jacobsen's result.

If $g^{(n)} = a_{n+1}$, $n \geq 0$, then $g^{(n)}$ is a sequence of tails for (2.2). Inserting this into the (truncated) sum (2.4), we find that

$$\sum_{k=0}^{m} \prod_{n=1}^{k} \left(-\frac{x}{a_{n+1}} \right) = a_1 f_m(x).$$

Entry 2 now follows from the theorem above.

We shall interpret Entries 3 and 4 formally. We emphasize that the arguments that we give are not rigorous. There is a slight misprint in the formulation of Entry 3 (p. 143).

Entry 3. *If $x, a_1, a_2, \ldots$ are arbitrary complex numbers, then*

$$x = a_1 + (x^2 + a_1(a_1 - 2a_2) - 2a_1(x^2 + a_2(a_2 - 2a_3) - 2a_2(\cdots)^{1/2})^{1/2})^{1/2}.$$

PROOF. It is easy to verify that

$$x - a_k = (x^2 + a_k(a_k - 2a_{k+1}) - 2a_k(x - a_{k+1}))^{1/2}, \qquad k \geq 1. \qquad (3.1)$$

Using (3.1) successively, we find that

$$x - a_1 = (x^2 + a_1(a_1 - 2a_2) - 2a_1(x - a_2))^{1/2}$$

$$= (x^2 + a_1(a_1 - 2a_2) - 2a_1(x^2 + a_2(a_2 - 2a_3) - 2a_2(x - a_3))^{1/2})^{1/2}$$

$$= \cdots,$$

and so the desired result follows.   □

Entry 4. *Let a, n, and x denote arbitrary complex numbers. Then*

$$f(x) := x + n + a$$

$$= (ax + (n + a)^2 + x(a(x + n) + (n + a)^2 + (x + n)(a(x + 2n)$$

$$+ (n + a)^2 + (x + 2n)(\cdots)^{1/2})^{1/2})^{1/2})^{1/2}.$$

PROOF. By successively substituting, we find that

$$f(x) = (ax + (n + a)^2 + xf(x + n))^{1/2}$$

$$= (ax + (n + a)^2 + x(a(x + n) + (n + a)^2 + (x + n)f(x + 2n))^{1/2})^{1/2}$$

$$= \cdots,$$

and therefore we obtain the proposed formula.   □

Examples. *We have*

(i) $3 = (1 + 2(1 + 3(1 + 4(1 + \cdots)^{1/2})^{1/2})^{1/2})^{1/2}$

and

(ii) $4 = (6 + 2(7 + 3(8 + 4(9 + \cdots)^{1/2})^{1/2})^{1/2})^{1/2}.$

Examples (i) and (ii) were submitted by Ramanujan [5], [16, p. 323] as a problem in the *Journal of the Indian Mathematical Society* and solutions were subsequently given by him. Example (i) appeared as a problem in the William Lowell Putnam competition in 1966 (J. H. McKay [1]).

T. Vijayaraghavan (Ramanujan [16, p. 348]) has shown that

$$(a_1 + (a_2 + (a_3 + \cdots(a_n)^{1/2})^{1/2})^{1/2})^{1/2}, \qquad a_n \geq 0,$$

tends to a limit as n tends to ∞ if and only if

$$\varlimsup_{n \to \infty} \frac{\mathrm{Log}\, a_n}{2^n} < \infty. \qquad (4.1)$$

See also Pólya and Szegö's book [1, pp. 37, 214]. Vijayaraghavan's theorem can be used to show that the infinite radicals in Examples (i) and (ii) are convergent (Ramanujan [16, p. 348]).

The literature on infinite radicals is rather scant, and so Herschfeld's paper [1] is to be particularly recommended. He points out that Ramanujan's proofs of (i) and (ii) are slightly incomplete, and he gives full rigorous solutions. This paper contains a good discussion on the convergence of infinite radicals. Elementary discussions of nested radicals have also been given by W. S. Sizer [1] and E. J. Allen [1].

We state Entry 5(i) as Ramanujan records it. But, as we shall see, Entry 5(i) is valid only for $\theta = 0$. We shall separate Entry 5(ii) into two parts. The first part will be proved rigorously; the second will be regarded as a formal identity. However, we shall indicate some values of θ for which the second part of Entry 5(ii) is rigorously true. We suggest to readers that they attempt to develop more thoroughly the theory of infinite radicals, so that perhaps concrete conditions may be imposed on the formal identities in Sections 3–5 to ensure their validity. Jacobsen's paper [4] is one in this direction.

Entry 5(i). *We have*

$$2 \cos \theta = (2 + 2 \cos 2\theta)^{1/2} = (2 + (2 + 2 \cos 4\theta)^{1/2})^{1/2}$$
$$= (2 + (2 + (2 + 2 \cos 8\theta)^{1/2})^{1/2})^{1/2} = \cdots .$$

PROOF. Repeatedly apply the identity

$$2 \cos(2^k \theta) = \pm (2 + 2 \cos(2^{k+1} \theta))^{1/2}, \qquad k \geq 0,$$

with the plus sign always chosen on the right side. However, unless $\theta = 0$, there clearly will be values of k when $\cos(2^k \theta) < 0$, and so we must choose the minus sign in such instances. If $\theta = 0$, Entry 5(i) implies that

$$2 = (2 + (2 + (2 + \cdots)^{1/2})^{1/2})^{1/2},$$

which is meaningful since (4.1) is easily seen to be satisfied. Furthermore, a direct proof may easily be given. (This last example appears in Zippin's book [1, p. 51].) $\square$

Entry 5(ii) (First Part). *Suppose that either* $|\theta| \leq \pi/6$ *or* $5\pi/6 \leq \theta \leq 7\pi/6$. *Then*

$$2 \cos \theta = (2 \cos 3\theta + 3(2 \cos 3\theta + 3(2 \cos 3\theta + \cdots)^{1/3})^{1/3})^{1/3}.$$

PROOF. For $n \geq 1$, let

$$R_n = (2 \cos 3\theta + 3(2 \cos 3\theta + 3(2 \cos 3\theta + \cdots)^{1/3})^{1/3})^{1/3},$$

where n cube roots are taken. Observe that

$$R_n = (2 \cos 3\theta + 3R_{n-1})^{1/3}, \qquad n \geq 2.$$

First suppose that $|\theta| \leq \pi/6$. Clearly, $R_{n-1} < R_n$ for each $n \geq 2$. Thus,

$$R_n^3 = 2 \cos 3\theta + 3R_{n-1} < 2 \cos 3\theta + 3R_n. \tag{5.1}$$

The polynomial $x^3 - 3x - 2 \cos 3\theta$ has three real roots, $2 \cos \theta$ and $-\cos \theta \pm \sqrt{3}|\sin \theta|$. For $|\theta| \le \pi/6$, $-\cos \theta \pm \sqrt{3}|\sin \theta| \le 0$. Therefore, $\{R_n\}$ is a non-negative, increasing sequence bounded above by the root $2 \cos \theta$. Thus, $\{R_n\}$ converges and, by (5.1), $\{R_n\}$ converges to a root of $x^3 - 3x - 2 \cos 3\theta$. As we have just seen, this root must be $2 \cos \theta$.

For $5\pi/6 \le \theta \le 7\pi/6$, consider $\alpha = \theta - \pi$. Thus, $|\alpha| \le \pi/6$. Using the foregoing analysis, we complete the proof. $\qquad\square$

We remark that if $\pi/2 < \theta < 5\pi/6$ or $7\pi/6 < \theta < 3\pi/2$, then $\{R_n\}$ converges to $-\cos \theta + \sqrt{3}|\sin \theta|$, while if $\pi/6 < \theta < \pi/2$ or $3\pi/2 < \theta < 11\pi/6$, $\{R_n\}$ converges to $-\cos \theta - \sqrt{3}|\sin \theta|$.

Entry 5(ii) (Second Part). *We have*

$$2 \cos \theta = (6 \cos \theta + (6 \cos 3\theta + (6 \cos 9\theta + \cdots)^{1/3})^{1/3})^{1/3}. \tag{5.2}$$

PROOF. Repeatedly employ the equality

$$2 \cos(3^k \theta) = (6 \cos(3^k \theta) + 2 \cos(3^{k+1} \theta))^{1/3}$$

for $k = 0, 1, 2, \ldots$.

We now indicate some special cases when the second part of Entry 5(ii) may be established rigorously.

If $\theta = 0$, then (5.2) becomes

$$2 = (6 + (6 + (6 + \cdots)^{1/3})^{1/3})^{1/3}. \tag{5.3}$$

To prove (5.3), define

$$R_n = (6 + (6 + \cdots 6^{1/3} \cdots)^{1/3})^{1/3}, \qquad n \ge 1,$$

where n cube roots are indicated. Observe that

$$R_n^3 = 6 + R_{n-1} < 6 + R_n, \qquad n \ge 2. \tag{5.4}$$

Now $x = 2$ is the only real root of the equation $x^3 - x - 6 = 0$. It follows that $R_{n-1} < R_n < 2$, $n \ge 2$. Thus, $\{R_n\}$ converges, and, by (5.4), the limit of $\{R_n\}$ equals 2.

If $\theta = \pi$, then (5.2) yields

$$-2 = (-6 + (-6 + (-6 + \cdots)^{1/3})^{1/3})^{1/3}$$
$$= -(6 + (6 + (6 + \cdots)^{1/3})^{1/3})^{1/3}, \tag{5.5}$$

which is valid by (5.3).

If $\theta = \pi/3$, the right side of (5.2) becomes

$$(3 + (-6 + (-6 + \cdots)^{1/3})^{1/3})^{1/3} = (3 - 2)^{1/3} = 1,$$

by (5.5). Hence, (5.2) is valid for $\theta = \pi/3$. In fact, by induction, it is easy to show that (5.2) holds for $\theta = \pi/3^k$, $k \ge 1$.

It may also be easily checked that (5.2) is valid if $\theta = \pi/2$ or $2\pi/3$, for example. If $\theta = \pi/4$, (5.2) holds, but the verification is more difficult. □

Entry 6. *Let $a > 0$ but $a \neq 1$. Suppose that n is a nonnegative integer. In the field of formal power series, put*

$$f_n(v) = \sum_{j=0}^{\infty} a_j(n)v^j,$$

where $a_0(n) = 1$, $a_1(n) = -a^{-n}$, and $a_j(n), j \geq 2$, is defined recursively by

$$a_j(n) = \frac{1}{2(a^{j-1} - 1)} \sum_{k=1}^{j-1} a_k(n)a_{j-k}(n). \tag{6.1}$$

Then for each nonnegative integer n,

$$\left(\frac{a(a-2)}{4} + \left(\frac{a(a-2)}{4} + \cdots + \left(\frac{a(a-2)}{4} + \frac{a}{2}f_0(v) \right)^{1/2} \cdots \right)^{1/2} \right)^{1/2} = \frac{a}{2}f_n(v), \tag{6.2}$$

where, on the left side, there are n iterated radicals. Furthermore,

$$f_n(v) = 1 - v/a^n + \frac{(v/a^n)^2}{2(a-1)} - \frac{(v/a^n)^3}{2(a-1)(a^2-1)}$$

$$+ \frac{(v/a^n)^4(a+5)}{8(a-1)(a^2-1)(a^3-1)} - \frac{(v/a^n)^5(2a^2+3a+7)}{8(a-1)(a^2-1)(a^3-1)(a^4-1)} + \cdots. \tag{6.3}$$

PROOF. If $n = 0$, (6.2) is trivial. Thus, assume that $n > 0$. Proceeding by induction and squaring both sides of (6.2), we find that we must show that

$$\frac{a(a-2)}{4} + \frac{a}{2}f_{n-1}(v) = \frac{a^2}{4}f_n^2(v), \qquad n \geq 1,$$

or, in other words,

$$\frac{a}{2} + \sum_{j=1}^{\infty} a_j(n-1)v^j = \frac{a}{2}f_n^2(v), \qquad n \geq 1,$$

Now, by (6.1) and induction, $a_j(n-1) = a^j a_j(n), j \geq 0, n \geq 1$, and so it suffices to show that

$$a^j a_j(n) = \frac{a}{2} \sum_{k=0}^{j} a_k(n)a_{j-k}(n), \qquad j \geq 2.$$

But the latter equality is equivalent to (6.1), and so the proof of (6.2) is complete.

The expansion (6.3) is easily determined by employing (6.1). □

Ramanujan's formulation of Entry 6 is slightly incorrect, for he claims that (p. 143)

$$a_j(n) = \frac{1}{2(a^{j-1} - 1)} \sum_{k=0}^{j-1} a_k(n) a_{j-1-k}(n),$$

which should be compared with (6.1).

Entry 7. *If x is not a negative integer, then*

$$1 = \frac{x+1}{x} \ \frac{x+2}{+\ x+1} \ \frac{x+3}{+\ x+2} + \cdots . \tag{7.1}$$

FIRST PROOF. We first derive a consequence of Entry 22 that we shall employ several times in the sequel. In Entry 22, replace x by x/α. Since the continued fraction converges uniformly with respect to α in a neighborhood of $\alpha = \infty$, we may let α tend to ∞ to deduce that, for $\beta \notin \{-1, -2, \cdots\}$,

$$\frac{{}_1F_1(\beta+1; \gamma+1; x)}{\gamma \, {}_1F_1(\beta; \gamma; x)} = \frac{1}{\gamma-x} \ \frac{(\beta+1)x}{+\ \gamma+1-x} \ \frac{(\beta+2)x}{+\ \gamma+2-x} + \cdots . \tag{7.2}$$

(An equivalent form of (7.2) was also found by Perron [3, p. 278, formula (8)].) By using Corollary 1 of Entry 21, we can show that (7.2) is also valid when β is a negative integer, provided that $\gamma \notin \{\beta, \beta - 1, \beta - 2, \ldots\}$.

   To prove (7.1), set $x = 1$ and $\beta = \gamma = x$ in (7.2). The result now easily follows. ☐

SECOND PROOF. The continued fraction (7.1) has tails $g^{(n)} \equiv 1$. The Nth partial sum of (2.4) is therefore equal to

$$\sum_{k=0}^{N} (-1)^k (x+1)_k, \tag{7.3}$$

which obviously cannot converge to a finite number. However, if x is not a nonpositive integer,

$$\frac{x+1}{x} \ \frac{x+2}{+\ x+1} \ \frac{x+3}{+\ x+2} + \cdots = \frac{\dfrac{x+1}{x}}{1} \ \frac{\dfrac{x+2}{x(x+1)}}{+\ 1} \ \frac{\dfrac{x+3}{(x+1)(x+2)}}{+\ 1} + \cdots ,$$

which converges by Worpitzky's theorem (Wall [1, p. 42]). Hence, by the theorem in Section 2, (7.3) tends to ∞ as N tends to ∞. So by the same theorem, (7.1) converges to $g^{(0)} = 1$. ☐

   It also should be remarked that Entry 7 follows from Entry 11 by setting $a = 1$ and $n = x + 1$.

   Ramanujan (p. 143) has written x instead of 1 on the left side of (7.1).

Corollary. *We have*

$$1 = \frac{2}{1} \ \frac{3}{+\ 2} \ \frac{4}{+\ 3} \ \frac{5}{+\ 4} + \cdots .$$

PROOF. Set $x = 1$ in Entry 7.   □

Entry 8. *Let n denote a positive integer and suppose that $x \neq -ka$, where k is a positive integer such that $1 \leq k \leq n$. Then*

$$\sum_{k=1}^{n} \frac{(-1)^{k+1}}{(x + a)(x + 2a) \cdots (x + ka)}$$

$$= \frac{1}{x + a} \frac{x + a}{+ x + 2a - 1} \frac{x + 2a}{+ x + 3a - 1 +} \cdots \frac{x + (n-1)a}{+ x + na - 1}. \quad (8.1)$$

FIRST PROOF. Denote the right side of (8.1) by A_n/B_n in the notation of Section 1. Then by (1.3),

$$A_n = (x + na - 1)A_{n-1} + (x + (n-1)a)A_{n-2}, \qquad n \geq 3,$$

or, upon iteration,

$$A_n - (x + na)A_{n-1} = -\{A_{n-1} - (x + (n-1)a)A_{n-2}\}$$
$$= \cdots = (-1)^n\{A_2 - (x + 2a)A_1\}$$
$$= (-1)^{n-1}, \qquad n \geq 3, \quad (8.2)$$

since $A_1 = 1$ and $A_2 = x + 2a - 1$.
   Similarly, by (1.4),

$$B_n - (x + na)B_{n-1} = -\{B_{n-1} - (x + (n-1)a)B_{n-2}\}$$
$$= \cdots = (-1)^n\{B_2 - (x + 2a)B_1\} = 0, \qquad n \geq 3,$$

since $B_1 = x + a$ and $B_2 = (x + a)(x + 2a)$. Hence,

$$B_n = (x + a)(x + 2a) \cdots (x + na), \qquad n \geq 1. \quad (8.3)$$

   On the other hand, let the left side of (8.1) be denoted by the rational function P_n/Q_n. Clearly,

$$Q_n = (x + a)(x + 2a) \cdots (x + na), \qquad n \geq 1. \quad (8.4)$$

Now, for $n \geq 2$,

$$\frac{P_n}{Q_n} = \frac{P_{n-1}}{Q_{n-1}} + \frac{(-1)^{n+1}}{Q_n} = \frac{(x + na)P_{n-1} + (-1)^{n+1}}{Q_n};$$

that is,

$$P_n = (x + na)P_{n-1} + (-1)^{n+1}, \qquad n \geq 2. \quad (8.5)$$

Hence, by (8.2) and (8.5), A_n and P_n satisfy the same recursion formula. Since $A_1 = P_1 = 1$ and $A_2 = P_2 = x + 2a - 1$, we conclude that $A_n = P_n$, $n \geq 1$. Also, by (8.3) and (8.4), $B_n = Q_n$, $n \geq 1$. Thus, the equality (8.1) has been established.   □

SECOND PROOF. We induct on n. For $n = 1$, (8.1) is trivially true.

Suppose that we denote the left side of (8.1) by $f_n(x)$. Proceeding by induction, we thus find that

$$(x + a)f_{n+1}(x) = 1 - f_n(x + a)$$

$$= 1 - \cfrac{1}{x + 2a + \cfrac{x + 2a}{x + 3a - 1 + \cdots + \cfrac{x + na}{x + (n + 1)a - 1}}}.$$

Letting

$$A = x + 3a - 1 + \cfrac{x + 3a}{x + 4a - 1 + \cdots + \cfrac{x + na}{x + (n + 1)a - 1}},$$

we then deduce that

$$(x + a)f_{n+1}(x) = 1 - \cfrac{1}{x + 2a + (x + 2a)/A}$$

$$= \frac{(x + 2a - 1) + (x + 2a)/A}{(x + 2a - 1) + (x + 2a)/A + 1}$$

$$= \cfrac{1}{1 + \cfrac{1}{(x + 2a - 1) + (x + 2a)/A}}.$$

Upon dividing both sides of the equality above by $x + a$, we arrive at (8.1), but with n replaced by $n + 1$. This completes the induction. $\square$

Corollary. *We have*

$$\frac{1}{e - 1} = \cfrac{1}{1} + \cfrac{1}{2} + \cfrac{2}{3} + \cdots.$$

PROOF. Let $x = 0$ and $a = 1$ in Entry 8 to obtain the equality

$$\sum_{k=1}^{n} \frac{(-1)^{k+1}}{k!} = \cfrac{1}{1} + \cfrac{1}{1} + \cfrac{2}{2} + \cfrac{3}{3} + \cdots + \cfrac{n - 1}{n - 1}.$$

Letting n tend to ∞ yields

$$1 - \frac{1}{e} = \cfrac{1}{1} + \cfrac{1}{1} + \cfrac{1}{2} + \cfrac{2}{3} + \cdots.$$

The desired formula now readily follows by inverting the equality above.

$\square$

The previous corollary is due to Euler [3].

Entry 9. *Let a and x be complex numbers such that either $x \neq -ka$ for $k \in \{1, 2, \ldots\}$ and $a \neq 0$, or that $a = 0$ and $|x| > 1$. Then*

$$\frac{x+a+1}{x+1} = \frac{x+a}{x-1} + \frac{x+2a}{x+a-1} + \frac{x+3a}{x+2a-1} + \cdots . \tag{9.1}$$

PROOF. We first indicate a formal nonrigorous argument. Observe that for each positive integer n,

$$\frac{x+na+1}{x+(n-1)a+1} = \frac{x+na}{x+(n-1)a-1 + \dfrac{x+(n+1)a+1}{x+na+1}}. \tag{9.2}$$

By applying this identity successively for $n = 1, 2, \ldots$, we formally derive (9.1).

We now give a rigorous proof based on (7.2). We first assume that $a \neq 0$. Putting $x = 1/a$, $\beta = x/a$, and $\gamma = (x-a)/a$ in (7.2), we find that, under the restriction $\beta = x/a \notin \{-1, -2, \ldots\}$,

$$\frac{1}{x-a} \frac{{}_1F_1\left(\dfrac{x+a}{a}; \dfrac{x}{a}; \dfrac{1}{a}\right)}{{}_1F_1\left(\dfrac{x}{a}; \dfrac{x-a}{a}; \dfrac{1}{a}\right)}$$

$$= \frac{1}{x-a-1} + \frac{x+a}{x-1} + \frac{x+2a}{x+a-1} + \frac{x+3a}{x+2a-1} + \cdots , \tag{9.3}$$

provided that $x \neq -ka$, where k is a positive integer. But,

$$\frac{{}_1F_1\left(\dfrac{x+a}{a}; \dfrac{x}{a}; \dfrac{1}{a}\right)}{{}_1F_1\left(\dfrac{x}{a}; \dfrac{x-a}{a}; \dfrac{1}{a}\right)} = \frac{x-a}{x} \frac{\dfrac{x}{a}e^{1/a} + \dfrac{1}{a}e^{1/a}}{\left(\dfrac{x}{a}-1\right)e^{1/a} + \dfrac{1}{a}e^{1/a}}$$

$$= \frac{(x-a)(x+1)}{x(x-a+1)}. \tag{9.4}$$

Substituting (9.4) into (9.3), taking the reciprocal of both sides, and simplifying, we arrive at (9.1).

Another proof for $a \neq 0$, depending on the theorem in Section 2, can be given. Equality (9.2) shows that

$$g^{(n-1)} = \frac{x+na+1}{x+(n-1)a+1}, \qquad n \geq 1,$$

is a sequence of tails for (9.1). Thus, a partial sum of (2.4) equals

$$\sum_{k=0}^{N} (-1)^k \prod_{k=1}^{N} \frac{\{x+(n-1)a+1\}\{x+na\}}{x+(n+1)a+1},$$

which must tend to ∞ by the same argument that was used in the second proof of Entry 7. Since $g^{(0)} = (x+a+1)/(x+1)$, the proof is complete by the theorem in Section 2.

For $a = 0$, the continued fraction (9.1) reduces to the periodic continued fraction $K(x/(x - 1))$. The convergence behavior of periodic continued fractions is well established. See, for instance, the text of Jones and Thron [1, pp. 47, 48]. Thus, $K(x/(x - 1))$ converges if and only if $x/(x - 1)^2$ does not lie on $(-\infty, -1/4)$. For $|x| > 1$, the continued fraction in (9.1) converges to 1, as claimed by Ramanujan. However, if $|x| < 1$, the continued fraction converges to $-x$.   □

Examples. *We have*

(i) $\dfrac{4}{3} = \dfrac{3}{1} + \dfrac{4}{2} + \dfrac{5}{3} + \dfrac{6}{4} + \cdots$

and

(ii) $\dfrac{5}{3} = \dfrac{4}{1} + \dfrac{6}{3} + \dfrac{8}{5} + \dfrac{10}{7} + \cdots$.

Proof. Set $x = 2$ and $a = 1$ in Entry 9 to deduce (i); similarly, set $x = 2$ and $a = 2$ to obtain (ii).   □

Entry 10. *If n is a positive integer, then*

$$n = \dfrac{1}{1 - n} + \dfrac{2}{2 - n} + \dfrac{3}{3 - n} + \cdots + \dfrac{n}{0} + \dfrac{n + 1}{1} + \dfrac{n + 2}{2} + \cdots .$$

First Proof. Putting $x = 1$, $\beta = 0$, and $\gamma = 1 - n$ in (7.2), we find that

$$0 = \frac{(1 - n)\, {}_1F_1(0; 1 - n; 1)}{{}_1F_1(1; 2 - n; 1)} = -n + \dfrac{1}{1 - n} + \dfrac{2}{2 - n} + \dfrac{3}{3 - n} + \cdots,$$

which completes the proof.   □

Second Proof. In (11.7), set $n = 1$ and replace a by n to deduce that

$$\dfrac{1}{1 - n} + \dfrac{2}{2 - n} + \cdots = 1 + \dfrac{n - 1}{3 - n} + \dfrac{n - 2}{4 - n} + \cdots .$$

We shall be finished if we can show that, for each positive integer n,

$$\dfrac{n}{2 - n} + \dfrac{n - 1}{3 - n} + \cdots = n. \qquad (10.1)$$

We prove (10.1) by inducting on n. If $n = 1$, (10.1) is trivial. Assuming that (10.1) holds with n replaced by $n - 1$, $n > 1$, we see that

$$\dfrac{n}{2 - n} + \dfrac{n - 1}{3 - n} + \cdots = \dfrac{n}{(2 - n) + (n - 1)} = n. \qquad □$$

The interpretation of Entry 11 was made difficult because Ramanujan left most of his notation undefined. Furthermore, some of his notation is unnecessary and so will not be given.

Entry 11. *Suppose that a is a positive integer and that* $n \notin \{0, -1, -2, \ldots\}$. *Define* N_a *and* D_a *by*

$$
{}_1F_1(1 - a; n + 2 - a; -1) = \frac{N_a}{(n + 2 - a)(n + 3 - a) \cdots n} \tag{11.1}
$$

and

$$
{}_1F_1(1 - a; n + 1 - a; -1) = \frac{D_a}{(n + 1 - a)(n + 2 - a) \cdots (n - 1)}, \tag{11.2}
$$

where if $a = 1$, *the denominators on the right sides of* (11.1) *and* (11.2) *are understood to be equal to 1. Then*

$$
\frac{N_a}{D_a} = \frac{n}{n - a} + \frac{n + 1}{n - a + 1} + \frac{n + 2}{n - a + 2} + \cdots \tag{11.3}
$$

and

$$
\frac{N_{a+1}}{N_a} = n + 2 - a + \frac{a - 1}{n + 3 - a} + \frac{a - 2}{n + 4 - a} + \cdots. \tag{11.4}
$$

PROOF. Since a is a positive integer, both ${}_1F_1(1 - a; n + 2 - a; -1)$ and ${}_1F_1(1 - a; n + 1 - a; -1)$ terminate, and so N_a and D_a are simply the numerators of the rational functions respectively obtained. In fact, N_a and D_a are polynomials in n of degree $a - 1$.

Setting $\beta = n$, $\gamma = n + 1 - a$, and $x = 1$ in (7.2), we find that

$$
\frac{n \, {}_1F_1(n + 1; n + 2 - a; 1)}{(n + 1 - a) \, {}_1F_1(n; n + 1 - a; 1)} = \frac{n}{n - a} + \frac{n + 1}{n - a + 1} + \frac{n + 2}{n - a + 2} + \ldots, \tag{11.5}
$$

where $n \notin \{0, -1, -2, \ldots\}$. But by Kummer's theorem (Entry 21 of Chapter 10),

$$
\frac{n \, {}_1F_1(n + 1; n + 2 - a; 1)}{(n + 1 - a) \, {}_1F_1(n; n + 1 - a; 1)} = \frac{n \, {}_1F_1(1 - a; n + 2 - a; -1)}{(n + 1 - a) \, {}_1F_1(1 - a; n + 1 - a; -1)}
$$

$$
= \frac{N_a}{D_a}. \tag{11.6}
$$

Thus, (11.3) follows from (11.5) and (11.6).

From (11.1),

$$
\frac{N_{a+1}}{N_a} = \frac{(n + 1 - a) \, {}_1F_1(-a; n + 1 - a; -1)}{{}_1F_1(1 - a; n + 2 - a; -1)}
$$

$$
= n + 2 - a + \frac{a - 1}{n + 3 - a} + \frac{a - 2}{n + 4 - a} + \cdots,
$$

where we have applied (7.2) with $\beta = -a$, $\gamma = n + 1 - a$, and $x = -1$. This application of (7.2) is valid by our remarks following (7.2). For if a is a positive integer, $\gamma = n + 1 - a \notin \{-a, -a - 1, -a - 2, \ldots\}$, since $n \notin \{0, -1, -2, \ldots\}$. This proves (11.4). $\qquad\square$

By generalizing the proof above, we can easily prove that

$$\frac{n}{n-a} + \frac{n+1}{n-a+1} + \frac{n+2}{n-a+2} + \cdots = 1 + \frac{a-1}{n+2-a} + \frac{a-2}{n+3-a} + \cdots,$$
(11.7)

provided that not both a and $-n$ are nonnegative integers.

Corollary 1. *If n is not a nonpositive integer, then*

$$\frac{n^2 + n + 1}{n^2 - n + 1} = \frac{n}{n-3} + \frac{n+1}{n-2} + \frac{n+2}{n-1} + \cdots.$$

PROOF. Let $a = 3$ in (11.3). $\qquad\square$

Corollary 2. *If n is not a nonpositive integer, then*

$$\frac{n^3 + 2n + 1}{(n-1)^3 + 2(n-1) + 1} = \frac{n}{n-4} + \frac{n+1}{n-3} + \frac{n+2}{n-2} + \cdots.$$

PROOF. Let $a = 4$ in (11.3). $\qquad\square$

Entry 12. *If $a \neq 0$ and $x \neq -ka$, where k is a positive integer,*

$$1 = \frac{x+a}{a} + \frac{(x+a)^2 - a^2}{a} + \frac{(x+2a)^2 - a^2}{a} + \frac{(x+3a)^2 - a^2}{a} + \cdots.$$
(12.1)

FIRST PROOF. In Entry 22, put $x = 1$, $\alpha = 0$, $\beta = (x - a)/a$, and $\gamma = (x + a)/a$. After simplification, we find that

$$\frac{x-a}{x+a} = \frac{x-a}{a} + \frac{(x+a)^2 - a^2}{a} + \frac{(x+2a)^2 - a^2}{a} + \cdots.$$

Multiplying both sides by $(x + a)/(x - a)$, we complete the proof. $\qquad\square$

SECOND PROOF. In Entry 27, let $x = 1$. Then set $y = 1 + 2x/a$ and $n = -4$. We then find that

$$1 + \frac{2x}{a} = 1 + \frac{4(x+a)^2/a^2 - 4}{2} + \frac{4(x+2a)^2/a^2 - 4}{2} + \frac{4(x+3a)^2/a^2 - 4}{2} + \cdots$$

$$= 1 + \frac{2}{a}\left\{ \frac{(x+a)^2 - a^2}{a} + \frac{(x+2a)^2 - a^2}{a} + \frac{(x+3a)^2 - a^2}{a} + \cdots \right\}$$

$$= 1 + \frac{2}{a}X,$$

say. Thus, $x = X$. Lastly,

$$1 = \frac{a + x}{a + x} = \frac{a + x}{a + X},$$

which is the desired formula.   □

THIRD PROOF. For $x = 0$, the result is trivial. Thus, assume that $x \neq -ka$, $0 \leq k < \infty$. A sequence of tails for (12.1) is given by

$$g^{(n)} = \begin{cases} 1, & \text{if } n = 0, \\ x + (n - 1)a. & \text{if } n \geq 1. \end{cases}$$

The sum in (2.4) is then equal to

$$\sum_{k=0}^{\infty} \prod_{n=1}^{k} \left(-\frac{a + x + (n - 1)a}{x + (n - 1)a} \right) = \sum_{k=0}^{\infty} (-1)^k \frac{x + ka}{x} = \infty.$$

By the theorem of Section 2, we conclude that the continued fraction in (12.1) converges to $g^{(0)} = 1$.   □

Entry 13. *Let a, b, and d be complex numbers such that either $d \neq 0$, $b \neq -kd$, where k is a nonnegative integer, and $\mathrm{Re}((a - b)/d) > 0$, or $d \neq 0$ and $a = b$, or $d = 0$ and $|a| < |b|$. Then*

$$a = \cfrac{ab}{a + b + d -} \cfrac{(a + d)(b + d)}{a + b + 3d} \cfrac{(a + 2d)(b + 2d)}{a + b + 5d} - \cdots. \tag{13.1}$$

FIRST PROOF. For this proof, we shall assume the first set of conditions on a, b, and d. We shall also need to assume that $(a + kd)(b + kd) \neq 0$, for each nonnegative integer k. Let $p_k = b + kd$, $k \geq 0$. Then

$$p_n = a + b + (2n + 1)d - \frac{(a + (n + 1)d)(b + (n + 1)d)}{p_{n+1}}, \qquad n \geq 0.$$

Writing $p_n = x_n/x_{n+1}$, $n \geq 0$, we may write the preceding formula in the form

$$x_n = (a + b + (2n + 1)d)x_{n+1} - (a + (n + 1)d)(b + (n + 1)d)x_{n+2}.$$

Setting $a + nd = y_n/y_{n+1}$, $n \geq 0$, we easily see that the same recurrence formula is satisfied by y_n.

Now if $x_0 = 1$,

$$x_{n+1} = \frac{x_{n+1}}{x_n} \frac{x_n}{x_{n-1}} \cdots \frac{x_1}{x_0} = \frac{1}{p_n} \frac{1}{p_{n-1}} \cdots \frac{1}{p_0}$$

$$= \frac{1}{(b + nd)(b + (n - 1)d) \cdots b}.$$

Similarly, if $y_0 = 1$,

$$y_{n+1} = \frac{1}{(a + nd)(a + (n - 1)d) \cdots a}.$$

Thus, under our assumptions, x_n/y_n tends to 0 as n tends to ∞.

We now apply a theorem in Perron's text [3, p. 97, Satz 2.46, C] to deduce that, under our hypotheses,

$$\frac{x_0}{x_1} = b = a + b + d - \frac{(a+d)(b+d)}{a+b+3d} - \frac{(a+2d)(b+2d)}{a+b+5d}$$

$$- \frac{(a+3d)(b+3d)}{a+b+7d} - \cdots .$$

Now take the reciprocal of both sides above and then multiply both sides by ab to obtain the proposed continued fraction representation. $\square$

SECOND PROOF. As in our first proof, we assume that the first set of hypotheses holds. In Entry 20, let $\alpha = b/(2d)$, $\beta = a/(2d)$, and $\gamma = (a + b + d)/(2d)$. By our hypotheses, each of the two hypergeometric series in Entry 20 converges at $x = -1$. By the remarks following Entry 20, we may let $x = -1$ in Entry 20. After a slight amount of manipulation, we find that

$$\frac{ab}{a+b+d} \frac{{}_2F_1\left(\dfrac{a+d}{2d}, \dfrac{a+2d}{2d}; \dfrac{a+b+3d}{2d}; 1\right)}{{}_2F_1\left(\dfrac{a+d}{2d}, \dfrac{a}{2d}; \dfrac{a+b+d}{2d}; 1\right)}$$

$$= \frac{ab}{a+b+d} - \frac{(a+d)(b+d)}{a+b+3d} - \frac{(a+2d)(b+2d)}{a+b+5d} - \cdots . \qquad (13.2)$$

If we now apply Gauss's theorem (Entry 8 of Chapter 10) to each of the hypergeometric series above, we find that the left side of (13.2) becomes

$$\frac{ab}{a+b+d} \frac{\Gamma\left(\dfrac{a+b+3d}{2d}\right)\Gamma\left(\dfrac{b}{2d}\right)}{\Gamma\left(\dfrac{b+2d}{2d}\right)\Gamma\left(\dfrac{a+b+d}{2d}\right)} = a,$$

which completes the proof. $\square$

THIRD PROOF. Assume that either of the first two sets of hypotheses is valid. Assume that $a + nd, b + nd \neq 0$ for each nonnegative integer n. Then

$$g^{(n)} = \begin{cases} a, & \text{if } n = 0, \\ -(a + nd), & \text{if } n \geq 1, \end{cases}$$

is a sequence of tails for (13.1). The series in (2.4) then becomes

$$\sum_{k=0}^{\infty} \prod_{n=1}^{k} \frac{b + (n-1)d}{a + nd} = {}_2F_1\left(\frac{b}{d}, 1; \frac{a}{d} + 1; 1\right),$$

which is known to diverge to ∞ if $\operatorname{Re}((b - a)/d) \geq 0$. Thus, by the aforementioned theorem, the continued fraction in (13.1) converges to $g^{(0)} = a$. $\square$

Our last proof is due to Jacobsen [3], who proves Entry 13 under all the given hypotheses on a, b, and d.

Entry 14. *If $a_1, a_2, \ldots, a_{2n}$ and x are arbitrary complex numbers, then*

$$\frac{a_1}{x} + \frac{a_2}{1} + \frac{a_3}{x} + \frac{a_4}{1} + \cdots + \frac{a_{2n}}{1}$$

$$= \frac{a_1}{x + a_2} - \frac{a_2 a_3}{x + a_3 + a_4} - \frac{a_4 a_5}{x + a_5 + a_6} - \cdots - \frac{a_{2n-2} a_{2n-1}}{x + a_{2n-1} + a_{2n}}. \quad (14.1)$$

PROOF. We shall induct on n. For $n = 1$, it is easy to verify that the proposed identity is valid. Now assume that (14.1) is true with n replaced by $n - 1$ for any fixed integer $n > 1$. Let

$$A = \frac{a_3}{x} + \frac{a_4}{1} + \frac{a_5}{x} + \frac{a_6}{1} + \cdots + \frac{a_{2n}}{1}$$

and

$$B = \frac{a_4 a_5}{x + a_5 + a_6} - \frac{a_6 a_7}{x + a_7 + a_8} - \cdots - \frac{a_{2n-2} a_{2n-1}}{x + a_{2n-1} + a_{2n}}.$$

Then, by induction,

$$\frac{a_1}{x} + \frac{a_2}{1} + \frac{a_3}{x} + \frac{a_4}{1} + \cdots + \frac{a_{2n}}{1} = \frac{a_1}{x} + \frac{a_2}{1 + A}$$

$$= \frac{a_1}{x} \cfrac{}{+ 1 + \cfrac{a_2}{\cfrac{a_3}{x + a_4 - B}}} = \cfrac{a_1}{x + a_2 - \cfrac{a_2 a_3}{x + a_3 + a_4 - B}},$$

which completes the proof. □

Entry 14 is actually a finite form of a special case of a classical result. Suppose that $K(a_n/b_n)$ has approximants f_n, $n \geq 1$. Then the even part of $K(a_n/b_n)$ is a continued fraction with approximants f_{2n}, $n \geq 1$. If $b_{2n} \neq 0, n \geq 1$, the even part is given by (up to equivalence transformations)

$$\frac{a_1 b_2}{a_2 + b_1 b_2} - \frac{a_2 a_3 b_4}{a_3 b_4 + b_2(a_4 + b_3 b_4)} - \frac{a_4 a_5 b_2 b_6}{a_5 b_6 + b_4(a_6 + b_5 b_6)}$$

$$- \cdots - \frac{a_{2n} a_{2n+1} b_{2n-2} b_{2n+2}}{a_{2n+1} b_{2n+2} + b_{2n}(a_{2n+2} + b_{2n+1} b_{2n+2})} - \cdots. \quad (14.2)$$

See, for example, the treatise of Jones and Thron [1, pp. 41, 42].

Preece [2] established a slight generalization of Entry 14 in infinite form, that is, a special case of (14.2). Rogers [2] also proved a corollary of (14.2).

We shall establish two renditions of Entry 15. We first regard (15.1) as a formal identity and provide a proof that is probably similar to that found by

Ramanujan. By a "formal identity" we mean that the two continued fractions in (15.1) below correspond to the same power series $\sum_{k=0}^{\infty} c_k x^{-k}$. In the second version, we offer conditions under which (15.1) is valid as an identity between two convergent continued fractions. We are very grateful to L. Jacobsen for the latter version.

Entry 15 (First Version). *As a formal identity,*

$$\frac{a_1 + h}{1} + \frac{a_1}{x} + \frac{a_2 + h}{1} + \frac{a_2}{x} + \cdots = h + \frac{a_1}{1} + \frac{a_1 + h}{x} + \frac{a_2}{1} + \frac{a_2 + h}{x} + \cdots. \tag{15.1}$$

PROOF. Let

$$F_k = x + \frac{a_k + h}{1} + \frac{a_k}{x} + \frac{a_{k+1} + h}{1} + \frac{a_{k+1}}{x} + \cdots, \qquad k \geq 2.$$

Denoting the left side of (15.1) by F, we find that

$$F = \frac{a_1 + h}{1 + a_1/F_2} = \frac{h(F_2 + a_1) + a_1(F_2 - h)}{F_2 + a_1}$$

$$= h + \frac{a_1(F_2 - h)}{F_2 + a_1} = h + \frac{a_1(F_2 - h)}{(F_2 - h) + (a_1 + h)}$$

$$= h + \frac{a_1}{1 + \dfrac{a_1 + h}{F_2 - h}}. \tag{15.2}$$

Next, for $k \geq 2$,

$$F_k - h = x - h + \frac{a_k + h}{1 + a_k/F_{k+1}} = x + \frac{a_k\left(1 - \dfrac{h}{F_{k+1}}\right)}{1 + \dfrac{a_k}{F_{k+1}}}$$

$$= x + \frac{a_k\left(1 - \dfrac{h}{F_{k+1}}\right)}{\left(1 - \dfrac{h}{F_{k+1}}\right) + \left(\dfrac{a_k + h}{F_{k+1}}\right)} = x + \frac{a_k}{1 + \dfrac{a_k + h}{F_{k+1} - h}}. \tag{15.3}$$

Now use (15.3) successively in (15.2) beginning with $k = 2$. This completes the proof since both continued fractions are regular C-fractions and thus correspond to uniquely determined power series (Jones and Thron [1, p. 222]).   □

Entry 15 (Second Version)

(i) *If the left side of (15.1) converges to F, say, then the odd part of the continued fraction on the right side of (15.1) converges to F, and conversely. In* *particular, this means that the identity holds in the usual sense if both continued fractions converge.*

(ii) *If the left side of (15.1) converges to F, then the right side of (15.1) converges to F, except possibly if h is a limit point of $\{-B_{2k}/B_{2k-1}\}$, where, in the notation (1.3) and (1.4), A_n/B_n denotes the nth approximant for the continued fraction on the left side, and conversely.*

PROOF. In the notation (1.3) and (1.4), let the left side of (15.1) have approximants A_n/B_n and the right side have approximants C_n/D_n. Then $A_0 = B_{-1} = 0$, $A_{-1} = B_0 = 1$, $D_{-1} = 0$, $C_{-1} = D_0 = 1$, and $C_0 = h$. Straightforward calculations show that

$$C_1 = A_1 = a_1 + h, \qquad D_1 = B_1 = 1,$$
$$A_2 = x(a_1 + h), \qquad B_2 = x + a_1,$$
$$C_2 = A_2 + A_1 h, \qquad D_2 = B_2 + B_1 h.$$

We shall now show, by induction, that for $k \geq 1$,

$$C_{2k-1} = A_{2k-1}, \qquad D_{2k-1} = B_{2k-1},$$
$$C_{2k} = A_{2k} + A_{2k-1} h, \qquad D_{2k} = B_{2k} + B_{2k-1} h. \tag{15.4}$$

For $k = 1$, each of the last four equalities has been demonstrated. Proceeding by induction, we find that

$$C_{2k-1} = C_{2k-2} + a_k C_{2k-3} = A_{2k-2} + A_{2k-3} h + a_k A_{2k-3}$$
$$= A_{2k-2} + (a_k + h) A_{2k-3} = A_{2k-1}$$

and

$$C_{2k} = xC_{2k-1} + (a_k + h)C_{2k-2} = xA_{2k-1} + (a_k + h)(A_{2k-2} + A_{2k-3} h)$$
$$= xA_{2k-1} + (a_k + h)A_{2k-2} + h(A_{2k-1} - A_{2k-2})$$
$$= (x + h)A_{2k-1} + a_k A_{2k-2} = A_{2k} + hA_{2k-1}.$$

The remaining two equalities in (15.4) may be established in a similar manner. Both conclusions of the second version of Entry 15 now follow from (15.4).

It also follows that the left side of (15.1) converges *generally* to F if and only if the right side converge *generally* to F. For the concept of general convergence, see the paper by L. Jacobsen [2].   □

Entry 16. *If neither m nor n is a negative integer, then*

$$\sum_{k=1}^{\infty} \frac{(-1)^{k+1}}{(m + k)(n + k)}$$

$$= \frac{1}{(m + 1)(n + 1) +} \frac{(m + 1)^2(n + 1)^2}{m + n + 3} + \frac{(m + 2)^2(n + 2)^2}{m + n + 5}$$

$$+ \frac{(m + 3)^2(n + 3)^2}{m + n + 7} + \cdots .$$

PROOF. We shall employ the corollary presented in Section 1. Letting $a_k = (-1)^{k+1}/(m+k)(n+k)$ and letting r tend to ∞, we find that

$$\sum_{k=1}^{\infty} \frac{(-1)^{k+1}}{(m+k)(n+k)}$$

$$= \frac{(m+1)^{-1}(n+1)^{-1}}{1} \quad \frac{(m+2)^{-1}(n+2)^{-1}}{+ (m+1)^{-1}(n+1)^{-1} - (m+2)^{-1}(n+2)^{-1}}$$

$$\frac{(m+1)^{-1}(n+1)^{-1}(m+3)^{-1}(n+3)^{-1}}{+ (m+2)^{-1}(n+2)^{-1} - (m+3)^{-1}(n+3)^{-1}} - \cdots$$

$$= \frac{1}{(m+1)(n+1)} \quad \frac{(m+1)^2(n+1)^2}{+ (m+2)(n+2) - (m+1)(n+1)}$$

$$\frac{(m+2)^2(n+2)^2}{+ (m+3)(n+3) - (m+2)(n+2) + \cdots},$$

from which the proposed identity readily follows formally.

The continued fraction converges for all m, n such that neither m nor n is a negative integer (Jacobsen [3, Theorem 2.3]). Since the series also converges in this domain, the identity is proved. ☐

The equality in Entry 17 refers only to the correspondence of the two sides; neither side need converge.

Entry 17. *Write*

$$\frac{1}{1+} \quad \frac{a_1 x}{1} + \frac{a_2 x}{1} + \frac{a_3 x}{1} + \cdots = \sum_{k=0}^{\infty} A_k(-x)^k, \tag{17.1}$$

where $A_0 = 1$. Let

$$P_n = a_1 a_2 \cdots a_{n-1}(a_1 + a_2 + \cdots + a_n), \qquad n \geq 1.$$

Then

$$P_1 = A_1,$$

$$P_2 = A_2,$$

$$P_3 = A_3 - a_1 A_2,$$

$$P_4 = A_4 - (a_1 + a_2)A_3,$$

$$P_5 = A_5 - (a_1 + a_2 + a_3)A_4 + a_1 a_3 A_3,$$

$$P_6 = A_6 - (a_1 + a_2 + a_3 + a_4)A_5 + (a_1 a_3 + a_2 a_4 + a_1 a_4)A_4.$$

In general, for $n \geq 1$,

$$P_n = \sum_{0 \leq k < n/2} (-1)^k \varphi_k(n) A_{n-k}, \tag{17.2}$$

where $\varphi_0(n) \equiv 1$ and $\varphi_r(n)$, $r \geq 1$, is defined recursively by

$$\varphi_r(n + 1) - \varphi_r(n) = a_{n-1}\varphi_{r-1}(n - 1). \tag{17.3}$$

FIRST PROOF. Let $C_n = C_n(x)$ and $B_n = B_n(x)$ denote the numerator and denominator, respectively, of the nth convergent of the continued fraction (17.1). Then, from (1.3) and (1.4),

$$\left.\begin{array}{ll} C_1 = C_2 = 1, & C_n = C_{n-1} + a_{n-1}xC_{n-2}, \\ B_1 = 1, B_2 = 1 + a_1x, B_n = B_{n-1} + a_{n-1}xB_{n-2}, & n \geq 3. \end{array}\right\} \tag{17.4}$$

By induction, it is easily seen that C_{2n-1}, C_{2n}, and B_{2n-1} are polynomials in x of degree $n - 1$, while B_{2n} is of degree n, where $n \geq 1$. Thus, for $n \geq 1$, set

$$B_n(x) = \sum_{k=0}^{[n/2]} \beta_k(n + 1)x^k. \tag{17.5}$$

We make the convention that $\beta_k(n + 1) = 0$ if $k > [n/2]$. From (17.4), it is obvious that $\beta_0(n + 1) = 1$ for each $n \geq 1$. Using (17.5) in the recursion formula for B_n given in (17.4) and equating coefficients of x^r, we readily deduce that

$$\beta_r(n + 1) - \beta_r(n) = a_{n-1}\beta_{r-1}(n - 1), \qquad r \geq 1. \tag{17.6}$$

Thus, by (17.3) and (17.6), we see that $\varphi_r(n)$ and $\beta_r(n)$ satisfy the same recursion formula. Since furthermore $\varphi_0(n) \equiv 1 \equiv \beta_0(n)$, we conclude that $\varphi_r(n) = \beta_r(n)$, $r \geq 0$. Also note that $\varphi_r(n) \equiv 0$ if $r \geq [n/2]$.

Put

$$E_n = \cfrac{a_n}{1} + \cfrac{a_{n+1}x}{1} + \cfrac{a_{n+2}x}{1} + \cdots, \tag{17.7}$$

where $n \geq 0$ and $a_0 = 1$. We shall show, by induction, that (see also Rogers' paper [2, p. 72, Eq. (1)])

$$E_0B_n - C_n = (-1)^n E_0E_1 \cdots E_n x^n, \qquad n \geq 1. \tag{17.8}$$

Since, by (17.7), $E_0 = 1/(1 + xE_1)$, (17.8) is easy to establish for $n = 1$. Assume now that (17.8) is valid for each nonnegative integer up to and including n. Then, by (17.4),

$$\begin{aligned} E_0B_{n+1} - C_{n+1} &= E_0B_n - C_n + a_n x(E_0B_{n-1} - C_{n-1}) \\ &= (-1)^n E_0E_1 \cdots E_n x^n + (-1)^{n-1} a_n x E_0E_1 \cdots E_{n-1}x^{n-1} \\ &= (-1)^n E_0E_1 \cdots E_{n-1}x^n(E_n - a_n) \\ &= (-1)^n E_0E_1 \cdots E_{n-1}x^n\left(\frac{a_n}{1 + xE_{n+1}} - a_n\right) \\ &= (-1)^{n+1} E_0E_1 \cdots E_n E_{n+1}x^{n+1}, \end{aligned}$$

and so the induction is complete.

Write

$$E_0 E_1 \cdots E_n = \sum_{k=0}^{\infty} e_k(n) x^k. \tag{17.9}$$

Setting $x = 0$, we find that

$$e_0(n) = a_1 a_2 \cdots a_n. \tag{17.10}$$

Next rewrite (17.9) in the form

$$\frac{1}{1 + xE_1} \frac{a_1}{1 + xE_2} \cdots \frac{a_n}{1 + xE_{n+1}} - a_1 a_2 \cdots a_n = \sum_{k=1}^{\infty} e_k(n) x^k.$$

Dividing both sides by x and then letting x tend to 0, we deduce that

$$e_1(n) = -a_1 a_2 \cdots a_n (a_1 + a_2 + \cdots + a_{n+1}) = -P_{n+1}, \tag{17.11}$$

for each nonnegative integer n.

In (17.8) replace n by $n - 1$. Then, by (17.1), (17.5), (17.9), (17.10), and (17.11),

$$\sum_{j=0}^{\infty} A_j (-x)^j \sum_{k=0}^{[(n-1)/2]} \varphi_k(n) x^k - C_{n-1}(x)$$

$$= E_0 B_{n-1}(x) - C_{n-1}(x)$$

$$= (-1)^{n-1} a_1 a_2 \cdots a_{n-1} x^{n-1} + (-1)^n P_n x^n + \cdots.$$

Equating coefficients of x^n, $n \geq 2$, yields

$$(-1)^n P_n = \sum_{k=0}^{[(n-1)/2]} (-1)^{n-k} A_{n-k} \varphi_k(n),$$

which is precisely (17.2). Since the case $n = 1$ of (17.2) is readily verified, the proof is complete. $\square$

Essentially the same proof that we have given above was independently and almost simultaneously discovered by Goulden and Jackson [2]. They [2] have also found a beautiful combinatorial proof of (17.3) by enumerating certain paths. Using a result of E. Frank [1], P. Achuthan and S. Ponnuswamy [1] have given a very short proof of Entry 17(i).

Before proceeding further, we shall find an exact formula for $\varphi_k(n)$, defined by (17.3).

First, it is not difficult to show that

$$\varphi_1(n) = \sum_{j=1}^{n-2} a_j$$

and

$$\varphi_2(n) = \sum_{\substack{1 \leq i \leq j-2 \\ 3 \leq j \leq n-2}} a_i a_j.$$

We shall show by induction on k that

$$\varphi_k(n) = \sum_{\substack{1 \le j_1 \le j_2 - 2 \\ 1 \le j_2 \le j_3 - 2 \\ \vdots \\ 1 \le j_k \le n-2}} a_{j_1} a_{j_2} \cdots a_{j_k}. \tag{17.12}$$

We have already indicated that (17.12) is true for $k = 1, 2$. Proceeding by induction and employing (17.3), we find that

$$\varphi_k(n) - \varphi_k(n-1) = a_{n-2} \sum_{\substack{1 \le j_1 \le j_2 - 2 \\ \vdots \\ 1 \le j_{r-i} \le n-4}} a_{j_1} a_{j_2} \cdots a_{j_{r-1}},$$

$$\varphi_k(n-1) - \varphi_k(n-2) = a_{n-3} \sum_{\substack{1 \le j_1 \le j_2 - 2 \\ \vdots \\ 1 \le j_{r-i} \le n-5}} a_{j_1} a_{j_2} \cdots a_{j_{r-1}},$$

$$\vdots$$

Adding together all the equalities above, we deduce (17.12). This completes the proof of the desired exact formula for $\varphi_k(n)$.

Rogers [2] has expressed $\varphi_k(n)$ by a determinant.

We are extremely grateful to G. E. Andrews for providing us with the following elegant, second proof of Entry 17. In fact, this proof was found prior to the proofs given and mentioned above. The first part of Andrews' argument was anticipated by De Morgan [1].

SECOND PROOF OF ENTRY 17. We first obtain a recursion formula for the coefficients A_k, $k \ge 0$. In order to do this, we introduce auxiliary coefficients $\bar{A}_k$, $k \ge 0$, which we now define. Of course, each coefficient A_k can be written in terms of $a_1, a_2, \ldots$. We define $\bar{A}_k$ by the same expression for A_k except that the subscript of each a_j appearing in A_k is increased by 1. For example, since $A_2 = a_1^2 + a_1 a_2$, we define $\bar{A}_2 = a_2^2 + a_2 a_3$.

Now, by (17.1),

$$\sum_{k=0}^{\infty} (-1)^k A_k x^k = \cfrac{1}{1 + a_1 x \left(\cfrac{1}{1} + \cfrac{a_2 x}{1} + \cfrac{a_3 x}{1} + \cdots \right)}$$

$$= \cfrac{1}{1 + a_1 x \sum_{k=0}^{\infty} (-1)^k \bar{A}_k x^k}$$

$$= \cfrac{1}{\sum_{k=0}^{\infty} (-1)^{k-1} a_1 \bar{A}_{k-1} x^k},$$

where $\bar{A}_{-1} = -1/a_1$. Multiply both sides of the extremal equation above by the denominator on the right side and equate coefficients of x^n on both sides to deduce that, for $n \ge 1$,

$$\sum_{k=-1}^{n-1} a_1 \bar{A}_k A_{n-k-1} = 0,$$

or

$$A_n = \sum_{k=0}^{n-1} a_1 \bar{A}_k A_{n-k-1}, \qquad n \geq 1, \tag{17.13}$$

which is the recurrence formula that we sought.

We now show that

> A_n is a homogeneous polynomial of degree n in the noncommutative variables $a_1, a_2, \ldots, a_n$, where the subscripts $j_1, j_2, \ldots, j_n$ of the monomials comprising A_n are precisely those sequences of positive integers starting at 1 for which $j_{k+1} - j_k \leq 1, j_k \geq 1$. $\qquad$ (17.14)

In order to make clearer the assertion above, we record the following examples:

$$A_1 = a_1,$$

$$A_2 = a_1 a_2 + a_1 a_1,$$

$$A_3 = a_1 a_2 a_2 + a_1 a_2 a_3 + a_1 a_2 a_1 + a_1 a_1 a_2 + a_1 a_1 a_1.$$

We now prove the assertion (17.14) by inducting on n. By using (17.13), we easily verify that (17.14) is true for $n = 1, 2, 3$, as indicated above. Assume that (17.14) is true up to but not including a specific integer n. Let A_n^* denote the polynomial described by (17.14). We shall show that A_n^* is equal to the right side of (17.13). Thus, $A_n^* = A_n$, which completes the induction. Let us divide the monomials comprising A_n^* into n classes. The kth class, $0 \leq k \leq n - 1$, consists of all monomials in A_n^* wherein the second appearance of a_1 is the $(k + 2)$nd term in the monomial. (Recall that a_1 begins each monomial.) Thus, the entries of the kth class are produced in the following manner. Start with a_1, adjoin a string of k a_j's, $j \geq 2$, that starts with a_2 and follows the appropriate subscript rules, and lastly adjoin a string of $n - k - 1$ a_j's that starts with a_1 and follows the prescribed subscript rules. But the entries for the string of k terms are generated by $\bar{A}_k$ and the entries for the remaining $n - k - 1$ terms are generated by A_{n-k-1}, by induction. Hence, the monomials in the kth class are generated by $a_1 \bar{A}_k A_{n-k-1}$. Summing on k, $0 \leq k \leq n - 1$, we find that

$$A_n^* = \sum_{k=0}^{n-1} a_1 \bar{A}_k A_{n-k-1},$$

which, by (17.13), completes the induction.

We now have a combinatorial interpretation (17.14) for A_n. After finding combinatorial interpretations for P_n and $\varphi_k(n)$, we shall use a sieving process to establish (17.2).

Let us say that a word of the type generated by A_n, that is, $a_{j_1} a_{j_2} \cdots a_{j_n}$, where $j_1 = 1$ and $j_{k+1} - j_k \leq 1$, with $j_k \geq 1$, has an "internal drop" if $j_{k+1} - j_k \neq 1$ for some k, $1 \leq k < n - 1$. Then we see that P_n is the polynomial in $a_1, a_2, \ldots, a_n$ composed of all words without internal drops.

From (17.12), observe that $\varphi_k(n)$ is a homogeneous polynomial of degree k in the noncommuting variables $a_1, a_2, \ldots, a_{n-2}$ wherein the subscripts of each monomial $a_{j_1} a_{j_2} \cdots a_{j_k}$ satisfy the inequalities $j_{i+1} - j_i \geq 2$, $1 \leq i \leq k - 1$.

We now begin the sieving procedure. We first examine A_n. Recall that an internal drop occurs when $j_{i+1} - j_i \neq 1$ and $1 \leq i < n - 1$. Let us call a_{j_k} the "top of the last internal drop" if k is maximal for internal drops; that is, if $j_{i+1} - j_i \neq 1$, $1 \leq i < n - 1$, then $j_{k+1} - j_k \neq 1$, $1 \leq k < n - 1$, and $i \leq k$. The top of the last internal drop must be one of the letters $a_1, a_2, \ldots, a_{n-2}$, since neither a_{n-1} nor a_n can be far enough to the left in a word to be at the top of an internal drop.

In order to eliminate all words from A_n with internal drops, we take the words from A_{n-1} and insert a_j, $1 \leq j \leq n - 2$, in the last position where it forms the top of an internal drop. Thus,

$$A_n - (a_1 + a_2 + \cdots + a_{n-2})A_{n-1} \tag{17.15}$$

does not possess any internal drops. (Note that we have written (17.15) commutatively; the correct noncommutative expression would have a_j, $1 \leq j \leq n - 2$, inserted as described above.) Unfortunately, there are words in (17.15) that were not originally in A_n. These words arose when the insertion of an a_j produced a subscript increase greater than or equal to 2 from the a_i immediately to the left of the inserted a_j. Of course, we must eliminate these undesirable words. We do this by taking the words of A_{n-2} and inserting pairs $a_i a_j$ with $j - i \geq 2$ so that a_j is at the top of the last internal drop. Hence,

$$A_n - \varphi_1(n)A_{n-1} + \varphi_2(n)A_{n-2} \tag{17.16}$$

does not possess internal drops. (Again note that (17.16) is a commutative representation of what is really a noncommutative polynomial in $a_1, a_2, \ldots, a_n$.) Unfortunately, we have now introduced some new words that were not originally under consideration. These new words have triples $a_i a_j a_k$ with $j - i \geq 2$ and $k - j \geq 2$ and with a_k at the top of the last internal drop.

We continue the process described above by induction. At each stage we must introduce a term

$$(-1)^k \varphi_k(n) A_{n-k}$$

to compensate for unwanted terms introduced at the previous stage. Fortunately, $\varphi_k(n) = 0$ for $k \geq n/2$, which is evident from (17.12). Thus, the sieving process terminates, and we reach the desired formula (17.2). $\square$

Among others, Muir [1] and Rogers [2] have studied the problem of deriving a continued fraction expansion from the coefficients of a power series. Both De Morgan [1] and Rogers [2] have commented on the fact that it is extremely more difficult to determine the power series coefficients A_k, $0 \leq k < \infty$, from a continued fraction of the form (17.1). Ramanujan's Entry 17 is a fascinating contribution to this more recondite converse problem.

By a theorem of Euler [1] (Jones and Thron [1, p. 37]) (see also (1.2)),

$$\frac{1}{1+} \frac{a_1 x}{1+} \frac{a_2 x}{1+} \cdots = \sum_{k=0}^{\infty} \frac{(-1)^k a_1 a_2 \cdots a_k}{B_k B_{k+1}} x^k,$$

where $B_k = B_k(x)$ is given by (1.4) and (17.5). Thus, $(-1)^n A_n$ is equal to the coefficient of x^n in

$$\sum_{k=0}^{n} \frac{(-1)^k a_1 \cdots a_k}{B_k(x) B_{k+1}(x)} x^k.$$

Obtaining a general formula for A_n in this manner seems hopeless.

However, a very complicated formula for A_n can be established combinatorially by counting planted plane trees with respect to their heights in two different ways. For a nice exposition of this proof, see the book of Goulden and Jackson [1]. See also a paper of Flajolet [1].

Corollary (i). *Write*

$$\frac{1}{1 + b_1 x +} \frac{a_1 x}{1 + b_2 x +} \frac{a_2 x}{1 + b_3 x +} \cdots = \sum_{k=0}^{\infty} A_k (-x)^k, \qquad (17.17)$$

where $A_0 = 1$. Define

$$P_n = a_1 a_2 \cdots a_{n-1}(a_1 + b_1 + a_2 + b_2 + \cdots + a_n + b_n), \qquad n \geq 1.$$

Then, for $n \geq 1$,

$$P_n = \sum_{k=0}^{n-1} (-1)^k \varphi_k(n) A_{n-k},$$

where $\varphi_0(n) \equiv 1$ and $\varphi_r(n)$, $r \geq 1$, is defined recursively by

$$\varphi_r(n + 1) - \varphi_r(n) = b_n \varphi_{r-1}(n) + a_{n-1} \varphi_{r-1}(n - 1).$$

As with Entry 17, Goulden and Jackson [2] independently and simultaneously discovered the proof that we found and which is recorded below. Goulden and Jackson [2] have also derived a combinatorial proof. Yet another proof of Corollary (i) has been found by Achuthan and Ponnuswamy [1]. McCabe [1] has established an identification of continued fractions of the type (17.17) with power series of the form $\sum_{k=0}^{\infty} B_k / x^k$. Since the proof below is very similar to the first proof of Entry 17, we give only a brief sketch.

PROOF. Let $C_n = C_n(x)$ and $B_n = B_n(x)$ denote the numerator and denominator, respectively, of the nth convergent of the continued fraction (17.17). Then

$$\left. \begin{array}{l} C_1 = 1, \qquad C_2 = 1 + b_2 x, \qquad C_n = (1 + b_n x)C_{n-1} + a_{n-1} x C_{n-2}, \\[2mm] B_1 = 1 + b_1 x, \qquad B_2 = 1 + (a_1 + b_1 + b_2)x + b_1 b_2 x^2, \\[2mm] B_n = (1 + b_n x)B_{n-1} + a_{n-1} x B_{n-2}, \end{array} \right\} \qquad (17.18)$$

where $n \geq 3$. Observe that $C_n(x)$ has degree $n - 1$ and $B_n(x)$ has degree n, $n \geq 1$. Thus, put

$$B_n(x) = \sum_{k=0}^{n} \beta_k(n + 1)x^k, \qquad n \geq 1. \qquad (17.19)$$

By substituting (17.19) into the recursion formula for $B_n(x)$ in (17.18) and equating coefficients of x^r, we deduce that $\beta_r(n) = \varphi_r(n)$, $r \geq 0$, $n \geq 2$.

The remainder of the proof is exactly parallel to that of the first proof of Entry 17. ☐

Corollary (ii). *Let $B_n(x)$ be defined as at the beginning of the proof of Entry 17. Then, for $n \geq 1$,*

$$B_n(x) = \sum_{k=0}^{[n/2]} \varphi_k(n+1)x^k.$$

PROOF. Corollary (ii) was established in the course of proving Entry 17. In particular, recall that $\beta_k(n) = \varphi_k(n)$ and consult (17.5). ☐

Example. *We have*

$$_2F_1^2(\tfrac{1}{2}, \tfrac{1}{2}; 1; x) = \frac{1}{1} - \frac{x}{2} - \frac{3x}{8} - \frac{5x}{2} - \frac{17x}{40} - \frac{23x}{2} - \frac{1395x}{3128} - \cdots. \qquad (17.22)$$

PROOF. Ramanujan evidently intends this example to be an illustration for Entry 17. In the notation of Entry 17,

$$a_1 = -\frac{1}{2}, \quad a_2 = -\frac{3}{16}, \quad a_3 = -\frac{5}{16}, \quad a_4 = -\frac{17}{80}, \quad a_5 = -\frac{23}{80}, \quad \text{and}$$

$$a_6 = -\frac{1395}{6256}.$$

Squaring $_2F_1(\tfrac{1}{2}, \tfrac{1}{2}; 1; x)$, we find, after some laborious computing, that

$$A_1 = -\frac{1}{2}, \quad A_2 = \frac{11}{32}, \quad A_3 = -\frac{17}{64}, \quad A_4 = \frac{1787}{2^{13}}, \quad A_5 = -\frac{3047}{2^{14}}, \quad \text{and}$$

$$A_6 = \frac{42631}{2^{18}}.$$

Lastly,

$$P_1 = -\frac{1}{2}, \quad P_2 = \frac{11}{32}, \quad P_3 = -\frac{3}{32}, \quad P_4 = \frac{291}{2^{13}}, \quad P_5 = -\frac{153}{2^{14}}, \quad \text{and}$$

$$P_6 = \frac{32337}{5 \cdot 2^{21}}.$$

All these calculations are in agreement with Entry 17, and so (17.22) is indeed correct. ☐

Entry 18. *Suppose that x and n are complex numbers such that either $x \notin [-1, 1]$, or $x = \pm 1$ and $\operatorname{Re} n \neq 0$, or n is an integer. Then*

$$\frac{(x+1)^n - (x-1)^n}{(x+1)^n + (x-1)^n} = \frac{n}{x} + \frac{n^2-1^2}{3x} + \frac{n^2-2^2}{5x} + \frac{n^2-3^2}{7x} + \cdots. \qquad (18.1)$$

FIRST PROOF. If we replace x by $1/x$ in Perron's book [3, p. 153, Eq. (9)], we obtain a continued fraction representation easily found to be equivalent to (18.1). By Perron's proof, (18.1) is valid for all complex numbers x outside $[-1, 1]$.

Now suppose that $x^2 = 1$ and $n \neq 0$. If Re $n > 0$, the left side of (18.1) is continuous for $|x| \geq 1$ and equals ± 1 at $x = \pm 1$, respectively. For Re $n > 0$, the continued fraction on the right side of (18.1) converges locally uniformly with respect to x for $x \geq 1$ and $x \leq -1$. Thus, by the uniform parabola theorem (Jacobsen [3]), (18.1) is valid for $x = \pm 1$ and Re $n > 0$. Since both sides of (18.1) are odd functions of n for $x = \pm 1$, (18.1) is valid for $x = \pm 1$ and Re $n < 0$ as well.

Lastly, suppose that n is an integer, and so both sides of (18.1) are rational functions of x. We already know that (18.1) holds for all $x \notin [-1, 1]$. Thus, by analytic continuation, (18.1) holds for all complex x.   □

Perron's derivation of Entry 18 arises from Entry 20.

SECOND PROOF. Let

$$g(m, n, x) = \frac{\Gamma(\frac{1}{2}(mx + m - n + 1))\Gamma(\frac{1}{2}(mx - m + n + 1))}{\Gamma(\frac{1}{2}(mx + m + n + 1))\Gamma(\frac{1}{2}(mx - m - n + 1))}.$$

Replacing x by mx in Entry 33, with $m > 0$, we find that, for Re $x > 0$,

$$\frac{1 - g(m, n, x)}{1 + g(m, n, x)} = \frac{mn}{mx +} \frac{(m^2 - 1^2)(n^2 - 1^2)}{3mx} + \frac{(m^2 - 2^2)(n^2 - 2^2)}{5mx} + \cdots$$

$$= \frac{n}{x +} \frac{(1 - 1/m^2)(n^2 - 1^2)}{3x} + \frac{(1 - 2^2/m^2)(n^2 - 2^2)}{5x} + \cdots .$$

$$(18.2)$$

Now let m tend to ∞ in (18.2). By using an asymptotic formula for the quotient of Γ-functions, Lemma 2, Section 24 of Chapter 11, or Stirling's formula, we find that

$$\lim_{m \to \infty} g(m, n, x) = (x + 1)^{-n}(x - 1)^n.$$

For x exterior to $(-\infty, 0]$, by the uniform parabola theorem (Jacobsen [3, Theorem 2.3]), the continued fraction on the right side of (18.2) converges uniformly with respect to m in a neighborhood of $m = \infty$. But by Perron's work [3, p. 153], or by the parabola theorem, the continued fraction in (18.1) converges for $x \notin [-1, 1]$. Thus, (18.1) holds for Re $x > 0$ and $x \notin (0, 1]$. By analytic continuation and our argument in the first proof, the domain of validity can be extended to that indicated.   □

Entry 18 is due to Euler [7] and easily implies a continued fraction expansion for $(x + 1)^n/(x - 1)^n$ due to Laguerre [1] (Perron [3, p. 153, Eq. (10)]).

If V_n denotes the left side of (18.1), then Ramanujan remarks that $V_n + 1/V_n = 2/V_{2n}$, a fact that is easily verified.

Corollary 1. *Let x be any complex number outside the cuts $(-i\infty, -i]$ and $[i, i\infty)$. Then*

$$\tan^{-1} x = \frac{x}{1} + \frac{x^2}{3} + \frac{(2x)^2}{5} + \frac{(3x)^2}{7} + \cdots.$$

For a proof, see Perron's text [3, p. 155]. Early proofs of Corollary 1 were given by Lambert [2], J. L. Lagrange [1], and Euler [7].

Corollary 2. *Let x be any complex number outside the cuts $(-\infty, -1]$ and $[1, \infty)$. Then*

$$\text{Log} \frac{1+x}{1-x} = \frac{2x}{1} - \frac{x^2}{3} - \frac{(2x)^2}{5} - \frac{(3x)^2}{7} - \cdots.$$

For a proof, see Perron's treatise [3, p. 154]. Corollary 2 is due to Euler [7]. For an application of Corollary 2 to product-weighted lead codes, see a paper of Jackson [1].

Corollary 3. *For any complex number x,*

$$\tan x = \frac{x}{1} - \frac{x^2}{3} - \frac{x^2}{5} - \frac{x^2}{7} - \cdots.$$

Corollary 3 was initially discovered by Lambert [1], [3]. A proof may be found in Perron's book [3, p. 157].

Corollary 4. *For any complex number x,*

$$\frac{e^x - 1}{e^x + 1} = \frac{x}{2} + \frac{x^2}{6} + \frac{x^2}{10} + \frac{x^2}{14} + \cdots.$$

Corollary 4 is due to Euler [5] and a proof may be found in Perron's text [3, p. 157].

Entry 19. *If n and x are arbitrary complex numbers, then*

$$\frac{x \, {}_0F_1(n+1; x)}{n \, {}_0F_1(n; x)} = \frac{\sqrt{x} \, J_n(2i\sqrt{x})}{iJ_{n-1}(2i\sqrt{x})} = \frac{x}{n} + \frac{x}{n+1} + \frac{x}{n+2} + \cdots,$$

where J_v denotes the ordinary Bessel function of order v.

FIRST PROOF. By a theorem of Euler [5] (Perron [3, p. 281, Satz 6.3]),

$$c + \frac{a}{c+d} + \frac{a}{c+2d} + \frac{a}{c+3d} + \cdots = c \frac{{}_0F_1(c/d; a/d^2)}{{}_0F_1(c/d+1; a/d^2)}, \qquad (19.1)$$

where $d \neq 0$. Let $c = n$, $a = x$, and $d = 1$ to find that

$$n + \frac{x}{n+1} + \frac{x}{n+2} + \frac{x}{n+3} + \cdots = n \frac{{}_0F_1(n; x)}{{}_0F_1(n+1; x)}.$$

Taking the reciprocal of both sides above and then multiplying both sides by x, we deduce the desired result.   □

SECOND PROOF. This proof is similar to the proof above, but employs a "finite" version of (19.1), namely, Entry 24. Simply let r tend to ∞ in Entry 24. After multiplying both sides by x/n, we complete the proof, since both sides converge for all x and n.   □

In fact, Entry 19 is classical; see, for example, the books of Wall [1, p. 349] or Jones and Thron [1, p. 168].

Entry 20. *If x is any complex number outside the interval $(-\infty, -1]$, or if α, $\beta, \gamma - \alpha$ or $\gamma - \beta \in \{0, -1, -2, \ldots\}$, then*

$$\frac{\alpha\beta x}{\gamma} \frac{{}_2F_1(\gamma - \alpha, \beta + 1; \gamma + 1; -x)}{{}_2F_1(\gamma - \alpha, \beta; \gamma; -x)}$$

$$= \frac{\alpha\beta x}{\gamma} + \frac{(\alpha - \gamma)(\beta - \gamma)x}{\gamma + 1} + \frac{(\alpha + 1)(\beta + 1)x}{\gamma + 2} + \frac{(\alpha - \gamma - 1)(\beta - \gamma - 1)x}{\gamma + 3}$$

$$+ \frac{(\alpha + 2)(\beta + 2)x}{\gamma + 4} + \cdots. \tag{20.1}$$

This result is very famous and is known as Gauss's continued fraction [1]. A proof may be found in the texts of Jones and Thron [1], Khovanskii [1], Perron [3], and Wall [1]. The cases when the continued fraction terminates are discussed by Perron [3, p. 151]. In the case $\gamma - \beta = \alpha + \frac{1}{2}$, Entry 20 may be extended to include $x = -1$.

It might be mentioned that Gauss's continued fraction may be found in Carr's book [1, p. 97], which was the most influential book in Ramanujan's development. Recent work on Gauss's continued fraction may be found in papers by Belevitch [1] and Ramanathan [1].

Entry 21. *We have*

$$\frac{\beta x}{\gamma} {}_2F_1(\beta + 1, 1; \gamma + 1; -x)$$

$$= \frac{\beta x}{\gamma} + \frac{\gamma(\beta + 1)x}{\gamma + 1} + \frac{1(\gamma - \beta)x}{\gamma + 2} + \frac{(\gamma + 1)(\beta + 2)x}{\gamma + 3} + \frac{2(\gamma + 1 - \beta)x}{\gamma + 4} + \cdots, \tag{21.1}$$

if either $x \notin (-\infty, -1]$, or β, γ or $\gamma - \beta \in \{0, -1, -2, \ldots\}$,

$$= \frac{\beta x}{\gamma} + \frac{(\beta + 1)x}{1} + \frac{1(1 + x)}{\gamma} + \frac{(\beta + 2)x}{1} + \frac{2(1 + x)}{\gamma} + \cdots, \tag{21.2}$$

if either Re $x > -\frac{1}{2}$ and not both $\beta + 1$ and $\gamma - \beta$ lie in $\{0, -1, -2, \ldots\}$, or $\beta \in \{0, -1, -2, \ldots\}$ and $\gamma - \beta \notin \{0, -1, -2, \ldots\}$,

$$= \frac{\beta x}{\gamma + x(\beta + 1) -} \; \frac{1(\beta + 1)x(x + 1)}{\gamma + 1 + x(\beta + 3) -} \; \frac{2(\beta + 2)x(x + 1)}{\gamma + 2 + x(\beta + 5) -} \cdots, \tag{21.3}$$

if either $\mathrm{Re}\, x > -\frac{1}{2}$ *and not both* $\beta + 1$ *and* $\gamma - \beta$ *lie in* $\{0, -1, -2, \ldots\}$, *or* $\beta \in \{0, -1, -2, \ldots\}$ *and* $\gamma - \beta \notin \{0, -1, -2, \ldots\}$.

PROOF. The expansion (21.1) follows from Entry 20 as follows. First, divide both sides of (20.1) by $\alpha\beta$. Set $\beta = 0$. Then replace α by $\gamma - \beta - 1$ and multiply both sides by β. The conditions on the parameters in Entry 20 are translated into the new conditions given for (21.1).

An indication of Ramanujan's proof is found in the first notebook (p. 217). Let

$$G = \frac{(\beta + 1)x}{\gamma + 1} \, {}_2F_1(\beta + 2, 1; \gamma + 2; -x).$$

Then

$$\frac{\beta x}{\gamma} \, {}_2F_1(\beta + 1, 1; \gamma + 1; -x) = \frac{\beta x}{\gamma}(1 - G) = \frac{\beta x}{\gamma} \; \frac{1}{1 + \dfrac{G}{1 - G}}. \tag{21.4}$$

Now in Entry 20, replace β by $\beta + 1$ and γ by $\gamma + 1$ and then set $\alpha = \gamma$. This yields

$$\frac{G}{1 - G} = \frac{(\beta + 1)x}{\gamma + 1} \; \frac{{}_2F_1(\beta + 2, 1; \gamma + 2; -x)}{{}_2F_1(\beta + 1, 1; \gamma + 1; -x)}$$

$$= \frac{(\beta + 1)x}{\gamma + 1} + \frac{1(\gamma - \beta)x}{\gamma + 2} + \frac{(\gamma + 1)(\beta + 2)x}{\gamma + 3} + \frac{2(\gamma - \beta + 1)x}{\gamma + 4}$$

$$+ \frac{(\gamma + 2)(\beta + 3)x}{\gamma + 5} + \cdots. \tag{21.5}$$

If we substitute (21.5) into (21.4), we complete the proof of (21.1).

We next prove (21.3). If $\mathrm{Re}\, x < \frac{1}{2}$, $\alpha + 1$ and $\gamma - \alpha$ are not both nonpositive integers, and $\beta + 1$ and $\gamma - \beta$ are not both nonpositive integers, then by a result of Nörlund [1] (Perron [3, p. 286, Eq. (10)]),

$$\frac{{}_2F_1(\alpha + 1, \beta + 1; \gamma + 1; x)}{\gamma \, {}_2F_1(\alpha, \beta; \gamma; x)}$$

$$= \frac{1}{\gamma - (1 + \alpha + \beta)x +} \; \frac{(\alpha + 1)(\beta + 1)(x - x^2)}{\gamma + 1 - (3 + \alpha + \beta)x}$$

$$+ \frac{(\alpha + 2)(\beta + 2)(x - x^2)}{\gamma + 2 - (5 + \alpha + \beta)x} + \cdots. \tag{21.6}$$

Setting $\alpha = 0$, replacing x by $-x$, and lastly multiplying both sides by βx, we complete the proof of (21.3). The cases when the continued fraction terminates are established by a familiar argument, since both sides are then rational functions of x.

Applying Entry 14 and letting n tend to ∞, or applying (14.2), we see that (21.3) is the even part of (21.2). Since (21.2) converges for Re $x > -\frac{1}{2}$, the identity follows.

Under certain conditions (21.1) can be extended to $x = -1$ and (21.2) and (21.3) to $x = -\frac{1}{2}$. $\square$

Corollary 1. *For every complex number x, we have*

$$\frac{x}{n} \, {}_1F_1(1; n + 1; x)$$

$$= \frac{x}{n} \, \frac{nx}{-\,n + 1} \, + \frac{x}{n + 2} \, - \frac{(n + 1)x}{n + 3} \, + \frac{2x}{n + 4} \, - \cdots$$

$$= \frac{x}{n - x} \, + \frac{x}{n + 1 - x} \, + \frac{2x}{n + 2 - x} \, + \frac{3x}{n + 3 - x} \, + \cdots \, .$$

PROOF. To prove the first equality, replace γ by n and x by $-x/\beta$ in (21.1). Letting β tend to ∞, we easily deduce the desired result.

To prove the second equality, employ (21.3) and proceed in precisely the same manner as above. $\square$

Corollary 2. *If x is any complex number, then*

$${}_1F_1(1; x + 1; x) = 1 + \frac{2x}{2} \, + \frac{3x}{3} \, + \frac{4x}{4} \, + \frac{5x}{5} \, + \cdots \, .$$

PROOF. Let $n = x$ in the second equality of Corollary 1. $\square$

Entry 22. *Assume that α, β, and γ are complex numbers such that not both $\beta + 1$ and $\gamma - \beta$ belong to $\{0, -1, -2, \ldots\}$ and not both $-\alpha$ and $\alpha + \gamma + 1$ are in $\{0, -1, -2, \ldots\}$. Suppose that either $|x| < 1$, or $\beta \in \{0, -1, -2, \ldots\}$, or $x = 1$ and Re$(\gamma - \alpha - \beta - 1) > 0$. Then*

$$\frac{\beta x}{\gamma} \, \frac{{}_2F_1(-\alpha, \beta + 1; \gamma + 1; -x)}{{}_2F_1(-\alpha, \beta; \gamma; -x)}$$

$$= \frac{\beta x}{\gamma - (\alpha + \beta + 1)x} \, + \frac{(\beta + 1)(\alpha + \gamma + 1)x}{\gamma + 1 - (\alpha + \beta + 2)x}$$

$$+ \frac{(\beta + 2)(\alpha + \gamma + 2)x}{\gamma + 2 - (\alpha + \beta + 3)x} \, + \cdots \, .$$

PROOF. Since by Entry 19 of Chapter 10,

$${}_2F_1(a, b; c; x) = (1 - x)^{c-a-b} \, {}_2F_1\left(c - a, b; c; \frac{x}{x - 1}\right),$$

we may write (21.6) in the form

$$\frac{x}{\gamma(1-x)} \frac{{}_2F_1\left(\gamma - \alpha, \beta + 1; \gamma + 1; \dfrac{x}{x-1}\right)}{{}_2F_1\left(\gamma - \alpha, \beta; \gamma; \dfrac{x}{x-1}\right)}$$

$$= \frac{x}{\gamma - (1 + \alpha + \beta)x} + \frac{(\alpha+1)(\beta+1)(x-x^2)}{\gamma + 1 - (3 + \alpha + \beta)x}$$

$$+ \frac{(\alpha+2)(\beta+2)(x-x^2)}{\gamma + 2 - (5 + \alpha + \beta)x} + \cdots,$$

provided that $\operatorname{Re} x < \frac{1}{2}$, not both $\beta + 1$ and $\gamma - \beta$ belong to $\{0, -1, -2, \ldots\}$, and not both $\alpha + 1$ and $\gamma - \alpha$ belong to $\{0, -1, -2, \ldots\}$. Letting $u = x/(1 - x)$, we find, after simplification, that

$$\frac{u \, {}_2F_1(\gamma - \alpha, \beta + 1; \gamma + 1; -u)}{\gamma \, {}_2F_1(\gamma - \alpha, \beta; \gamma; -u)}$$

$$= \frac{u}{\gamma(u + 1) - (1 + \alpha + \beta)u} + \frac{(\alpha + 1)(\beta + 1)u}{(\gamma + 1)(u + 1) - (3 + \alpha + \beta)u}$$

$$+ \frac{(\alpha + 2)(\beta + 2)u}{(\gamma + 2)(u + 1) - (5 + \alpha + \beta)u} + \cdots,$$

provided that $|u| < 1$. Replacing α by $\alpha + \gamma$ in the foregoing equality, we readily complete the proof of Entry 22 for $|x| < 1$.

Lastly, observe that the left side of Entry 22 is analytic in a neighborhood of $x = 1$. For $x = 1$, the continued fraction converges to an analytic function of α, β, and γ provided that $\operatorname{Re}(\gamma - \alpha - \beta - 1) > 0$, by the uniform parabola theorem (Jacobsen [3, Theorem 2.3]). This then completes the proof of Entry 22 for $x = 1$. □

Perron [3, p. 306] attributes Entry 22 to Andoyer [1], and, as we have seen, Entry 22 is equivalent to Nörlund's result (21.6). However, R. Askey [2] has pointed out that Entry 22 is really due to Euler [6], [2]. A somewhat more detailed discussion of Entries 20–22, along with the associated contiguous relations, may be found in a paper by K. G. Ramanathan [1].

As with Entry 17, the equality in (23.1) below refers only to the correspondence of the two sides, for neither side needs to converge.

Entry 23. *Write, for each nonnegative integer n,*

$$\frac{a_n}{b_n x} + \frac{a_{n+1}}{b_{n+1} x} + \frac{a_{n+2}}{b_{n+2} x} + \cdots = c_n \sum_{k=0}^{\infty} A_n(k)(-x)^k, \qquad (23.1)$$

where $A_n(0) = 1$. Then $c_n c_{n+1} = a_n$,

$$A_n(1) + A_{n+1}(1) = \frac{b_n}{c_{n+1}} = \frac{b_n c_n}{a_n}, \tag{23.2}$$

$$A_n(2) + A_{n+1}(2) = A_n^2(1),$$

$$A_n(3) + A_{n+1}(3) = A_n(1)\{A_n(2) - A_{n+1}(2)\}, \tag{23.3}$$

$$A_n(4) + A_{n+1}(4) = A_n(1)\{A_n(3) - A_{n+1}(3)\} - A_n(2)A_{n+1}(2),$$

and, in general, for $k \geq 3$,

$$A_n(k) + A_{n+1}(k) = A_n(1)\{A_n(k-1) - A_{n+1}(k-1)\} - \sum_{j=2}^{k-2} A_n(j)A_{n+1}(k-j).$$
$$\tag{23.4}$$

PROOF. From (23.1),

$$\frac{a_n}{b_n x + c_{n+1} \sum_{k=0}^{\infty} A_{n+1}(k)(-x)^k} = c_n \sum_{k=0}^{\infty} A_n(k)(-x)^k,$$

or

$$a_n = c_n \left(b_n x + c_{n+1} \sum_{k=0}^{\infty} A_{n+1}(k)(-x)^k \right) \sum_{k=0}^{\infty} A_n(k)(-x)^k$$

$$= -c_n b_n \sum_{k=1}^{\infty} A_n(k-1)(-x)^k$$

$$+ c_n c_{n+1} \sum_{k=0}^{\infty} \sum_{j=0}^{k} A_n(j)A_{n+1}(k-j)(-x)^k.$$

Now equate coefficients of x^k, $k \geq 0$, on both sides. For $k = 0$, we find that $c_n c_{n+1} = a_n$, and, for $k \geq 1$, we deduce that

$$A_n(k) + A_{n+1}(k) = \frac{b_n}{c_{n+1}} A_n(k-1) - \sum_{j=1}^{k-1} A_n(j)A_{n+1}(k-j). \tag{23.5}$$

Letting $k = 1$, we immediately deduce (23.2). Using (23.2) in (23.5), we find that, for $k \geq 2$,

$$A_n(k) + A_{n+1}(k) = \{A_n(1) + A_{n+1}(1)\}A_n(k-1) - \sum_{j=1}^{k-1} A_n(j)A_{n+1}(k-j).$$

Upon simplifying the equality above, we deduce both (23.3) and (23.4). □

Example. We have

$$\lim_{x \to +\infty} \left\{ \sqrt{\frac{2x}{\pi}} - \frac{x}{1} + \frac{2x}{2} + \frac{3x}{3} + \frac{4x}{4} + \cdots \right\} = \frac{2}{3\pi}.$$

PROOF. From Entry 47, for $x > 0$,

$$1 + \frac{e^x \Gamma(x+1)}{x^x} - \int_0^{\infty} e^{-t} \left(1 + \frac{t}{x} \right)^x dt = 1 + \frac{2x}{2} + \frac{3x}{3} + \frac{4x}{4} + \cdots.$$

Taking the reciprocal of both sides, we find that

$$L(x) := \frac{x}{1 + \dfrac{e^x \Gamma(x+1)}{x^x} - \displaystyle\int_0^\infty e^{-t}\left(1 + \frac{t}{x}\right)^x dt} = \frac{x}{1} + \frac{2x}{2} + \frac{3x}{3} + \frac{4x}{4} + \cdots .$$

It therefore remains to show that

$$L(x) = \sqrt{\frac{2x}{\pi}} - \frac{2}{3\pi} + o(1), \tag{23.6}$$

as x tends to ∞.

Write

$$L(x) = \frac{x}{\frac{1}{2}e^x x^{-x}\Gamma(x+1) + \theta_x}.$$

When x is a positive integer, Ramanujan [4], [16, p. 324] derived an asymptotic expansion for θ_x as x tends to ∞. Watson [3] later established the expansion for general $x > 0$. See also the corollary to Entry 48 below and Entry 6 of Chapter 13. Using this asymptotic series (48.4) and Stirling's formula, we find that

$$L(x) = \frac{x}{\sqrt{\dfrac{\pi x}{2}} + O(x^{-1/2}) + \dfrac{1}{3} + O(1/x)}$$

$$= \sqrt{\frac{2x}{\pi}} - \frac{2}{3\pi} + O(x^{-1/2}),$$

as x tends to ∞. Thus, (23.6) is established, and the proof is completed. $\qquad\square$

Entry 24. *Let n and x be any complex numbers, and let r be any positive integer. Let*

$$f(n, r, x) = \sum_{k=0}^\infty \frac{(-r+k-1)_k x^k}{(n)_k(-n-r)_k k!}.$$

Then

$$\frac{n}{n} + \frac{x}{n+1} + \frac{x}{n+2} + \frac{x}{n+3} + \cdots + \frac{x}{n+r} = \frac{f(n+1, r-1, x)}{f(n, r, x)}. \tag{24.1}$$

PROOF. We shall induct on r. For $r = 1$,

$$\frac{n}{n} + \frac{x}{n+1} = \frac{n}{n + x/(n+1)} = \frac{1}{1 + \dfrac{x}{n(n+1)}},$$

and so (24.1) is established for $r = 1$.

Now assume that (24.1) is true when r is replaced by $r - 1$ for some fixed integer r, $r \geq 2$. Then applying the induction hypothesis with n replaced by $n + 1$, we find that

$$\frac{n}{n+} \frac{x}{n+1+} \frac{x}{n+2+} \cdots + \frac{x}{n+r} = \frac{n}{n + \dfrac{x}{n+1} \dfrac{f(n+2, r-2, x)}{f(n+1, r-1, x)}}$$

$$= \frac{nf(n+1, r-1, x)}{nf(n+1, r-1, x) + \dfrac{x}{n+1} f(n+2, r-2, x)}. \tag{24.2}$$

We are thus led to examine, for $k \geq 1$,

$$\frac{n(-r+k)_k}{(n+1)_k(-n-r)_k k!} + \frac{(-r+k)_{k-1}}{(n+1)_k(-n-r)_{k-1}(k-1)!}$$

$$= \frac{(-r+k)_{k-1}}{(n+1)_k(-n-r)_{k-1}(k-1)!} \left\{ \frac{n(-r+2k-1)}{(-n-r+k-1)k} + 1 \right\}$$

$$= \frac{(-r+k)_{k-1}}{(n+1)_k(-n-r)_{k-1}(k-1)!} \frac{(n+k)(-r+k-1)}{(-n-r+k-1)k}$$

$$= \frac{(-r+k-1)_k}{(n+1)_{k-1}(-n-r)_k k!} = \frac{n(-r+k-1)_k}{(n)_k(-n-r)_k k!}.$$

Hence,

$$nf(n+1, r-1, x) + \frac{x}{n+1} f(n+2, r-2, x) = nf(n, r, x). \tag{24.3}$$

Substituting (24.3) into (24.2), we complete the induction.   □

Entry 24 is a rather remarkable result, for it gives a continued fraction expansion for the quotient of hypergeometric polynomials,

$$\frac{{}_2F_3\left(\dfrac{1-r}{2}, -\dfrac{r}{2}; -r, n+1, -r-n; x\right)}{{}_2F_3\left(\dfrac{-1-r}{2}, -\dfrac{r}{2}; -r-1, n, -r-n; x\right)}.$$

Entry 25. *Suppose that either n is an odd integer and x is any complex number or that n is any complex number and* Re $x > 0$. *Then*

$$\frac{\Gamma(\tfrac{1}{4}(x+n+1))\Gamma(\tfrac{1}{4}(x-n+1))}{\Gamma(\tfrac{1}{4}(x+n+3))\Gamma(\tfrac{1}{4}(x-n+3))}$$

$$= \frac{4}{x-} \frac{n^2-1^2}{2x-} \frac{n^2-3^2}{2x-} \frac{n^2-5^2}{2x} - \cdots. \tag{25.1}$$

Entry 25 is originally due to Euler [6, Sec. 67]. Stieltjes [2], [4, pp. 329–394] derived Entry 25 from Entry 22. Other proofs have been found by Perron [3, p. 35] and Ramanathan [1]. The hypotheses in these proofs are stronger than those we have given. Proofs under the stated hypotheses have been derived by Jacobsen [3] and Masson [3].

FIRST PROOF. We offer another proof which is based upon Entry 39, for Re $x > 0$, or for either n or ℓ in $\{\pm 1, \pm 3, \pm 5, \ldots\}$. First, rewrite Entry 39 in the form

$$\frac{8}{P} + \frac{1}{2}(x^2 + \ell^2 - n^2 - 1)$$

$$= x^2 - 1 + \cfrac{1^2 - n^2}{1} + \cfrac{1^2 - \ell^2}{x^2 - 1} + \cfrac{3^2 - n^2}{1} + \cfrac{3^2 - \ell^2}{x^2 - 1} + \cdots,$$

or

$$\frac{1}{8/P + \frac{1}{2}(x^2 + \ell^2 - n^2 - 1)}$$

$$= \cfrac{1}{x^2 - 1} + \cfrac{1^2 - n^2}{1} + \cfrac{1^2 - \ell^2}{x^2 - 1} + \cfrac{3^2 - n^2}{1} + \cfrac{3^2 - \ell^2}{x^2 - 1} + \cdots$$

$$= \cfrac{1}{x^2 - n^2} - \cfrac{(1^2 - n^2)(1^2 - \ell^2)}{x^2 - \ell^2 - n^2 + 9} - \cfrac{(3^2 - n^2)(3^2 - \ell^2)}{x^2 - \ell^2 - n^2 + 33} - \cdots,$$

by Entry 14. Now take the reciprocal of both sides above and then solve for P, which again involves taking reciprocals. Hence,

$$P = \cfrac{8}{\frac{1}{2}(x^2 - \ell^2 - n^2 + 1)} - \cfrac{(1^2 - n^2)(1^2 - \ell^2)}{x^2 - \ell^2 - n^2 + 9} - \cfrac{(3^2 - n^2)(3^2 - \ell^2)}{x^2 - \ell^2 - n^2 + 33} - \cdots.$$

Replacing x by $x + \ell$, we find that, either for Re$(x + \ell) > 0$, or for n or ℓ belonging to $\{\pm 1, \pm 3, \pm 5, \ldots\}$,

$$\frac{\Gamma(\frac{1}{4}(x + n + 1))\Gamma(\frac{1}{4}(x - n + 1))}{\Gamma(\frac{1}{4}(x + n + 3))\Gamma(\frac{1}{4}(x - n + 3))} \ell \frac{\Gamma(\frac{1}{4}(x + 2\ell + n + 1))\Gamma(\frac{1}{4}(x + 2\ell - n + 1))}{\Gamma(\frac{1}{4}(x + 2\ell + n + 3))\Gamma(\frac{1}{4}(x + 2\ell - n + 3))}$$

$$= \cfrac{8\ell}{\frac{1}{2}(x^2 + 2x\ell - n^2 + 1)} - \cfrac{(1^2 - n^2)(1^2 - \ell^2)}{x^2 + 2x\ell - n^2 + 9}$$

$$- \cfrac{(3^2 - n^2)(3^2 - \ell^2)}{x^2 + 2x\ell - n^2 + 33} - \cdots$$

$$= \cfrac{8}{\frac{1}{2}(2x + (x^2 - n^2 + 1)/\ell)} - \cfrac{(1^2 - n^2)(1/\ell^2 - 1)}{2x + (x^2 - n^2 + 9)/\ell}$$

$$- \cfrac{(3^2 - n^2)(3^2/\ell^2 - 1)}{2x + (x^2 - n^2 + 33)/\ell} - \cdots.$$

Now let ℓ tend to ∞. By using the reflection formula for the gamma function and Stirling's formula, we deduce that

$$\lim_{\ell \to \infty} \ell \frac{\Gamma(\frac{1}{4}(x + 2\ell + n + 1))\Gamma(\frac{1}{4}(x + 2\ell - n + 1))}{\Gamma(\frac{1}{4}(x + 2\ell + n + 3))\Gamma(\frac{1}{4}(x + 2\ell - n + 3))} = 2,$$

and so Entry 25 readily follows, since the continued fraction above converges uniformly in a neighborhood of $\ell = \infty$ under the stated hypotheses.   □

We next offer another proof of Entry 25 that is due to D. Masson [3]. In [1], [2], and [3], Masson employs second-order linear recurrence relations and a theorem of Pincherle [1] to represent a general class of continued fractions by quotients of hypergeometric functions. He also determines the rate of convergence of the continued fractions and establishes connections with several types of orthogonal polynomials. However, we shall not discuss the latter topics here.

Consider the recurrence relation

$$X_{n+1} - b_n X_n - a_n X_{n-1} = 0, \tag{25.2}$$

where $a_n = -(an^2 + bn + c)$ and $b_n = z - dn$; here, $a, b, c,$ and d are constants. A solution $X_n^{(s)}$ is said to be subdominant if for any other linearly independent solution $X_n^{(d)}$ of (25.2),

$$\lim_{n \to \infty} X_n^{(s)}/X_n^{(d)} = 0.$$

Pincherle's theorem [1] then states that $K(a_n/b_n)$ converges if and only if there exist linearly independent solutions $X_n^{(s)}$ and $X_n^{(d)}$ of (25.2), as described above. Moreover,

$$K(a_n/b_n) = -X_1^{(s)}/X_0^{(s)}. \tag{25.3}$$

We now quote Masson's primary theorem [2], [3] for us.

Theorem 1. *Let* $a, d^2 - 4a \neq 0$. *Then* (25.2) *has linearly independent solutions*

$$X_{n-1}^{\pm}(a, b, c, d; z)$$

$$= \left(\pm \frac{a}{\mu} \right)^n \frac{\Gamma(n + \alpha)\Gamma(n + \beta)}{\Gamma(n + \gamma^{\pm})} {}_2F_1(n + \alpha, n + \beta; n + \gamma^{\pm}; \delta^{\pm}),$$

where

$$\mu = (d^2 - 4a)^{1/2}, \qquad -\pi/2 < \arg \mu \leq \pi/2,$$

$$\delta^{\pm} = \tfrac{1}{2}(1 \pm d/\mu),$$

$$\gamma^{\pm} = \left(\frac{a + b}{a} \right) \delta^{\pm} \pm z/\mu,$$

and α *and* β *are defined by*

$$a(n + \alpha)(n + \beta) = an^2 + bn + c.$$

Moreover, there exists a subdominant solution if and only if

$$\left| \text{Re}\left(\frac{d}{\mu}\right) \right| + \left| \text{Re}\left(\left(\frac{a+b}{2a}\right)\frac{d}{\mu} + \frac{z}{\mu}\right) \right| \neq 0,$$

and it is given by

$$X_n^{(s)} = \begin{cases} X_n^+, & \text{if } \text{Re}(d/\mu) < 0 \text{ or if } \text{Re}(d/\mu) = 0 \text{ and } \text{Re}(\gamma^+ - \gamma^-) > 0, \\ X_n^-, & \text{if } \text{Re}(d/\mu) > 0 \text{ or if } \text{Re}(d/\mu) = 0 \text{ and } \text{Re}(\gamma^+ - \gamma^-) < 0. \end{cases}$$

We shall apply Theorem 1 to prove the following theorem from which Entry 25 follows as a corollary.

Theorem 2. *If $\pm \text{Im } z > 0$ and $n \in \mathbb{C}$, then*

$$CF := \frac{1}{z-} \frac{(1^2 - n^2)/4}{z} \frac{(3^2 - n^2)/4}{z} - \cdots$$

$$= 2\left(z \pm 4i \frac{\Gamma\left(\dfrac{3+n \mp iz}{4}\right)\Gamma\left(\dfrac{3-n \mp iz}{4}\right)}{\Gamma\left(\dfrac{1+n \mp iz}{4}\right)\Gamma\left(\dfrac{1-n \mp iz}{4}\right)}\right)^{-1}. \tag{25.4}$$

PROOF. For brevity, we set $\delta = \delta^\pm$ and $\gamma = \gamma^\pm$.

Comparing the left sides of (25.3) and (25.4), we see that $a = 1$, $b = -1$, $c = (1 - n^2)/4$, and $d = 0$. In the notation of Theorem 1, we then find that $\mu = 2i$, $\delta = \frac{1}{2}$, $\gamma = \pm z/(2i)$, $\alpha = (n - 1)/2$, and $\beta = -(n + 1)/2$. If $\pm \text{Im } z > 0$, then by Pincherle's theorem and Theorem 1, there exists a subdominant solution of (25.2) such that

$$\frac{-(1^2 - n^2)/4}{z} \frac{(3^2 - n^2)/4}{z} - \cdots = -\frac{X_1^{(s)}}{X_0^{(s)}}. \tag{25.5}$$

Since $X_1 = zX_0 - \alpha\beta X_{-1}$, we may rewrite (25.5) in the form

$$CF = \frac{X_0^{(s)}}{\alpha\beta X_{-1}^{(s)}}, \tag{25.6}$$

where, by Theorem 1,

$$\frac{X_0^{(s)}}{\alpha\beta X_{-1}^{(s)}} = \pm \frac{{}_2F_1(\alpha + 1, \beta + 1; \gamma + 1; \frac{1}{2})}{2i\gamma \, {}_2F_1(\alpha, \beta; \gamma; \frac{1}{2})}. \tag{25.7}$$

It remains to evaluate this quotient of hypergeometric functions.

Recall that (Erdélyi [1, p. 104, Eq. (51)], Bailey [4, p. 11, Eq. (3)])

$$_2F_1(a, 1 - a; c; \tfrac{1}{2}) = \frac{2^{1-c}\Gamma(c)\Gamma(\frac{1}{2})}{\Gamma(\frac{1}{2}a + \frac{1}{2}c)\Gamma(\frac{1}{2}c - \frac{1}{2}a + \frac{1}{2})}. \tag{25.8}$$

It follows immediately that

$$_2F_1(\alpha + 1, \beta + 1; \gamma + 1; \tfrac{1}{2}) = \frac{2^{-\gamma}\Gamma(\gamma + 1)\Gamma(\tfrac{1}{2})}{\Gamma\left(\dfrac{3}{4} + \dfrac{n}{4} + \dfrac{\gamma}{2}\right)\Gamma\left(\dfrac{3}{4} - \dfrac{n}{4} + \dfrac{\gamma}{2}\right)}. \qquad (25.9)$$

In order to evaluate $_2F_1(\alpha, \beta; \gamma; \tfrac{1}{2})$, we must use contiguous relations to obtain functions evaluable by (25.8). Using a common abbreviated notation in Erdélyi's compendium [1, p. 103], we solve (31) there for $F(a - 1)$ and then replace F by an expression for F obtained from (32). Accordingly, after simplification, we find that

$$F(a - 1)$$

$$= \frac{1}{(b - a)(c - a)}(b\{c - 2a - (b - a)z\}F(b + 1) + a(a + b - c)F(a + 1)).$$

We apply this formula with $F(a - 1) = {}_2F_1(\alpha, \beta; \gamma; \tfrac{1}{2})$. After considerable simplification, we deduce that

$$_2F_1(\alpha, \beta; \gamma; \tfrac{1}{2})$$

$$= \tfrac{1}{2} {}_2F_1(\alpha + 1, \beta + 1; \gamma; \tfrac{1}{2}) + \frac{\gamma}{2\left(\gamma - \dfrac{1}{2} - \dfrac{n}{2}\right)} {}_2F_1(\alpha + 2, \beta; \gamma; \tfrac{1}{2})$$

$$= \frac{2^{-\gamma}\Gamma(\gamma)\Gamma(\tfrac{1}{2})}{\Gamma\left(\dfrac{1}{4} + \dfrac{n}{4} + \dfrac{\gamma}{2}\right)\Gamma\left(\dfrac{1}{4} - \dfrac{n}{4} + \dfrac{\gamma}{2}\right)} + \frac{2^{-\gamma-1}\gamma\Gamma(\gamma)\Gamma(\tfrac{1}{2})}{\Gamma\left(\dfrac{3}{4} + \dfrac{n}{4} + \dfrac{\gamma}{2}\right)\Gamma\left(\dfrac{3}{4} - \dfrac{n}{4} + \dfrac{\gamma}{2}\right)},$$

by (25.8). Putting this and (25.9) into (25.7), employing (25.7) in conjunction with (25.6), and lastly taking reciprocals of both sides, we conclude that

$$\frac{1}{CF} = \pm 2i \frac{\Gamma\left(\dfrac{3}{4} + \dfrac{n}{4} + \dfrac{\gamma}{2}\right)\Gamma\left(\dfrac{3}{4} - \dfrac{n}{4} + \dfrac{\gamma}{2}\right)}{\Gamma\left(\dfrac{1}{4} + \dfrac{n}{4} + \dfrac{\gamma}{2}\right)\Gamma\left(\dfrac{1}{4} - \dfrac{n}{4} + \dfrac{\gamma}{2}\right)} + \frac{z}{2}.$$

Taking the reciprocal once again, we deduce (25.4). $\square$

SECOND PROOF OF ENTRY 25. It is now easy to deduce (25.1). We remark at the outset that each of the two equalities in (25.4) yields (25.1).

Let $x = iz$, where $\mathrm{Im}\, z < 0$. Then $\mathrm{Re}\, x > 0$. From (25.4), we easily deduce that

$$\frac{1}{2x} - \frac{n^2 - 1^2}{2x} - \frac{n^2 - 3^2}{2x} - \cdots$$

$$= -i\left(-ix - 4i\frac{\Gamma(\tfrac{1}{4}(3 + n + x))\Gamma(\tfrac{1}{4}(3 - n + x))}{\Gamma(\tfrac{1}{4}(1 + n + x))\Gamma(\tfrac{1}{4}(1 - n + x))}\right)^{-1}.$$

After taking the reciprocal of each side, we find that

$$x - \frac{n^2 - 1^2}{2x} - \frac{n^2 - 3^2}{2x} - \cdots = 4\frac{\Gamma(\frac{1}{4}(3 + n + x))\Gamma(\frac{1}{4}(3 - n + x))}{\Gamma(\frac{1}{4}(1 + n + x))\Gamma(\frac{1}{4}(1 - n + x))}.$$

Taking the reciprocal of both sides once again, we complete the proof. □

Masson's proof is very interesting because it brings the hypergeometric functions out of the closet. Thus, there is a connection with the previous entries that was not heretofore noticed. Although Ramanujan probably did not derive in this way the many continued fractions for gamma functions that appear in this chapter, we have gained some insight into why these continued fractions exist.

Corollary 1. *If* Re $x > 0$, *then*

$$\frac{\Gamma^2(\frac{1}{4}(x + 1))}{\Gamma^2(\frac{1}{4}(x + 3))} = \frac{4}{x} + \frac{1^2}{2x} + \frac{3^2}{2x} + \frac{5^2}{2x} + \cdots.$$

PROOF. Set $n = 0$ in Entry 25. □

Corollary 1 was first proved by Bauer [2] in 1872 and was communicated by Ramanujan [16, p. xxvii] in his first letter to Hardy. Corollary 1 was also recently submitted as a problem by W. B. Jordan [1].

If we put $x = 1$ in Corollary 1, we obtain Lord Brouncker's continued fraction for π,

$$\pi = \frac{4}{1} + \frac{1^2}{2} + \frac{3^2}{2} + \frac{5^2}{2} + \cdots.$$

For a very interesting historical account of Brouncker's continued fraction, see Dutka's paper [2].

Corollary 2. *If* Re $x > 0$, *then*

$$\frac{\Gamma(\frac{1}{8}(x + 3))\Gamma(\frac{1}{8}(x + 1))}{\Gamma(\frac{1}{8}(x + 7))\Gamma(\frac{1}{8}(x + 5))} = \frac{8}{x} + \frac{1 \cdot 3}{2x} + \frac{5 \cdot 7}{2x} + \frac{9 \cdot 11}{2x} + \cdots.$$

PROOF. Replace x by $x/2$ and n by $\frac{1}{2}$ in Entry 25. □

Entry 26. *Suppose that n is an odd integer and x is any complex number or that n is an arbitrary complex number and* Re $x > 0$. *Then*

$$\frac{\Gamma^2(\frac{1}{4}(x + n + 1))\Gamma^2(\frac{1}{4}(x - n + 1))}{\Gamma^2(\frac{1}{4}(x + n + 3))\Gamma^2(\frac{1}{4}(x - n + 3))}$$

$$= \frac{8}{\frac{1}{2}(x^2 + n^2 - 1) +} \frac{1^2 - n^2}{1} + \frac{1^2}{x^2 - 1 +} \frac{3^2 - n^2}{1} + \frac{3^2}{x^2 - 1 +} \cdots$$

$$= \frac{8}{\frac{1}{2}(x^2 - n^2 - 1) + 1 +} \frac{1}{x^2 - 1} + \frac{1^2 - n^2}{1} + \frac{3^2}{x^2 - 1} + \frac{3^2 - n^2}{1} + \cdots.$$

PROOF. To obtain the first equality, set $\ell = 0$ in Entry 39. The second equality follows from Entry 39 by letting $n = 0$ and replacing ℓ by n.

Alternatively, the two continued fractions can formally be shown to be equal by an application of Entry 15. Let $h = -n^2$ and $a_k = (2k - 1)^2$, $k \geq 1$, and also replace x by $x^2 - 1$ in Entry 15. The desired equality easily follows, since both continued fractions terminate if n is an odd integer, and since both continued fractions converge for Re $x > 0$, otherwise.   □

Corollary. *If* Re $x > 0$, *then*

$$\frac{\Gamma^4(\frac{1}{4}(x + 1))}{\Gamma^4(\frac{1}{4}(x + 3))} = \frac{8}{\frac{1}{2}(x^2 - 1) +} \frac{1^2}{1 +} \frac{1^2}{x^2 - 1 +} \frac{3^2}{1 +} \frac{3^2}{x^2 - 1 +} \cdots.$$

PROOF. Set $n = 0$ in Entry 26.   □

The next theorem is found in Ramanujan's [16, p. xxix] second letter to Hardy. The first proof in print was provided by Preece [1]. Entry 27 can also be found in Perron's book [3, p. 37, Eq. (31)]. A very instructive proof of Entry 27 has been derived by Ramanathan [1].

Entry 27. *Suppose that $x, y > 0$. Then*

$$x + \frac{(1 + y)^2 + n}{2x} + \frac{(3 + y)^2 + n}{2x} + \frac{(5 + y)^2 + n}{2x} + \cdots$$

$$= y + \frac{(1 + x)^2 + n}{2y} + \frac{(3 + x)^2 + n}{2y} + \frac{(5 + x)^2 + n}{2y} + \cdots.$$

For an improved version of Entry 27, see Jacobsen's paper [3].

Entry 28. *Let* Re $x > 0$ *and* $|\arg n| \leq \pi/2 - \delta$, *for some positive number δ. Then*

$$\lim_{n \to \infty} \frac{x + \dfrac{n^2 + 1^2}{2x} + \dfrac{n^2 + 3^2}{2x} + \cdots}{n + \dfrac{x^2 - 1^2}{2n} + \dfrac{x^2 - 3^2}{2n} + \cdots} = 1. \tag{28.1}$$

PROOF. Apply Entry 25 with n replaced by in to find that, for Re $x > 0$,

$$\frac{\Gamma(\frac{1}{4}(x + in + 1))\Gamma(\frac{1}{4}(x - in + 1))}{\Gamma(\frac{1}{4}(x + in + 3))\Gamma(\frac{1}{4}(x - in + 3))} = \frac{4}{x +} \frac{n^2 + 1^2}{2x} + \frac{n^2 + 3^2}{2x} + \cdots,$$

or

$$4\frac{\Gamma(\frac{1}{4}(x + in + 3))\Gamma(\frac{1}{4}(x - in + 3))}{\Gamma(\frac{1}{4}(x + in + 1))\Gamma(\frac{1}{4}(x - in + 1))} = x + \frac{n^2 + 1^2}{2x} + \frac{n^2 + 3^2}{2x} + \cdots.$$

Next, apply Entry 25 with x and n interchanged to obtain, for Re $n > 0$,

$$4\frac{\Gamma(\frac{1}{4}(n + x + 3))\Gamma(\frac{1}{4}(n - x + 3))}{\Gamma(\frac{1}{4}(n + x + 1))\Gamma(\frac{1}{4}(n - x + 1))} = n - \frac{x^2 - 1^2}{2n} - \frac{x^2 - 3^2}{2n} - \cdots.$$

Now, for $|\arg n| \le \pi/2 - \delta$,

$$\lim_{n \to \infty} \frac{\Gamma(\frac{1}{4}(x + in + 3))\Gamma(\frac{1}{4}(x - in + 3))\Gamma(\frac{1}{4}(n + x + 1))\Gamma(\frac{1}{4}(n - x + 1))}{\Gamma(\frac{1}{4}(x + in + 1))\Gamma(\frac{1}{4}(x - in + 1))\Gamma(\frac{1}{4}(n + x + 3))\Gamma(\frac{1}{4}(n - x + 3))}$$

$$= 1,$$

where we have applied Stirling's formula for the quotient of two gamma functions (Lemma 2, Section 24, Chapter 11). Thus, we have shown that

$$\lim_{n \to \infty} \frac{x + \dfrac{n^2 + 1^2}{2x} + \dfrac{n^2 + 3^2}{2x} + \cdots}{n - \dfrac{x^2 - 1^2}{2n} - \dfrac{x^2 - 3^2}{2n} - \cdots} = 1. \tag{28.2}$$

However, for Re $n > 0$,

$$\lim_{n \to \infty} \frac{n - \dfrac{x^2 - 1^2}{2n} - \dfrac{x^2 - 3^2}{2n} - \cdots}{n + \dfrac{x^2 - 1^2}{2n} + \dfrac{x^2 - 3^2}{2n} + \cdots} = 1, \tag{28.3}$$

because the numerator and denominator above are both of the form $n + O(1/n)$ as n tends to ∞. Combining (28.2) and (28.3), we deduce (28.1). □

In his first notebook (p. 160), Ramanujan states a more precise version of Entry 28,

$$\frac{x + \dfrac{n^2 + 1^2}{2x} + \dfrac{n^2 + 3^2}{2x} + \cdots}{n + \dfrac{x^2 - 1^2}{2n} + \dfrac{x^2 - 3^2}{2n} + \cdots} = \frac{1 - e^{-\pi n}}{1 - 2e^{-\pi n/2}\sin(\pi x/2) + e^{-\pi n}}.$$

Ramanujan probably intends the right side to be an approximation to the left side for n large. However, the right side is $1 + O(e^{-\pi n/2})$ as n tends to ∞. A close analysis of our proof of Entry 28 shows that the left side of (28.1) is of the form $1 + O(1/n)$ as n tends to ∞ and that the estimate $O(1/n)$ cannot be improved. Thus, Ramanujan's claim does not appear to have a valid interpretation.

Entry 29. *Let n be an odd integer and x complex, or let n be complex and Re $x > 0$. Then*

$$\sum_{k=1}^{\infty} \left\{ \frac{(-1)^{k+1}}{x+n+2k-1} + \frac{(-1)^{k+1}}{x-n+2k-1} \right\}$$

$$= \frac{1}{x+} \; \frac{1^2-n^2}{x} \; +\frac{2^2}{x} \; +\frac{3^2-n^2}{x} \; +\frac{4^2}{x} \; +\frac{5^2-n^2}{x} \; +\cdots .$$

We provide two proofs under more restrictive hypotheses than what we have given. Jacobsen [3] has proved Entry 29 with the stated conditions.

FIRST PROOF. Our first proof merely consists of a reformulation of a result found in Perron's book [3, p. 33, Eq. (12)]),

$$\frac{4}{x+} \; \frac{1^2-n^2}{x} \; +\frac{2^2}{x} \; +\frac{3^2-n^2}{x} \; +\frac{4^2}{x} \; +\frac{5^2-n^2}{x} \; +\cdots$$

$$= \psi\left(\frac{x+n+3}{4}\right) + \psi\left(\frac{x-n+3}{4}\right) - \psi\left(\frac{x+n+1}{4}\right) - \psi\left(\frac{x-n+1}{4}\right),$$

(29.1)

where $x > 0$ and $n^2 < 1$. Now employ (0.1) and simplify to complete the proof. $\square$

In fact, Entry 29 was first proved in print in 1953 under these stronger hypotheses by Perron [2] who derived it from Entry 34 below.

SECOND PROOF. Since

$$\frac{1}{1+t^2} = \sum_{k=0}^{\infty} (-1)^k t^{2k}, \qquad |t| < 1,$$

we find that, for Re $x > -1$,

$$H(x) := \int_0^1 \frac{t^x}{1+t^2} \, dt = \sum_{k=0}^{\infty} \frac{(-1)^k}{x+2k+1} .$$

(29.2)

Then for Re$(x \pm n) > -1$,

$$H(x+n) + H(x-n) = \int_0^1 \frac{t^x(t^n + t^{-n})}{1+t^2} \, dt$$

$$= \int_0^{\infty} e^{-ux} \frac{\cosh(nu)}{\cosh u} \, du,$$

(29.3)

where we have made the change of variable $t = e^{-u}$. But for $x > 0$ and $n^2 < 1$, Rogers [3] has shown that

$$\int_0^{\infty} e^{-xu} \frac{\cosh(nu)}{\cosh u} \, du = \frac{1}{x+} \; \frac{1^2-n^2}{x} \; +\frac{2^2}{x} \; +\frac{3^2-n^2}{x} \; +\frac{4^2}{x} \; +\cdots .$$

(29.4)

Employing (29.2) and (29.4) in (29.3), we arrive at the desired formula. $\square$

Corollary. *If* Re $x > 0$, *then*

$$2 \sum_{k=1}^{\infty} \frac{(-1)^{k+1}}{x + 2k - 1} = \frac{1}{x} + \frac{1^2}{x} + \frac{2^2}{x} + \frac{3^2}{x} + \frac{4^2}{x} + \cdots.$$

PROOF. Set $n = 0$ in Entry 29.   □

Entry 30. *Suppose either that n is an integer or that* Re $x > 0$. *Then*

$$\sum_{k=0}^{\infty} \left\{ \frac{1}{x - n + 2k + 1} - \frac{1}{x + n + 2k + 1} \right\}$$

$$= \frac{n}{x} + \frac{1^2(1^2 - n^2)}{3x} + \frac{2^2(2^2 - n^2)}{5x} + \frac{3^2(3^2 - n^2)}{7x} + \cdots. \quad (30.1)$$

FIRST PROOF. Letting

$$R = \Gamma(\tfrac{1}{2}(x + m + n + 1))\Gamma(\tfrac{1}{2}(x - m - n + 1))$$

and

$$T = \Gamma(\tfrac{1}{2}(x + m - n + 1))\Gamma(\tfrac{1}{2}(x - m + n + 1)),$$

we first write Entry 33 in the form

$$\frac{R - T}{R + T} = \frac{mn}{x} + \frac{(m^2 - 1^2)(n^2 - 1^2)}{3x} + \frac{(m^2 - 2^2)(n^2 - 2^2)}{5x} + \cdots,$$

where x, m, and n are complex numbers such that Re $x > 0$, or either m or n is an integer. Thus,

$$\lim_{m \to 0} \frac{1}{m} \frac{R - T}{R + T} = \frac{n}{x} + \frac{1^2(1^2 - n^2)}{3x} + \frac{2^2(2^2 - n^2)}{5x} + \cdots. \quad (30.2)$$

On the other hand, a direct calculation with the use of L'Hospital's rule shows that

$$\lim_{m \to 0} \frac{1}{m} \frac{R - T}{R + T} = \frac{1}{2}\psi\left(\frac{x + n + 1}{2}\right) - \frac{1}{2}\psi\left(\frac{x - n + 1}{2}\right)$$

$$= \sum_{k=0}^{\infty} \left\{ \frac{1}{x - n + 2k + 1} - \frac{1}{x + n + 2k + 1} \right\}. \quad (30.3)$$

Combining (30.2) and (30.3), we finish the proof.   □

SECOND PROOF. This proof requires that $n^2 < 1$ and $x > 0$. Proceeding in somewhat the same way as in the second proof of Entry 29, we find that, for Re$(x \pm n) > -1$,

$$\int_0^1 \frac{t^{x-n} - t^{x+n}}{1 - t^2} \, dt = \sum_{k=0}^{\infty} \left\{ \frac{1}{x - n + 2k + 1} - \frac{1}{x + n + 2k + 1} \right\}. \quad (30.4)$$

On the other hand, letting $t = e^{-u}$ and using a theorem of Stieltjes [1],

[4, pp. 378–391], which was also proved by Rogers [3], we find that, for $x > 0$ and $n^2 < 1$,

$$\int_0^1 \frac{t^{x-n} - t^{x+n}}{1 - t^2} \, dt = \int_0^\infty e^{-xu} \frac{\sinh(nu)}{\sinh u} \, du$$

$$= \frac{n}{x +} \frac{1^2(1^2 - n^2)}{3x} + \frac{2^2(2^2 - n^2)}{5x} + \cdots. \qquad (30.5)$$

Combining (30.4) and (30.5), we complete the second proof. □

Corollary. *If* Re $x > 0$, *then*

$$2 \sum_{k=0}^\infty \frac{1}{(x + 2k + 1)^2} = \frac{1}{x +} \frac{1^4}{3x} + \frac{2^4}{5x} + \frac{3^4}{7x} + \cdots.$$

PROOF. Divide both sides of (30.1) by n and let n tend to 0. □

If we set $x = 1$ in the corollary above, we deduce that

$$\frac{1}{2}\zeta(2) = \frac{\pi^2}{12} = \frac{1}{1 +} \frac{1^4}{3} + \frac{2^4}{5} + \frac{3^4}{7} + \cdots.$$

For a simple proof of this expansion, see a note by Madhava [1].

In fact, the corollary above is due to Stieltjes [1, Eq. (22)], [4, p. 387].

Entry 31. *Suppose that n is an even integer or that* Re $x > 0$ *and n is any complex number. Then*

$$\sum_{k=0}^\infty \left\{ \frac{(-1)^k}{x - n + 2k + 1} - \frac{(-1)^k}{x + n + 2k + 1} \right\}$$

$$= \frac{n}{x^2 - 1 +} \frac{2^2 - n^2}{1} + \frac{2^2}{x^2 - 1 +} \frac{4^2 - n^2}{1} + \frac{4^2}{x^2 - 1 +} \cdots. \qquad (31.1)$$

FIRST PROOF. From Entry 36, if Re $x > 0$ or if n is an even integer,

$$\lim_{\ell \to 0} \frac{1}{\ell} \frac{1 - P}{1 + P} = \frac{n}{x^2 - 1 +} \frac{2^2 - n^2}{1} + \frac{2^2}{x^2 - 1 +} \frac{4^2 - n^2}{1} + \frac{4^2}{x^2 - 1 +} \cdots. \qquad (31.2)$$

On the other hand, a direct calculation with the use of L'Hospital's rule gives

$$\lim_{\ell \to 0} \frac{1}{\ell} \frac{1 - P}{1 + P} = \frac{1}{4} \left\{ \psi\left(\frac{x + n + 1}{4}\right) + \psi\left(\frac{x - n + 3}{4}\right) \right.$$

$$\left. - \psi\left(\frac{x - n + 1}{4}\right) - \psi\left(\frac{x + n + 3}{4}\right) \right\}$$

$$= \sum_{k=1}^{\infty} \left\{ -\frac{1}{x+n+4k-3} - \frac{1}{x-n+4k-1} \right.$$

$$\left. + \frac{1}{x-n+4k-3} + \frac{1}{x+n+4k-1} \right\}. \qquad (31.3)$$

Equalities (31.2) and (31.3) taken together yield (31.1).   □

SECOND PROOF. This proof requires more severe restrictions on x and n. As in the second proofs of Entries 29 and 30, we easily find that, for $\operatorname{Re}(x \pm n) > -1$,

$$\sum_{k=0}^{\infty} \left\{ \frac{(-1)^k}{x-n+2k+1} - \frac{(-1)^k}{x+n+2k+1} \right\} = \int_0^{\infty} e^{-xu} \frac{\sinh(nu)}{\cosh u} \, du.$$

But Stieltjes [3], [4, pp. 402–566] and later Rogers [3] have shown that, for $x > 0$ and $n^2 < 1$,

$$\int_0^{\infty} e^{-xu} \frac{\sinh(nu)}{\cosh u} \, du$$

$$= \frac{n}{x^2-1} + \frac{2^2-n^2}{1} + \frac{2^2}{x^2-1} + \frac{4^2-n^2}{1} + \frac{4^2}{x^2-1} + \cdots.$$

The foregoing two equalities imply (31.1).   □

Corollary. *If* $\operatorname{Re} x > 0$, *then*

$$2 \sum_{k=0}^{\infty} \frac{(-1)^k}{(x+2k+1)^2} = \frac{1}{x^2-1} + \frac{2^2}{1} + \frac{2^2}{x^2-1} + \frac{4^2}{1} + \frac{4^2}{x^2-1} + \cdots.$$

PROOF. Divide both sides of (31.1) by n and then let n tend to 0.   □

If we put $x = 2$ in the foregoing corollary, we obtain the following elegant continued fraction for Catalan's constant G:

$$2G := 2 \sum_{k=0}^{\infty} \frac{(-1)^k}{(2k+1)^2} = 2 - \frac{1}{3} + \frac{2^2}{1} + \frac{2^2}{3} + \frac{4^2}{1} + \frac{4^2}{3} + \cdots.$$

Of course, similar continued fraction expansions for G can be obtained by setting $x = 2n$, where n is any positive integer, in the corollary above. This same infinite set of continued fractions for G was independently found by H. Cohen (personal communication) who obtained them from a *different* formula.

Entry 32(i). *If* $\operatorname{Re} x > 0$, *then*

$$1 + 2x \sum_{k=1}^{\infty} \frac{(-1)^k}{x+2k} = \frac{1}{x} + \frac{1 \cdot 2}{x} + \frac{2 \cdot 3}{x} + \frac{3 \cdot 4}{x} + \cdots. \qquad (32.1)$$

PROOF. Let

$$P = P(x, n) = \frac{\Gamma(\frac{1}{4}(x + n + 3))\Gamma(\frac{1}{4}(x - n + 3))}{\Gamma(\frac{1}{4}(x + n + 1))\Gamma(\frac{1}{4}(x - n + 1))}.$$

Then by Entry 25, for Re $x > 0$ and $n \neq 1$,

$$4P = x + \frac{1^2 - n^2}{2x} + \frac{3^2 - n^2}{2x} + \frac{5^2 - n^2}{2x} + \cdots,$$

or

$$\frac{4P - x}{1 - n} = \frac{1 + n}{2x} + \frac{3^2 - n^2}{2x} + \frac{5^2 - n^2}{2x} + \cdots. \tag{32.2}$$

Note that $P(x, 1) = x/4$. We now let n tend to 1 in (32.2) and apply L'Hospital's rule on the left side. We then find that

$$\frac{x}{4}\left\{2\psi\left(\frac{x + 2}{4}\right) - \psi\left(\frac{x}{4} + 1\right) - \psi\left(\frac{x}{4}\right)\right\} = \frac{2}{2x} + \frac{2 \cdot 4}{2x} + \frac{4 \cdot 6}{2x} + \frac{6 \cdot 8}{2x} + \cdots.$$

Simplifying each side above, we arrive at (32.1). □

Entry 32(ii). *If* Re $x > 0$, *then*

$$1 + 2x^2 \sum_{k=1}^{\infty} \frac{(-1)^k}{(x + k)^2} = \frac{1}{x} + \frac{1^2}{x} + \frac{1 \cdot 2}{x} + \frac{2^2}{x} + \frac{2 \cdot 3}{x} + \frac{3^2}{x} + \cdots.$$

PROOF. Let

$$P = P(x, n) = \psi\left(\frac{x + n + 3}{4}\right) + \psi\left(\frac{x - n + 3}{4}\right)$$
$$- \psi\left(\frac{x + n + 1}{4}\right) - \psi\left(\frac{x - n + 1}{4}\right).$$

Then from (29.1), we find that, for Re $x > 0$ and $n \neq 1$,

$$\frac{4/P - x}{1 - n} = \frac{1 + n}{x} + \frac{2^2}{x} + \frac{3^2 - n^2}{x} + \frac{4^2}{x} + \frac{5^2 - n^2}{x} + \cdots. \tag{32.3}$$

Observe that $P(x, 1) = 4/x$. Letting n tend to 1 in (32.3) and employing L'Hospital's rule, we find that

$$\frac{4\frac{\partial}{\partial n}P(x, n)\Big|_{n=1}}{P^2(x, 1)} = x^2 \sum_{k=1}^{\infty} \left\{\frac{1}{(x + 4k)^2} - \frac{2}{(x + 4k - 2)^2} + \frac{1}{(x + 4k - 4)^2}\right\}$$

$$= \frac{2}{x} + \frac{2^2}{x} + \frac{2 \cdot 4}{x} + \frac{4^2}{x} + \frac{4 \cdot 6}{x} + \cdots.$$

Replacing x by $2x$, we deduce that

$$1 + 2x^2 \sum_{k=1}^{\infty} \frac{(-1)^k}{(x+k)^2} = \frac{2}{2x} + \frac{2^2}{2x} + \frac{2 \cdot 4}{2x} + \frac{4^2}{2x} + \frac{4 \cdot 6}{2x} + \cdots.$$

Simplifying the right side, we complete the proof.   □

If we set $x = 1$ in Entry 32(ii), we deduce that

$$\zeta(2) = 1 + \frac{1}{1} + \frac{1^2}{1} + \frac{1 \cdot 2}{1} + \frac{2^2}{1} + \frac{2 \cdot 3}{1} + \frac{3^2}{1} + \cdots.$$

Putting $x = \frac{1}{2}$ in Entry 32(ii) yields another continued fraction for G,

$$2G = 1 + \frac{1}{\frac{1}{2}} + \frac{1^2}{\frac{1}{2}} + \frac{1 \cdot 2}{\frac{1}{2}} + \frac{2^2}{\frac{1}{2}} + \frac{2 \cdot 3}{\frac{1}{2}} + \frac{3^2}{\frac{1}{2}} + \cdots.$$

Entry 32(iii). *If* $\operatorname{Re} x > -\frac{1}{2}$, *then*

$$\zeta(3, x + 1) := \sum_{k=1}^{\infty} \frac{1}{(x+k)^3}$$

$$= \frac{1}{2x(x+1)} + \frac{1^3}{1} + \frac{1^3}{6x(x+1)} + \frac{2^3}{1} + \frac{2^3}{10x(x+1)} + \cdots$$

$$= \frac{1}{2x^2 + 2x + 1} - \frac{1^6}{3(2x^2 + 2x + 3)}$$

$$- \frac{2^6}{5(2x^2 + 2x + 7)} - \frac{3^6}{7(2x^2 + 2x + 13)} - \cdots. \qquad (32.4)$$

PROOF. In Entry 35, replace x by $2x + 1$. Then $y = 4x(x + 1) + 2m - m^2$, and we need to require that $\operatorname{Re} x > -\frac{1}{2}$. Also let $\ell = n = m$. Noting that $t = 0$ and using the second continued fraction of Entry 35, we find that

$$\frac{1-P}{1+P} = \cfrac{1 - \dfrac{\Gamma(\frac{1}{2}(2x+2+3m))\Gamma^3(\frac{1}{2}(2x+2-m))}{\Gamma(\frac{1}{2}(2x+2-3m))\Gamma^3(\frac{1}{2}(2x+2+m))}}{1 + \dfrac{\Gamma(\frac{1}{2}(2x+2+3m))\Gamma^3(\frac{1}{2}(2x+2-m))}{\Gamma(\frac{1}{2}(2x+2-3m))\Gamma^3(\frac{1}{2}(2x+2+m))}}$$

$$= \frac{2m^3}{y - 2m^3} + \frac{2(1-m)(1^2-m^2)}{1} + \frac{2(1+m)(1^2-m^2)}{3y}$$

$$+ \frac{2(2-m)(2^2-m^2)}{1} + \frac{2(2+m)(2^2-m^2)}{5y} + \cdots. \qquad (32.5)$$

Now divide both sides of (32.5) by m^3 and let m tend to 0. On the right side, we arrive at

$$\frac{2}{4x(x+1)} + \frac{2 \cdot 1^3}{1} + \frac{2 \cdot 1^3}{12x(x+1)} + \frac{2 \cdot 2^3}{1} + \frac{2 \cdot 2^3}{20x(x+1)} + \cdots.$$

Simplifying above, we obtain the former continued fraction of (32.4).

Next, write the aforementioned continued fraction in the equivalent form

$$\frac{1}{2x(x+1)} + \frac{1^3}{1} + \frac{1^3/3}{2x(x+1)} + \frac{2^3/3}{1} + \frac{2^3/5}{2x(x+1)} + \frac{3^3/5}{1} + \cdots.$$

Applying Entry 14 to this continued fraction, we deduce the equality between the continued fractions of (32.4).

For brevity, set $z = x + 1$. For $\operatorname{Re} z > \frac{1}{2}$, it remains to examine, by (32.5),

$$\lim_{m \to 0} \frac{1}{m^3} \frac{1-P}{1+P}$$

$$= \lim_{m \to 0} \frac{1}{2m^3 \Gamma^4(z)} \left(\left\{ \Gamma(z) - \frac{3m}{2}\Gamma'(z) + \frac{3^2 m^2}{2^3}\Gamma''(z) - \frac{3^2 m^3}{2^4}\Gamma'''(z) + \cdots \right\} \right.$$

$$\times \left\{ \Gamma(z) + \frac{m}{2}\Gamma'(z) + \frac{m^2}{2^3}\Gamma''(z) + \frac{m^3}{2^4 \cdot 3}\Gamma'''(z) + \cdots \right\}^3$$

$$- \left\{ \Gamma(z) + \frac{3m}{2}\Gamma'(z) + \frac{3^2 m^2}{2^3}\Gamma''(z) + \frac{3^2 m^3}{2^4}\Gamma'''(z) + \cdots \right\}$$

$$\left. \times \left\{ \Gamma(z) - \frac{m}{2}\Gamma'(z) + \frac{m^2}{2^3}\Gamma''(z) - \frac{m^3}{2^4 \cdot 3}\Gamma'''(z) + \cdots \right\}^3 \right)$$

$$= \frac{1}{\Gamma^4(z)} \left(\left\{ -\frac{3^2}{2^4} + \frac{1}{2^4} \right\} \Gamma'''(z)\Gamma^3(z) + \left\{ \frac{3^3}{2^4} - \frac{3^2}{2^4} + \frac{6}{2^4} \right\} \Gamma''(z)\Gamma'(z)\Gamma^2(z) \right.$$

$$\left. + \left\{ \frac{1}{2^3} - \frac{3^2}{2^3} \right\} \Gamma'(z)^3\Gamma(z) \right)$$

$$= -\frac{\Gamma'''(z)}{2\Gamma(z)} + \frac{3\Gamma''(z)\Gamma'(z)}{2\Gamma^2(z)} - \frac{\Gamma'(z)^3}{\Gamma^3(z)}$$

$$= -\frac{1}{2}\frac{d^2}{dz^2}\left(\frac{\Gamma'(z)}{\Gamma(z)}\right) = -\frac{1}{2}\psi''(z) = \sum_{k=0}^{\infty} \frac{1}{(z+k)^3}.$$

The proof is now complete.   □

Ramanujan's second continued fraction in Entry 32(iii) is slightly in error (p. 149).

We might compare Entry 32(iii) with another continued fraction for $\zeta(3, x)$,

$$4x^3\zeta(3, x) = 2x + 2 + \frac{1}{x} + \frac{p_1}{x} + \frac{q_1}{x} + \frac{p_2}{x} + \frac{q_2}{x} + \cdots,$$

where, for $k \geq 1$,

$$p_k = \frac{k^2(k+1)}{4k+2} \quad \text{and} \quad q_k = \frac{k(k+1)^2}{4k+2}.$$

The last result was discovered by Stieltjes [1], [4, pp. 378–391].

Setting $x = 1$ in Entry 32(iii), we deduce the following beautiful continued fraction for $\zeta(3)$:

$$\zeta(3) = 1 + \cfrac{1}{2\cdot2 +} \cfrac{1^3}{1 +} \cfrac{1^3}{6\cdot2 +} \cfrac{2^3}{1 +} \cfrac{2^3}{10\cdot2 +} \cdots.$$

This continued fraction also follows from work of Apéry [1] and was of crucial importance in his famous proof that $\zeta(3)$ is irrational.

Entry 33. *Let x, m, and n be complex. If either m or n is an integer or if* Re $x > 0$, *then*

$$\frac{\Gamma(\frac{1}{2}(x+m+n+1))\Gamma(\frac{1}{2}(x-m-n+1)) - \Gamma(\frac{1}{2}(x+m-n+1))\Gamma(\frac{1}{2}(x-m+n+1))}{\Gamma(\frac{1}{2}(x+m+n+1))\Gamma(\frac{1}{2}(x-m-n+1)) + \Gamma(\frac{1}{2}(x+m-n+1))\Gamma(\frac{1}{2}(x-m+n+1))}$$

$$= \cfrac{mn}{x +} \cfrac{(m^2-1^2)(n^2-1^2)}{3x +} \cfrac{(m^2-2^2)(n^2-2^2)}{5x}$$

$$+ \cfrac{(m^2-3^2)(n^2-3^2)}{7x} + \cdots.$$

PROOF. Set

$$R(m) = \frac{\Gamma(\frac{1}{2}(x+\ell+n+1)+m)\Gamma(\frac{1}{2}(x-\ell-n+1)+m)}{\Gamma(\frac{1}{2}(x-\ell+n+1)+m)\Gamma(\frac{1}{2}(x+\ell-n+1)+m)}$$

and

$$T = \frac{\Gamma(\frac{1}{2}(x+\ell-n+1))\Gamma(\frac{1}{2}(x-\ell+n+1))}{\Gamma(\frac{1}{2}(x+\ell+n+1))\Gamma(\frac{1}{2}(x-\ell-n+1))}.$$

Suppose that m is a positive integer in Entry 35. Replacing x by $x + m$ in Entry 35, we find that

$$\frac{1 - R(m)T}{1 + R(m)T} = \cfrac{2\ell mn}{x^2+2mx-\ell^2-n^2+1 +} \cfrac{4(\ell^2-1^2)(m^2-1^2)(n^2-1^2)}{3(x^2+2mx-\ell^2-n^2+5)}$$

$$+ \cfrac{4(\ell^2-2^2)(m^2-2^2)(n^2-2^2)}{5(x^2+2mx-\ell^2-n^2+13)} + \cdots$$

$$= \cfrac{\ell n}{x+(x^2-\ell^2-n^2+1)/2m +} \cfrac{(\ell^2-1^2)(n^2-1^2)(1-1/m^2)}{3(x+(x^2-\ell^2-n^2+5)/2m)}$$

$$+ \cfrac{(\ell^2-2^2)(n^2-2^2)(1-2^2/m^2)}{5(x+(x^2-\ell^2-n^2+13)/2m)} + \cdots. \tag{33.1}$$

Now let m tend to ∞ in (33.1). By Stirling's formula, $R(m)$ tends to 1 as m tends to ∞. The continued fraction converges uniformly with respect to m in a neighborhood of $m = \infty$ if Re $x > 0$. Hence,

$$\frac{1-T}{1+T} = \frac{\ell n}{x +} \; \frac{(\ell^2-1^2)(n^2-1^2)}{3x} \; + \; \frac{(\ell^2-2^2)(n^2-2^2)}{5x} \; + \cdots .$$

Replacing ℓ by m above, we complete the proof.   □

In fact, Entry 33 was first proved in print by Nörlund [1] under more restrictive hypotheses.

The continued fraction in Entry 33 is a special case of a more general continued fraction for a quotient of two integrals involving hypergeometric functions that was discovered by Stieltjes [1], [4, p. 389, Eq. (29)].

Entry 34. *Suppose that n is an odd integer or ℓ is an even integer, or assume that Re $x > 0$ with ℓ and n arbitrary complex numbers. Define*

$$P = \frac{\Gamma(\tfrac{1}{4}(x+\ell+n+1))\Gamma(\tfrac{1}{4}(x+\ell-n+1))\Gamma(\tfrac{1}{4}(x-\ell+n+3))\Gamma(\tfrac{1}{4}(x-\ell-n+3))}{\Gamma(\tfrac{1}{4}(x-\ell+n+1))\Gamma(\tfrac{1}{4}(x-\ell-n+1))\Gamma(\tfrac{1}{4}(x+\ell+n+3))\Gamma(\tfrac{1}{4}(x+\ell-n+3))}.$$

Then

$$\frac{1-P}{1+P} = \frac{\ell}{x +} \; \frac{1^2-n^2}{x} \; + \; \frac{2^2-\ell^2}{x} \; + \; \frac{3^2-n^2}{x} \; + \; \frac{4^2-\ell^2}{x} \; + \cdots .$$

Entry 34 was stated by Ramanujan [16, p. 350] in his first letter to Hardy. The first published proof was provided by Preece [2]. Another proof has been devised by Perron [1], [3, p. 34, Eq. (15)]. These two proofs require stronger hypotheses. Jacobsen [3] has shown how to establish Entry 34 under the given assumptions on the parameters.

Corollary. *Suppose that $\mathrm{Re}(x/\gamma) \neq 0$. Put*

$$F(\alpha, \beta) = \tan^{-1}\left\{\frac{\alpha}{x +} \; \frac{\beta^2+\gamma^2}{x} \; + \; \frac{\alpha^2+(2\gamma)^2}{x} \; + \; \frac{\beta^2+(3\gamma)^2}{x} \; + \cdots \right\}.$$

Then

$$F(\alpha, \beta) + F(\beta, \alpha) = 2F\{\tfrac{1}{2}(\alpha+\beta), \tfrac{1}{2}(\alpha+\beta)\}.$$

The corollary was communicated by Ramanujan [16, p. 353] in his second letter to Hardy. Again, the first published proof was given by Preece [2], and indeed this result is a corollary of Entry 34.

Entry 35. *Let $x, \ell, m,$ and n denote complex numbers and put $y = x^2 - (1-m)^2$ and $t = (n^2 - \ell^2)(1 - 2m)$. Define*

$$P = \frac{\Gamma(\tfrac{1}{2}(x+\ell+m+n+1))\Gamma(\tfrac{1}{2}(x+\ell-m-n+1))\Gamma(\tfrac{1}{2}(x-\ell+m-n+1))\Gamma(\tfrac{1}{2}(x-\ell-m+n+1))}{\Gamma(\tfrac{1}{2}(x-\ell-m-n+1))\Gamma(\tfrac{1}{2}(x-\ell+m+n+1))\Gamma(\tfrac{1}{2}(x+\ell-m+n+1))\Gamma(\tfrac{1}{2}(x+\ell+m-n+1))}.$$

Then if either $\ell, m,$ or n is an integer or if Re $x > 0$,

$$\frac{1-P}{1+P} = \frac{2\ell mn}{x^2 - \ell^2 - m^2 - n^2 + 1 +} \frac{4(\ell^2 - 1^2)(m^2 - 1^2)(n^2 - 1^2)}{3(x^2 - \ell^2 - m^2 - n^2 + 5)}$$

$$+ \frac{4(\ell^2 - 2^2)(m^2 - 2^2)(n^2 - 2^2)}{5(x^2 - \ell^2 - m^2 - n^2 + 13) +} \cdots$$

$$= \frac{2\ell mn}{y + t - 2\ell^2 m +} \frac{2(1-m)(1^2 - n^2)}{1} + \frac{2(1+m)(1^2 - \ell^2)}{3y + t}$$

$$+ \frac{2(2-m)(2^2 - n^2)}{1} + \frac{2(2+m)(2^2 - \ell^2)}{5y + t} + \cdots. \tag{35.1}$$

PROOF. The first equality was shown by Watson [8] to be a corollary of Entry 40. If either ℓ, m, or n is an integer, Watson's limiting process is trivially justified. If ℓ, m, and n are nonintegral, then the limiting process is more difficult to justify. We refer the reader to Jacobsen's paper [3], where this justification is carefully presented.

To prove the second equality, we employ the following generalization of Entry 14 but special case of (14.2). If the former continued fraction converges, then

$$\frac{a_1}{x_1 +} \frac{a_2}{1 +} \frac{a_3}{x_3 +} \frac{a_4}{1 +} \cdots + \frac{a_{2k-1}}{x_{2k-1} +} \frac{a_{2k}}{1} + \cdots$$

$$= \frac{a_1}{x_1 + a_2 -} \frac{a_2 a_3}{x_3 + a_3 + a_4 -} \frac{a_4 a_5}{x_5 + a_5 + a_6 -} \cdots$$

$$- \frac{a_{2k-2} a_{2k-1}}{x_{2k-1} + a_{2k-1} + a_{2k} +} \cdots.$$

Thus, with $a_1 = 2\ell mn$, $a_2 = 2(1-m)(1-n^2)$, ... and $x_1 = y + t - 2\ell^2 m$, $x_3 = 3y + t, \ldots$, we find that

$$\frac{2\ell mn}{y + t - 2\ell^2 m +} \frac{2(1-m)(1-n^2)}{1} + \frac{2(1+m)(1-\ell^2)}{3y + t} + \frac{2(2-m)(2^2 - n^2)}{1}$$

$$+ \frac{2(2+m)(2^2 - \ell^2)}{5y + t} + \cdots + \frac{2(k-m)(k^2 - n^2)}{1} + \cdots$$

$$= \frac{2\ell mn}{x^2 - \ell^2 - m^2 - n^2 + 1 -} \frac{4(1-m^2)(1-\ell^2)(1-n^2)}{3(x^2 - \ell^2 - m^2 - n^2 + 5)}$$

$$- \frac{4(2^2 - m^2)(2^2 - \ell^2)(2^2 - n^2)}{5(x^2 - \ell^2 - m^2 - n^2 + 13)} - \cdots$$

$$- \frac{4((k-1)^2 - m^2)((k-1)^2 - l^2)((k-1)^2 - n^2)}{(2k-1)(x^2 - \ell^2 - m^2 - n^2 + 2k^2 - 2k + 1)} + \cdots, \tag{35.2}$$

where we have used the easily proved identity

$$(2j + 1)y + t + 2(j + m)(j^2 - \ell^2) + 2(j + 1 - m)((j + 1)^2 - n^2)$$
$$= (2j + 1)(x^2 - \ell^2 - m^2 - n^2 + 2j^2 + 2j + 1).$$

This establishes the second equality in (35.1).   □

Entry 36. *Suppose either that n or ℓ is an even integer or that* Re $x > 0$ *and n and ℓ are arbitrary complex numbers. Let*

$$P = \frac{\Gamma(\frac{1}{4}(x + \ell + n + 3))\Gamma(\frac{1}{4}(x - \ell - n + 3))\Gamma(\frac{1}{4}(x + \ell - n + 1))\Gamma(\frac{1}{4}(x - \ell + n + 1))}{\Gamma(\frac{1}{4}(x + \ell + n + 1))\Gamma(\frac{1}{4}(x - \ell - n + 1))\Gamma(\frac{1}{4}(x + \ell - n + 3))\Gamma(\frac{1}{4}(x - \ell + n + 3))}.$$

Then

$$\frac{1 - P}{1 + P} = \frac{\ell n}{x^2 - 1 - \ell^2 +} \; \frac{2^2 - n^2}{1} \; \frac{2^2 - \ell^2}{+ \; x^2 - 1 +} \; \frac{4^2 - n^2}{1} \; \frac{4^2 - \ell^2}{+ \; x^2 - 1 +} \cdots.$$

PROOF. In the second equality of Entry 35, let $m = \frac{1}{2}$ and replace x, n, and ℓ by $x/2$, $n/2$, and $\ell/2$, respectively. After simplification, the proposed identity follows.   □

Entry 37. *Suppose that either ℓ or n is an integer or that* Re $x > 0$. *Then*

$$\frac{1}{2}\left\{ \psi\left(\frac{x + \ell - n + 1}{2}\right) + \psi\left(\frac{x - \ell + n + 1}{2}\right) \right.$$
$$\left. - \psi\left(\frac{x + \ell + n + 1}{2}\right) - \psi\left(\frac{x - \ell - n + 1}{2}\right) \right\}$$
$$= \frac{2\ell n}{x^2 - 1 + n^2 - \ell^2 +} \; \frac{2(1^2 - n^2)}{1} \; \frac{2(1^2 - \ell^2)}{+ \; 3(x^2 - 1) + n^2 - \ell^2}$$
$$+ \frac{4(2^2 - n^2)}{1} \; \frac{4(2^2 - \ell^2)}{+ \; 5(x^2 - 1) + n^2 - \ell^2 +} \cdots. \tag{37.1}$$

PROOF. Taking the second equality in (35.1), divide both sides by m and then let m tend to 0. Applying L'Hospital's rule on the left side, we readily deduce the desired formula with no difficulty.   □

Entry 38. *Assume that either n is an integer or that* Re $x > 0$. *Then*

$$\sum_{k=0}^{\infty} \frac{1}{(x - n + 2k + 1)^2} - \sum_{k=0}^{\infty} \frac{1}{(x + n + 2k + 1)^2}$$
$$= \frac{n}{x^2 - 1 + n^2 +} \; \frac{2(1^2 - n^2)}{1} \; \frac{2 \cdot 1^2}{+ \; 3(x^2 - 1) + n^2}$$
$$+ \frac{4(2^2 - n^2)}{1} \; \frac{4 \cdot 2^2}{+ \; 5(x^2 - 1) + n^2 +} \cdots$$
$$= \frac{n}{x^2 - n^2 + 1 -} \; \frac{4(1^2 - n^2)1^4}{3(x^2 - n^2 + 5) -} \; \frac{4(2^2 - n^2)2^4}{5(x^2 - n^2 + 13) -} \cdots. \tag{38.1}$$

PROOF. To prove the first equality in (38.1), divide both sides of (37.1) by 2ℓ and let ℓ tend to 0. Applying L'Hospital's rule on the left side, we easily achieve the desired equality.

The second equality in (38.1) is also easily established. First, divide both sides of the first equality in (35.1) by m and then let m tend to 0. Of course, this gives a second continued fraction for the left side of (37.1). Now divide both sides by ℓ and let ℓ tend to 0. $\square$

Entry 39. *Let ℓ and n denote arbitrary complex numbers. Suppose that x is complex with* Re $x > 0$ *or that either n or ℓ is an odd integer. Then*

$$P := \frac{\Gamma(\tfrac{1}{4}(x + \ell + n + 1))\Gamma(\tfrac{1}{4}(x - \ell + n + 1))\Gamma(\tfrac{1}{4}(x + \ell - n + 1))\Gamma(\tfrac{1}{4}(x - \ell - n + 1))}{\Gamma(\tfrac{1}{4}(x + \ell + n + 3))\Gamma(\tfrac{1}{4}(x - \ell + n + 3))\Gamma(\tfrac{1}{4}(x + \ell - n + 3))\Gamma(\tfrac{1}{4}(x - \ell - n + 3))}$$

$$= \frac{8}{(x^2 - \ell^2 + n^2 - 1)/2 +} \; \frac{1^2 - n^2}{1} \; \frac{1^2 - \ell^2}{+ \; x^2 - 1 +} \; \frac{3^2 - n^2}{1} \; \frac{3^2 - \ell^2}{+ \; x^2 - 1 + \cdots} .$$
(39.1)

PROOF. We shall prove Entry 39 for $-\infty < \ell^2, n^2 < 1$ and $x > 1$. An argument of Jacobsen [3] can then be used to extend the domains of convergence for ℓ, n, and x to those indicated.

To prove Entry 39, we employ the following theorem found in Perron's text [3, p. 27, Satz 1.13]. Suppose that all the elements are positive in both continued fractions below. Assume also that each continued fraction converges. Then

$$b_0 + \frac{a_1}{b_1 +} \; \frac{a_2}{b_2 +} \; \frac{a_3}{b_3 + \cdots}$$

$$= b_0 + r_0 + \frac{\varphi_1}{b_1 + r_1 +} \; \frac{a_1\varphi_2/\varphi_1}{b_2 + r_2 - r_0\varphi_2/\varphi_1 +} \; \frac{a_2\varphi_3/\varphi_2}{b_3 + r_3 - r_1\varphi_3/\varphi_2 + \cdots} ,$$
(39.2)

where

$$\varphi_k = a_k - r_{k-1}(b_k + r_k), \qquad k \geq 1.$$
(39.3)

(The parameters r_k, $k \geq 0$, have no restrictions other than those imposed above.)

Let

$$F(x) = F(x, \ell, n) = \frac{x^2 - \ell^2 + n^2 - 1}{2} + \frac{1^2 - n^2}{1} \; \frac{1^2 - \ell^2}{+ \; x^2 - 1}$$

$$+ \frac{3^2 - n^2}{1} \; \frac{3^2 - \ell^2}{+ \; x^2 - 1 + \cdots} .$$
(39.4)

In the notation above, $a_{2k} = (2k - 1)^2 - \ell^2$, $a_{2k-1} = (2k - 1)^2 - n^2$, $b_{2k} = x^2 - 1$, and $b_{2k-1} = 1$, where $k \geq 1$. Write

$$r_{2k} = d_1k + c_1 \quad \text{and} \quad r_{2k-1} = d_2k + c_2, \qquad k \geq 1.$$
(39.5)

Our first goal is to determine c_1, c_2, d_1, and d_2 so that φ_k is constant for $k \geq 1$.

From (39.3), (39.5), and the aforementioned formula for a_{2k}, it follows that

$$d_1 d_2 = 4. \tag{39.6}$$

Thus, from (39.3), we find that

$$\varphi_{2k} = (2k-1)^2 - \ell^2 - (d_2 k + c_2)(x^2 - 1 + d_1 k + c_1)$$
$$= -\{4 + d_2(x^2 - 1 + c_1) + c_2 d_1\}k + 1 - \ell^2 - c_2(x^2 - 1 + c_1) \tag{39.7}$$

and

$$\varphi_{2k-1} = (2k-1)^2 - n^2 - \{d_1(k-1) + c_1\}\{1 + d_2 k + c_2\}$$
$$= -\{4 + d_1(1 + c_2) - d_2(d_1 - c_1)\}k$$
$$\quad + 1 - n^2 + (d_1 - c_1)(1 + c_2), \tag{39.8}$$

where $k \geq 1$. By our prescriptions, we require that

$$d_2(x^2 - 1 + c_1) + c_2 d_1 = -4 = d_1(1 + c_2) - d_2(d_1 - c_1). \tag{39.9}$$

Using (39.6) and simplifying the extremal equality above, we find that

$$d_1^2 - 4d_1 + 4(1 - x^2) = 0.$$

We shall choose the positive root $d_1 = 2x + 2$. Thus, by (39.6), $d_2 = 2/(x + 1)$.

Since we wish φ_k to be constant, by (39.7) and (39.8), we need to stipulate that

$$1 - \ell^2 - c_2(x^2 - 1 + c_1) = 1 - n^2 + (d_1 - c_1)(1 + c_2).$$

Simplifying, we find that

$$c_1 - c_2(x + 1)^2 = \ell^2 - n^2 + 2(x + 1). \tag{39.10}$$

On the other hand, from (39.9),

$$c_1 + c_2(x + 1)^2 = -(x + 1)^2. \tag{39.11}$$

Adding (39.10) and (39.11), we deduce that

$$c_1 = \tfrac{1}{2}(\ell^2 - n^2 - x^2 + 1),$$

and so

$$c_2 = -1 - \frac{\ell^2 - n^2 - x^2 + 1}{2(x + 1)^2}.$$

Hence, we have determined the parameters c_1, c_2, d_1, and d_2 so that φ_k is constant, namely, from (39.5),

$$r_{2k} = 2(x + 1)k + \tfrac{1}{2}(\ell^2 - n^2 - x^2 + 1), \qquad k \geq 1, \tag{39.12}$$

and

$$r_{2k-1} = \frac{2}{x+1}k - 1 - \frac{\ell^2 - n^2 - x^2 + 1}{2(x+1)^2}. \tag{39.13}$$

Let us set $\varphi_k = \alpha$. By (39.8) and our determinations above,

$$\alpha = 1 - n^2 + (d_1 - c_1)(1 + c_2)$$

$$= \frac{4(1 - n^2)(x+1)^2 - 4(x+1)(\ell^2 - n^2 - x^2 + 1) + (\ell^2 - n^2 - x^2 + 1)^2}{4(x+1)^2}$$

$$= \frac{-4n^2(x+1)^2 + \{\ell^2 - n^2 - x^2 + 1 - 2(x+1)\}^2}{4(x+1)^2}$$

$$= \frac{\{\ell^2 - n^2 - x^2 + 1 - 2(1+n)(x+1)\}\{\ell^2 - n^2 - x^2 + 1 - 2(1-n)(x+1)\}}{4(x+1)^2}.$$

The numerator above is a polynomial in x of degree 4. It is easily checked that the four roots of this polynomial are $x + 1 = \pm \ell \pm n$, where all four possible combinations of signs are taken. Hence,

$$\alpha = \frac{(x + 1 + \ell + n)(x + 1 - \ell - n)(x + 1 + \ell - n)(x + 1 - \ell + n)}{4(x+1)^2}. \tag{39.14}$$

Recalling the definition (39.4), applying (39.2), and employing (39.12) and (39.13), we have shown that

$$F(x) = \cfrac{\alpha}{\cfrac{2}{x+1} - \cfrac{\ell^2 - n^2 - x^2 + 1}{2(x+1)^2} + \cfrac{1^2 - n^2}{x^2 - 1 + 2x + 2}}$$

$$+ \frac{1^2 - \ell^2}{1 + 2/(x+1)} + \frac{3^2 - n^2}{x^2 - 1 + 2x + 2} + \frac{3^2 - \ell^2}{1 + 2/(x+1)} + \cdots$$

$$= \cfrac{\alpha(x+1)^2}{\{(x+2)^2 - \ell^2 + n^2 - 1\}/2 + \cfrac{1^2 - n^2}{1} + \cfrac{1^2 - \ell^2}{(x+1)(x+3)}}$$

$$+ \frac{3^2 - n^2}{1} + \frac{3^2 - \ell^2}{(x+1)(x+3)} + \cdots$$

$$= \frac{\alpha(x+1)^2}{F(x+2)}. \tag{39.15}$$

For brevity, set, for any function f,

$$\prod_{\pm} f(x + k \pm \ell \pm n)$$

$$= f(x + k + \ell + n)f(x + k - \ell - n)f(x + k + \ell - n)f(x + k - \ell + n).$$

Hence, from (39.14) and (39.15),

$$F(x)F(x+2) = \frac{1}{4}\prod_{\pm}(x + 1 \pm \ell \pm n),$$

and so

$$\frac{F(x)F(x+2)}{F(x+2)F(x+4)} = \prod_{\pm} \left(\frac{x+1\pm\ell\pm n}{x+3\pm\ell\pm n}\right).$$

By iteration of this formula, we find that, for each positive integer m,

$$\frac{F(x)}{F(x+4m)} = \prod_{k=0}^{m-1} \prod_{\pm} \left(\frac{x+4k+1\pm\ell\pm n}{x+4k+3\pm\ell\pm n}\right)$$

$$= \frac{1}{m^2} \prod_{\pm} \prod_{k=0}^{m-1} \frac{(\frac{1}{4}(x+1\pm\ell\pm n)+k)m!\,m^{(x-1\pm\ell\pm n)/4}}{(\frac{1}{4}(x+3\pm\ell\pm n)+k)m!\,m^{(x-3\pm\ell\pm n)/4}}.$$

Hence,

$$\lim_{m\to\infty} \frac{F(x)m^2}{F(x+4m)} = \prod_{\pm} \frac{\Gamma(\frac{1}{4}(x+3\pm\ell\pm n))}{\Gamma(\frac{1}{4}(x+1\pm\ell\pm n))} = \frac{1}{P}. \tag{39.16}$$

From the definition of $F(x)$ in (39.4), we easily see that

$$\lim_{m\to\infty} \frac{m^2}{F(x+4m)} = \frac{1}{8}. \tag{39.17}$$

Combining (39.16) and (39.17), we deduce (39.1).   □

H. Cohen has communicated to us a similar proof of Entry 39. His proof is based on Apéry's method for accelerating the convergence of a continued fraction. For a complete description of this method, see Cohen's seminar notes [1]. Accounts are also given in papers by Apéry [1] and Batut and Olivier [1].

The equality (39.2) is called the Bauer–Muir transformation. Jacobsen [5] has shown that the conditions for its validity can be considerably relaxed.

We might note an interesting consequence of Entry 39. From Malmstén's integral representation for Log $\Gamma(z)$ (Whittaker and Watson [1, p. 249]), we find that

$$\text{Log } P = \int_0^\infty \left(\frac{\sum_{\pm} e^{-(x\pm\ell\pm n+1)t/4} - \sum_{\pm} e^{-(x\pm\ell\pm n+3)t/4}}{1-e^{-t}} - 2e^{-t}\right)\frac{dt}{t},$$

where $\sum_{\pm}$ indicates a sum of four terms with each possible combination of signs taken. Simplifying, we find that

$$\text{Log } P = 2\int_0^\infty \left(\frac{e^{-tx/4}\cosh(\ell t/4)\cosh(nt/4)}{\cosh(t/4)} - e^{-t}\right)\frac{dt}{t}$$

$$= 2\int_0^\infty e^{-tx}\left(\frac{\cosh(\ell t)\cosh(nt)}{\cosh t} - 1\right)\frac{dt}{t} - 2\int_0^\infty \frac{e^{-4t}-e^{-xt}}{t}\,dt$$

$$= 2\int_0^\infty e^{-tx}\left(\frac{\cosh(\ell t)\cosh(nt)}{\cosh t} - 1\right)\frac{dt}{t} + 2\text{ Log }\left(\frac{4}{x}\right), \tag{39.18}$$

by Frullani's theorem (Edwards [2, pp. 337–342] or Part I [9, p. 313, Eq.

(2.15)]). Exponentiating (39.18) and combining the result with (39.1), we deduce that

$$\exp\left(2\int_0^\infty e^{-tx}\left(\frac{\cosh(\ell t)\cosh(nt)}{\cosh t}-1\right)\frac{dt}{t}\right)$$

$$=\frac{x^2/2}{(x^2-\ell^2+n^2-1)/2\,+}\,\frac{1^2-n^2}{1\,+}\,\frac{1^2-\ell^2}{x^2-1\,+}\,\frac{3^2-n^2}{1\,+}\,\frac{3^2-\ell^2}{x^2-1\,+}\cdots,$$

$$(39.19)$$

where $0\le|\ell|,|n|<1$ and $x>1$.

The expansion (39.19) appears to be new. It generalizes a result of Rogers [3] and is similar to results of both Rogers [3] and Stieltjes [1], [4, pp. 378–391].

Entry 40. *Let*

$$P=\prod\Gamma(\tfrac{1}{2}(\alpha\pm\beta\pm\gamma\pm\delta\pm\varepsilon+1)),$$

where the product contains eight gamma functions and where the argument of each gamma function contains an even number of minus signs. Let

$$Q=\prod\Gamma(\tfrac{1}{2}(\alpha\pm\beta\pm\gamma\pm\delta\pm\varepsilon+1)),$$

where the product contains eight gamma functions and where the argument of each gamma function contains an odd number of minus signs. Suppose that at least one of the parameters $\beta,\gamma,\delta,\varepsilon$ is equal to a nonzero integer. Then

$$\frac{P-Q}{P+Q}$$

$$=\frac{8\alpha\beta\gamma\delta\varepsilon}{1\{2(\alpha^4+\beta^4+\gamma^4+\delta^4+\varepsilon^4+1)-(\alpha^2+\beta^2+\gamma^2+\delta^2+\varepsilon^2-1)^2-2^2\}}$$

$$+\frac{64(\alpha^2-1^2)(\beta^2-1^2)(\gamma^2-1^2)(\delta^2-1^2)(\varepsilon^2-1^2)}{3\{2(\alpha^4+\beta^4+\gamma^4+\delta^4+\varepsilon^4+1)-(\alpha^2+\beta^2+\gamma^2+\delta^2+\varepsilon^2-5)^2-6^2\}}$$

$$+\frac{64(\alpha^2-2^2)(\beta^2-2^2)(\gamma^2-2^2)(\delta^2-2^2)(\varepsilon^2-2^2)}{5\{2(\alpha^4+\beta^4+\gamma^4+\delta^4+\varepsilon^4+1)-(\alpha^2+\beta^2+\gamma^2+\delta^2+\varepsilon^2-13)^2-14^2\}}+\cdots.$$

$$(40.1)$$

Entry 40 is certainly one of Ramanujan's crowning achievements in the theory of continued fractions. Watson [8] has given the only published proof of Entry 40.

In an address before the London Mathematical Society in 1931, Watson [7] discussed Entry 40 but incorrectly wrote 9 and 10 instead of 13 and 14, respectively, in the last recorded denominator above. In a footnote of [8], Watson remarked: "Through an error in copying which occurred when I previously published an enunciation of the theorem...." However, Watson did copy the result faithfully; Ramanujan had made the same error (p. 152).

Throughout the notebooks, Ramanujan normally did not completely state identities involving sequences, but he did usually give enough terms to determine the sequence. In particular, if a sequence is linear, Ramanujan often gave only two terms, while if a sequence is quadratic, he would give three. In the first notebook, he only stated two terms of the sequences $2n^2 + 2n + 1$ and $2n^2 + 2n + 2$; that is, 1, 5 and 2, 6, respectively, that occur on the right side of (40.1). This was probably carelessness on his part for he most likely knew the quadratic patterns of the sequences. When he wrote his second notebook, a revised enlargement of the first, he decided to add one more term. However, he evidently did not rederive his identity and erroneously assumed that the two sequences are linear. Ironically, Watson's statement of Entry 40 in [8] also contains a misprint. Watson [8] also obtained a q-analogue of Entry 40.

It is natural to ask if the hypotheses on β, γ, δ, and ε can be relaxed. Jacobsen [3] has answered this by proving the following theorem.

Theorem. *The continued fraction on the right side of* (40.1) *converges to a meromorphic function* $F(\alpha, \beta, \gamma, \delta, \varepsilon)$ *in* $\mathscr{C}^5$. *Furthermore,* $F \neq (P - Q)/(P + Q)$.

The identity of F is not known.

Entry 41. *Let* x *and* γ *be complex numbers such that either* $|x + 1| > 1$ *or* γ *is a nonnegative integer. Then*

$$
{}_2F_1(-\beta, 1; \gamma + 1; -x) = \frac{\Gamma(\beta + 1)\Gamma(\gamma + 1)(1 + x)^{\beta + \gamma}}{\Gamma(\beta + \gamma + 1)x^\gamma}
$$

$$
- \cfrac{\gamma}{(\beta + 1)x + 1 - \gamma - \cfrac{1(1 - \gamma)(x + 1)}{(\beta + 2)x + 3 - \gamma}}
$$

$$
- \cfrac{2(2 - \gamma)(x + 1)}{(\beta + 3)x + 5 - \gamma} - \cdots. \tag{41.1}
$$

PROOF. From Erdélyi's treatise [1, p. 108, formula (2)],

$$
{}_2F_1(-\beta, 1; \gamma + 1; -x)
$$

$$
= \frac{\Gamma(-\beta - 1)\Gamma(\gamma + 1)}{\Gamma(-\beta)\Gamma(\gamma)x} {}_2F_1(1, 1 - \gamma; \beta + 2; -1/x)
$$

$$
+ \frac{\Gamma(\beta + 1)\Gamma(\gamma + 1)x^\beta}{\Gamma(\beta + \gamma + 1)} {}_2F_1(-\beta, -\beta - \gamma; -\beta; -1/x)
$$

$$
= -\frac{\gamma}{(\beta + 1)x} {}_2F_1(1, 1 - \gamma; \beta + 2; -1/x) + \frac{\Gamma(\beta + 1)\Gamma(\gamma + 1)(x + 1)^{\beta + \gamma}}{\Gamma(\beta + \gamma + 1)x^\gamma}.
$$

$$
\tag{41.2}
$$

Now apply (21.3) with β, γ, and x replaced by $-\gamma, \beta + 1$, and $1/x$, respectively, to deduce that

$$-\frac{\gamma}{(\beta + 1)x}\,_2F_1(1, 1 - \gamma; \beta + 2; -1/x)$$

$$= -\frac{\gamma/x}{\beta + 1 + (1 - \gamma)/x} - \frac{1(1 - \gamma)(1 + 1/x)/x}{\beta + 2 + (3 - \gamma)/x} - \frac{2(2 - \gamma)(1 + 1/x)/x}{\beta + 3 + (5 - \gamma)/x} - \cdots.$$

$$(41.3)$$

Translating the conditions under which (21.3) is valid, we find that (41.3) holds if $\mathrm{Re}\ 1/x > -\frac{1}{2}$ with not both $1 - \gamma$ and $\beta + \gamma + 1$ belonging to $\{0, -1, -2, \ldots\}$, or if γ is a nonnegative integer and $\beta + \gamma + 1 \notin \{0, -1, -2, \ldots\}$. Combining (41.2) and (41.3), we obtain (41.1) under the conditions given in the previous sentence. Now $\mathrm{Re}\ 1/x > -\frac{1}{2}$ if and only if $|x + 1| > 1$. Lastly, Jacobsen [3] has employed the uniform parabola theorem to remove the extraneous conditions on β and γ given above. $\qquad\square$

Entry 42. *If n is a nonnegative integer, or if $x \notin (-\infty, 0]$, then*

$$_1F_1(1; n + 1; x)$$

$$= \frac{e^x\Gamma(n + 1)}{x^n} - \frac{n}{x +} \frac{1 - n}{1} \frac{1}{+ x +} \frac{2 - n}{1} \frac{2}{+ x +} \frac{3 - n}{1} \frac{3}{+ x + \cdots}$$

$$= \frac{e^x\Gamma(n + 1)}{x^n} - \frac{n}{x + 1 - n} - \frac{1(1 - n)}{x + 3 - n} - \frac{2(2 - n)}{x + 5 - n} - \frac{3(3 - n)}{x + 7 - n} - \cdots.$$

$$(42.1)$$

PROOF. In Entry 41, replace x by x/β and γ by n. If $|1 + x/\beta| > 1$, or if n is a nonnegative integer, we find that

$$_2F_1(-\beta, 1; n + 1; -x/\beta) = \frac{\Gamma(\beta + 1)\Gamma(n + 1)(1 + x/\beta)^{\beta + n}}{\Gamma(\beta + n + 1)(x/\beta)^n}$$

$$- \frac{\gamma}{(\beta + 1)x/\beta + 1 - n} - \frac{1(1 - n)(1 + x/\beta)}{(\beta + 2)x/\beta + 3 - n}$$

$$- \frac{2(2 - n)(1 + x/\beta)}{(\beta + 3)x/\beta + 5 - n} - \cdots.$$

Since the continued fraction above converges uniformly with respect to β in a neighborhood of $\beta = \infty$, we may let β tend to ∞ to deduce the second equality in (42.1).

To obtain the first equality in (42.1), apply Entry 14. $\qquad\square$

Entry 42 was first discovered by Legendre [1]. See also Nielsen's book [1, p. 217] for a proof.

Corollary. *If either x is exterior to $(-\infty, 0]$ or if n is a positive integer, then*

$$\sum_{k=0}^{\infty} \frac{(-x)^k}{k!(n + k)} = \frac{\Gamma(n)}{x^n} - \frac{e^{-x}}{x +} \frac{1 - n}{1} \frac{1}{+ x +} \frac{2 - n}{1} \frac{2}{+ x + \cdots}.$$

PROOF. Multiplying both sides of (42.1) by e^{-x}/n and comparing the resulting equality with that above, we see that we must show that

$$e^{-x} \sum_{k=0}^{\infty} \frac{x^k}{(n)_{k+1}} = \sum_{k=0}^{\infty} \frac{(-x)^k}{k!(n+k)}. \tag{42.2}$$

Applying Entry 21 of Chapter 10 with x replaced by $-x$, n replaced by $n+1$, and $m = n$, we deduce (42.2). □

Entry 43. *If x is any complex number outside $(-\infty, 0]$, then*

$$\sum_{k=0}^{\infty} \frac{x^k}{1 \cdot 3 \cdots (2k+1)} = \sqrt{\frac{\pi}{2x}} \, e^{x/2} - \frac{1}{x+1} + \frac{1}{x+1} + \frac{2}{x+1} + \frac{3}{x+1} + \frac{4}{x+1} + \frac{5}{x+1} + \cdots$$

$$= \sqrt{\frac{\pi}{2x}} \, e^{x/2} - \frac{1}{x+1} - \frac{1 \cdot 2}{x+5} + \frac{3 \cdot 4}{x+9} - \frac{5 \cdot 6}{x+13} - \cdots.$$
$$\tag{43.1}$$

PROOF. Putting $n = \frac{1}{2}$ in Entry 42, we find that

$$\sum_{k=0}^{\infty} \frac{(2x)^k}{1 \cdot 3 \cdots (2k+1)} = \sqrt{\frac{\pi}{4x}} \, e^x - \frac{1/2}{x} + \frac{1/2}{1} + \frac{1}{x} + \frac{3/2}{1} + \frac{2}{x} + \frac{5/2}{1} + \cdots.$$

Replacing x by $x/2$, we obtain an equivalent form of the first continued fraction of (43.1).

The second continued fraction in (43.1) follows in the same way from the second continued fraction of (42.1). Alternatively, apply Entry 14 to the first continued fraction in (43.1). □

Corollary 1. *For Re $x > 0$,*

$$F(x) := \int_0^x e^{-t^2} \, dt = \frac{\sqrt{\pi}}{2} - \frac{e^{-x^2}}{2x} + \frac{1}{x} + \frac{2}{2x} + \frac{3}{x} + \frac{4}{2x} + \cdots.$$

PROOF. By (42.2), for $n > 0$,

$$\int_0^x e^{-t} t^{n-1} \, dt = \sum_{k=0}^{\infty} \frac{(-1)^k x^{n+k}}{k!(n+k)} = e^{-x} \sum_{k=0}^{\infty} \frac{x^{n+k}}{(n)_{k+1}}. \tag{43.2}$$

Let $n = \frac{1}{2}$ and replace t by t^2 and x by x^2. Applying Entry 43, we then find that, for x^2 exterior to $(-\infty, 0]$,

$$F(x) = xe^{-x^2} \sum_{k=0}^{\infty} \frac{x^{2k}}{(\frac{3}{2})_k} = xe^{-x^2} \sum_{k=0}^{\infty} \frac{(2x^2)^k}{1 \cdot 3 \cdots (2k+1)}$$

$$= xe^{-x^2} \left\{ \sqrt{\frac{\pi}{4x^2}} \, e^{x^2} - \frac{1}{2x^2} + \frac{1}{1} + \frac{2}{2x^2} + \frac{3}{1} + \frac{4}{2x^2} + \frac{5}{1} + \cdots \right\},$$

which is equivalent to the proposed formula. □

Corollary 2. *Let x be real. Then as x tends to* ∞,

$$\int_0^x \frac{F(t)}{t}\, dt = \frac{\sqrt{\pi}}{2}\left(\frac{\gamma}{2} + \text{Log}(2x)\right) + o(1), \tag{43.3}$$

where F is defined in Corollary 1 and γ *denotes Euler's constant.*

PROOF. Integrating by parts, we find that

$$\int_0^x \frac{F(t)}{t}\, dt = F(x)\,\text{Log}\,x - \int_0^x e^{-t^2}\,\text{Log}\,t\, dt$$

$$= \left(\int_0^\infty e^{-t^2}\, dt - \int_x^\infty e^{-t^2}\, dt\right)\text{Log}\,x$$

$$- \left(\int_0^\infty e^{-t^2}\,\text{Log}\,t\, dt - \int_x^\infty e^{-t^2}\,\text{Log}\,t\, dt\right)$$

$$= \frac{\sqrt{\pi}}{2}\,\text{Log}\,x - \int_0^\infty e^{-t^2}\,\text{Log}\,t\, dt + o(1), \tag{43.4}$$

as x tends to ∞.

From the integral definition of $\Gamma(x)$, for $x > 0$,

$$\Gamma'(x) = 4\int_0^\infty e^{-t^2} t^{2x-1}\,\text{Log}\,t\, dt.$$

In particular,

$$\Gamma'(\tfrac{1}{2}) = 4\int_0^\infty e^{-t^2}\,\text{Log}\,t\, dt = -\sqrt{\pi}(\gamma + 2\,\text{Log}\,2), \tag{43.5}$$

which was established by Ramanujan (p. 92) in Chapter 8. (See our book [9, p. 184, Cor. 3(i)].) Employing (43.5) in (43.4), we deduce (43.3) at once. □

Entry 44. *For x > 0, define*

$$\varphi(x) = \int_0^\infty \frac{e^{-t}}{x+t}\, dt.$$

Then for x > 0,

$$\int_0^x \frac{1 - e^{-t}}{t}\, dt = \sum_{k=1}^\infty \frac{(-1)^{k-1} x^k}{k!k} = \gamma + \text{Log}\,x + e^{-x}\varphi(x), \tag{44.1}$$

where γ *denotes Euler's constant.*

PROOF. At the outset, we remark that essentially the same calculations are made in slightly more detail in our edited version of Chapter 4 [9, p. 103].

The first equality in (44.1) is readily established by writing the integrand as a Maclaurin series and inverting the order of summation and integration.

Next, making a simple change of variable in the definition of φ and using a well-known integral representation for γ (Olver [1, p. 40]), we find that

$$e^{-x}\varphi(x) + \gamma + \operatorname{Log} x = \int_x^\infty \frac{e^{-t}}{t}\, dt + \int_0^1 \frac{1 - e^{-t}}{t}\, dt - \int_1^\infty \frac{e^{-t}}{t}\, dt + \int_1^x \frac{dt}{t}.$$

Upon simplification, we complete the proof of the second equality in (44.1).

$\square$

Entry 44(i). *Let x be real. Then as x approaches ∞,*

$$\varphi(x) \sim \sum_{k=0}^\infty \frac{(-1)^k k!}{x^{k+1}}.$$

Entry 44(i) was established by Euler, and a rigorous discussion of it can be found in Hardy's book [5, pp. 26, 27]. Ramanujan also stated this result in Chapter 4 (p. 44); see our book [9, pp. 101–102].

For Entry 44(ii), we quote Ramanujan (p. 153).

Entry 44(ii). *$\varphi(x)$ lies between $1/x$ and $1/(x + 1)$ and very nearly equals $\sqrt{\varphi(x + 1)/x}$.*

PROOF. Letting n tend to 0 in the corollary of Section 42, we find that, for $x > 0$,

$$\sum_{k=1}^\infty \frac{(-1)^{k+1} x^k}{k!\,k} = \lim_{n \to 0}\left(\frac{1}{n} - \frac{\Gamma(n)}{x^n}\right) + e^{-x}f(x)$$

$$= \gamma + \operatorname{Log} x + e^{-x}f(x), \tag{44.2}$$

where

$$f(x) = \frac{1}{x} + \frac{1}{1} + \frac{1}{x} + \frac{2}{1} + \frac{2}{x} + \cdots. \tag{44.3}$$

Comparing (44.1) and (44.2), we deduce that $f(x) = \varphi(x)$.

Now from (44.3), it is immediate that $\varphi(x) < 1/x$. Next, if

$$F = x + \frac{2}{1} + \frac{2}{x} + \frac{3}{1} + \frac{3}{x} + \cdots,$$

we can write (44.3) as

$$\varphi(x) = \frac{1}{x + 1/(1 + 1/F)} = \frac{1}{x + \dfrac{F}{1 + F}} > \frac{1}{x + 1}.$$

Thus, Ramanujan's upper and lower bounds for $\varphi(x)$ are established.

Squaring the asymptotic series from Entry 44(i), we find that, as x tends to ∞,

$$\varphi^2(x) \sim \frac{1}{x^2} - \frac{2}{x^3} + \frac{5}{x^4} - \frac{16}{x^5} + \cdots .$$

On the other hand, also from Entry 44(i), as x tends to ∞,

$$\frac{\varphi(x+1)}{x} \sim \frac{1}{x(x+1)} - \frac{1}{x(x+1)^2} + \frac{2}{x(x+1)^3} + \cdots$$

$$= \frac{1}{x^2}\left(1 - \frac{1}{x} + \frac{1}{x^2} - \frac{1}{x^3}\right) - \frac{1}{x^3}\left(1 - \frac{2}{x} + \frac{3}{x^2}\right)$$

$$+ \frac{2}{x^4}\left(1 - \frac{3}{x}\right) + O\left(\frac{1}{x^6}\right)$$

$$= \frac{1}{x^2} - \frac{2}{x^3} + \frac{5}{x^4} - \frac{10}{x^5} + O\left(\frac{1}{x^6}\right).$$

Thus, the initial three terms of the asymptotic expansions for $\varphi^2(x)$ and $\varphi(x+1)/x$ agree. Hence, Ramanujan's approximation for $\varphi(x)$ is reasonable.

$\square$

Entry 44(iii). *For $x > 0$,*

$$\varphi(x) = \frac{1}{x} \, \frac{1}{+} \, \frac{1}{x} \, \frac{1}{+} \, \frac{1}{x} \, \frac{2}{+} \, \frac{2}{x} \, \frac{2}{+} \, \frac{3}{x} \, \frac{3}{+} \, \frac{3}{x} + \cdots$$

$$= \frac{1}{x+1} - \frac{1^2}{x+3} - \frac{2^2}{x+5} - \frac{3^2}{x+7} - \cdots .$$

PROOF. The former continued fraction was established in the course of proving Entry 44(ii) (see (44.3)). To obtain the latter continued fraction, apply Entry 14.

$\square$

In fact, Entry 44(iii) is valid for all complex x outside $(-\infty, 0)$ (Jacobsen [3]).

The second continued fraction above was first derived by Tschebyscheff [1].

Entry 44(iv). *Let x be any complex number exterior to $(-\infty, 0]$, and let n be a natural number. Then*

$$\varphi(x) = \sum_{k=0}^{n-1} \frac{(-1)^k k!}{x^{k+1}} + \frac{(-1)^n n!}{x^n}$$

$$\times \left(\frac{1}{x+n+1} - \frac{n+1}{x+n+3} - \frac{2(n+2)}{x+n+5} - \frac{3(n+3)}{x+n+7} - \cdots \right).$$

PROOF. Integrating by parts n times, we find that

$$\varphi(x) = \sum_{k=0}^{n-1} \frac{(-1)^k k!}{x^{k+1}} + (-1)^n n! \int_0^\infty \frac{e^{-t}}{(x+t)^{n+1}} \, dt$$

$$= \sum_{k=0}^{n-1} \frac{(-1)^k k!}{x^{k+1}} + \frac{(-1)^n}{x^n} \int_0^\infty \frac{e^{-t} t^n}{x+t} \, dt, \qquad (44.4)$$

where we have used the equality (Perron [3, p. 219])

$$\frac{1}{\Gamma(b)} \int_0^\infty \frac{e^{-t} t^{b-1}}{(1+t)^a} \, dt = \frac{1}{\Gamma(a)} \int_0^\infty \frac{e^{-t} t^{a-1}}{(1+t)^b} \, dt, \qquad \text{Re } a, \text{ Re } b > 0.$$

However, for $x \notin (-\infty, 0]$ (Perron [3, p. 219, Eq. (12)], Khovanskii [1, p. 148, Eq. (11.17)]),

$$\frac{1}{n!} \int_0^\infty \frac{e^{-t} t^n}{x+t} \, dt = \frac{1}{x+n+1} - \frac{n+1}{x+n+3} - \frac{2(n+2)}{x+n+5} - \frac{3(n+3)}{x+n+7} - \cdots .$$

$$(44.5)$$

Substituting (44.5) into (44.4), we deduce the proposed identity.   □

Corollary 1. *Let*

$$H_n = \sum_{k=1}^n \frac{1}{k}.$$

Then if $x > 0$,

$$\sum_{k=1}^\infty \frac{H_k x^k}{k!} = e^x (\text{Log } x + \gamma) + \varphi(x).$$

Corollary 1 is also given by Ramanujan in Chapter 4 (p. 44). See the author's book [9, p. 103] for a proof.

Our formulation of Corollary 2 corrects that given by Ramanujan (p. 153).

Corollary 2. *For $|h| < 1$ and $n > 0$, define $f(h, n)$ by*

$$\int_0^{n(1-h)} \frac{1 - e^{-t}}{t} \, dt = \gamma + \text{Log } n + e^{-n} \varphi(n) - e^{-n} f(h, n). \qquad (44.6)$$

Then

$$f(h, n) = \sum_{k=1}^\infty \frac{1}{k} \left(e^n - \sum_{j=0}^{k-1} \frac{n^j}{j!} \right) h^k.$$

PROOF. First, if $h = 0$, we see from Entry 44 that $f(0, n) = 0$. For brevity, set $g(h) = f(h, n)$. Clearly, we shall be finished if we can show that

$$g^{(k+1)}(0) = k! \left(e^n - \sum_{j=0}^k \frac{n^j}{j!} \right), \qquad k \geq 0. \qquad (44.7)$$

First, differentiating (44.6), we find that

$$e^{-n}g'(h) = \frac{e^{nh-n} - 1}{h - 1}.$$ (44.8)

Setting $h = 0$ in (44.8), we deduce (44.7) in the case $k = 0$. For $k > 0$, we apply Leibniz's rule to (44.8) to find that

$$e^{-n}g^{(k+1)}(h) = \sum_{j=0}^{k} \binom{k}{j} \frac{d^j}{dh^j} \left(\frac{1}{h-1}\right) \frac{d^{k-j}}{dh^{k-j}}(e^{nh-n} - 1)$$

$$= \sum_{j=0}^{k} \binom{k}{j} \frac{(-1)^j j!}{(h-1)^{j+1}} \{n^{k-j}e^{nh-n} - \delta_j\},$$

where $\delta_k = 1$ and $\delta_j = 0$, $0 \le j \le k - 1$. Thus,

$$g^{(k+1)}(0) = k!e^n - \sum_{j=0}^{k} \binom{k}{j} j! n^{k-j}.$$

Equality (44.7) now follows upon replacing j by $k - j$ above. ☐

Ramanujan concludes Section 44 by recording the values $\varphi(1) = 0.5963474$ and $\varphi(\frac{1}{2}) = 0.9229106$. From (44.1),

$$\varphi(1) = e\left(\sum_{k=1}^{\infty} \frac{(-1)^{k-1}}{k!k} - \gamma\right)$$

and

$$\varphi(\tfrac{1}{2}) = \sqrt{e}\left(\sum_{k=1}^{\infty} \frac{(-1)^{k-1}}{2^k k!k} - \gamma - \text{Log } 2\right).$$

Using calculated values for γ, e, $\sqrt{e}$, and Log 2 (Abramowitz and Stegun [1, pp. 2, 3]) and 11 and 9 terms, respectively, from the two sums above, we can readily verify that Ramanujan's calculations are correct.

Entries 45(i), (ii). *Consider the continued fraction*

$$\frac{1}{1} + \frac{x}{1} + \frac{x}{1} + \frac{2x}{1} + \frac{2x}{1} + \frac{3x}{1} + \frac{3x}{1} + \cdots + \frac{(n-1)x}{1} + \frac{nx}{1}.$$

Then in the notation of (1.4), for $n \ge 1$,

$$B_{2n}(x) := B_{2n} = \sum_{k=0}^{n} \frac{(-n)_k^2 x^k}{k!}$$ (45.1)

and

$$B_{2n-1}(x) := B_{2n-1} = \sum_{k=0}^{n-1} \frac{(-n)_k^2}{k!}\left(1 - \frac{k}{n}\right)x^k.$$ (45.2)

PROOF. We shall induct on n. For $n = 1$, both (45.1) and (45.2) are easily seen to be correct.

We shall thus assume that both (45.1) and (45.2) are true up to a specific positive integer n. By (1.4),

$$B_{2n+1}(x) = B_{2n}(x) + nxB_{2n-1}(x)$$

$$= \sum_{k=0}^{n} \frac{(-n)_k^2 x^k}{k!} + \sum_{k=1}^{n} \frac{(-n)_{k-1}^2 (n-k+1)x^k}{(k-1)!}.$$

But, for $1 \le k \le n$,

$$\frac{(-n)_k^2}{k!} + \frac{(-n)_{k-1}^2 (n-k+1)}{(k-1)!} = \frac{(-n)_{k-1}^2 (n-k+1)}{k!} \{(n-k+1) + k\}$$

$$= \frac{(-n-1)_k^2}{k!} \left(1 - \frac{k}{n+1} \right).$$

Hence, we have established (45.2) with n replaced by $n + 1$.

By (1.4) and the proof just completed above,

$$B_{2n+2}(x) = B_{2n+1}(x) + (n+1)xB_{2n}(x)$$

$$= \sum_{k=0}^{n+1} \frac{(-n-1)_k^2}{k!} \left(1 - \frac{k}{n+1} \right) x^k + (n+1) \sum_{k=1}^{n+1} \frac{(-n)_{k-1}^2}{(k-1)!} x^k.$$

But, for $1 \le k \le n + 1$,

$$\frac{(-n-1)_k^2}{k!} \left(1 - \frac{k}{n+1} \right) + (n+1) \frac{(-n)_{k-1}^2}{(k-1)!}$$

$$= \frac{(-n-1)_k^2}{k!} \left(1 - \frac{k}{n+1} + \frac{k}{n+1} \right) = \frac{(-n-1)_k^2}{k!}.$$

Hence, (45.1) is established with n replaced by $n + 1$.   □

We have slightly rearranged the ordering of the formulas in Section 46.

Entry 46(i). *For $|x| < 1$, set*

$$\Gamma(x+1) = \sum_{k=0}^{\infty} \frac{A_k(-x)^k}{k!}. \tag{46.1}$$

Define $\varphi_n(x)$ as the constant term in the Laurent expansion of $x^p\Gamma(1-p)/p^n$, $0 < |p| < 1$, where n is a nonnegative integer. Then, if $x \ne 0$,

$$\varphi_n(x) = \frac{1}{n!} \sum_{k=0}^{n} \binom{n}{k} A_{n-k} \operatorname{Log}^k x. \tag{46.2}$$

Furthermore, define $\psi_n(x)$, $n \ge 0$, by

$$\sum_{k=1}^{\infty} \frac{(-1)^{k-1} x^k}{k^n k!} = \varphi_n(x) + (-1)^{n-1} e^{-x} \psi_n(x). \tag{46.3}$$

Then, for $n \ge 1$,

$$\psi_n(x) - \psi_n'(x) = \frac{\psi_{n-1}(x)}{x}. \tag{46.4}$$

PROOF. First, for $|p| < 1$, by (46.1),

$$\frac{x^p \Gamma(1-p)}{p^n} = \frac{1}{p^n} \sum_{k=0}^{\infty} \frac{p^k \text{Log}^k x}{k!} \sum_{j=0}^{\infty} \frac{A_j p^j}{j!}$$

$$= \frac{1}{p^n} \sum_{r=0}^{\infty} \frac{1}{r!} \left(\sum_{k=0}^{r} \binom{r}{k} A_{r-k} \text{Log}^k x \right) p^r.$$

Equality (46.2) is now immediate.

Using (46.2) in (46.3) and differentiating both sides with respect to x, we find that, for $n \geq 1$,

$$(-1)^n e^{-x} \psi_n(x) + (-1)^{n-1} e^{-x} \psi_n'(x)$$

$$= \sum_{k=1}^{\infty} \frac{(-1)^{k-1} x^{k-1}}{k^{n-1} k!} - \frac{1}{n!} \sum_{k=1}^{n} \binom{n}{k} A_{n-k} k \frac{\text{Log}^{k-1} x}{x}$$

$$= \frac{1}{x} \left(\sum_{k=1}^{\infty} \frac{(-1)^{k-1} x^k}{k^{n-1} k!} - \frac{1}{(n-1)!} \sum_{k=0}^{n-1} \binom{n-1}{k} A_{n-1-k} \text{Log}^k x \right)$$

$$= \frac{1}{x} (-1)^{n-2} e^{-x} \psi_{n-1}(x).$$

The proof of (46.4) is now complete. □

Entry 46(ii). *For $n \geq 1$,*

$$A_n = \sum_{k=1}^{n} \frac{(n-1)!}{(n-k)!} S_k A_{n-k}, \tag{46.5}$$

where A_k is defined by (46.1), $S_1 = \gamma$, and $S_k = \zeta(k)$, $k \geq 2$, where ζ denotes the Riemann zeta-function.

PROOF. Entry 46(ii) is a reformulation of a well-known result that can be found in Luke's book [1, p. 27]. Namely, if

$$\Gamma(x+1) = \sum_{k=0}^{\infty} b_k x^k, \qquad |x| < 1,$$

then, for $n \geq 1$,

$$n b_n = \sum_{k=1}^{n} (-1)^k S_k b_{n-k}. \tag{46.6}$$

Translating the recursion formula (46.6) in terms of the coefficients A_k, we readily obtain (46.5). □

We state Entry 46(iii) as recorded by Ramanujan. Afterward, we discuss the accuracy of his numerical calculations.

Entry 46(iii). *In the notation* (46.6),

$$b_1 = -0.5772156649,$$
$$b_2 = \quad 0.9890560173,$$
$$b_3 = -0.9074790803,$$
$$b_4 = \quad 0.9817280965.$$

Furthermore, if we write

$$\Gamma(x+1) = 1 + b_1 x + b_2 x^2 + b_3 x^3 + b_4 \frac{x^4}{1 + \theta_x x}, \qquad (46.7)$$

then

$$\theta_0 = 1.00027,$$
$$\theta_1 = 51/52,$$
$$\theta_2 = 77/82,$$
$$\theta_6 = 5/68,$$
$$\theta_7 = -1/38$$

"nearly."

The coefficient b_1 is equal to $-\gamma$, and the numerical value that is given is correct. The given values for b_2, b_3, and b_4 do not seem to be correct. We have employed (46.6) along with values of S_k given in Abramowitz and Stegun's tables [1, p. 811] and have found that

$$b_2 = \quad 0.9890559953,$$
$$b_3 = -0.9074790762,$$
$$b_4 = \quad 0.9817280865.$$

Evidently, we are to interpret θ_x to be that unique number yielding an equality in (46.7). The values given by Ramanujan are rational approximations. The value for θ_0 is enigmatic, because, for $x = 0$, θ_0 is not well defined. In the table below, we give the calculated values of the right side of (46.7) using Ramanujan's determinations and also our determinations of b_2, b_3, and b_4.

x	θ_x	$\Gamma(x+1)$	Ramanujan's Value	Our Value
1	51/52	1	0.999990949	0.999990967
2	77/82	2	1.999702292	1.999702625
6	5/68	720	719.9611865	719.9612493
7	−1/38	5040	2623.541808	2623.542013

Thus, the values for θ_1, θ_2, and θ_6 give good approximations, but the value for θ_7 certainly does not.

We are very grateful to Henri Cohen for motivating the proof of Entry 46(iv) below. In particular, he informed us of formula (46.20). As in Entry 17, the equality below refers only to the correspondence between the two sides. The left side is a power series, and the continued fraction on the right side is the (unique) C-fraction corresponding to the power series.

Entry 46(iv). *If n is a nonnegative integer, then*

$$\psi_n(x) = \cfrac{x}{\left(x + \cfrac{n}{2 +} \cfrac{5n + 10}{6x +} \cfrac{41n + 58}{10 +} \cdots\right)^{n+1}}. \qquad (46.8)$$

PROOF. From (46.4), it is clear that $\psi_n(x)$ can be expressed as a power series in $1/x$. Putting

$$\psi_n(x) = \sum_{k=0}^{\infty} \frac{a_k(n)}{x^k}, \qquad n \geq 0,$$

we then write (46.4) in the form

$$\sum_{k=0}^{\infty} \frac{a_k(n)}{x^k} + \sum_{k=2}^{\infty} \frac{(k-1)a_{k-1}(n)}{x^k} = \sum_{k=1}^{\infty} \frac{a_{k-1}(n-1)}{x^k}, \qquad (46.9)$$

where $n \geq 1$. It follows immediately that $a_0(n) = 0$ if $n \geq 1$, $a_1(n) = 0$ if $n \geq 2$, and

$$a_k(n) + (k-1)a_{k-1}(n) = a_{k-1}(n-1), \qquad (46.10)$$

for $k \geq 2$ and $n \geq 1$. Now assume that, up to some fixed integer $k - 1$, $a_{k-1}(n) = 0$ if $n \geq k$. Thus, $a_{k-1}(n-1) = 0$ if $n \geq k + 1$. It follows from (46.10) and our inductive assumption that $a_k(n) = 0$ if $n \geq k + 1$. Hence, we shall rewrite (46.9) in the form

$$\sum_{k=0}^{\infty} \frac{b_k(n)}{x^{n+k}} + \sum_{k=1}^{\infty} \frac{(n+k-1)b_{k-1}(n)}{x^{n+k}} = \sum_{k=0}^{\infty} \frac{b_k(n-1)}{x^{n+k}}.$$

Hence, for $n \geq 1$,

$$b_0(n) = b_0(n-1) \qquad (46.11)$$

and, for $k, n \geq 1$,

$$b_k(n) + (n+k-1)b_{k-1}(n) = b_k(n-1). \qquad (46.12)$$

From the definition (46.3) of $\psi_n(x)$, it is easy to see that $\psi_0(x) \equiv 1$. Hence, by (46.11) and induction, we find that

$$b_0(n) = 1, \qquad n \geq 0. \qquad (46.13)$$

Next, in (46.12), let $k = 1$ and replace n by j. Since $b_0(j) = 1, j \geq 0$, we find that

$$b_1(j) + j = b_1(j - 1), \qquad j \geq 1. \tag{46.14}$$

Summing both sides of (46.14) for $1 \leq j \leq n$ and recalling that $b_1(0) = a_1(0) = 0$, we deduce that

$$b_1(n) + \sum_{j=1}^{n} j = 0,$$

or

$$b_1(n) = -\tfrac{1}{2}n(n + 1). \tag{46.15}$$

Put $k = 2$ and $n = j$ in (46.12) to obtain the equality

$$b_2(j) + (j + 1)b_1(j) = b_2(j - 1). \tag{46.16}$$

Sum both sides of (46.16) on j, $1 \leq j \leq n$. Using the fact that $b_2(0) = 0$ as well as (46.15), we find that

$$b_2(n) = \frac{1}{2} \sum_{j=1}^{n} (j^3 + 2j^2 + j)$$

$$= \tfrac{1}{24}n(n + 1)(n + 2)(3n + 5). \tag{46.17}$$

Lastly, we set $k = 3$ and $n = j$ in (46.12) and find that

$$b_3(j) + (j + 2)b_2(j) = b_3(j - 1). \tag{46.18}$$

Summing both sides of (46.18) for $1 \leq j \leq n$ and employing (46.17), we find that

$$b_3(n) = -\frac{1}{24} \sum_{j=1}^{n} j(j + 1)(j + 2)^2(3j + 5)$$

$$= -\tfrac{1}{48}n(n + 1)(n + 2)^2(n + 3)^2, \tag{46.19}$$

after a lengthy calculation. (Formulas for summing $\sum_{1 \leq j \leq n} j^k$, $1 \leq k \leq 5$, may be found in Gradshteyn and Ryzhik's tables [1, pp. 1, 2].)

In conclusion, from (46.13), (46.15), (46.17), and (46.19), we have demonstrated that, for x sufficiently large,

$$\psi_n(x) = \frac{1}{x^n}\left(1 - \frac{n(n + 1)}{2x} + \frac{n(n + 1)(n + 2)(3n + 5)}{24x^2}\right.$$

$$\left. - \frac{n(n + 1)(n + 2)^2(n + 3)^2}{48x^3} + \cdots\right). \tag{46.20}$$

Now, by (46.8), we wish to prove that

$$\{x^{-1}\psi_n(x)\}^{-1/(n+1)} = x\{x^n\psi_n(x)\}^{-1/(n+1)}$$

$$= x + \frac{n}{2} + \cfrac{\frac{5n + 10}{6x}}{1} + \cfrac{\frac{41n + 58}{10}}{1} + \cdots$$

$$= x\left(1 + \cfrac{\frac{n}{2x}}{1} + \cfrac{\frac{5n + 10}{12x}}{1} + \cfrac{\frac{41n + 58}{60x}}{1} + \cdots\right).$$

In fact, it will be slightly more convenient to show that the reciprocals of the expressions above ae equal. Hence, we shall prove that

$$\{x^n\psi_n(x)\}^{1/(n+1)} = \frac{1}{1} + \frac{\dfrac{n}{2x}}{1} + \frac{\dfrac{5n+10}{12x}}{1} + \frac{\dfrac{41n+58}{60x}}{1} + \cdots. \qquad (46.21)$$

In order to establish (46.21), we shall first compute the power series for $\{x^n\psi_n(x)\}^{1/(n+1)}$ in powers of $1/x$. By (46.20) and the binomial theorem, we find that, for x sufficiently large,

$$\{x^n\psi_n(x)\}^{1/(n+1)}$$

$$= 1 + \frac{1}{n+1}\left(-\frac{n(n+1)}{2x} + \frac{n(n+1)(n+2)(3n+5)}{24x^2}\right.$$

$$\left. - \frac{n(n+1)(n+2)^2(n+3)^2}{48x^3} + \cdots\right)$$

$$- \frac{n}{2(n+1)^2}\left(-\frac{n(n+1)}{2x} + \frac{n(n+1)(n+2)(3n+5)}{24x^2} + \cdots\right)^2$$

$$+ \frac{n(2n+1)}{6(n+1)^3}\left(-\frac{n(n+1)}{2x} + \cdots\right)^3 + \cdots.$$

We now compute the coefficients $c_1(n)$, $c_2(n)$, and $c_3(n)$ of $1/x$, $1/x^2$, and $1/x^3$, respectively. Clearly, $c_1(n) = -n/2$. Second,

$$c_2(n) = \frac{n(n+2)(3n+5)}{24} - \frac{n^3}{8} = \frac{n(11n+10)}{24}.$$

Third,

$$c_3(n) = -\frac{n(n+2)^2(n+3)^2}{48} + \frac{n^3(n+2)(3n+5)}{48} - \frac{n^4(2n+1)}{48}$$

$$= -\frac{n(9n^2+20n+12)}{16}.$$

Hence,

$$\{x^n\psi_n(x)\}^{1/(n+1)} = 1 - \frac{n}{2x} + \frac{n(11n+10)}{24x^2} - \frac{n(9n^2+20n+12)}{16x^3} + \cdots. \qquad (46.22)$$

We now employ Entry 17 to compute the continued fraction representation (46.21). In the notation of Entry 17, by (46.22),

$$A_1 = \frac{n}{2}, \quad A_2 = \frac{n(11n+10)}{24}, \quad \text{and} \quad A_3 = \frac{n(9n^2+20n+12)}{16}.$$

First,

$$a_1 = A_1 = \frac{n}{2}. \qquad (46.23)$$

Second,

$$P_2 = a_1(a_1 + a_2) = A_2 = \frac{n(11n + 10)}{24}.$$

Using (46.23) and solving for a_2, we readily find that

$$a_2 = \frac{5n + 10}{12}. \tag{46.24}$$

Lastly,

$$P_3 = a_1 a_2 (a_1 + a_2 + a_3) = A_3 - a_1 A_2 = \frac{n(9n^2 + 20n + 12)}{16} - \frac{n^2(11n + 10)}{24}.$$

Solving for a_3 and employing (46.23) and (46.24), we find, after a mild calculation, that

$$a_3 = \frac{41n + 58}{60}. \tag{46.25}$$

Employing (46.23)–(46.25) in Entry 17, we complete the proof of (46.21).   □

Example. *For $x > 0$, let*

$$F(x) = \int_0^x \frac{1 - e^{-t}}{t}\, dt.$$

Then

$$\lim_{x \to \infty} \left(\int_0^x \frac{F(t)}{t}\, dt - \tfrac{1}{2} F^2(x) \right) = \frac{\pi^2}{12}.$$

PROOF. First, from Entry 44,

$$\tfrac{1}{2} F^2(x) = \tfrac{1}{2}\gamma^2 + \tfrac{1}{2} \operatorname{Log}^2 x + \gamma \operatorname{Log} x + o(1), \tag{46.26}$$

as x tends to ∞.

Next, integrating by parts twice and using Entry 44, we find that, as x tends to ∞,

$$\int_0^x \frac{F(t)}{t}\, dt$$

$$= F(x) \operatorname{Log} x - \int_0^x \frac{1 - e^{-t}}{t} \operatorname{Log} t\, dt$$

$$= (\gamma + \operatorname{Log} x) \operatorname{Log} x + o(1) - \tfrac{1}{2}(1 - e^{-x}) \operatorname{Log}^2 x + \tfrac{1}{2}\int_0^x e^{-t} \operatorname{Log}^2 t\, dt. \tag{46.27}$$

Combining (46.26) and (46.27), we deduce that

$$\int_0^x \frac{F(t)}{t}\, dt - \tfrac{1}{2}F^2(x) = -\tfrac{1}{2}\gamma^2 + \frac{1}{2}\int_0^\infty e^{-t}\, \mathrm{Log}^2\, t\, dt + o(1)$$

$$= -\tfrac{1}{2}\gamma^2 + \tfrac{1}{2}\Gamma''(1) + o(1), \tag{46.28}$$

as x tends to ∞.

By Entry 26 of Chapter 7 (see the author's book [9, p. 176]),

$$\mathrm{Log}\,\Gamma(x+1) = -\gamma x + \sum_{k=2}^\infty \frac{\zeta(k)(-x)^k}{k}, \qquad |x| < 1.$$

Hence, after two differentiations,

$$\frac{\Gamma''(x+1)}{\Gamma(x+1)} - \psi^2(x+1) = \sum_{k=2}^\infty (-1)^k(k-1)\zeta(k)x^{k-2},$$

and so

$$\Gamma''(1) = \psi^2(1) + \zeta(2) = \gamma^2 + \pi^2/6.$$

Substituting the value for $\Gamma''(1)$ found above into (46.28) and letting x tend to ∞, we complete the proof. $\qquad\square$

Entry 47. *If n is any complex number outside of $(-\infty, 0]$, then*

$$\int_0^\infty e^{-x}(1 + x/n)^n\, dx$$

$$= 1 + \frac{n}{1+} \frac{1(n-1)}{3} \frac{2(n-2)}{+\ \ 5} \frac{3(n-3)}{+\ \ 7} + \cdots \tag{47.1}$$

$$= 2 + \frac{n-1}{2} \frac{1(n-2)}{+\ \ 4} \frac{2(n-3)}{+\ \ 6} \frac{3(n-4)}{+\ \ 8} + \cdots \tag{47.2}$$

$$= \frac{e^n\Gamma(n+1)}{n^n} - \frac{2n}{2} \frac{3n}{+\ 3} \frac{4n}{+\ 4} \frac{5n}{+\ 5} + \cdots. \tag{47.3}$$

PROOF. In (21.2), let $x = \gamma/n$ and $\beta = -n$. Thus, under certain restrictions on γ and n arising from (21.2),

$${}_2F_1(1-n, 1; \gamma+1; -\gamma/n)$$

$$= \frac{\gamma}{\gamma+} \frac{(1-n)\gamma/n}{1} \frac{1(1+\gamma/n)}{+\ \ \ \gamma} \frac{(2-n)\gamma/n}{+\ \ \ 1} \frac{2(1+\gamma/n)}{+\ \ \ \gamma} + \cdots. \tag{47.4}$$

Now, for $\mathrm{Re}(\gamma/n) > 0$ (Bailey [4, p. 4]),

$${}_2F_1(1-n, 1; \gamma+1; -\gamma/n) = \gamma \int_0^1 (1-t)^{\gamma-1}(1 + t\gamma/n)^{n-1}\, dt$$

$$= \int_0^\gamma (1 - u/\gamma)^{\gamma-1}(1 + u/n)^{n-1}\, du. \tag{47.5}$$

Thus, letting γ tend to ∞ in (47.4) and (47.5), we find that, for n exterior to $(-\infty, 0]$,

$$\int_0^\infty e^{-u}(1 + u/n)^{n-1}\, du = \frac{1}{1+} \; \frac{(1-n)/n}{1} \; + \frac{1/n}{1} + \; \frac{(2-n)/n}{1} \; + \frac{2/n}{1} + \cdots .$$

Integrating by parts once, adding 1 to both sides, and writing the right side above in an equivalent form, we see that

$$\int_0^\infty e^{-u}(1 + u/n)^n\, du = 1 + \frac{n}{n+} \; \frac{1-n}{1} \; + \frac{1}{n+} \; \frac{2-n}{1} \; + \frac{2}{n+} \cdots$$

$$= 1 + \frac{n}{1+} \; \frac{n-1}{3} \; + \frac{2(n-2)}{5} \; + \frac{3(n-3)}{7} + \cdots ,$$

by Entry 14. This completes the proof of (47.1).

Second, let $x = \gamma/n$ and $\beta = 1 - n$ in (21.2). Then, for $\mathrm{Re}(\gamma/n) > 0$,

$$\frac{n-1}{n}\,{}_2F_1(2 - n, 1; \gamma + 1; -\gamma/n)$$

$$= \frac{(n-1)\gamma/n}{\gamma} \; + \frac{(2-n)\gamma/n}{1} \; + \frac{1(1+\gamma/n)}{\gamma} \; + \frac{(3-n)\gamma/n}{1} \; + \frac{2(1+\gamma/n)}{\gamma} + \cdots .$$

Now proceed as above and let γ tend to ∞ to find that, if n is outside $(-\infty, 0]$,

$$\frac{n-1}{n}\int_0^\infty e^{-t}(1 + t/n)^{n-2}\, dt$$

$$= \frac{(n-1)/n}{1} \; + \frac{(2-n)/n}{1} \; + \frac{1/n}{1} + \; \frac{(3-n)/n}{1} \; + \frac{2/n}{1} + \cdots$$

$$= \frac{n-1}{n} \; + \frac{2-n}{1} \; + \frac{1}{n+} \; \frac{3-n}{1} \; + \frac{2}{n+} \; \frac{4-n}{1} + \cdots$$

$$= \frac{n-1}{2} \; + \frac{n-2}{4} \; + \frac{2(n-3)}{6} \; + \frac{3(n-4)}{8} + \cdots , \tag{47.6}$$

by Entry 14.

Assuming that n is any complex number outside $(-\infty, 0]$ and integrating by parts twice, we find that

$$\frac{n-1}{n}\int_0^\infty e^{-t}(1 + t/n)^{n-2}\, dt = -2 + \int_0^\infty e^{-t}(1 + t/n)^n\, dt.$$

Substituting the formula above into (47.6), we establish (47.2).

Third, setting $x = t - n$, we find that

$$\int_0^\infty e^{-x}(1 + x/n)^n\, dx = \frac{e^n}{n^n}\int_n^\infty e^{-t}t^n\, dt$$

$$= \frac{e^n\Gamma(n+1)}{n^n} - \frac{e^n}{n^n}\int_0^n e^{-t}t^n\, dt$$

$$= \frac{e^n \Gamma(n+1)}{n^n} - \sum_{k=0}^{\infty} \frac{n^{k+1}}{(n+1)_{k+1}}$$

$$= \frac{e^n \Gamma(n+1)}{n^n} + 1 - {}_1F_1(1; n+1; n),$$

where in the penultimate line we employed (43.2). Applying Corollary 2 in Section 21, we complete the proof of (47.3). □

In essence, Entry 47 is due to Nielsen [1], [2]. Equality (47.1) may be derived from [2, p. 46, Eq. (6)]. Equality (47.2) can be deduced from [2, p. 47, Eq. (11)]. Lastly, equality (47.3) can be proved by using [1, p. 219, Eq. (8)]. Note that, by Corollary 2 in Section 21, the continued fraction in (47.3) actually converges for all complex n.

Entry 48. *As n tends to ∞,*

$$\int_0^{\infty} e^{-x}(1 + x/n)^n \, dx = \frac{e^n \Gamma(n+1)}{2n^n} + \frac{2}{3} - \frac{4}{135n} + \frac{8}{2835n^2}$$

$$+ \frac{16}{8505n^3} - \frac{8992}{3^8 \cdot 5^2 \cdot 7 \cdot 11 n^4} + \cdots. \quad (48.1)$$

The asymptotic expansion given above first appeared in Ramanujan's solution to an ultimately famous problem proposed by Ramanujan [4], [16, pp. 323, 324] in the *Journal of the Indian Mathematical Society*. In addition to Ramanujan's (formal) solution, later proofs were given by Watson [3] and Szegö [1]. In fact, the last displayed term on the right side of (48.1) has not been recorded by any of the aforementioned authors. Further coefficients have been calculated by Bowman et al. [1] and Marsaglia [1].

The corollary below is similar to the aforementioned problem posed by Ramanujan [4], [16, pp. 323, 324]. A version of this corollary was also communicated by Ramanujan [16, p. xxvi] in his first letter to Hardy.

Corollary. *Define $\theta = \theta_n$ by*

$$\sum_{k=0}^{n-1} \frac{n^k}{k!} + \frac{n^n}{n!} \theta = \frac{e^n}{2}. \quad (48.2)$$

Then

$$\theta \approx \theta^* = \theta_n^* := \frac{4 + 15n}{8 + 45n}. \quad (48.3)$$

PROOF. As Ramanujan [4], [16, p. 324] easily demonstrated,

$$\theta = \frac{e^n \Gamma(n+1)}{2n^n} + 1 - \int_0^{\infty} e^{-x}(1 + x/n)^n \, dx, \quad (48.4)$$

and so (48.1) may be reformulated as

$$\theta = \frac{1}{3} + \frac{4}{135n} - \frac{8}{2835n^2} + \cdots, \tag{48.5}$$

as n tends to ∞. On the other hand,

$$\frac{4 + 15n}{8 + 45n} = \frac{1}{3} + \frac{4}{135n} - \frac{32}{6075n^2} + \cdots, \tag{48.6}$$

as n tends to ∞. Thus, θ^* is a fairly good approximation to θ.   □

In 1983, a problem similar to the corollary above was published, and a lengthy discussion, with three solutions, was given in a later issue of the *Mathematical Gazette* [2]. In particular, suppose that each of the n independent random variables X_k, $1 \le k \le n$, has a Poisson distribution with parameter 1. Then $S_n := \sum_{k=1}^{n} X_k$ has a Poisson distribution with parameter n. Thus,

$$P(S_n \le n) = e^{-n} \sum_{k=0}^{n} \frac{n^k}{k!}.$$

After applying the central limit theorem, we conclude that

$$\lim_{n \to \infty} P(S_n \le n) = \tfrac{1}{2}.$$

For further connections of the aforementioned corollary to probability, see the papers by Bowman et al. [1] and Lawden [1].

The integral of Entry 48, as well as a generalization, arises in a solution of the famous "birthday surprise" problem. See the delightful paper by Blaum et al. [1] where earlier work of Klamkin and Newman [1] is corrected and greatly extended.

A result analogous to (48.5) has been obtained by Copson [1] for e^{-n}. More precisely, if φ_n is defined by

$$e^{-n} = \sum_{k=0}^{n-1} \frac{(-n)^k}{k!} + \frac{(-n)^n}{n!} \varphi_n,$$

then

$$\varphi_n = \frac{1}{2} + \frac{1}{8n} + \frac{1}{32n^2} + \cdots,$$

as n tends to ∞.

Generalizations of Ramanujan's and Copson's theorems have been established by Buckholtz [1] and Paris [1]. The commentary in Szegö's *Collected Papers* [2, pp. 151, 152] provides a good summary of the literature on generalizations and related problems. Another proof of Ramanujan's result (48.5) as well as some related results may be found in Knuth's book [1, pp. 112–117]. Carlitz [1] has examined a class of functions arising in the work of Ramanujan, Copson, and Buckholtz. Jogdeo and Samuels [1] considered a binomial analogue of (48.2).

Ramanujan concludes Section 48 with the following table.

n	θ_n	θ_n^*
0	0.50000	0.50000
$\frac{1}{2}$	0.37750	0.37705
1	0.35914	0.35849
$\frac{3}{2}$	0.35146	0.35099
2	0.34726	0.34694
∞	0.33333	0.33333

Of course, when $n = 0$, it is trivial that $\theta_0 = \theta_0^* = \frac{1}{2}$. From (48.5) and (48.6), it is clear that $\theta_\infty = \theta_\infty^* = \frac{1}{3}$. The proposed values for θ_1, θ_1^*, θ_2, θ_2^*, $\theta_{1/2}^*$, and $\theta_{3/2}^*$ are easily corroborated by using the definitions of θ_n and θ_n^* given in (48.2) and (48.3). It remains to examine the values of $\theta_{1/2}$ and $\theta_{3/2}$.

In order to calculate $\theta_{1/2}$ and $\theta_{3/2}$, we shall employ (48.4) and the continued fraction (47.3). Hence,

$$\theta_n = 1 - \frac{e^n \Gamma(n+1)}{2n^n} + \frac{2n}{2} + \frac{3n}{3} + \frac{4n}{4} + \frac{5n}{5} + \cdots, \qquad n > 0. \qquad (48.7)$$

In the notation of (1.3) and (1.4), when $n = \frac{1}{2}$,

$$A_k = (k+1)A_{k-1} + \tfrac{1}{2}(k+1)A_{k-2}, \qquad k \geq 1,$$

and

$$B_k = (k+1)B_{k-1} + \tfrac{1}{2}(k+1)B_{k-2}, \qquad k \geq 1.$$

By successive calculations, we eventually find that

$$\frac{A_5}{B_5} = 0.4106925, \qquad \frac{A_6}{B_6} = 0.4106857, \qquad \frac{A_7}{B_7} = 0.4106862.$$

Thus,

$$\frac{2/2}{2} + \frac{3/2}{3} + \frac{4/2}{4} + \frac{5/2}{5} + \cdots = 0.410686.$$

Since

$$\frac{1}{2}\sqrt{\frac{e\pi}{2}} = 1.033182838,$$

we conclude from (48.7) that Ramanujan's proposed value for $\theta_{1/2}$ is correct.

If $n = \frac{3}{2}$, again, from (1.3) and (1.4),

$$A_k = (k+1)A_{k-1} + \tfrac{3}{2}(k+1)A_{k-2}, \qquad k \geq 1,$$

and

$$B_k = (k+1)B_{k-1} + \tfrac{3}{2}(k+1)B_{k-2}, \qquad k \geq 1.$$

Iterated calculations yield

$$\frac{A_7}{B_7} = 0.972952, \qquad \frac{A_8}{B_8} = 0.972930, \qquad \frac{A_9}{B_9} = 0.972933.$$

Proceeding as above, we find that $\theta_{3/2} = 0.35145$, which differs slightly from the value given by Ramanujan.

Ramanujan [4], [16, p. 324] conjectured, probably partially on the basis of his calculations above, that θ_n always lies between $\frac{1}{2}$ and $\frac{1}{3}$. This conjecture was proved by both Watson [3] and Szegö [1].

Entry 49. *For each integer $n \geq 2$, define $\theta = \theta_n$ by*

$$\gamma + \text{Log}\, n + \sum_{k=1}^{\infty} \frac{n^k}{k!k} = e^n \left(\sum_{k=0}^{n-2} \frac{k!}{n^{k+1}} + \frac{(n-1)!}{n^n} \theta \right),$$

where γ denotes Euler's constant. Then, as n tends to ∞,

$$\theta = \frac{2}{3} + \frac{4}{135n} + \frac{8}{2835n^2} + \cdots.$$

We are very grateful to F. W. J. Olver for providing us the following solution based on material from his book [1].

PROOF. First, observe that, for $n > 0$,

$$\sum_{k=1}^{\infty} \frac{n^k}{k!k} = \int_0^n \frac{e^t - 1}{t} \, dt. \tag{49.1}$$

By combining (49.1) with a familiar formula for γ (Olver [1, p. 40]), we readily find that

$$\gamma + \text{Log}\, n + \sum_{k=1}^{\infty} \frac{n^k}{k!k} = PV \int_{-\infty}^{n} \frac{e^t}{t} \, dt =: Ei(n),$$

where $n > 0$. Olver has calculated an asymptotic series for $Ei(n)$, and in the notation of his text [1, p. 529, Eq. (4.06)], $\theta = C_{n-1}(n)$. By [1, p. 529, formula (4.07)],

$$\theta = C_{n-1}(n) \sim \sum_{k=0}^{\infty} \frac{\gamma_k(1)}{(n-1)^k}, \tag{49.2}$$

as n tends to ∞, where the first three values for $\gamma_k(1)$ are given by (see [1, p. 530])

$$\gamma_0(1) = \frac{2}{3}, \quad \gamma_1(1) = \frac{4}{135}, \quad \text{and} \quad \gamma_2(1) = -\frac{76}{2835}.$$

Putting these values in (49.2), we deduce that

$$\theta = \frac{2}{3} + \frac{4}{135(n-1)} - \frac{76}{2835(n-1)^2} + \cdots$$

$$= \frac{2}{3} + \frac{4}{135n}\left(1 + \frac{1}{n}\right) - \frac{76}{2835n^2} + O\left(\frac{1}{n^3}\right),$$

from which the proposed asymptotic expansion follows.   □

For much of the theory of $Ei(n)$, see Nielsen's book [2].

CHAPTER 13

Integrals and Asymptotic Expansions

In assessing the content of Ramanujan's first letter to him, Hardy [9, p. 9] judged that "on the whole, the integral formulae seemed the least impressive." Later he added that Ramanujan's definite integral formulae "are still interesting and will repay a careful analysis" [9, p. 186]. Indeed, a dismissal of Ramanujan's contributions to integration would have been decidedly premature. First, we might recall that this first letter contained several remarkable formulas on series and continued fractions. In evaluating infinite series and deriving series identities, Ramanujan had no peers, except for possibly Euler and Jacobi. Ramanujan's work on continued fraction expansions of analytic functions ranks as one of his most brilliant achievements. Thus, if Ramanujan's contributions to integrals dim slightly in comparison, it is only because the glitter of diamonds surpasses that of rubies. Indeed, there are many elegant and important integrals that bear Ramanujan's name. (See, for example, Entry 22.)

Chapter 13 is largely devoted to integrals. In this chapter, we find some of Ramanujan's more prominent integral evaluations. In particular, many of the integrals from [8], [16, pp. 53–58] are found here. But much more importantly, Chapter 13 contains some absolutely remarkable results not heretofore observed. Entry 6 gives an asymptotic expansion of a certain integral and provides a generalization of a famous question posed by Ramanujan [4], [16, pp. 323, 324] in the *Journal of the Indian Mathematical Society*. The latter problem and related asymptotic expansions may be found at the end of Chapter 12. Entry 7 is a highlight of Chapter 13 and a truly remarkable formula. Ramanujan offers here an asymptotic expansion of a certain integral as two parameters tend to ∞. From both theoretical and computational standpoints, Entry 7 was very difficult for us to prove. As a by-product of Entry 7, we obtain an asymptotic expansion for the hypergeometric function

$_2F_1\left(1, m; m - n; \dfrac{m - n}{m}\right)$ as m, n, and $m - n$ tend to ∞. Such an expansion does not appear to have been previously given in the literature. Another elegant asymptotic formula for an integral appears in Entry 8. This expansion is related to the confluent hypergeometric functions $\Phi(a, c; z)$ and $\Psi(a, c; z)$ (Lebedev [1, pp. 260, 263]). We have proved a generalization of Entry 8 in Section 10 (see (10.22)). Entry 5 is a very unusual integral formula that has its roots in a favorite theorem of Ramanujan, an interpolation formula in the theory of integral transforms. Special cases of Entry 5 are formulas for K-Bessel and confluent hypergeometric functions.

In addition to theorems on integrals, Chapter 13 contains material on infinite series. Undoubtedly, the most impressive results on series appear in Section 10. Entry 10 offers an extraordinarily beautiful asymptotic expansion for series that are remindful of hypergeometric series. We know of nothing like it in the literature. Corollary (i) is also a very interesting result which, in a special case, is related to Entry 8 and therefore to confluent hypergeometric functions.

It should be remarked that none of Ramanujan's integral evaluations or asymptotic expansions is accompanied by conditions of validity. Particularly in Entries 5, 7, and 10, the determination of these conditions was not an easy task.

For an enlightening discussion of several of Ramanujan's asymptotic expansions and for some further generalizations, see Evans' paper [1].

As might be expected, several of Ramanujan's integral evaluations are classical. It would be very difficult to determine the original discoverers of these results, and so we usually content ourselves with just pointing out their appearances in the tables of Gradshteyn and Ryzhik [1].

Occasionally, we shall write expressions such as

$$f(x) \sim g(x)h\left(a_0 + \frac{a_1}{x} + \frac{a_2}{x^2} + \cdots\right).$$

By this we mean that

$$f(x) = g(x)h(F(x)),$$

where $F(x)$ has the asymptotic expansion

$$F(x) \sim a_0 + \frac{a_1}{x} + \frac{a_2}{x^2} + \cdots,$$

as x tends to ∞.

Entry 1. *Let $n \geq 0$ and put $N = [n + 1]$. Then*

$$\int_0^\infty x^{-n-1} \sum_{k=0}^\infty A_k(-x)^k \, dx = (-1)^N \int_0^\infty x^{-n+N-1} \sum_{k=0}^\infty A_{N+k}(-x)^k \, dx,$$

when the right side is meaningful.

Ramanujan does not intend Entry 1 to be a theorem, but instead he is *defining* the integral on the left side by the expression on the right side. To illustrate Entry 1, Ramanujan gives the example

$$\int_0^\infty \frac{e^{-x^2}}{x^4} \, dx = \tfrac{2}{3}\sqrt{\pi},$$

which is to be interpreted as

$$\int_0^\infty \frac{e^{-x^2} - 1 + x^2}{x^4} \, dx = \tfrac{2}{3}\sqrt{\pi}. \tag{1.1}$$

This result is easy to establish either directly or by using the general formula (Whittaker and Watson [1, p. 243])

$$\Gamma(z) = \int_0^\infty t^{z-1} \left(e^{-t} - \sum_{k=0}^n \frac{(-t)^k}{k!} \right) dt \tag{1.2}$$

due to Cauchy and Saalschütz, where the integer n is chosen so that $-n - 1 < \mathrm{Re}(z) < -n$. Hence, employing (1.2), we find that

$$\int_0^\infty \frac{e^{-x^2} - 1 + x^2}{x^4} \, dx = \frac{1}{2} \int_0^\infty t^{-5/2}(e^{-t} - 1 + t) \, dt$$

$$= \tfrac{1}{2}\Gamma(-\tfrac{3}{2}) = \tfrac{2}{3}\Gamma(\tfrac{1}{2}) = \tfrac{2}{3}\sqrt{\pi},$$

which establishes (1.1).

Corollary. *If $a, n > 0$ and b is real, then*

$$\int_0^\infty e^{-ax} x^{n-1} \, \frac{\cos(bx)}{\sin(bx)} \, dx = \frac{\Gamma(n)}{(a^2 + b^2)^{n/2}} \, \frac{\cos(n \tan^{-1}(b/a))}{\sin(n \tan^{-1}(b/a))}.$$

These two formulas are well known (Gradshteyn and Ryzhik [1, p. 490]). Ramanujan furthermore remarks that the integrals above "for negative values of n are known." Indeed, Ramanujan's definition in Entry 1 assigns a meaning to these integrals for negative values of n. In fact, these same formulas still hold if $n < 0$, provided that n is not a negative integer. To that end, using Ramanujan's definition from Entry 1 and (1.2) and defining the nonnegative integer m by $-m - 1 < n < -m$, we find that

$$\int_0^\infty e^{-ax} x^{n-1} \cos(bx) \, dx = \frac{1}{2} \int_0^\infty x^{n-1}(e^{x(-a+bi)} + e^{x(-a-bi)}) \, dx$$

$$= \tfrac{1}{2}\{(a - bi)^{-n} + (a + bi)^{-n}\} \int_0^\infty x^{n-1} e^{-x} \, dx$$

$$= \tfrac{1}{2}\{(a - bi)^{-n} + (a + bi)^{-n}\} \int_0^\infty x^{n-1} \left(e^{-x} - \sum_{k=0}^m \frac{(-x)^k}{k!} \right) dx$$

$$= (a^2 + b^2)^{-n/2} \cos(n \tan^{-1}(b/a))\Gamma(n).$$

A similar argument holds for $\sin(bx)$ in place of $\cos(bx)$.

Entry 2. *Let φ have $m + 1$ continuous derivatives. Then*

(i) $$\int \varphi(x)e^{-nx}\,dx = -e^{-nx}\sum_{k=0}^{m}\frac{\varphi^{(k)}(x)}{n^{k+1}} + \frac{1}{n^{m+1}}\int \varphi^{(m+1)}(x)e^{-nx}\,dx,$$

(ii) $$\int \varphi(x)\cos(nx)\,dx$$

$$= \sin(nx)\sum_{k=0}^{m/2}\frac{(-1)^{k}\varphi^{(2k)}(x)}{n^{2k+1}} + \cos(nx)\sum_{k=0}^{m/2-1}\frac{(-1)^{k}\varphi^{(2k+1)}(x)}{n^{2k+2}}$$

$$+ \frac{(-1)^{m/2+1}}{n^{m+1}}\int \varphi^{(m+1)}(x)\sin(nx)\,dx, \quad \textit{if m is even,}$$

$$= \sin(nx)\sum_{k=0}^{(m-1)/2}\frac{(-1)^{k}\varphi^{(2k)}(x)}{n^{2k+1}} + \cos(nx)\sum_{k=0}^{(m-1)/2}\frac{(-1)^{k}\varphi^{(2k+1)}(x)}{n^{2k+2}}$$

$$+ \frac{(-1)^{(m+1)/2}}{n^{m+1}}\int \varphi^{(m+1)}(x)\cos(nx)\,dx, \quad \textit{if m is odd,}$$

(iii) $$\int \varphi(x)\sin(nx)\,dx$$

$$= \sin(nx)\sum_{k=0}^{m/2-1}\frac{(-1)^{k}\varphi^{(2k+1)}(x)}{n^{2k+2}} - \cos(nx)\sum_{k=0}^{m/2}\frac{(-1)^{k}\varphi^{(2k)}(x)}{n^{2k+1}}$$

$$+ \frac{(-1)^{m/2}}{n^{m+1}}\int \varphi^{(m+1)}(x)\cos(nx)\,dx, \quad \textit{if m is even,}$$

$$= \sin(nx)\sum_{k=0}^{(m-1)/2}\frac{(-1)^{k}\varphi^{(2k+1)}(x)}{n^{2k+2}} - \cos(nx)\sum_{k=0}^{(m-1)/2}\frac{(-1)^{k}\varphi^{(2k)}(x)}{n^{2k+1}}$$

$$+ \frac{(-1)^{(m+1)/2}}{n^{m+1}}\int \varphi^{(m+1)}(x)\sin(nx)\,dx, \quad \textit{if m is odd.}$$

All the equalities above may be established by successively integrating by parts.

Entry 3. *Let $n, x > 0$ and define θ and r by $\theta = \tan^{-1}(n/x)$ and $r = (n^2 + x^2)^{1/2}$. Suppose that m is any positive integer. Then as x tends to ∞,*

$$\int_{x}^{\infty} e^{-t^2}\cos(2nt)\,dt$$

$$= \frac{e^{-x^2}}{2}\sum_{k=0}^{m-1}\frac{(-1)^{k}(\frac{1}{2})_k\cos(2nx + (2k+1)\theta)}{r^{2k+1}} + O(r^{-2m-1}).$$

PROOF. Upon successively integrating by parts, we find that, for x sufficiently large,

$$\int_x^\infty e^{-t^2} \cos(2nt)\, dt$$

$$= \frac{e^{-n^2}}{2} \int_x^\infty e^{-(t-in)^2}\, dt + \frac{1}{2} e^{-n^2} \int_x^\infty e^{-(t+in)^2}\, dt$$

$$= \frac{e^{-n^2}}{4}\left(\int_{(x-in)^2}^\infty \frac{e^{-t}}{\sqrt{t}}\, dt + \int_{(x+in)^2}^\infty \frac{e^{-t}}{\sqrt{t}}\, dt \right)$$

$$= \frac{e^{-n^2}}{4}\left(\frac{e^{-(x-in)^2}}{x-in} + \frac{e^{-(x+in)^2}}{x+in} - \frac{1}{2}\int_{(x-in)^2}^\infty \frac{e^{-t}}{t^{3/2}}\, dt - \frac{1}{2}\int_{(x+in)^2}^\infty \frac{e^{-t}}{t^{3/2}}\, dt \right)$$

$$= \frac{1}{4}\left(\frac{e^{-x^2+2inx}}{x-in} + \frac{e^{-x^2-2inx}}{x+in} - \frac{e^{-x^2+2inx}}{2(x-in)^3} - \frac{e^{-x^2-2inx}}{2(x+in)^3} \right.$$

$$\left. + \frac{3}{2^2} e^{-n^2}\left\{ \int_{(x-in)^2}^\infty \frac{e^{-t}}{t^{5/2}}\, dt + \int_{(x+in)^2}^\infty \frac{e^{-t}}{t^{5/2}}\, dt \right\} \right)$$

$$= \frac{e^{-x^2}}{4}\left(\frac{e^{2inx+i\theta}}{r} + \frac{e^{-2inx-i\theta}}{r} - \frac{e^{2inx+3i\theta}}{2r^3} - \frac{e^{-2inx-3i\theta}}{2r^3} \right.$$

$$\left. + \frac{3e^{2inx+5i\theta}}{2^2 r^5} + \frac{3e^{-2inx-5i\theta}}{2^2 r^5} \right)$$

$$- \frac{15}{2^5} e^{-n^2}\left\{ \int_{(x-in)^2}^\infty \frac{e^{-t}}{t^{7/2}}\, dt + \int_{(x+in)^2}^\infty \frac{e^{-t}}{t^{7/2}}\, dt \right\}.$$

It is now clear that, after m integrations by parts, we may easily deduce the desired formula. $\quad\square$

Entry 4. *Suppose that φ is entire, n is real, and that the integrals and series below converge. Then*

$$\int_0^\infty e^{-x^2}\{e^{2nx}\varphi(x) + e^{-2nx}\varphi(-x)\}\, dx = \int_0^\infty e^{n^2-x^2}\{\varphi(n+x) + \varphi(n-x)\}\, dx$$

$$= \sqrt{\pi}\, e^{n^2} \sum_{k=0}^\infty \frac{\varphi^{(2k)}(n)}{2^{2k} k!}.$$

PROOF. Letting I denote the integral at the far left side, we find that

$$I = \int_{-\infty}^\infty e^{-x^2+2nx}\varphi(x)\, dx = e^{n^2}\int_{-\infty}^\infty e^{-(x-n)^2}\varphi(x)\, dx$$

$$= e^{n^2}\int_{-\infty}^\infty e^{-x^2}\varphi(n+x)\, dx = \int_0^\infty e^{n^2-x^2}\{\varphi(n+x) + \varphi(n-x)\}\, dx,$$

and so the first equality of Entry 4 is established.

Expanding $\varphi(n+x)$ and $\varphi(n-x)$ in power series, simplifying, and inverting the order of summation and integration by a theorem in Titchmarsh's book

[1, p. 47], we find that

$$\int_0^\infty e^{-x^2}\{\varphi(n+x) + \varphi(n-x)\}\, dx = 2\int_0^\infty e^{-x^2} \sum_{k=0}^\infty \frac{\varphi^{(2k)}(n)x^{2k}}{(2k)!}\, dx$$

$$= 2\sum_{k=0}^\infty \frac{\varphi^{(2k)}(n)}{(2k)!} \int_0^\infty e^{-x^2} x^{2k}\, dx$$

$$= \sum_{k=0}^\infty \frac{\varphi^{(2k)}(n)}{(2k)!} \Gamma(k+\tfrac{1}{2}),$$

from which the second equality of Entry 4 easily follows.   □

As an example, if we put $\varphi(x) = e^x$ in Entry 4, we find that

$$\int_0^\infty e^{-x^2}\cosh((2n+1)x)\, dx = \int_0^\infty e^{n^2+n-x^2}\cosh x\, dx = \frac{\sqrt{\pi}\, e^{n^2+n+1/4}}{2}.$$

In order to state Entry 5, we first need to enunciate a theorem due to Hardy [9, p. 186, formula (A)]. See also Part I [9, p. 299]. Let $s = \sigma + it$ with σ and t both real. Let $H(\delta) = \{s : \sigma \geq -\delta\}$, where $0 < \delta < 1$. Suppose that $\psi(s)$ is analytic on $H(\delta)$ and that there exist constants C, P, and A with $A < \pi$ such that

$$|\psi(s)| \leq Ce^{P\sigma + A|t|}, \tag{5.1}$$

for all $s \in H(\delta)$. For $x > 0$ and $0 < c < \delta$, define

$$\Psi(x) = \frac{1}{2\pi i} \int_{c-i\infty}^{c+i\infty} \frac{\pi}{\sin(\pi s)} \psi(-s) x^{-s}\, ds. \tag{5.2}$$

If $0 < x < e^{-P}$, an application of the residue theorem yields (Hardy [9, p. 189])

$$\Psi(x) = \sum_{k=0}^\infty \psi(k)(-x)^k.$$

Finally, if $0 < \sigma < \delta$ (Hardy [9, pp. 189, 190]),

$$\int_0^\infty \Psi(x) x^{s-1}\, dx = \frac{\pi}{\sin(\pi s)} \psi(-s). \tag{5.3}$$

Entry 5. *Let $\psi(s)$ satisfy the hypotheses of Hardy's theorem given above for some $\delta > \tfrac{1}{2}$. Put $\psi(s) = A_{2s+1}/\Gamma(s+1)$, and so, in the notation above,*

$$\Psi(x) = \sum_{k=0}^\infty \frac{A_{2k+1}(-x)^k}{k!}, \qquad 0 < x < e^{-P}.$$

Suppose that for $a = 2\delta > 1$, $x^{a-1/2}\Psi(x^2) \in L^2(0,\infty)$. Then

$$\int_0^\infty e^{-1/x^2}\Psi(x^2)\, dx = \frac{\sqrt{\pi}}{2} \sum_{k=0}^\infty \frac{(-2)^k A_k}{k!}.$$

Ramanujan (p. 156) states Entry 5 in the form

$$\int_0^\infty e^{-1/x^2} \sum_{k=0}^\infty \frac{(-1)^k A_{2k+1} x^{2k}}{k!} \, dx = \frac{\sqrt{\pi}}{2} \sum_{k=0}^\infty \frac{(-2)^k A_k}{k!}.$$

Although $\Psi(x)$ has been defined for $x > 0$ by (5.2), there is no guarantee that its power series converges for all x.

PROOF. First, for $0 < \sigma < 2\delta$,

$$\int_0^\infty x^{s-1} \Psi(x^2) \, dx = \frac{1}{2} \int_0^\infty u^{s/2-1} \Psi(u) \, du = \tfrac{1}{2} \Gamma(s/2) A_{-s+1}, \qquad (5.4)$$

by (5.3).

Second, for $\sigma < 0$,

$$\int_0^\infty x^{s-1} e^{-1/x^2} \, dx = \frac{1}{2} \int_0^\infty u^{-s/2-1} e^{-u} \, du = \tfrac{1}{2} \Gamma(-s/2). \qquad (5.5)$$

We now apply Parseval's theorem for Mellin transforms (Titchmarsh [2, p. 95]). Using (5.4) and (5.5), we find that, for $a > 1$,

$$\int_0^\infty e^{-1/x^2} \Psi(x^2) \, dx = \frac{1}{2\pi i} \int_{a-i\infty}^{a+i\infty} \frac{1}{4} \Gamma\left(\frac{s}{2}\right) A_{-s+1} \Gamma\left(\frac{s-1}{2}\right) ds. \qquad (5.6)$$

In order to evaluate the integral on the right side above, we examine

$$I_{M,N} := \frac{1}{2\pi i} \int_{C_{M,N}} \frac{1}{4} \Gamma\left(\frac{s}{2}\right) A_{-s+1} \Gamma\left(\frac{s-1}{2}\right) ds = \frac{\sqrt{\pi}}{2\pi i} \int_{C_{M,N}} A_{-s+1} 2^{-s} \Gamma(s-1) \, ds,$$
$$(5.7)$$

where $C_{M,N}$ is a positively oriented rectangle with vertices $a \pm iM$ and $-N \pm iM$, where $M, N > 0$ and $N \equiv \tfrac{1}{2} \pmod 1$. By hypothesis, the only singularities of the integrand for $\sigma \le a$ are at $s = 1 - k$, where k is a non-negative integer. Thus, by (5.7) and the residue theorem,

$$I_{M,N} = \sqrt{\pi} \sum_{0 \le k < N+1} A_k 2^{k-1} \frac{(-1)^k}{k!}. \qquad (5.8)$$

By (5.1), for $\sigma \le a$,

$$\left| \Gamma\left(\frac{s}{2}\right) A_{-s+1} \Gamma\left(\frac{s-1}{2}\right) \right| = \pi^2 \left| \frac{\csc \dfrac{\pi s}{2} \csc \dfrac{\pi(s-1)}{2}}{\Gamma\left(\dfrac{3-s}{2}\right) \Gamma\left(1 - \dfrac{s}{2}\right)} A_{1-s} \right|$$

$$\le C \pi^2 \left| \frac{\csc \dfrac{\pi s}{2} \csc \dfrac{\pi(s-1)}{2}}{\Gamma\left(\dfrac{3-s}{2}\right)} \right| e^{-P\sigma/2 + A|t|/2}.$$

From the upper bound above and from Stirling's formula, we easily see, by first letting M tend to ∞ and then letting N tend to ∞, that

$$\lim_{M,N\to\infty} I_{M,N} = \frac{1}{2\pi i} \int_{a-i\infty}^{a+i\infty} \frac{1}{4} \Gamma\left(\frac{s}{2}\right) A_{-s+1} \Gamma\left(\frac{s-1}{2}\right) ds$$

$$= \sqrt{\pi} \sum_{k=0}^{\infty} A_k 2^{k-1} \frac{(-1)^k}{k!}, \tag{5.9}$$

by (5.8). Substituting (5.9) into (5.6), we complete the proof.   □

As a first illustration of Entry 5, we note that (Gradshteyn and Ryzhik [1, p. 307])

$$\int_0^\infty e^{-x^2 - 1/x^2} dx = \frac{\sqrt{\pi}}{2} e^{-2}.$$

For a second example, take $\Psi(x) = (1 + x)^{-\mu}$, where $\mu > \frac{1}{2}$. Then

$$A_s = \frac{\Gamma(\mu + \frac{1}{2}s - \frac{1}{2})}{\Gamma(\mu)}.$$

An application of Entry 5 then yields

$$\int_0^\infty e^{-1/x^2}(1 + x^2)^{-\mu} dx = \frac{\sqrt{\pi}}{2\Gamma(\mu)} \sum_{k=0}^{\infty} \frac{(-2)^k \Gamma(\mu + \frac{1}{2}k - \frac{1}{2})}{k!}$$

$$= \frac{\Gamma(\mu - \frac{1}{2})}{2} \Psi(\mu - \frac{1}{2}, \frac{1}{2}; 1),$$

where in the last line $\Psi(a, c; z)$ denotes the confluent hypergeometric function mentioned in the introduction to this chapter.

The theorem of Hardy that we quoted above is a rigorous reformulation of one of Ramanujan's favorite theorems. It is Entry 11 of Chapter 4 and also appears as Theorem I in his quarterly reports. See our first volume [9, pp. 105, 298] on Ramanujan's notebooks, where many applications of Ramanujan's theorem are also found. Hardy's book [9, Chapter 12] also contains several applications. According to J. Edwards [2, p. 213], a special case of Ramanujan's theorem, or the case $s = \frac{1}{2}$ of (5.3), was established by J. W. L. Glaisher.

An alternative approach to Entry 5 is now sketched. Suppose that we expand $\exp(-1/x^2)$ in a power series, invert the order of summation and integration, and apply the aforementioned favorite theorem of Ramanujan. Accordingly, we find that

$$\int_0^\infty e^{-1/x^2} \Psi(x^2) dx = \frac{1}{2} \sum_{j=0}^{\infty} \frac{(-1)^j}{j!} \int_0^\infty u^{-j-1/2} \Psi(u) du$$

$$= \frac{1}{2} \sum_{j=0}^{\infty} \frac{(-1)^j}{j!} \Gamma(-j + \frac{1}{2}) A_{2j}$$

$$= \frac{\sqrt{\pi}}{2} \sum_{j=0}^{\infty} \frac{2^{2j} A_{2j}}{(2j)!}.$$

Thus, we obtain the "wrong" answer; the odd indexed terms do not appear! Now, in fact, Ramanujan used this same type of argument in many similar instances; see our account of the quarterly reports in [9]. Despite the non-rigorous nature of the procedure, Ramanujan possessed extraordinary intuition in determining when the process leads to the correct formula and when it leads to an incorrect formula.

The case $h = 0$ of the asymptotic expansion in Entry 6 below is essentially a famous problem that Ramanujan [4], [16, pp. 323, 324] submitted to the *Journal of the Indian Mathematical Society*. See also Entry 48 of Chapter 12 for the case $h = 0$. Watson [3] has made a more detailed study of this asymptotic expansion, and we shall use some of his analysis in our proof of the generalization below.

It should be remarked that the first integral below is equal to $n\Psi(1, n + 2 - h; n)$ (Lebedev [1, p. 268, formula (9.11.6)]), where $\Psi(a, c; z)$ denotes the confluent hypergeometric function.

Entry 6. *Let $n > 0$ and suppose that m is a positive integer. Then*

$$\int_0^\infty e^{-x}\left(1 + \frac{x}{n}\right)^{n-h} dx = \sum_{k=0}^{m-1} \frac{(-1)^k(-n+h)_k}{n^k}$$

$$+ \frac{(-1)^m(-n+h)_m}{n^m} \int_0^\infty e^{-x}\left(1 + \frac{x}{n}\right)^{n-h-m} dx$$

$$= \frac{e^n\Gamma(n-h+1)}{2n^{n-h}} + A_0 - \frac{A_1}{n} + \frac{A_2}{n^2} + \cdots,$$

as n tends to ∞. Here,

$$A_0 = \frac{2}{3} - h, \quad A_1 = \frac{4}{135} - \frac{h^2(1-h)}{3}, \quad and$$

$$A_2 = \frac{8}{2835} + \frac{2h(1-h)}{135} - \frac{h(1-h^2)(2-3h^2)}{45}.$$

PROOF. The first equality in Entry 6 follows by successively integrating by parts m times.

We now establish the asymptotic expansion. Putting $x = (U - 1)n$ and $x = un$, respectively, in the two integrals below, we find that

$$\int_0^\infty e^{-x}\left(1 + \frac{x}{n}\right)^{n-h} dx - \frac{e^n\Gamma(n-h+1)}{2n^{n-h}}$$

$$= \int_0^\infty e^{-x}\left(1 + \frac{x}{n}\right)^{n-h} dx - \frac{e^n}{2n^{n-h}} \int_0^\infty e^{-x}x^{n-h} dx$$

$$= n\int_1^\infty e^{n(1-U)}U^{n-h} dU - \frac{n}{2}\int_0^\infty e^{n(1-u)}u^{n-h} du$$

$$= \frac{n}{2} \int_1^\infty e^{n(1-U)} U^{n-h} \, dU - \frac{n}{2} \int_0^1 e^{n(1-u)} u^{n-h} \, du$$

$$= \frac{n}{2} \int_0^\infty e^{-nt} \left(U^{-h} \frac{dU}{dt} + u^{-h} \frac{du}{dt} \right) dt, \tag{6.1}$$

where we have made the changes of variables $e^{1-U} U = e^{-t}$ and $e^{1-u} u = e^{-t}$, respectively, in the two foregoing integrals.

From Watson's paper [3], for t sufficiently small,

$$U(t) = 1 + (2t)^{1/2} + \frac{2t}{3} + \frac{(2t)^{3/2}}{36} - \frac{2t^2}{135} + \frac{(2t)^{5/2}}{4320} + \frac{4t^3}{8505} + \cdots.$$

It follows that

$$\frac{dU}{dt} = \frac{1}{(2t)^{1/2}} + \frac{2}{3} + \frac{(2t)^{1/2}}{12} - \frac{4t}{135} + \frac{(2t)^{3/2}}{864} + \frac{4t^2}{2835} + \cdots$$

and

$$U(t)^{-h} = 1 - h \left\{ (2t)^{1/2} + \frac{2t}{3} + \frac{(2t)^{3/2}}{36} - \frac{2t^2}{135} + \frac{(2t)^{5/2}}{4320} + \cdots \right\}$$

$$+ \frac{h(h+1)}{2} \left\{ (2t)^{1/2} + \frac{2t}{3} + \frac{(2t)^{3/2}}{36} - \frac{2t^2}{135} + \cdots \right\}^2$$

$$- \frac{h(h+1)(h+2)}{6} \left\{ (2t)^{1/2} + \frac{2t}{3} + \frac{(2t)^{3/2}}{36} + \cdots \right\}^3$$

$$+ \frac{h(h+1)(h+2)(h+3)}{24} \left\{ (2t)^{1/2} + \frac{2t}{3} + \cdots \right\}^4$$

$$- \frac{h(h+1)(h+2)(h+3)(h+4)}{120} \left\{ (2t)^{1/2} + \cdots \right\}^5.$$

The expansion for $u(t)$ in ascending powers of $\sqrt{t}$ is the same as that for $U(t)$, except that the coefficients of odd powers of $\sqrt{t}$ are of opposite signs. Omitting all the algebraic calculations, we find that

$$U^{-h} \frac{dU}{dt} = \frac{1}{(2t)^{1/2}} + \left(\frac{2}{3} - h \right) + c_1 t^{1/2} + \left(-\frac{4}{135} - \frac{2h}{3} + \frac{4h(h+1)}{3} \right.$$

$$\left. - \frac{h(h+1)(h+2)}{3} \right) t + c_2 t^{3/2} + \left(\frac{4}{2835} + \frac{2h}{135} + \frac{44h(h+1)}{135} \right.$$

$$- \frac{7h(h+1)(h+2)}{9} + \frac{h(h+1)(h+2)(h+3)}{3}$$

$$\left. - \frac{h(h+1)(h+2)(h+3)(h+4)}{30} \right) t^2 + \cdots,$$

where $c_1, c_2, \ldots$ are certain constants, depending on h but not on t. The expansion for $u^{-h}\, du/dt$ is the same as that above, except that the coefficients of odd powers of $\sqrt{t}$ are of opposite signs. By the same justification as in Watson's proof [3], we thus obtain the following asymptotic expansion as n tends to ∞:

$$
\frac{n}{2} \int_0^\infty e^{-nt} \left(U^{-h} \frac{dU}{dt} + u^{-h} \frac{du}{dt} \right) dt
$$

$$
= \frac{n}{2} \int_0^\infty e^{-nt} \left\{ \left(\frac{4}{3} - 2h \right) + \left(-\frac{8}{135} - \frac{4h}{3} + \frac{8h(h+1)}{3} - \frac{2h(h+1)(h+2)}{3} \right) t \right.
$$

$$
+ \left(\frac{8}{2835} + \frac{4h}{135} + \frac{88h(h+1)}{135} - \frac{14h(h+1)(h+2)}{9} \right.
$$

$$
+ \frac{2h(h+1)(h+2)(h+3)}{3}
$$

$$
\left. - \frac{h(h+1)(h+2)(h+3)(h+4)}{15} \right) t^2 + \cdots \right\} dt
$$

$$
= \frac{2}{3} - h + \frac{1}{n} \left(-\frac{4}{135} + \frac{h^2}{3} - \frac{h^3}{3} \right)
$$

$$
+ \frac{1}{n^2} \left(\frac{8}{2835} - \frac{4h}{135} - \frac{2h^2}{135} + \frac{h^3}{9} - \frac{h^5}{15} \right) + \cdots .
$$

By (6.1), this completes the proof.   □

Entry 7. Let $m > n + 1$. If m and n tend to ∞ while $m - n$ remains bounded, then

$$
I := I(m, n) := (m - n - 1) \int_0^\infty \frac{(1 + x/n)^n}{(1 + x/m)^m} \, dx = n \sum_{k=0}^\infty \frac{(m - n)^k}{(m - n)_k} + O(1). \quad (7.1)
$$

Put

$$
R = \frac{n(m - n)}{2m}. \quad (7.2)
$$

If m, n, and $m - n$ tend to ∞, implying that also R tends to ∞, then we have the asymptotic expansion

$$
I = \frac{m^{m+1} \Gamma(n+1) \Gamma(m-n+1)}{2n^n \Gamma(m+1)(m-n)^{m-n}} + A_1 + A_2 + A_3 + A_4 + \cdots, \quad (7.3)
$$

where A_k, $1 \le k < \infty$, is a rational function of m and n such that

$$
A_k = O(mR^{1-k}), \quad (7.4)
$$

as m, n, and $m - n$ tend to ∞. Moreover,

$$A_1 = \frac{2(m + n)}{3}, \tag{7.5}$$

$$A_2 = -\frac{4(m + n)(m - 2n)(m - \frac{1}{2}n)}{135mn(m - n)}, \tag{7.6}$$

$$A_3 = \frac{8(m^3 + n^3)(m - 2n)(m - \frac{1}{2}n)}{2835m^2 n^2 (m - n)^2}, \tag{7.7}$$

and

$$A_4 = \frac{16(m^3 + n^3)(m - 2n)(m - \frac{1}{2}n)(m^2 - mn + n^2)}{8505m^3 n^3 (m - n)^3}. \tag{7.8}$$

PROOF. Replacing x by nx in (7.1), we find that

$$I = (m - n - 1)n \int_0^\infty (1 + x)^n (1 + nx/m)^{-m} \, dx. \tag{7.9}$$

Using a standard integral representation for the hypergeometric function $_2F_1(a, b; c; z)$ (Luke [1, p. 57, Eq. (2)]), we deduce that

$$I = n \, _2F_1\left(1, m; m - n; \frac{m - n}{m}\right). \tag{7.10}$$

We first suppose that as m and n tend to ∞, $m - n < B$ for some constant B. By (7.10),

$$I = n \sum_{k=0}^\infty \frac{(m - n)^k}{(m - n)_k} \frac{(m)_k}{m^k},$$

and so

$$\frac{mI}{n} - m \sum_{k=0}^\infty \frac{(m - n)^k}{(m - n)_k} = \sum_{k=0}^\infty \frac{(m - n)^k((m)_k - m^k)}{(m - n)_k m^{k-1}}. \tag{7.11}$$

To prove (7.1), we shall show that the left side of (7.11) is bounded as m and n tend to ∞, by proving that

$$T_k := \frac{(m - n)^k((m)_k - m^k)}{(m - n)_k m^{k-1}} < 2^{-k}, \tag{7.12}$$

for all m and k sufficiently large.

Clearly,

$$\frac{(m - n)^k}{(m - n)_k} < \frac{B^k}{k!}. \tag{7.13}$$

Next, by the mean value theorem,

$$\frac{(m)_k - m^k}{m^{k-1}} < \frac{(m + k)^k - m^k}{m^{k-1}} \le k^2 \left(1 + \frac{k}{m}\right)^{k-1}. \tag{7.14}$$

Thus, by (7.13), (7.14), and Stirling's formula, for k and m sufficiently large,

$$T_k < \frac{B^k}{k!} k^2 2^{k-1} < 2^{-k}, \quad \text{if } k \leq m,$$

and

$$T_k < \frac{B^k}{k!} k^2 \left(\frac{2k}{m}\right)^{k-1} < 2^{-k}, \quad \text{if } m < k.$$

Hence, (7.12) is established, and therefore the proof of (7.1) is complete.

Second, we suppose that m, n, and $m - n$ tend to ∞. For brevity, set

$$S = \frac{m^{m+1} \Gamma(n+1) \Gamma(m-n+1)}{2n^n \Gamma(m+1)(m-n)^{m-n}}.$$

Employing a basic integral representation for the beta function (Gradshteyn and Ryzhik [1, p. 948, formula 3]), we see that

$$
\begin{aligned}
S &= \frac{m^m(m-n-1)\Gamma(m-n-1)\Gamma(n+1)}{2n^n(m-n)^{m-n-1}\Gamma(m)} \\
&= \frac{m^m(m-n-1)}{2n^n(m-n)^{m-n-1}} \int_0^\infty \frac{x^n\,dx}{(1+x)^m} \\
&= \frac{m^m(m-n-1)n}{2(m-n)^m} \int_0^\infty \frac{t^n\,dt}{\left(1 + \dfrac{n}{m-n}t\right)^m} \\
&= \frac{(m-n-1)n}{2} \int_{-1}^\infty (u+1)^n \left(\frac{n}{m}u+1\right)^{-m} du. \tag{7.15}
\end{aligned}
$$

Combining (7.9) and (7.15), we obtain the representation

$$
\begin{aligned}
I - S = \frac{(m-n-1)n}{2} \Bigg\{ &\int_0^\infty (U+1)^n \left(\frac{n}{m}U+1\right)^{-m} dU \\
&- \int_{-1}^0 (u+1)^n \left(\frac{n}{m}u+1\right)^{-m} du \Bigg\}. \tag{7.16}
\end{aligned}
$$

The former integrand in (7.16) is decreasing on $(0, \infty)$, while the latter integrand is increasing on $(-1, 0)$. In order to see this, define

$$Q(z) = (z+1)^n \left(\frac{n}{m}z+1\right)^{-m}, \tag{7.17}$$

and observe that

$$\frac{d}{dz} \operatorname{Log} Q(z) = \frac{n}{z+1} - \frac{n}{\dfrac{n}{m}z+1} = -\frac{2Rz}{(z+1)\left(\dfrac{n}{m}z+1\right)}, \tag{7.18}$$

where R is defined by (7.2). Now replace both integrands in (7.16) by e^{-t} to obtain

$$I - S = \frac{(m - n - 1)n}{2} \int_0^\infty e^{-t} \{U'(t) + u'(t)\} \, dt. \tag{7.19}$$

By the inverse function theorem, for $t > 0$ and t sufficiently small,

$$u(t) = \sum_{k=1}^\infty a_k t^{k/2} \quad \text{and} \quad U(t) = \sum_{k=1}^\infty (-1)^k a_k t^{k/2}, \tag{7.20}$$

where the coefficients a_k, $1 \le k < \infty$, are functions of m and n with

$$a_1 = -R^{-1/2}. \tag{7.21}$$

Recalling (7.17), we observe that, for $|u| < 1$,

$$t = f(u) := -\operatorname{Log} Q(u) = m \operatorname{Log}\left(1 + \frac{n}{m} u\right) - n \operatorname{Log}(1 + u) = Ru^2 \sum_{k=0}^\infty c_k u^k, \tag{7.22}$$

where

$$c_k = \frac{2(-1)^k (m^{k+1} - n^{k+1})}{(k + 2)m^k(m - n)}, \qquad k \ge 0. \tag{7.23}$$

Note that

$$c_0 = 1, \quad c_1 = -\frac{2(m + n)}{3m}, \quad \text{and} \quad |c_k| \le \tfrac{4}{3}, \qquad k \ge 0. \tag{7.24}$$

Thus, for $|z| \le \tfrac{1}{3}$,

$$\left| \sum_{k=0}^\infty c_k z^k \right| \ge 1 - \sum_{k=1}^\infty |c_k| |z|^k \ge 1 - \frac{4}{3} \sum_{k=1}^\infty 3^{-k} = \frac{1}{3}. \tag{7.25}$$

We next proceed to show how the coefficients a_k in (7.20) are related to the coefficients c_k defined in (7.22) and (7.23).

For $t > 0$ and t sufficiently small, let

$$g(t) = \sum_{k=1}^\infty a_k t^k.$$

From (7.22), $t^2 = f(u(t^2)) = f(g(t))$, and so $t = \sqrt{f(g(t))}$. Applying g to the last equality, we find that

$$u = g(\sqrt{f(u)}), \tag{7.26}$$

for $u < 0$ and u sufficiently close to zero. Let $R(F)$ denote the residue of a function $F(z)$ at a pole $z = 0$. Then by the Lagrange inversion formula, (36.8) of Chapter 11, for $k \ge 1$,

$$a_k = R(z^{-k-1}g(z)) = R\left(f(z)^{-(k+1)/2} g(\sqrt{f(z)}) \frac{d}{dz} \sqrt{f(z)} \right). \tag{7.27}$$

By (7.26) and (7.27), for $k \geq 1$,

$$a_k = -R\left(\frac{z}{k}\frac{d}{dz}f(z)^{-k/2}\right) = \frac{1}{k}R(f(z)^{-k/2}).\tag{7.28}$$

Now by (7.22) and (7.25), for $0 < |z| \leq \frac{1}{3}$,

$$|f(z)| \geq R|z|^2/3 > 0.\tag{7.29}$$

Hence, by (7.28), (7.29), and the residue theorem,

$$a_k = \frac{1}{2\pi i k}\int_{|z|=1/3} f(z)^{-k/2}\,dz,\qquad k \geq 1.\tag{7.30}$$

Finally, by (7.29) and (7.30), for $k \geq 1$,

$$|a_k| \leq \frac{1}{2\pi k}\frac{2\pi}{3}\sup_{|z|=1/3}|f(z)|^{-k/2} \leq \frac{1}{3k}\left(\frac{27}{R}\right)^{k/2}.\tag{7.31}$$

It follows that the expansions for $u(t)$ and $U(t)$ given in (7.20) are valid for $0 < t \leq R/30$.

By (7.20), (7.21), and (7.31), there exists a positive number $\delta < \frac{1}{30}$ such that

$$u(\delta R) < -\delta \quad \text{and} \quad U(\delta R) > \delta,\tag{7.32}$$

since $|u(\delta R)|$ and $|U(\delta R)|$ both exceed

$$\sqrt{\delta} - \sum_{k=2}^{\infty}|a_k|(\delta R)^{k/2} \geq \sqrt{\delta} - \sum_{k=2}^{\infty}(27\delta)^{k/2}$$

$$= \sqrt{\delta} - \frac{27\delta}{1 - \sqrt{27\delta}} > \delta.$$

Now return to (7.19) and write

$$I - S = H + J,\tag{7.33}$$

where

$$H = \frac{(m - n - 1)n}{2}\int_0^{\delta R} e^{-t}\{U'(t) + u'(t)\}\,dt$$

and

$$J = \frac{(m - n - 1)n}{2}\int_{\delta R}^{\infty} e^{-t}\{U'(t) + u'(t)\}\,dt.$$

Fix a positive integer K. By (7.20),

$$H = (m - n - 1)n\int_0^{\delta R} e^{-t}\sum_{k=1}^{\infty}ka_{2k}t^{k-1}\,dt = H_1 + H_2,\tag{7.34}$$

where

$$H_1 = (m - n - 1)n\sum_{k=1}^{K}ka_{2k}\int_0^{\delta R}e^{-t}t^{k-1}\,dt\tag{7.35}$$

and

$$H_2 = (m - n - 1)n \int_0^{\delta R} e^{-t} \sum_{k=K+1}^{\infty} k a_{2k} t^{k-1} \, dt. \tag{7.36}$$

By (7.31),

$$H_2 < (m - n)n \int_0^{\delta R} \frac{e^{-t}}{t} \sum_{k=K+1}^{\infty} \left(\frac{27t}{R}\right)^k dt$$

$$= 2mR \int_0^{\delta R} \frac{e^{-t}}{t} \left(\frac{27t}{R}\right)^{K+1} \left(1 - \frac{27t}{R}\right)^{-1} dt$$

$$< 2mR \left(\frac{27}{R}\right)^{K+1} (1 - 27\delta)^{-1} \int_0^{\delta R} e^{-t} t^K \, dt$$

$$= O(mR^{-k}), \tag{7.37}$$

as m, n, and $m - n$ tend to ∞. Thus, by (7.34)–(7.37),

$$H = H_1 + O(mR^{-K}) = (m - n - 1)n \sum_{k=1}^{K} a_{2k} k! + O(mR^{-K}), \tag{7.38}$$

as m, n, and $m - n$ tend to ∞.

Define, for $k \geq 1$,

$$A_k = n(m - n)a_{2k}k! - na_{2k-2}(k - 1)!, \tag{7.39}$$

where $a_0 = 0$. Then, by (7.2) and (7.31), for $k \leq K$,

$$A_k = O(mR \cdot R^{-k}) + O(nR^{1-k}) = O(mR^{1-k}),$$

as m, n, and $m - n$ approach ∞. Thus, (7.4) holds.

By (7.31), (7.38), and (7.39),

$$H = \sum_{k=1}^{K} A_k + O(mR^{-K}), \tag{7.40}$$

as m, n, and $m - n$ tend to ∞. In order to prove (7.3), it suffices, by (7.33) and (7.40), to prove that

$$J = O(me^{-Rg}),$$

for some fixed positive constant g. Since $(m - n - 1)n = O(mR)$, it suffices to show that, for some constant $g > 0$,

$$\int_{\delta R}^{\infty} e^{-t} U'(t) \, dt, \quad \int_{\delta R}^{\infty} e^{-t} u'(t) \, dt = O(e^{-Rg}),$$

as m, n, and $m - n$ tend to ∞. Changing variables, using (7.32), and recalling the remark made after (7.16), we see that it suffices to show that, for some $g > 0$,

$$\int_{\delta}^{\infty} (U + 1)^n \left(\frac{n}{m} U + 1\right)^{-m} dU, \quad \int_{-1}^{-\delta} (u + 1)^n \left(\frac{n}{m} u + 1\right)^{-m} du = O(e^{-Rg}). \tag{7.41}$$

Now let $\mu = \pm\delta$. By (7.29),

$$|f(\mu)| \geq R\delta^2/3. \tag{7.42}$$

By (7.17) and the aforementioned remark prior to (7.17), $0 < Q(\mu) < 1$. Thus, by (7.22), $f(\mu) > 0$. Thus, by (7.42), $f(\mu) \geq Rg$, with $g = \delta^2/3$; that is,

$$Q(\mu) = e^{-f(\mu)} \leq e^{-Rg}.$$

Hence,

$$\int_{-1}^{-\delta} Q(u)\, du < e^{-Rg},$$

since the integrand is increasing on $(-1, -\delta)$. Similarly,

$$\int_{\delta}^{3} Q(U)\, dU < 3e^{-Rg}.$$

Thus, by the last two inequalities, to complete the proof of (7.41), it suffices to prove that

$$Q(U) < U^{-R/2}, \tag{7.43}$$

when $U \geq 1$, for then

$$\int_{3}^{\infty} Q(U)\, dU < \int_{3}^{\infty} U^{-R/2}\, dU = \frac{6 \cdot 3^{-R/2}}{R - 2} = O(e^{-R/2}).$$

By (7.18), for $U \geq 1$,

$$\frac{d}{dU} \operatorname{Log}(U^{R/2}Q(U)) = \frac{R}{2U} - \frac{2RU}{(U + 1)\left(\dfrac{n}{m}U + 1\right)}$$

$$< \frac{R}{2U}\left(1 - \frac{4U^2}{(U + 1)^2}\right) \leq 0.$$

Thus, $U^{R/2}Q(U)$ is decreasing for $U \geq 1$. Moreover, with $w = m/n$,

$$Q(1) = \left(\frac{2}{(1 + 1/w)^w}\right)^n < 1,$$

since $(1 + 1/w)^w$ is increasing for $w \geq 1$. This completes the proof of (7.43) and consequently of (7.3) as well.

In order to calculate A_1, A_2, A_3, and A_4, by (7.39), we need to determine a_2, a_4, a_6, and a_8. To do this, we employ (7.28). From (7.28) and the value of c_1 given in (7.24), it is easy to see that

$$a_2 = \frac{2(m + n)}{3n(m - n)}.$$

However, the calculations of a_4, a_6, and a_8 rapidly increase in difficulty. After very many hours of excruciatingly laborious calculation, we found that

$$a_4 = -\frac{2(m + n)(m^2 - 25mn + n^2)}{135mn^2(m - n)^2},$$

$$a_6 = \frac{4(m + n)(m^4 - 14m^3n + 267m^2n^2 - 14mn^3 + n^4)}{8505m^2n^3(m - n)^3},$$

and

$$a_8 = \frac{2(m + n)(m^6 - 3m^5n - 12m^4n^2 + 389m^3n^3 - 12m^2n^4 - 3mn^5 + n^6)}{25515m^3n^4(m - n)^4}.$$

The values (7.5)–(7.8) now follow from (7.39) and the evaluations given above.

□

Customarily, Ramanujan provides no hypotheses for Entry 7. Only the expansion (7.3) is given, and (7.1) is not found in the notebooks. Although Ramanujan was very familiar with the Lagrange inversion formula, it is very doubtful that our proof is substantially like that found by Ramanujan. In particular, our calculations of A_3 and A_4 were so involved that Ramanujan must have had a proof wherein the coefficients A_k arise more naturally with less computation.

By combining (7.3) and (7.10) with Stirling's formula, we obtain an asymptotic expansion for $_2F_1(1, m; m - n; (m - n)/m)$, as m, n and $m - n$ tend to ∞. The asymptotic behavior of this $_2F_1$ function for general $m > n > 0$ with m tending to ∞ is discussed in the paper by Evans [1, Theorems 15–17]. A vast literature on asymptotic expansions of hypergeometric functions exists, but this asymptotic expansion appears to be new.

Entry 8. *As n tends to ∞,*

$$\int_0^\infty \left\{ \frac{n^x\Gamma(n + 1)}{\Gamma(n + x + 1)} + e^{-x}\left(1 + \frac{x}{n}\right)^n \right\} dx = \frac{e^n\Gamma(n + 1)}{n^n} + \frac{6n}{12n + 1} + O(n^{-3/2}).$$

Before proving Entry 8, we indicate its connection with the confluent hypergeometric functions $\Psi(a, c; z)$ and $\Phi(a, c; z)$. As mentioned prior to Entry 6,

$$n\Psi(1, n + 2; n) = \int_0^\infty e^{-x}\left(1 + \frac{x}{n}\right)^n dx.$$

Also from Lebedev's text [1, p. 263, Eq. (9.10.3)] and the definition of Φ [1, p. 260, Eq. (9.9.1)],

$$n\Psi(1, n + 2; n) = n\frac{\Gamma(-n - 1)}{\Gamma(-n)}\Phi(1, n + 2; n) + \frac{\Gamma(n + 1)}{n^n}\Phi(-n, -n; n)$$

$$= -\frac{n}{n + 1}\sum_{k=0}^\infty \frac{n^k}{(n + 2)_k} + \frac{e^n\Gamma(n + 1)}{n^n}$$

$$= -\sum_{k=1}^{\infty} \frac{n^k}{(n+1)_k} + \frac{e^n \Gamma(n+1)}{n^n}$$

$$= -\sum_{k=0}^{\infty} \frac{n^k \Gamma(n+1)}{\Gamma(n+k+1)} + \frac{e^n \Gamma(n+1)}{n^n} + 1.$$

Thus, Entry 8 may be rewritten in the form

$$\int_0^{\infty} \frac{n^x \Gamma(n+1)}{\Gamma(n+x+1)} dx - \sum_{k=0}^{\infty} \frac{n^k \Gamma(n+1)}{\Gamma(n+k+1)} = -\frac{6n+1}{12n+1} + O(n^{-3/2}). \quad (8.1)$$

PROOF. From Stirling's formula, as n tends to ∞,

$$\frac{n^x \Gamma(n+1)}{\Gamma(n+x+1)} = \frac{e^x n^{n+x+1/2}}{(n+x)^{n+x+1/2}} \left\{ 1 + \frac{1}{12n} - \frac{1}{12(n+x)} + O\left(\frac{1}{n^2}\right) \right\},$$

uniformly for $0 \le x < \infty$. Thus,

$$\int_0^{\infty} \frac{n^x \Gamma(n+1)}{\Gamma(n+x+1)} dx = \int_0^{\infty} \frac{e^x n^{n+x+1/2}}{(n+x)^{n+x+1/2}} \left\{ 1 + \frac{1}{12n} - \frac{1}{12(n+x)} + O\left(\frac{1}{n^2}\right) \right\}$$

$$= \left\{ 1 + \frac{1}{12n} + O\left(\frac{1}{n^2}\right) \right\} n \int_0^{\infty} (1+t)^{-1/2} \left\{ \frac{e^t}{(1+t)^{1+t}} \right\}^n dt$$

$$- \frac{1}{12} \int_0^{\infty} (1+t)^{-3/2} \left\{ \frac{e^t}{(1+t)^{1+t}} \right\}^n dt$$

$$= I_1 + I_2, \quad (8.2)$$

say.

As t increases from 0 to ∞, $e^t/(1+t)^{1+t}$ decreases monotonically from 1 to 0. To apply Watson's lemma (Copson [3, p. 49], Olver [1, p. 113]), set

$$v = (1+t) \operatorname{Log}(1+t) - t$$

$$= \sum_{k=2}^{\infty} \frac{(-t)^k}{k(k-1)}, \quad |t| < 1. \quad (8.3)$$

For v sufficiently small and nonnegative, let

$$t = \sum_{k=1}^{\infty} c_k v^{k/2}. \quad (8.4)$$

Now substitute (8.4) into (8.3) and solve for $c_1, \ldots, c_4$. After a lengthy calculation, we find that

$$t = (2v)^{1/2} + \frac{1}{3} v - \frac{\sqrt{2}}{26} v^{3/2} + \frac{2}{135} v^2 + \cdots, \quad (8.5)$$

and so

$$\frac{dt}{dv} = \frac{1}{(2v)^{1/2}} + \frac{1}{3} - \frac{(2v)^{1/2}}{24} + \frac{4}{135} v + \cdots, \quad (8.6)$$

for v sufficiently small and $v \ge 0$.

Again, from (8.5), for v sufficiently small and nonnegative,

$$(1 + t)^{-1/2} = 1 - \frac{1}{2}\left\{(2v)^{1/2} + \frac{1}{3}v - \frac{\sqrt{2}}{36}v^{3/2} + \cdots\right\} + \frac{3}{8}\left\{(2v)^{1/2} + \frac{1}{3}v + \cdots\right\}^2$$

$$- \frac{5}{16}\{(2v)^{1/2} + \cdots\}^3 + \cdots$$

$$= 1 - \frac{(2v)^{1/2}}{2} + \frac{7}{12}v - \frac{13\sqrt{2}}{36}v^{3/2} + \cdots. \qquad (8.7)$$

Hence, from (8.6) and (8.7), for $v \geq 0$ and v sufficiently small,

$$(1 + t)^{-1/2}\frac{dt}{dv} = \frac{1}{(2v)^{1/2}} - \frac{1}{6} + \frac{(2v)^{1/2}}{12} - \frac{103}{1080}v + \cdots.$$

Thus, by Watson's lemma,

$$I_1 = \left(n + \frac{1}{12} + O\left(\frac{1}{n}\right)\right)\int_0^\infty (1 + t)^{-1/2}\frac{dt}{dv}e^{-nv}\,dv$$

$$= \left(n + \frac{1}{12} + O\left(\frac{1}{n}\right)\right)\int_0^\infty e^{-nv}\left\{\frac{1}{(2v)^{1/2}} - \frac{1}{6} + \frac{(2v)^{1/2}}{12} - \frac{103}{1080}v + \cdots\right\}\,dv$$

$$= \left(n + \frac{1}{12} + O\left(\frac{1}{n}\right)\right)\left\{\sqrt{\frac{\pi}{2n}} - \frac{1}{6n} + \frac{\sqrt{2\pi}}{24n^{3/2}} - \frac{103}{1080n^2} + O\left(\frac{1}{n^{5/2}}\right)\right\}$$

$$= \sqrt{\frac{\pi n}{2}} - \frac{1}{6} + \frac{1}{12}\sqrt{\frac{2\pi}{n}} - \frac{118}{1080n} + O\left(\frac{1}{n^{3/2}}\right), \qquad (8.8)$$

as n tends to ∞.

Next, from (8.5) and (8.6),

$$(1 + t)^{-3/2}\frac{dt}{dv} = \left(1 - \frac{3(2v)^{1/2}}{2} + \cdots\right)\left(\frac{1}{(2v)^{1/2}} + \frac{1}{3} + \cdots\right)$$

$$= \frac{1}{(2v)^{1/2}} - \frac{7}{6} + \cdots,$$

for $v \geq 0$ and v sufficiently small. Hence, by Watson's lemma,

$$I_2 = -\frac{1}{12}\int_0^\infty (1 + t)^{-3/2}\frac{dt}{dv}e^{-nv}\,dv$$

$$= -\frac{1}{12}\int_0^\infty e^{-nv}\left\{\frac{1}{(2v)^{1/2}} - \frac{7}{6} + \cdots\right\}\,dv$$

$$= -\frac{1}{24}\sqrt{\frac{2\pi}{n}} + \frac{7}{72n} + O\left(\frac{1}{n^{3/2}}\right), \qquad (8.9)$$

as n tends to ∞.

Putting (8.8) and (8.9) in (8.2), we conclude that

$$\int_0^\infty \frac{n^x \Gamma(n+1)}{\Gamma(n+x+1)} \, dx = \sqrt{\frac{\pi n}{2}} - \frac{1}{6} + \frac{1}{24}\sqrt{\frac{2\pi}{n}} - \frac{13}{1080n} + O\left(\frac{1}{n^{3/2}}\right)$$

$$= \frac{e^n \Gamma(n+1)}{2n^n} - \frac{1}{6} - \frac{13}{1080n} + O\left(\frac{1}{n^{3/2}}\right), \qquad (8.10)$$

as n tends to ∞.

Next, from Entry 6, as n tends to ∞,

$$\int_0^\infty e^{-x}(1 + x/n)^n \, dx = \frac{e^n \Gamma(n+1)}{2n^n} + \frac{2}{3} - \frac{4}{135n} + O\left(\frac{1}{n^2}\right). \qquad (8.11)$$

Combining (8.10) and (8.11), we deduce that

$$\int_0^\infty \left\{\frac{n^x \Gamma(n+1)}{\Gamma(n+x+1)} + e^{-x}(1 + x/n)^n\right\} dx = \frac{e^n \Gamma(n+1)}{n^n} + \frac{1}{2} - \frac{1}{24n} + O\left(\frac{1}{n^{3/2}}\right),$$

as n tends to ∞. Since, as n tends to ∞,

$$\frac{6n}{12n+1} = \frac{1}{2} - \frac{1}{24n} + O\left(\frac{1}{n^2}\right),$$

we conclude the proof of Ramanujan's approximation.   □

For a generalization of Entry 8, see (10.22).

Entry 9. *If*

$$\varphi(m) = \int_0^\infty \frac{e^{-m^2 x^2}}{1+x^2} \, dx$$

and if $|m| \geq |n|$, *where m and n are real, then*

$$\int_0^\infty \frac{e^{-m^2 x^2}}{1+x^2} \cos(2mnx) \, dx = \frac{e^{-n^2}}{2}\{\varphi(m+n) + \varphi(m-n)\}. \qquad (9.1)$$

PROOF. First, note that (9.1) is trivial for $n = 0$. Assume next that $0 < n < m$. Then

$$\int_0^\infty \frac{e^{-m^2 x^2}}{1+x^2} \cos(2mnx) \, dx = \frac{i}{4} \int_{-\infty}^\infty \left(\frac{1}{x+i} - \frac{1}{x-i}\right) e^{-m^2 x^2} \cos(2mnx) \, dx$$

$$= \frac{ie^{-n^2}}{4} \int_{-\infty}^\infty \left(\frac{1}{x+i} - \frac{1}{x-i}\right) e^{-(mx-in)^2} \, dx$$

$$= \frac{ie^{-n^2}}{4}(I_1 - I_2), \qquad (9.2)$$

say.

Let $p = n/m$, so that $0 < p < 1$. By integrating $e^{-(mz-in)^2}/(z + i)$ around a rectangle with vertices $\pm N$ and $\pm N + ip$, applying Cauchy's theorem, and letting N tend to ∞, we find that

$$I_1 = \int_{-\infty}^{\infty} \frac{e^{-m^2x^2}}{x + i(1 + p)} \, dx = \frac{m + n}{m} \int_{-\infty}^{\infty} \frac{e^{-(m+n)^2u^2}}{\dfrac{m + n}{m}u + \dfrac{m + n}{m}i} \, du$$

$$= \int_{-\infty}^{\infty} \frac{e^{-(m+n)^2u^2}}{u + i} \, du = \int_{-\infty}^{\infty} \frac{(u - i)e^{-(m+n)^2u^2}}{1 + u^2} \, du$$

$$= -2i\varphi(m + n). \tag{9.3}$$

Proceeding in the same fashion as above and setting $x = (m - n)u/m$, we find that

$$I_2 = \int_{-\infty}^{\infty} \frac{e^{-m^2x^2}}{x - i(1 - p)} \, dx = \int_{-\infty}^{\infty} \frac{(u + i)e^{-(m-n)^2u^2}}{1 + u^2} \, du = 2i\varphi(m - n). \tag{9.4}$$

Substituting (9.3) and (9.4) into (9.2), we easily deduce (9.1) for $0 < n < m$.

Observe that both sides of (9.1) are even functions of n. Hence, (9.1) holds for $-m < n < m$. Since the left side of (9.1) is an even function of m and since $\varphi(r)$ is an even function of r, we see that (9.1) is valid for $|n| < |m|$. By continuity, (9.1) holds for $|m| = |n|$ as well. This completes the proof. $\square$

We now find the analogue of (9.1) when $|n| > |m|$. Suppose that $0 < m < n$. As before, I_1 is given by (9.3). But, letting $R(i)$ denote the residue of $e^{-(mz-in)^2}/(z - i)$ at the simple pole $z = i$, we find that

$$I_2 = \int_{-\infty}^{\infty} \frac{e^{-m^2x^2}}{x + i(p - 1)} \, dx + 2\pi i R(i)$$

$$= \int_{-\infty}^{\infty} \frac{e^{-(n-m)^2u^2}}{u + i} \, du + 2\pi i e^{(m-n)^2}$$

$$= \int_{-\infty}^{\infty} \frac{(u - i)e^{-(n-m)^2u^2}}{1 + u^2} \, du + 2\pi i e^{(m-n)^2}$$

$$= -2i\varphi(n - m) + 2\pi i e^{(m-n)^2}. \tag{9.5}$$

Hence, substituting (9.3) and (9.5) into (9.2), we easily find that

$$\int_0^{\infty} \frac{e^{-m^2x^2}}{1 + x^2} \cos(2mnx) \, dx = \tfrac{1}{2}e^{-n^2}\{\varphi(m + n) - \varphi(m - n)\} + \tfrac{1}{2}\pi e^{m^2 - 2mn}. \tag{9.6}$$

By the same arguments are before, (9.6) is valid in general for $|n| \geq |m| \geq 0$.

Another proof of Entry 9 can be given by combining a result of Binet (Burkhardt [1, p. 1154]) with some formulas in Nielsen's book [2, pp. 18, 19,

Eqs. (5), (13)]. Entry 9 can also be derived by an appropriate application of Parseval's theorem.

Entry 10. *Let α, β, γ, and δ be fixed real numbers with $\gamma > \delta \geq 0$. Assume that for some fixed $d > 0$, $\varphi(x)$ is analytic and nonzero in the disk $|x| \leq d$; $\varphi(x)$ and $\varphi'(x)$ are positive for $x \geq -d$; and there exists a constant $M > 0$ such that*

$$x\varphi'(x) \geq M\varphi(x) \qquad (10.1)$$

for all $x \geq d$. Let $h > 0$. Then as h tends to 0,

$$S = S(h) := \sum_{k=0}^{\infty} \prod_{j=1}^{k} \frac{\varphi(h\alpha + hj\delta)}{\varphi(h\beta + hj\gamma)}$$

$$= \sqrt{\frac{\pi\varphi(0)}{2h(\gamma - \delta)\varphi'(0)}} + \frac{1}{3}\frac{\gamma + \delta}{\gamma - \delta}\left\{1 - \frac{\varphi(0)\varphi''(0)}{\varphi'(0)^2}\right\} + \frac{\alpha - \beta}{\gamma - \delta} + O(\sqrt{h}). \qquad (10.2)$$

Two functions $\varphi(x)$ that satisfy Entry 10 are e^x and $(1 + x)^n$, $n > 0$ (see Corollary (i) below). Observe that if φ satisfies Entry 10, so do e^φ and φ^c, for any $c > 0$. Also, if φ_1 and φ_2 obey the hypotheses of Entry 10, then $\varphi_1\varphi_2$ does as well. Entry 10 is truly a remarkable theorem, and there does not appear to be anything like it in the literature. The form of this asymptotic formula is reminiscent of the asymptotic formulas that arise in the method of stationary phase and in other asymptotic estimates of integrals.

PROOF. Let $L(x) = \text{Log } \varphi(x)$ and $w = [h^{-3/5}]$. Write

$$S = S_1 + S_2,$$

where

$$S_1 = \sum_{k \leq w} \prod_{j=1}^{k} \frac{\varphi(h\alpha + hj\delta)}{\varphi(h\beta + hj\gamma)} \quad \text{and} \quad S_2 = \sum_{k > w} \prod_{j=1}^{k} \frac{\varphi(h\alpha + hj\delta)}{\varphi(h\beta + hj\gamma)}.$$

We first examine S_1. Choose h so small that

$$|h\alpha + hj\delta|, |h\beta + hj\gamma| \leq d,$$

for each j, $1 \leq j \leq w$. Since, for $|x| \leq d$,

$$L(x) = L(0) + L'(0)x + \tfrac{1}{2}L''(0)x^2 + O(x^3),$$

we find that

$$S_1 = \sum_{k \leq w} \exp\left(\sum_{j=1}^{k} \{L(h\alpha + hj\delta) - L(h\beta + hj\gamma)\}\right)$$

$$= \sum_{k \leq w} \exp\left(\sum_{j=1}^{k} \{L'(0)h(\alpha - \beta + j(\delta - \gamma)) + \tfrac{1}{2}L''(0)h^2((\delta^2 - \gamma^2)j^2\right.$$

$$\left. + O(j)) + h^3 O(j^3)\}\right)$$

$$= \sum_{k \leq w} \exp(L'(0)h((\alpha - \beta)k + \tfrac{1}{2}(\delta - \gamma)(k^2 + k))$$

$$+ \tfrac{1}{2}L''(0)h^2(\tfrac{1}{3}(\delta^2 - \gamma^2)k^3 + O(k^2)) + h^3 O(k^4))$$

$$= \sum_{k \leq w} \exp(-Ahk^2 + Bhk + Ch^2k^3 + O(h^2k^2) + O(h^3k^4)),$$

where

$$A = \tfrac{1}{2}(\gamma - \delta)L'(0) > 0, \qquad B = \tfrac{1}{2}L'(0)(2\alpha - 2\beta + \delta - \gamma),$$
$$C = \tfrac{1}{6}L''(0)(\delta^2 - \gamma^2). \tag{10.3}$$

Since $e^x = 1 + x + O(x^2)$, whenever $x = O(1)$, we deduce that

$$S_1 = \sum_{k \leq w} e^{-Ahk^2}\{1 + (Bhk + Ch^2k^3 + O(h^2k^2) + O(h^3k^4)) + O(h^4k^6)\}$$

$$= T_0 + BhT_1 + Ch^2T_3 + O(h^2T_2 + h^3T_4 + h^4T_6), \tag{10.4}$$

where

$$T_r = \sum_{k=0}^{w} e^{-Ahk^2}k^r, \qquad r \geq 0.$$

Furthermore, define

$$V_r = \int_0^\infty e^{-Aht^2}t^r\,dt, \qquad r \geq 0.$$

Then

$$V_0 = \sqrt{\frac{\pi}{4Ah}}, \qquad V_1 = \frac{1}{2Ah},$$

and, for $r \geq 2$,

$$V_r = \frac{r-1}{2Ah}V_{r-2} = O(h^{-(r+1)/2}).$$

Recall now the Euler–Maclaurin summation formula (Olver [1, p. 285]). Let a and b denote nonnegative integers with $b > a$. Suppose that $f^{(2m)}(t)$ is absolutely integrable over $[a, b]$, where m is a fixed positive integer. Then

$$\sum_{k=a}^{b} f(k) = \int_a^b f(t)\,dt + \tfrac{1}{2}\{f(a) + f(b)\} + \sum_{k=1}^{m-1}\frac{B_{2k}}{(2k)!}\{f^{(2k-1)}(b)$$

$$- f^{(2k-1)}(a)\} + R_m, \tag{10.5}$$

where

$$R_m = \int_a^b \frac{B_{2m} - B_{2m}(t - [t])}{(2m)!}f^{(2m)}(t)\,dt. \tag{10.6}$$

Here B_j denotes the jth Bernoulli number and $B_j(x)$ denotes the jth Bernoulli polynomial, $0 \leq j < \infty$. The Euler–Maclaurin summation formula was the

focus of much of Ramanujan's work. In particular, see Chapters 6–8 of Part I
[9] and Chapter 15 in this book.

Applying the Euler–Maclaurin formula (10.5) with $f(t) = \exp(-Aht^2)t^r$,
$a = 0$, $b = w$, and $m = 1$, we easily find that, as h tends to 0,

$$T_0 = V_0 + \tfrac{1}{2} + O(\sqrt{h})$$

and, for $r \geq 1$,

$$T_r = V_r + O(h^{-r/2}).$$

Thus, by (10.4),

$$S_1 = \sqrt{\frac{\pi}{4Ah}} + \frac{1}{2} + \frac{B}{2A} + \frac{C}{2A^2} + O(\sqrt{h}). \tag{10.7}$$

Comparing the right sides of (10.2) and (10.7) with the help of (10.3), we find
that they agree. Thus, it remains to show that $S_2 = O(\sqrt{h})$, as h tends to 0.

Let $N + 1$ denote the smallest integer j, $j \geq 1$, for which $\alpha + j\delta < \beta + j\gamma$.
Then

$$S_2 = \sum_{k>w} \exp\left(\sum_{j=1}^{k} \{L(h\alpha + hj\delta) - L(h\beta + hj\gamma)\}\right)$$

$$\ll \sum_{k>w} \exp\left(\sum_{j=N+1}^{k} \{L(h\alpha + hj\delta) - L(h\beta + hj\gamma)\}\right)$$

$$= \sum_{k>w} \exp\left(-\sum_{j=N+1}^{k} L'(\theta_j)h(\beta - \alpha + j(\gamma - \delta))\right),$$

where we have applied the mean value theorem, and so

$$h(\alpha + j\delta) < \theta_j < h(\beta + j\gamma). \tag{10.8}$$

Since $L'(x)$ is continuous for $|x| \leq d$ and $L'(x) > 0$ for $x \geq -d$, there exists
a constant $Q > 0$ such that $L'(x) \geq Q$ whenever $|x| \leq d$. The terms with
$h(\alpha + j\delta) \leq d < h(\beta + j\gamma)$ make a total contribution that is less than 1 to each
summand on k. Hence,

$$S_2 \ll \sum_{k>w} \exp\left(-\sum_{\substack{j=N+1 \\ h(\beta+j\gamma) \leq d}}^{k} Qh(\beta - \alpha + j(\gamma - \delta))\right.$$

$$\left. - \sum_{\substack{j=N+1 \\ h(\alpha+j\delta)>d}}^{k} L'(\theta_j)h(\beta - \alpha + j(\gamma - \delta))\right)$$

$$= \sum_{k>w} \exp\left(-\sum_{j=N+1}^{\min(k,g)} Qh(\beta - \alpha + j(\gamma - \delta))\right.$$

$$\left. - \sum_{j=f+1}^{k} L'(\theta_j)h(\beta - \alpha + j(\gamma - \delta))\right),$$

where $g = [(dh^{-1} - \beta)/\gamma]$ and $f = [(dh^{-1} - \alpha)/\delta]$ with the understanding that
$f = \infty$ if $\delta = 0$.

Now by (10.1), $xL'(x) \geq M > 0$, for $x \geq d$. Thus,

$$L'(\theta_j)(\beta - \alpha + j(\gamma - \delta)) \geq \frac{Mh(\beta - \alpha + j(\gamma - \delta))}{\theta_j}$$

$$\geq \frac{Mh(\beta - \alpha + j(\gamma - \delta))}{h(\beta + j\gamma)}$$

$$\geq R,$$

for some constant $R > 0$, where we have used (10.8). Thus,

$$S_2 \ll \sum_{k>w} \exp\left(-Qh \sum_{j=N+1}^{\min(k,g)} (\beta - \alpha + j(\gamma - \delta)) - R \sum_{j=f+1}^{k} 1\right)$$

$$= P_1 + P_2 + P_3,$$

where

$$P_1 = \sum_{w<k\leq g} \exp(-Qh\{(\beta - \alpha)(k - N) + \tfrac{1}{2}(\gamma - \delta)(k^2 + k - N^2 - N)\}),$$

$$P_2 = \sum_{g<k\leq f} \exp(-Qh\{(\beta - \alpha)(g - N) + \tfrac{1}{2}(\gamma - \delta)(g^2 + g - N^2 - N)\}),$$

and

$$P_3 = \sum_{k>f} \exp(-Qh\{(\beta - \alpha)(g - N)$$

$$+ \tfrac{1}{2}(\gamma - \delta)(g^2 + g - N^2 - N)\} - R(k - f)).$$

It is not difficult to see that there exist positive constants Q_1, Q_2, and Q_3 such that, as h tends to 0,

$$P_1 \ll \sum_{k>w} \exp(-Q_1 hk^2) = O(\sqrt{h}),$$

$$P_2 \ll f\exp(-Q_2 hg^2) \ll fe^{-Q_3/h} = O(\sqrt{h}),$$

and

$$P_3 \ll e^{-Q_3/h} \sum_{k=1}^{\infty} e^{-Rk} \ll e^{-Q_3/h} = O(\sqrt{h}).$$

Thus,

$$S_2 \ll P_1 + P_2 + P_3 = O(\sqrt{h}),$$

as h tends to 0. This completes the proof. $\square$

Corollary (i). *Let $n > 0$. Then as x tends to ∞,*

$$\sum_{k=0}^{\infty} \left\{\frac{x^k \Gamma(x + 1)}{\Gamma(x + k + 1)}\right\}^n = \sqrt{\frac{\pi x}{2n}} + \frac{1}{3n} + O(x^{-1/2}).$$

We first offer a short proof for the case when $n = 1$ and x is a positive integer. Using the corollary and (48.5) in Section 48 of Chapter 12 and

Stirling's formula, we find that

$$\sum_{k=0}^{\infty} \frac{x^k \Gamma(x+1)}{\Gamma(x+k+1)} = \frac{x!}{x^x} \sum_{k=0}^{\infty} \frac{x^{x+k}}{(x+k)!} = \frac{x!}{x^x} \sum_{k=x}^{\infty} \frac{x^k}{k!}$$

$$= \frac{x!}{x^x} \left\{ e^x - \sum_{k=0}^{x-1} \frac{x^k}{k!} \right\}$$

$$= \frac{e^x x!}{2x^x} + \frac{1}{3} + O\left(\frac{1}{x}\right)$$

$$= \sqrt{\frac{\pi x}{2}} + O\left(\frac{1}{\sqrt{x}}\right) + \frac{1}{3} + O\left(\frac{1}{x}\right),$$

as x tends to ∞. The result now follows in the case that $n = 1$ and x is a positive integer.

Second, we remark that a more precise version of Corollary (i) in the case $n = 1$ has essentially already been proved in this chapter. By combining (8.1) and (8.10), we deduce that

$$\sum_{k=0}^{\infty} \frac{x^k \Gamma(x+1)}{\Gamma(x+k+1)} = \sqrt{\frac{\pi x}{2}} + \frac{1}{3} + \frac{1}{24}\sqrt{\frac{2\pi}{x}} + \frac{4}{135x} + O\left(\frac{1}{x^{3/2}}\right),$$

as x tends to ∞.

We next give two proofs of Corollary (i), in general. The first uses Entry 10; the second is *ab initio*.

FIRST PROOF. In Entry 10, let $\varphi(t) = (1 + t)^n$, $\alpha = \beta = \delta = 0$, $\gamma = 1$, and $x = 1/h$. Brief calculations of the expressions on the right side of (10.2) complete the proof. □

SECOND PROOF. For $u \geq 0$, set

$$f(u) = \left(\frac{x^u \Gamma(x+1)}{\Gamma(x+u+1)} \right)^n. \tag{10.9}$$

By Stirling's formula, as x tends to ∞,

$$f(u) = e^{un}\left(1 + \frac{u}{x}\right)^{-n(x+u+1/2)} \left\{ 1 + \frac{1}{12x} + \frac{1}{288x^2} - \cdots \right\}^n$$

$$\times \left\{ 1 + \frac{1}{12(x+u)} + \frac{1}{288(x+u)^2} - \cdots \right\}^{-n}. \tag{10.10}$$

Hence, with $t = u/x$,

$$\int_0^{\infty} f(u)\, du = (1 + O(x^{-1})) \int_0^{\infty} e^{un}\left(1 + \frac{u}{x}\right)^{-n(x+u+1/2)} du$$

$$= x(1 + O(x^{-1})) \int_0^{\infty} (1 + t)^{-n/2} \left\{ \frac{e^t}{(1+t)^{1+t}} \right\}^{nx} dt. \tag{10.11}$$

To apply Watson's lemma (Olver [1, p. 113]), we set $v = (1 + t) \text{Log}(1 + t)$ $- t$ and proceed as in the proof of Entry 8. Using (8.5) and (8.6), we find that for $v \geq 0$ and sufficiently small,

$$(1 + t)^{-n/2} \frac{dt}{dv} = (2v)^{-1/2} + \left(\frac{1}{3} - \frac{n}{2}\right) + \cdots . \tag{10.12}$$

Hence, from (10.11) and (10.12), as x tends to ∞,

$$\int_0^\infty f(u)\, du = (x + O(1)) \int_0^\infty e^{-nxv} (1 + t)^{-n/2} \frac{dt}{dv}\, dv$$

$$= (x + O(1)) \int_0^\infty e^{-nxv} \left\{(2v)^{-1/2} + \left(\frac{1}{3} - \frac{n}{2}\right) + \cdots\right\} dv$$

$$= \sqrt{\frac{\pi x}{2n}} + \left(\frac{1}{3n} - \frac{1}{2}\right) + O(x^{-1/2}). \tag{10.13}$$

For each pair of nonnegative integers k, r, let $A_{k,r}(z)$ denote a function with an asymptotic expansion

$$A_{k,r}(z) = a_0 + \frac{a_1}{z} + \frac{a_2}{z^2} + \cdots, \tag{10.14}$$

as z tends to ∞, where the coefficients a_i, $i \geq 0$, may depend on k and r, and where, for each positive integer j, (10.14) becomes an asymptotic expansion of $A_{k,r}^{(j)}(z)$ after j-fold term by term differentiation with respect to z. Using (10.10) and induction on r, it can be shown that, for each positive integer r, $f^{(r)}(u)$ has the form

$$f^{(r)}(u) = f(u) \sum_{k=0}^r A_{k,r}(x + u)(x + u)^{-[(k+1)/2]} \text{Log}^{r-k}\left(1 + \frac{u}{x}\right), \tag{10.15}$$

as x tends to ∞. In particular,

$$f^{(r)}(0) = O(x^{-[(r+1)/2]}), \tag{10.16}$$

as x tends to ∞, and

$$f^{(r)}(u) \to 0, \tag{10.17}$$

as u tends to ∞.

Applying the Euler–Maclaurin formula (10.5) with $f(u)$ defined by (10.9), $a = 0$, and $b = \infty$, we find that, in view of (10.17),

$$\sum_{k=0}^\infty f(k) = \int_0^\infty f(u)\, du + \frac{1}{2} - \sum_{k=1}^{m-1} \frac{B_{2k}}{(2k)!} f^{(2k-1)}(0) + R_m, \tag{10.18}$$

where

$$R_m = \int_0^\infty \frac{B_{2m} - B_{2m}(t - [t])}{(2m)!} f^{(2m)}(t)\, dt. \tag{10.19}$$

By (10.13) and (10.18) with $m = 1$, it remains to show that $R_1 = O(x^{-1/2})$, as x tends to ∞. We shall show more generally that, for each integer $m \geq 1$,

$$R_m = O(x^{1/2-m}), \tag{10.20}$$

as x tends to ∞. Observe that (10.16) and (10.18)–(10.20) imply the interesting infinite asymptotic expansion

$$\sum_{k=0}^{\infty} f(k) - \int_0^{\infty} f(u)\, du \sim \frac{1}{2} - \sum_{k=1}^{\infty} \frac{B_{2k}}{(2k)!} f^{(2k-1)}(0), \tag{10.21}$$

as x tends to ∞.

By (10.15) and (10.19), as x tends to ∞,

$$R_m \ll \int_0^{\infty} |f^{(2m)}(u)|\, du \ll \int_0^{\infty} f(u) \sum_{k=0}^{2m} (x+u)^{-[(k+1)/2]} \operatorname{Log}^{2m-k}\left(1 + \frac{u}{x}\right) du$$

$$\ll \sum_{k=0}^{2m} x^{-[(k+1)/2]} \int_0^{\infty} f(u) \operatorname{Log}^{2m-k}\left(1 + \frac{u}{x}\right) du.$$

Set $t = u/x$ and apply (10.10) to deduce that

$$R_m \ll \sum_{k=0}^{2m} x^{1-[(k+1)/2]} \int_0^{\infty} (1+t)^{-n/2} \left\{\frac{e^t}{(1+t)^{1+t}}\right\}^{nx} \operatorname{Log}^{2m-k}(1+t)\, dt.$$

Setting $v = (1+t) \operatorname{Log}(1+t) - t$, we then obtain

$$R_m \ll \sum_{k=0}^{2m} x^{1-[(k+1)/2]} \int_0^{\infty} e^{-nxv}(1+t)^{-n/2} \frac{dt}{dv} \operatorname{Log}^{2m-k}(1+t)\, dv.$$

By (10.12), (8.5), and Watson's lemma,

$$R_m \ll \sum_{k=0}^{2m} x^{1-[(k+1)/2]} \int_0^{\infty} e^{-nxv}(2v)^{-1/2}(2v)^{(2m-k)/2}\, dv$$

$$\ll \sum_{k=0}^{2m} x^{1-[(k+1)/2]} \int_0^{\infty} e^{-nxv} v^{m-(k+1)/2}\, dv$$

$$\ll \sum_{k=0}^{2m} x^{(k+1)/2-[(k+1)/2]-m} = O(x^{1/2-m}),$$

as x tends to ∞. This completes the proof of (10.20). $\qquad\square$

The second proof above is substantially due to F. W. J. Olver (personal communication), who established (10.20) in the case $m = 1$. By an extension of his ideas, we have proved (10.20) for all m in order to obtain the asymptotic formula (10.21). As an application of (10.21), we demonstrate that

$$\sum_{k=0}^{\infty} \left\{\frac{x^k \Gamma(x+1)}{\Gamma(x+k+1)}\right\}^n - \int_0^{\infty} \left\{\frac{x^t \Gamma(x+1)}{\Gamma(x+t+1)}\right\}^n dt = \frac{1}{2} + \frac{n}{24x} + O(x^{-2}), \tag{10.22}$$

where $n > 0$ and x tends to ∞. Observe that (10.22) generalizes Entry 8, since in the case $n = 1$, (10.22) implies (8.1). To verify (10.22), logarithmically differentiate with respect to u in (10.10) to obtain

$$\frac{f'(u)}{f(u)} = -\frac{n}{2(u+x)} - n \log\left(1 + \frac{u}{x}\right) - n\left\{1 + \frac{1}{12(x+u)}\right.$$

$$\left. + \frac{1}{288(x+u)^2} - \cdots\right\}^{-1}\left\{-\frac{1}{12(x+u)^2} - \frac{1}{144(x+u)^3} + \cdots\right\}.$$

Thus,

$$f'(0) = -\frac{n}{2x} + O(x^{-2})$$

as x tends to ∞. Since $B_2/2! = \frac{1}{12}$, (10.22) therefore follows from (10.16) and (10.21).

Corollary (ii). *If n is a positive integer, as x tends to ∞,*

$$\sum_{k=0}^{\infty}\left(\frac{x^k}{k!}\right)^n \sim \frac{\exp\left\{nx + \frac{n^2-1}{24}\left(\frac{1}{nx} + \frac{1}{2n^2x^2} + \cdots\right)\right\}}{\sqrt{n}(2\pi x)^{(n-1)/2}}. \tag{10.23}$$

It is tempting to conclude that the sum in the exponent on the right side is equal to $-\log(1 - 1/(nx))$. However, then we would have an exact formula rather than an asymptotic formula, and it is clear that this exact formula could not possibly be true for $n > 1$.

For $n = 1$, (10.23) is trivial. For $n = 2$, the left side of (10.23) is equal to $I_0(2x)$, where I_0 is the Bessel function of imaginary argument of order 0. In this case, the first three terms

$$\frac{e^{2x}}{2\sqrt{\pi x}}\left(1 + \frac{1}{16x} + \frac{9}{512x^2} + \cdots\right)$$

agree with the asymptotic expansion for $I_0(2x)$ found in Watson's treatise [9, p. 203, Eq. (2)]. The case $n = 5$ was communicated by Ramanujan in his first letter to Hardy [16, p. xxvi] and was proved by Watson [2].

PROOF. Ramanujan's result follows easily from a general result proved by Barnes [1, p. 115]. Accordingly, Barnes showed that (see also Watson's paper [2])

$$\sum_{k=0}^{\infty}\left(\frac{x^k}{k!}\right)^n \sim \frac{e^{nx}}{\sqrt{n}(2\pi x)^{(n-1)/2}}\left(1 + \frac{n^2-1}{24nx} + \frac{(n^2-1)(n^2+23)}{1152n^2x^2} + \cdots\right),$$

as x tends to ∞. Expanding the exponential on the right side of (10.23), we find that Ramanujan's result is in agreement with that of Barnes (for the first three terms). $\qquad\square$

For another approach to Corollary (ii), when n is any positive number, see the text by Olver [1, pp. 307–309].

Entry 11(i). *As x tends to* ∞,

$$\sum_{k=0}^{\infty} \left(\frac{ex}{k}\right)^k \sim \sqrt{2\pi x}\, \exp\left(x - \frac{1}{24x} - \frac{1}{48x^2} - \left(\frac{1}{36} + \frac{1}{5760}\right)\frac{1}{x^3} + \cdots\right).$$

PROOF. We shall apply a general asymptotic formula

$$e^{-x}\sum_{k=0}^{\infty} \frac{\varphi(k)x^k}{k!} \sim \varphi(x) + \frac{x}{2}\varphi''(x) + \frac{x}{6}\varphi'''(x) + \frac{x^2}{8}\varphi^{(4)}(x)$$

$$+ \frac{x}{24}\varphi^{(4)}(x) + \frac{x^2}{12}\varphi^{(5)}(x) + \frac{x^3}{48}\varphi^{(6)}(x) + \cdots, \quad (11.1)$$

as x tends to ∞, that is found in Chapter 3, Entry 10 of the second notebook. The function $\varphi(x) = e^x\Gamma(x+1)/x^x$ is easily seen to satisfy the hypotheses of a rigorous formulation of this theorem (Part I [9, pp. 57, 58]). Thus, by (11.1) and Stirling's formula,

$$e^{-x}\sum_{k=0}^{\infty} \left(\frac{ex}{k}\right)^k \sim \sqrt{2\pi}\left(x^{1/2} + \frac{1}{12x^{1/2}} + \frac{1}{288x^{3/2}} - \frac{139}{51840x^{5/2}} + \cdots\right)$$

$$+ \frac{x}{2}\sqrt{2\pi}\left(-\frac{1}{4x^{3/2}} + \frac{1}{16x^{5/2}} + \frac{5}{384x^{7/2}} + \cdots\right)$$

$$+ \frac{x}{6}\sqrt{2\pi}\left(\frac{3}{8x^{5/2}} - \frac{5}{32x^{7/2}} + \cdots\right)$$

$$+ \frac{x^2}{8}\sqrt{2\pi}\left(-\frac{15}{16x^{7/2}} + \frac{35}{64x^{9/2}} + \cdots\right)$$

$$+ \frac{x}{24}\sqrt{2\pi}\left(-\frac{15}{16x^{7/2}} + \cdots\right) + \frac{x^2}{12}\sqrt{2\pi}\left(\frac{105}{32x^{9/2}} + \cdots\right)$$

$$+ \frac{x^3}{48}\sqrt{2\pi}\left(-\frac{945}{64x^{11/2}} + \cdots\right) + \cdots$$

$$= \sqrt{2\pi x}\left(1 - \frac{1}{24x} - \frac{23}{2^7\cdot 3^2 x^2} - \frac{11237}{2^{10}\cdot 3^4\cdot 5x^3} + \cdots\right), \quad (11.2)$$

as x tends to ∞.

On the other hand,

$$\exp\left(-\frac{1}{24x} - \frac{1}{48x^2} - \left(\frac{1}{36} + \frac{1}{5760}\right)\frac{1}{x^3} + \cdots\right)$$

$$= 1 - \frac{1}{24x} - \frac{1}{48x^2} - \left(\frac{1}{36} + \frac{1}{5760}\right)\frac{1}{x^3} + \cdots + \frac{1}{2\cdot 24^2 x^2}$$

$$+ \frac{1}{24\cdot 48x^3} + \cdots - \frac{1}{6\cdot 24^3 x^3} + \cdots$$

$$= 1 - \frac{1}{24x} - \frac{23}{2^7\cdot 3^2 x^2} - \frac{11237}{2^{10}\cdot 3^4\cdot 5x^3} + \cdots.$$

Comparing the two asymptotic expansions found above with that in Entry 11(i), we complete the proof. $\square$

Ramanujan's asymptotic formula (11.1) is very useful and powerful. In addition to Part I, see the paper by R. J. Evans [1] for several applications. Corollary 14 of his paper provides a solution to a previously unsolved problem of Appledorn [1].

Entry 11(ii). *As n tends to ∞,*

$$I_n := \int_0^\infty \frac{x^{n-1}\, dx}{\displaystyle\sum_{k=0}^\infty (x/k)^k} \sim n^n\left(\frac{1}{n} + \frac{1}{2n^2} + \frac{1}{3n^3} + \frac{3}{8n^4} + \cdots\right).$$

PROOF. By (11.2),

$$I_n = e^n \int_0^\infty \frac{x^{n-1}\, dx}{\displaystyle\sum_{k=0}^\infty (ex/k)^k}$$

$$= e^n \int_0^\infty \frac{x^{n-1}\, dx}{e^x\sqrt{2\pi x}\left(1 - \dfrac{1}{24x} - \dfrac{23}{2^7\cdot 3^2 x^2} - \dfrac{11237}{2^{10}\cdot 3^4\cdot 5x^3} + \cdots\right)}$$

$$= \frac{e^n}{\sqrt{2\pi}} \int_0^\infty e^{-x} x^{n-3/2}\left\{1 + \frac{1}{24x} + \frac{25}{2^7\cdot 3^2 x^2} + \frac{11957}{2^{10}\cdot 3^4\cdot 5x^3} + \cdots\right\} dx$$

$$= \frac{e^n}{\sqrt{2\pi}}\left\{\Gamma(n - \tfrac{1}{2}) + \frac{1}{24}\Gamma(n - \tfrac{3}{2}) + \frac{25}{2^7\cdot 3^2}\Gamma(n - \tfrac{5}{2})\right.$$

$$\left. + \frac{11957}{2^{10}\cdot 3^4\cdot 5}\Gamma(n - \tfrac{7}{2}) + \cdots\right\}.$$

It seems convenient to express each of the gamma functions above in terms of $\Gamma(n + \tfrac{1}{2})$ and then use the asymptotic series (Olver [1, p. 295]),

$$\Gamma(n + \tfrac{1}{2}) \sim \sqrt{2\pi}\, n^n e^{-n}\left(1 - \frac{1}{24n} + \frac{1}{2^7\cdot 3^2 n^2} + \frac{1003}{2^{10}\cdot 3^4\cdot 5n^3} + \cdots\right),$$

as n tends to ∞. Hence, as n tends to ∞,

$$I_n = \frac{e^n\Gamma(n + \tfrac{1}{2})}{\sqrt{2\pi}(n - \tfrac{1}{2})}\left\{1 + \frac{1}{24(n - \tfrac{3}{2})} + \frac{25}{2^7\cdot 3^2(n - \tfrac{3}{2})(n - \tfrac{5}{2})}\right.$$

$$\left. + \frac{11957}{2^{10}\cdot 3^4\cdot 5(n - \tfrac{3}{2})(n - \tfrac{5}{2})(n - \tfrac{7}{2})} + \cdots\right\}$$

$$= n^n \left\{ 1 - \frac{1}{24n} + \frac{1}{2^7 \cdot 3^2 n^2} + \frac{1003}{2^{10} \cdot 3^4 \cdot 5n^3} + \cdots \right\}$$

$$\times \frac{1}{n} \left\{ 1 + \frac{1}{2n} + \frac{1}{4n^2} + \frac{1}{8n^3} + \cdots \right\} \left\{ 1 + \frac{1}{24n} + \frac{1}{16n^2} + \frac{3}{32n^3} + \cdots \right.$$

$$+ \frac{25}{2^7 \cdot 3^2 n^2} + \frac{25}{2^5 \cdot 3^2 n^3} + \cdots + \frac{11957}{2^{10} \cdot 3^4 \cdot 5n^3} + \cdots \right\}.$$

Collecting together the coefficients of $1/n^k$, $1 \le k \le 4$, we complete the proof. $\square$

Entry 11(iii).

$$S := \text{Log } 2 \sum_{k=2}^{\infty} \frac{(-1)^k}{k \text{ Log } k} + \text{Log}^2 \, 2 \sum_{k=2}^{\infty} \frac{1}{k \text{ Log } k \text{ Log}(2k)} = 1.$$

Entry 11(iii) was, in fact, submitted as a problem by Ramanujan to the *Journal of the Indian Mathematical Society* [12], [16, p. 333].

PROOF. We shall show by induction on n that, for $n \ge 0$,

$$S = \sum_{k=1}^{n} \frac{1}{k(k+1)} + \sum_{k=2}^{\infty} \frac{\text{Log}^2 \, 2}{k \text{ Log}(2^n k) \text{ Log}(2^{n+1} k)} + \sum_{k=2}^{\infty} \frac{(-1)^k \text{ Log } 2}{k \text{ Log}(2^n k)}. \quad (11.3)$$

By definition of S, (11.3) is valid for $n = 0$. Now,

$$\sum_{k=2}^{\infty} \frac{\text{Log}^2 \, 2}{k \text{ Log}(2^n k) \text{ Log}(2^{n+1} k)} + \sum_{k=2}^{\infty} \frac{(-1)^k \text{ Log } 2}{k \text{ Log}(2^n k)}$$

$$= \sum_{k=2}^{\infty} \frac{\{(-1)^k + 1\} \text{ Log}^2 \, 2 + (-1)^k \text{ Log } 2 \text{ Log}(2^n k)}{k \text{ Log}(2^n k) \text{ Log}(2^{n+1} k)}$$

$$= \sum_{k=1}^{\infty} \frac{\text{Log}^2 \, 2}{k \text{ Log}(2^{n+1} k) \text{ Log}(2^{n+2} k)} + \sum_{k=2}^{\infty} \frac{(-1)^k \text{ Log } 2}{k \text{ Log}(2^{n+1} k)}$$

$$= \frac{1}{(n+1)(n+2)} + \sum_{k=2}^{\infty} \frac{\text{Log}^2 \, 2}{k \text{ Log}(2^{n+1} k) \text{ Log}(2^{n+2} k)} + \sum_{k=2}^{\infty} \frac{(-1)^k \text{ Log } 2}{k \text{ Log}(2^{n+1} k)},$$

which completes the induction. Letting n tend to ∞ in (11.3), we easily conclude that

$$S = \sum_{k=1}^{\infty} \frac{1}{k(k+1)} = 1. \qquad \square$$

Ramanujan begins Section 12 by briefly describing Entry 10 of Chapter 3. He concludes this section by giving an example that is an elaboration of Example 2, Section 10 of Chapter 3. Pollak and Shepp [1] have proposed an equivalent asymptotic expansion, but with less terms.

Example. *As x tends to ∞,*

$$\sum_{k=1}^{\infty} \frac{\text{Log}(k+1)x^k}{k!} \sim e^x \left(\text{Log } x + \frac{1}{2x} + \frac{1}{12x^2} + \frac{1}{12x^3} + \frac{19}{120x^4} + \frac{9}{20x^5} + \cdots \right).$$

PROOF. As in the proof of Entry 11(i), we apply Entry 10 of Chapter 3. However, in addition to the seven terms displayed in (11.1), nine more terms are needed. Thus, as x tends to ∞,

$$e^{-x} \sum_{k=1}^{\infty} \frac{\text{Log}(k+1)x^k}{k!} \sim \text{Log}(x+1) - \frac{x}{2(x+1)^2} + \frac{x}{3(x+1)^3} - \frac{3x^2}{4(x+1)^4}$$

$$- \frac{x}{4(x+1)^4} + \frac{2x^2}{(x+1)^5} - \frac{5x^3}{2(x+1)^6} + \frac{x}{5(x+1)^5}$$

$$- \frac{25x^2}{6(x+1)^6} + \frac{15x^3}{(x+1)^7} - \frac{105x^4}{8(x+1)^8} - \frac{x}{6(x+1)^6}$$

$$+ \frac{8x^2}{(x+1)^7} - \frac{245x^3}{4(x+1)^8} + \frac{140x^4}{(x+1)^9}$$

$$- \frac{189x^5}{2(x+1)^{10}} + \cdots$$

$$= \text{Log } x + \frac{1}{x} - \frac{1}{2x^2} + \frac{1}{3x^3} - \frac{1}{4x^4} + \frac{1}{5x^5} + \cdots$$

$$- \frac{1}{2x} \left(1 - \frac{2}{x} + \frac{3}{x^2} - \frac{4}{x^3} + \frac{5}{x^4} + \cdots \right) + \frac{1}{3x^2} \left(1 - \frac{3}{x} \right.$$

$$+ \frac{6}{x^2} - \frac{10}{x^3} + \cdots \bigg) - \frac{3}{4x^2} \left(1 - \frac{4}{x} + \frac{10}{x^2} - \frac{20}{x^3} + \cdots \right)$$

$$- \frac{1}{4x^3} \left(1 - \frac{4}{x} + \frac{10}{x^2} + \cdots \right) + \frac{2}{x^3} \left(1 - \frac{5}{x} + \frac{15}{x^2} + \cdots \right)$$

$$- \frac{5}{2x^3} \left(1 - \frac{6}{x} + \frac{21}{x^2} + \cdots \right) + \frac{1}{5x^4} \left(1 - \frac{5}{x} + \cdots \right)$$

$$- \frac{25}{6x^4} \left(1 - \frac{6}{x} + \cdots \right) + \frac{15}{x^4} \left(1 - \frac{7}{x} + \cdots \right)$$

$$- \frac{105}{8x^4} \left(1 - \frac{8}{x} + \cdots \right) - \frac{1}{6x^5} + \frac{8}{x^5} - \frac{245}{4x^5} + \frac{140}{x^5}$$

$$- \frac{189}{2x^5} + \cdots .$$

Collecting the coefficients of x^{-k}, $1 \le k \le 5$, we arrive at Ramanujan's asymptotic expansion.   □

Entry 13. *Let* α, β, γ, *and* δ *be any complex numbers. Then*

$$\int_0^\infty \frac{dx}{(x^2 + \alpha^2)(x^2 + \beta^2)(x^2 + \gamma^2)(x^2 + \delta^2)}$$

$$= \frac{\pi}{6} \frac{(\alpha + \beta + \gamma + \delta)^3 - (\alpha^3 + \beta^3 + \gamma^3 + \delta^3)}{\alpha\beta\gamma\delta(\alpha + \beta)(\beta + \gamma)(\gamma + \alpha)(\alpha + \delta)(\beta + \delta)(\gamma + \delta)}. \qquad (13.1)$$

Corollary. *If* α, β, γ, *and* δ *are the roots of the polynomial* $x^4 - px^3 + qx^2 - rx + s$, *then*

$$\int_0^\infty \frac{dx}{(x^2 + \alpha^2)(x^2 + \beta^2)(x^2 + \gamma^2)(x^2 + \delta^2)} = \frac{\pi}{2s} \frac{1}{r - \dfrac{ps}{q - r/p}}. \qquad (13.2)$$

Ramanujan's formula (13.2) is the same as an evaluation in Gradshteyn and Ryzhik's tables [1, p. 218, formula (5)]. Since $p = \alpha + \beta + \gamma + \delta$, $q = \alpha\beta + \alpha\gamma + \alpha\delta + \beta\gamma + \beta\delta + \gamma\delta$, $r = \alpha\beta\gamma + \alpha\beta\delta + \alpha\gamma\delta + \beta\gamma\delta$, and $s = \alpha\beta\gamma\delta$, formula (13.2) can be rewritten in the form (13.1), after a tedious calculation.

Entry 14. *If* x *is arbitrary and* $a \neq 0, -1, -2, \ldots$, *then*

$$\sum_{k=0}^\infty \frac{(-1)^k (2a)_k (a + k)}{k! \{(a + k)^2 + x^2\}} = \frac{\Gamma^2(a)}{2\Gamma(2a) \prod_{k=0}^\infty \left\{1 + \left(\dfrac{x}{a + k}\right)^2\right\}}.$$

A proof of Entry 14 was published by Ramanujan [8], [16, p. 53].

R. Askey has pointed out the following observation. Letting $b = ix$ and $c = -ix$ and using a value for ${}_5F_4$ found in Wilson's paper [1, Eq. (2.4)], we find that the sum in Entry 14 equals

$$\frac{a}{a^2 + x^2} {}_4F_3 \left[\begin{matrix} 2a, a + 1, a + ix, a - ix \\ a, a + 1 + ix, a + 1 - ix \end{matrix} ; -1 \right]$$

$$= \frac{a}{(a + b)(a + c)} \lim_{d \to -\infty} {}_5F_4 \left[\begin{matrix} 2a, a + 1, a + b, a + c, a + d \\ a, a + 1 - b, a + 1 - c, a + 1 - d \end{matrix} ; 1 \right]$$

$$= \frac{a}{(a + b)(a + c)}$$

$$\times \lim_{d \to -\infty} \frac{\Gamma(a + 1 - b)\Gamma(a + 1 - c)\Gamma(a + 1 - d)\Gamma(1 - a - b - c - d)}{\Gamma(2a + 1)\Gamma(1 - b - c)\Gamma(1 - b - d)\Gamma(1 - c - d)}.$$

Entry 14 now follows after computing the limit above.

Example. *For* $n, a > 0$,

$$\int_0^\infty \frac{\cos(nx)\, dx}{a^2 + x^2} = \frac{\pi}{2a} e^{-na}.$$

This formula is well known and was established by Ramanujan in his quarterly reports (Part I [9, p. 322]) via the Fourier cosine inversion formula.

Entry 15. *For a > 0 and n real,*

$$\int_0^\infty |\Gamma(a + ix)|^2 \cos(2nx)\, dx = \tfrac{1}{2}\sqrt{\pi}\,\Gamma(a)\Gamma(a + \tfrac{1}{2}) \operatorname{sech}^{2a} n.$$

Entry 15 was proved by Ramanujan in [8], [16, pp. 53, 54].

We note the following generalization of Entry 15. If $a > 0$ and $|\operatorname{Re} y| < \tfrac{1}{2}$, then

$$\int_{-\infty}^\infty |\Gamma(a + ix)|^2 e^{2yx}\, dx = \sqrt{\pi}\,\Gamma(a)\Gamma(a + \tfrac{1}{2}) \sec^{2a} y.$$

For this and substantial ramifications, see Wilson's paper [1].

Entry 16. *For a and n both real, and n integral in (iv),*

(i) $\displaystyle \int_0^\infty \frac{\sinh(ax)}{\sinh(\pi x)} \cos(nx)\, dx = \frac{1}{2} \frac{\sin a}{\cosh n + \cos a}$, $\qquad |a| < \pi$,

(ii) $\displaystyle \int_0^\infty \frac{\cosh(ax)}{\sinh(\pi x)} \sin(nx)\, dx = \frac{1}{2} \frac{\sinh n}{\cosh n + \cos a}$, $\qquad |a| < \pi$,

(iii) $\displaystyle \int_0^\infty \frac{\sin(nx)}{e^{2\pi x} - 1}\, dx = \frac{1}{2}\left(\frac{1}{e^n - 1} + \frac{1}{2} - \frac{1}{n}\right)$, $\qquad n > 0$,

(iv) $\displaystyle \int_0^\infty \frac{x^{2n-1}}{e^{2\pi x} - 1}\, dx = \frac{(-1)^{n-1} B_{2n}}{4n}$, $\qquad n > 0$,

$\displaystyle \int_0^\infty \frac{x^{2n}}{\cosh(\pi x/2)}\, dx = (-1)^n E_{2n}$, $\qquad n \geq 0$,

where B_k and E_k, $0 \leq k < \infty$, denote the kth Bernoulli and Euler numbers, respectively.

In each case below, [1] refers to the tables of Gradshteyn and Ryzhik.

Both (i) and (ii) can be found in [1, p. 504]. Ramanujan has stated (iii) in [8], [16, p. 56] but does not give a proof. Formula (iii), however, is easily derived from [1, p. 481, formula 3.911, No. 2]. Both integrals in (iv) are classical [1, pp. 1076, 349].

Entry 17. *Let $\varphi(z)$ be analytic for $a \leq \operatorname{Re}(z) \leq n$, where a is a nonnegative integer. Suppose that*

$$\lim_{y \to \infty} |\varphi(x \pm iy)| e^{-2\pi y} = 0,$$

uniformly for $a \leq x \leq n$. Then

$$\sum_{k=a}^{n} \varphi(k) = \int_{a}^{n} \varphi(u)\, du + \tfrac{1}{2}\{\varphi(a) + \varphi(n)\}$$

$$- i \int_{0}^{\infty} \frac{\varphi(n + iu) - \varphi(n - iu) - \varphi(a + iu) + \varphi(a - iu)}{e^{2\pi u} - 1}\, du. \quad (17.1)$$

Entry 17 is the famous Abel–Plana summation formula (Henrici [1, p. 274], Whittaker and Watson [1, p. 145]). For the history of this formula and some of its applications, see Lindelöf's book [1, Chapter 3]. Ramanujan's formulation of Entry 17 is not as precise as that given above, because all those expressions that are independent of n are not explicitly given.

Corollary. *For each positive integer n,*

$$\text{Log } n! = n \text{ Log } n - n + \tfrac{1}{2} \text{Log}(2\pi n) + 2 \int_{0}^{\infty} \frac{\tan^{-1}(x/n)}{e^{2\pi x} - 1}\, dx.$$

This corollary is easily established by setting $\varphi(x) = \text{Log } x$ in Entry 17. Details may be found in Lindelöf's text [1, pp. 69, 70]. Whittaker and Watson [1, pp. 250, 251] give another proof and attribute the result to Binet in 1839.

Entry 18(i). *Let $t > 0$, and fix a positive integer n. Set $x = tn$, and put $\varphi(z) = f(t + tz) - f(tz)$ for a given function f. Suppose that $\varphi(z)$ satisfies the hypotheses of Entry 17 with $a = 0$. Then*

$$f(x) + \tfrac{1}{2}\varphi(n) = \tfrac{1}{2}\{f(0) + f(t)\} + \int_{0}^{n} \varphi(u)\, du$$

$$- i \int_{0}^{\infty} \frac{\varphi(n + iu) - \varphi(n - iu) - \varphi(iu) + \varphi(-iu)}{e^{2\pi u} - 1}\, du.$$

PROOF. Apply Entry 17 to $\varphi(z)$ with $a = 0$. Now observe that the left side of (17.1) is equal to

$$\sum_{k=0}^{n} \varphi(k) = \varphi(n) + f(x) - f(0).$$

After some rearrangement, we deduce the desired result. □

Our formulation of Entry 18(i) is rather different from that of Ramanujan since he does not record those parts of the formula that do not depend on x. Furthermore, there are two misprints in his statement (p. 159). In order to prove Entry 18(ii), which is likewise not properly stated by Ramanujan, we need to establish a lemma that is similar in character to Entry 17.

Lemma. *Let $n = 2m$ be an even positive integer. Suppose that $\varphi(z)$ is analytic on $0 \le \text{Re}(z) \le n$ and that*

$$\lim_{y \to \infty} |\varphi(x \pm iy)| e^{-\pi y/2} = 0,$$

uniformly for $0 \leq x \leq n$. *Then, provided that the integrals below exist,*

$$2 \sum_{k=1}^{m} (-1)^k \varphi(2k - 1)$$

$$= (-1)^m \int_0^\infty \frac{\varphi(n + iu) + \varphi(n - iu)}{e^{\pi u/2} + e^{-\pi u/2}} \, du - \int_0^\infty \frac{\varphi(iu) + \varphi(-iu)}{e^{\pi u/2} + e^{-\pi u/2}} \, du.$$

PROOF. Let C_N denote the positively oriented rectangle with vertices $\pm iN$ and $n \pm iN$. By the residue theorem,

$$\frac{1}{2\pi i} \int_{C_N} \frac{\pi \varphi(z) \, dz}{\cos(\pi z/2)} = 2 \sum_{k=1}^{m} (-1)^k \varphi(2k - 1).$$

If we let N tend to ∞ and invoke our hypotheses, we find that

$$2 \sum_{k=1}^{m} (-1)^k \varphi(2k - 1) = \frac{1}{2i} \int_{n-i\infty}^{n+i\infty} \frac{\varphi(z) \, dz}{\cos(\pi z/2)} - \frac{1}{2i} \int_{-i\infty}^{i\infty} \frac{\varphi(z) \, dz}{\cos(\pi z/2)}. \quad (18.1)$$

Letting $z = n + iu$ and recalling that $n = 2m$, we find that

$$\frac{1}{2i} \int_{n-i\infty}^{n+i\infty} \frac{\varphi(z) \, dz}{\cos(\pi z/2)} = (-1)^m \int_{-\infty}^\infty \frac{\varphi(n + iu) \, du}{e^{\pi u/2} + e^{-\pi u/2}}$$

$$= (-1)^m \int_0^\infty \frac{\varphi(n + iu) + \varphi(n - iu)}{e^{\pi u/2} + e^{-\pi u/2}} \, du.$$

The remaining integral in (18.1) can be transformed in a similar fashion. The desired result now follows. □

Entry 18(ii). *Let* $t > 0$, *and fix an even positive integer* $n = 2m$. *Set* $x = tn$ *and define* $\varphi(z) = f(tz + t) + f(tz - t)$ *for a given function* f. *Suppose that* $\varphi(z)$ *satisfies the hypotheses of the previous lemma. Then*

$$2f(x) = 2(-1)^m f(0) + \int_0^\infty \frac{\varphi(n + iu) + \varphi(n - iu)}{e^{\pi u/2} + e^{-\pi u/2}} \, du$$

$$- (-1)^m \int_0^\infty \frac{\varphi(iu) + \varphi(-iu)}{e^{\pi u/2} + e^{-\pi u/2}} \, du,$$

provided that the integrals above exist.

PROOF. Apply the previous lemma and observe that

$$2 \sum_{k=1}^{m} (-1)^k \varphi(2k - 1) = 2(-1)^m f(x) - 2f(0).$$

The desired equality now follows. □

Entry 19. *Let $n > 0$. If*

$$\psi(n) = \int_0^h \varphi(x) \cos(nx)\, dx,$$

then

(i)
$$\int_0^\infty \psi(x) \cos(mx)\, dx = \begin{cases} \dfrac{\pi}{2}\varphi(m), & m < h, \\[2mm] \dfrac{\pi}{4}\varphi(m), & m = h, \\[2mm] 0, & m > h; \end{cases}$$

if

$$\psi(n) = \int_0^h \varphi(x) \sin(nx)\, dx,$$

then

(ii)
$$\int_0^\infty \psi(x) \sin(mx)\, dx = \begin{cases} \dfrac{\pi}{2}\varphi(m), & m < h, \\[2mm] \dfrac{\pi}{4}\varphi(m), & m = h, \\[2mm] 0, & m > h. \end{cases}$$

Entry 19 follows easily from the Fourier integral theorem (Titchmarsh [1, pp. 432–435], [2, pp. 16, 17]) and is valid when φ is continuous and of bounded variation on $[0, h]$. Entry 19 is also given in Ramanujan's quarterly reports (Part I [9, p. 333]).

Corollary. *If $a > 0$ and n is real, then*

$$\int_0^\infty \operatorname{sech}^{2a} x \cos(2nx)\, dx = \frac{\sqrt{\pi}\,|\Gamma(a + in)|^2}{2\Gamma(a)\Gamma(a + \tfrac{1}{2})}.$$

This result was proved by Ramanujan [8], [16, p. 54] by means of the Fourier inversion formula and Entry 15.

We note the following generalization of the previous corollary. If $a > |\operatorname{Re} y|$, then

$$\int_{-\infty}^\infty \operatorname{sech}^{2a} x\, e^{2yx}\, dx = 2^{2a-1}\frac{\Gamma(a + y)\Gamma(a - y)}{\Gamma(2a)}.$$

To see this, observe that

$$\int_{-\infty}^\infty \operatorname{sech}^{2a} x\, e^{2yx}\, dx = \int_{-\infty}^\infty \left(\frac{2}{e^x + e^{-x}}\right)^{2a} e^{2yx}\, dx$$

$$= \int_0^\infty \left(\frac{2}{t + 1/t}\right)^{2a} t^{2y-1}\, dt$$

$$= 2^{2a} \int_0^\infty t^{2y+2a-1}(1 + t^2)^{-2a} \, dt$$

$$= 2^{2a-1} \int_0^\infty u^{a+y-1}(1 + u)^{-2a} \, du$$

$$= 2^{2a-1} \frac{\Gamma(a + y)\Gamma(a - y)}{\Gamma(2a)},$$

where we have employed a familiar integral representation for the beta-function.

Entry 20. *If $n > 0$ and $0 \le a < \pi$, then*

$$\int_0^\infty \frac{\sinh(ax)}{\sinh(\pi x)} \frac{dx}{1 + n^2 x^2} = \sum_{k=1}^\infty \frac{(-1)^{k+1} \sin(ka)}{1 + nk}.$$

This result is classical (Gradshteyn and Ryzhik [1, p. 352]) and is easily established by contour integration.

Entry 21. *Let p, q, and n be real. Suppose that $\varphi_j(p, x)$ and $F(nx)$ are continuous for $\alpha_j \le x \le \beta_j$, where $j = 1, 2$. Define $\psi_1(p, n)$ and $\psi_2(p, n)$ by*

$$\int_{\alpha_1}^{\beta_1} \varphi_1(p, x)F(nx) \, dx = \psi_1(p, n) \quad and \quad \int_{\alpha_2}^{\beta_2} \varphi_2(p, x)F(nx) \, dx = \psi_2(p, n).$$

Then

$$\int_{\alpha_1}^{\beta_1} \varphi_1(p, x)\psi_2(q, nx) \, dx = \int_{\alpha_2}^{\beta_2} \varphi_2(q, x)\psi_1(p, nx) \, dx.$$

Entry 21 is easily established by inverting the order of integration.

The following corollary, which is Parseval's theorem for cosine transforms, is formally a special case of Entry 21. However, since the intervals of integration are not finite, different hypotheses, which we have taken from Titchmarsh's book [2, p. 54], must be assumed. Both Entry 21 and the corollary below were proved formally by Ramanujan in [8], [16, pp. 55, 56].

Corollary. *Let p, q, l, and n be real. Suppose that $\varphi(p, x) \in L(0, \infty)$ in the variable x and that $\lim_{x \to 0+} \varphi(p, x)$ exists. Define*

$$\psi(p, n) = \int_0^\infty \varphi(p, x) \cos(nx) \, dx,$$

which we assume is integrable over any finite interval in $0 \le n < \infty$. Also suppose that $\psi(p, n)$ tends to 0 as n tends to ∞. Then

$$\frac{\pi}{2} \int_0^\infty \varphi(p, x)\varphi(q, lx) \, dx = \int_0^\infty \psi(q, x)\psi(p, lx) \, dx.$$

The corollary above and the example below were communicated in Ramanujan's first letter to Hardy [16, p. 350]. Earlier, Ramanujan [6] had submitted this example as a problem to the *Journal of the Indian Mathematical Society*. Ramanujan also established this example in [8], [16, p. 55].

Example. *If* $\alpha\beta = \pi/4$, *then*

$$\sqrt{\alpha}\int_0^\infty \frac{e^{-x^2}\,dx}{\cosh(\alpha x)} = \sqrt{\beta}\int_0^\infty \frac{e^{-x^2}\,dx}{\cosh(\beta x)}.$$

Ramanujan's next statement is enigmatic. He says that the example above can be derived from the formula

$$\sqrt{\alpha}\sum_{k=0}^\infty \frac{(-1)^k E_{2k+1}\alpha^{2k}}{k!} = \sqrt{\beta}\sum_{k=0}^\infty \frac{(-1)^k E_{2k+1}\beta^{2k}}{k!}, \tag{21.1}$$

"which is obtained from the theorem"

$$\sum_{k=1}^\infty (-1)^{k+1}\varphi(k) = \sum_{k=0}^\infty (-1)^k\varphi(-k). \tag{21.2}$$

Equality (21.1) is really just a very special case of the Poisson summation formula (see Corollary (i) in Section 31) when the functions appearing in the formula are self-reciprocal Fourier transforms of a special type. Formula (21.2) was stated by Ramanujan in Chapter 4, Section 9, Example 2 and, as to be expected, is valid only under severe restrictions (Part I, p. 97).

Entry 22. *If* $a, b > 0$, *then*

(i) $$\int_0^\infty |\Gamma(a+ix)\Gamma(b+ix)|^2\,dx = \frac{\sqrt{\pi}\,\Gamma(a)\Gamma(a+\tfrac{1}{2})\Gamma(b)\Gamma(b+\tfrac{1}{2})\Gamma(a+b)}{2\Gamma(a+b+\tfrac{1}{2})};$$

if $0 < a < b + \tfrac{1}{2}$, *then*

(ii) $$\int_0^\infty \left|\frac{\Gamma(a+ix)}{\Gamma(b+1+ix)}\right|^2\,dx = \frac{\sqrt{\pi}\,\Gamma(a)\Gamma(a+\tfrac{1}{2})\Gamma(b-a+\tfrac{1}{2})}{2\Gamma(b+1)\Gamma(b+\tfrac{1}{2})\Gamma(b-a+1)}.$$

These two beautiful formulas were derived by Ramanujan in [8], [16, pp. 57, 54] and are perhaps his most famous integral evaluations. It should be mentioned, however, that Barnes [2, pp. 154, 155] established an extension of (i) at roughly the same time that Ramanujan discovered Entry 22. R. Roy [1] has employed Mellin transforms to give a proof of (ii). For ramifications of these results, see papers of Wilson [1] and Askey and Wilson [1].

Entry 22(i) can be generalized in the following way. If we apply Parseval's theorem (Titchmarsh [2, p. 5]), the corollary in Section 19, and Legendre's duplication formula, we find that

$$2 \int_0^\infty |\Gamma(a + ix)\Gamma(b + ix)|^2 \cos(xy) \, dx$$

$$= \frac{\pi}{4^{a+b-1}} \Gamma(2a)\Gamma(2b) \int_{-\infty}^\infty \operatorname{sech}^{2a} \frac{u}{2} \operatorname{sech}^{2b} \frac{u + y}{2} \, du,$$

where $y \geq 0$. Glasser [1] has shown how to evaluate integrals like that on the right side above.

Entry 23. *Let $a > 0$, $m < 1$, and $m + n > 0$. Then*

$$\int_0^\infty \frac{x^{-m}\Gamma(x + a)}{\Gamma(x + a + n + 1)} \, dx = \frac{\pi \csc(\pi m)}{\Gamma(n + 1)} \sum_{k=0}^\infty \frac{(-n)_k}{k!(a + k)^m}.$$

We have not been able to find this result in the literature. Ramanujan has also obtained this integral formula in his quarterly reports, and a complete proof may be found in Part I, pp. 303, 304.

Entry 24(i) offers the triviality

$$\sum_{k=0}^n A_k = \sum_{k=0}^\infty A_{n-k} - \sum_{k=1}^\infty A_{-k},$$

which is followed by a corollary in which A_k above is replaced by $A_k/\Gamma(k + 1)$. The intent of Entry 24(ii),

$$\lim_{N \to \infty} \sum_{k=-N}^N \varphi(x + k) = \lim_{N \to \infty} \sum_{k=-N}^N \varphi(y + k),$$

is indeed unclear. What can be said?

Ramanujan, in a corollary, claims that

$$\frac{x^h}{h!} + \sum_{k=1}^\infty \left(\frac{x^{h+kn}}{\Gamma(h + kn + 1)} + \frac{x^{h-kn}}{\Gamma(h - kn + 1)} \right)$$

$$= 1 + \sum_{k=1}^\infty \left(\frac{x^{kn}}{\Gamma(kn + 1)} + \frac{x^{-kn}}{\Gamma(-kn + 1)} \right) = \frac{e^x}{n},$$

where $n \leq 1$ and x and h are arbitrary. Although these equalities are true for $h = 0$ and $n = 1$, they certainly are false in general, because the far left side is a nonconstant function of h and the expressions to the right are not. Moreover, the series diverge if n is not an integer.

In Entry 24(iii), Ramanujan offers the equality

$$\int_{-\infty}^\infty \frac{\varphi(x)}{\Gamma(x + 1)} \, dx = \sum_{n=0}^\infty \frac{\varphi(n)}{\Gamma(n + 1)}. \tag{24.1}$$

Instances when an integral is equal to the corresponding sum are rare. For examples of this phenomenon, see papers by Boas and Pollard [1], Krishnan [1], and Forrester [1]. See also Entries 5(i), (ii) and Entries 16(i), (ii) in Chapter 14.

In Corollary (i), Ramanujan claims that

$$\int_{-\infty}^{\infty} \frac{a^x}{\Gamma(x + 1)} \, dx = e^a,$$

which follows formally from (24.1) by setting $\varphi(x) = a^x$. However, if $a = 0$, Ramanujan's claim is clearly false, and if a is real and nonzero, the integral diverges.

In Corollary (ii), Ramanujan asserts that

$$\int_{-\infty}^{\infty} \frac{a^x \Gamma(n + 1) \, dx}{\Gamma(x + 1)\Gamma(n - x + 1)} = (1 + a)^n, \qquad (24.2)$$

which follows formally from (24.1) by letting $\varphi(x) = a^x \Gamma(n + 1)/\Gamma(n - x + 1)$. Again, (24.2) is false for $a = 0$, while the integral diverges for real $a \neq 0, \pm 1$. If $a = e^{i\alpha}$, $|\alpha| < \pi$, with α real, then (24.2) is valid and, in fact, was proved by Ramanujan in his paper [14], [16, pp. 216–229, Eq. (1.2)].

Entry 25(i) is the special case $b = 0$ of Entry 25(ii).

In Entry 25(ii), Ramanujan writes

$$A_n := \int_0^{\infty} \left(\frac{a^{b+x}}{\Gamma(b + x + 1)} + \frac{a^{b-x}}{\Gamma(b - x + 1)} \right) \cos(nx) \, dx$$

$$= e^{a \cos n} \cos(a \sin n - nb)$$

and

$$B_n := \int_0^{\infty} \left(\frac{a^{b+x}}{\Gamma(b + x + 1)} - \frac{a^{b-x}}{(b - x + 1)} \right) \sin(nx) \, dx$$

$$= e^{a \cos n} \sin(a \sin n - nb),$$

where presumably a is real.

It is easy to see, by Stirling's formula, that both of these integrals diverge if $a \neq 0$. But let us discern how Ramanujan reasoned. By simple changes of variable and (24.1),

$$A_n + iB_n = \int_{-\infty}^{\infty} \frac{a^{b+x}}{\Gamma(b + x + 1)} e^{inx} \, dx$$

$$= \int_{-\infty}^{\infty} \frac{a^u}{\Gamma(u + 1)} e^{in(u-b)} \, du$$

$$= \sum_{k=0}^{\infty} \frac{a^k e^{in(k-b)}}{k!}$$

$$= \exp(ae^{in} - inb).$$

Equating real and imaginary parts, we complete Ramanujan's formal derivation.

The content of Entries 23–25 perhaps served as the seed for Ramanujan's beautiful paper [14] on integrals involving the gamma function.

Example (i). *The maximum value of* $a^x/\Gamma(x + 1)$ *is equal to*

$$\frac{a^{a-1/2}}{\Gamma(a + \frac{1}{2})}\left\{1 + \frac{1}{1152a^3} + O\left(\frac{1}{a^4}\right)\right\}$$

when a is large.

PROOF. Differentiating $a^x/\Gamma(x + 1)$ with respect to x, we find that $a^x/\Gamma(x + 1)$ achieves its maximum when

$$\psi(x + 1) - \text{Log } a = 0, \tag{25.1}$$

where, as usual, $\psi(x) = \Gamma'(x)/\Gamma(x)$. Now in Entry 15 of Chapter 8 (p. 95), (Part I, p. 194), Ramanujan derived an asymptotic expansion for the root x of (25.1) in descending powers of a as a tends to ∞; namely,

$$x + \frac{1}{2} = a - \frac{1}{24a} + \frac{3}{640a^3} + \cdots.$$

Letting

$$\varepsilon = \varepsilon(a) = -\frac{1}{24a} + \frac{3}{640a^3},$$

we find that, for a large,

$$\frac{a^x}{\Gamma(x + 1)} = \frac{a^{a-1/2+\varepsilon+O(a^{-4})}}{\Gamma(a + \frac{1}{2} + \varepsilon + O(a^{-4}))} = \frac{a^{a-1/2+\varepsilon}}{\Gamma(a + \frac{1}{2} + \varepsilon)}\left\{1 + O\left(\frac{1}{a^4}\right)\right\}. \tag{25.2}$$

From Lemma 2, Section 24 of Chapter 11,

$$\frac{\Gamma(a + \frac{1}{2} + \varepsilon)}{\Gamma(a + \frac{1}{2})} = a^\varepsilon\left(1 + \frac{\varepsilon^2}{2a} + \frac{\varepsilon}{24a^2} + O\left(\frac{1}{a^4}\right)\right)$$

$$= a^\varepsilon\left(1 - \frac{1}{1152a^3} + O\left(\frac{1}{a^4}\right)\right),$$

as a tends to ∞. Using the expansion above in (25.2), we deduce the desired approximation. □

Our version of Example (i) is different from that of Ramanujan, who writes that the maximum value of $a^x/\Gamma(x + 1)$ is

$$\frac{a^{a-1/2}}{\Gamma(a + \frac{1}{2})} \exp\left(\frac{1}{1152a^3 + 323.2a}\right)$$

"very nearly." This agrees with our statement, except for the appearance of the expression $323.2a$, which is apparently incorrect.

In Example (ii), Ramanujan states a version of the Euler–Maclaurin summation formula (10.5) and remarks that it "is very useful in evaluating definite integrals."

Entry 26(i). *Let $n > 0$ and suppose that m is a nonnegative integer. Then*

$$\int_0^\infty \frac{\cos(2nx)\,dx}{(1+x^2)^{m+1}} = \frac{\pi n^m e^{-2n}}{2m!} \sum_{k=0}^m \frac{(m+k)!}{(4n)^k(m-k)!k!}.$$

This result is classical (Gradshteyn and Ryzhik [1, p. 413]) and can be established by contour integration.

Entry 26(ii). *Let $p > 0$ and suppose that m and n are nonnegative integers with $m \le n$. Then*

$$I(m, n) := \int_0^\infty \frac{x^{2m}\cos(px)}{(1+x^2)^{n+1}}\,dx = \frac{(-1)^m \pi e^{-p}}{2^{n+1}n!} \sum_{r=0}^n A_r p^{n-r}, \qquad (26.1)$$

where, for $r \ge 0$,

$$A_r = \frac{(n+r)!}{2^r r!(n-r)!} \sum_{k=0}^{\min(r,\,m)} \frac{4^k(-r)_k(-m)_k(-n)_k}{(-n-r)_{2k}k!}.$$

PROOF. First, for $n = 0$, the proposed formula is readily established, for example, by the calculus of residues. Thus, in the sequel, we assume that $n > 0$.

We shall induct on m. For $m = 0$, formula (26.1) is seen to be valid by Entry 26(i). Now it is easy to see that

$$I(m+1, n) = I(m, n-1) - I(m, n), \qquad (26.2)$$

where $m \ge 0, n \ge 1$. Inducting on m, we shall employ (26.2) to show that (26.1) is true with m replaced by $m + 1$. To that end,

$$I(m+1, n) = \frac{(-1)^m \pi e^{-p}}{2^n(n-1)!} \sum_{r=0}^{n-1} p^{n-1-r} \frac{(n-1+r)!}{2^r r!(n-1-r)!}$$

$$\times \sum_{k=0}^{\min(r,\,m)} \frac{4^k(-r)_k(-m)_k(-n+1)_k}{(-n+1-r)_{2k}k!} - \frac{(-1)^m \pi e^{-p}}{2^{n+1}n!}$$

$$\times \sum_{r=0}^n p^{n-r} \frac{(n+r)!}{2^r r!(n-r)!} \sum_{k=0}^{\min(r,\,m)} \frac{4^k(-r)_k(-m)_k(-n)_k}{(-n-r)_{2k}k!}$$

$$= \frac{(-1)^m \pi e^{-p}}{2^{n+1}n!} \sum_{r=0}^n \frac{p^{n-r}}{(n-r)!} \left\{ \frac{2n(n-2+r)!}{2^{r-1}(r-1)!} \right.$$

$$\times \sum_{k=0}^{\min(r-1,\,m)} \frac{4^k(-r+1)_k(-m)_k(-n+1)_k}{(-n-r+2)_{2k}k!} - \frac{(n+r)!}{2^r r!}$$

$$\left. \times \sum_{k=0}^{\min(r,\,m)} \frac{4^k(-r)_k(-m)_k(-n)_k}{(-n-r)_{2k}k!} \right\}$$

$$= \frac{(-1)^{m+1}\pi e^{-p}}{2^{n+1}n!} \sum_{r=0}^{n} p^{n-r} \frac{(n+r)!}{(n-r)!2^r r!}$$

$$\times \left\{ \sum_{k=0}^{\min(r,m)} \frac{4^k(-r)_k(-m)_k(-n)_k}{(-n-r)_{2k}k!} - \frac{4nr}{(n+r)(n+r-1)} \right.$$

$$\times \left. \sum_{k=1}^{\min(r,m+1)} \frac{4^{k-1}(-r+1)_{k-1}(-m)_{k-1}(-n+1)_{k-1}}{(-n-r+2)_{2k-2}(k-1)!} \right\}$$

$$= \frac{(-1)^{m+1}\pi e^{-p}}{2^{n+1}n!} \sum_{r=0}^{n} p^{n-r} \frac{(n+r)!}{(n-r)!2^r r!}$$

$$\times \left\{ \sum_{k=0}^{\min(r,m+1)} \frac{4^k(-r)_k(-m)_k(-n)_k}{(-n-r)_{2k}k!} \right.$$

$$\left. - \sum_{k=1}^{\min(r,m+1)} \frac{4^k(-r)_k(-m)_k(-n)_k}{(-n-r)_{2k}(k-1)!} \right\}.$$

Since

$$\frac{(-m)_k}{k!} - \frac{(-m)_{k-1}}{(k-1)!} = \frac{(-m-1)_k}{k!},$$

the desired formula, (26.1) with m replaced by $m+1$, follows. $\square$

Entry 27. *If n is an even positive integer, then*

$$\prod_{k=1}^{\infty} \left\{ 1 + \left(\frac{x}{k}\right)^n \right\}^2$$

$$= \prod_{k=1}^{n/2} \frac{\cosh(2\pi x \sin((2k-1)\pi/n)) - \cos(2\pi x \cos((2k-1)\pi/n))}{2\pi^2 x^2}. \quad (27.1)$$

PROOF. We have

$$\prod_{j=1}^{n/2} \frac{\cosh(2\pi x \sin((2j-1)\pi/n)) - \cos(2\pi x \cos((2j-1)\pi/n))}{2\pi^2 x^2}$$

$$= \prod_{j=1}^{n/2} \left(\frac{\sinh(i\pi x e^{-\pi i(2j-1)/n})}{i\pi x e^{-\pi i(2j-1)/n}} \frac{\sinh(-i\pi x e^{\pi i(2j-1)/n})}{-i\pi x e^{\pi i(2j-1)/n}} \right)$$

$$= \prod_{k=1}^{\infty} \prod_{j=1}^{n/2} \left(1 - \frac{x^2 e^{-2\pi i(2j-1)/n}}{k^2} \right) \left(1 - \frac{x^2 e^{2\pi i(2j-1)/n}}{k^2} \right). \quad (27.2)$$

Comparing (27.1) with (27.2) and replacing x/k by x, we find that it remains to show that

$$(1 + x^n)^2 = \prod_{j=1}^{n/2} (1 - x^2 e^{-2\pi i(2j-1)/n})(1 - x^2 e^{2\pi i(2j-1)/n}).$$

It is easily checked that the $2n$ roots on the left side are precisely the same as the $2n$ roots on the right side, and so the proof is complete. $\square$

Corollary (i). *If n is arbitrary, then*

$$\prod_{k=1}^{\infty}\left\{1+\left(\frac{2n}{n+k}\right)^{3}\right\}=\frac{\Gamma^{3}(n+1)\sinh(\pi n\sqrt{3})}{\Gamma(3n+1)\pi n\sqrt{3}}.$$

Corollary (ii). *If n is arbitrary, then*

$$\prod_{k=1}^{\infty}\left\{1+\left(\frac{2n+1}{n+k}\right)^{3}\right\}=\frac{\Gamma^{3}(n+1)\cosh\{\pi(n+\tfrac{1}{2})\sqrt{3}\}}{\Gamma(3n+2)\pi}.$$

The latter two formulas were proven by Ramanujan in [9], [16, p. 51].

Entry 28. *If m and n are positive integers and x is arbitrary, then*

$$mn\sum_{k=0}^{\infty}\frac{x^{nk}}{(nk)!}=\sum_{k=0}^{mn-1}e^{x\cos(2\pi k/n)}\cos(x\sin(2\pi k/n)).$$

PROOF. Letting $k=jn+r, 0\leq j\leq m-1, 0\leq r\leq n-1$, below, we find that

$$\sum_{k=0}^{mn-1}e^{x\cos(2\pi k/n)}\cos(x\sin(2\pi k/n))=\sum_{k=0}^{mn-1}\exp(xe^{2\pi ik/n})$$

$$=\sum_{r=0}^{n-1}\sum_{j=0}^{m-1}\exp(xe^{2\pi ir/n})$$

$$=m\sum_{r=0}^{n-1}\sum_{j=0}^{\infty}\frac{(xe^{2\pi ir/n})^{j}}{j!}$$

$$=m\sum_{j=0}^{\infty}\frac{x^{j}}{j!}\sum_{r=0}^{n-1}e^{2\pi irj/n}.$$

The last inner sum is equal to 0 unless $n\,|\,j$ in which case it is equal to n. The proof is now complete. $\square$

Entries 29(i), (ii). *Suppose that $p\geq 0$, l is a nonnegative integer, and n is a positive integer with $n>l$. Then*

$$\int_{0}^{\infty}\frac{(-x^{2})^{l}}{1+x^{2n}}\cos(px)\,dx$$

$$=\begin{cases}\dfrac{\pi}{2n}e^{-p}+\dfrac{\pi}{n}\displaystyle\sum_{k=1}^{(n-1)/2}e^{-p\cos(\pi k/n)}\cos\left(\dfrac{(2l+1)\pi k}{n}-p\sin\dfrac{\pi k}{n}\right), & \text{if n is odd,}\\[4mm]\dfrac{\pi}{n}\displaystyle\sum_{k=1}^{n/2}e^{-p\cos((2k-1)\pi/2n)}\cos\left(\dfrac{(2l+1)(2k-1)\pi}{2n}-p\sin\dfrac{(2k-1)\pi}{2n}\right), \\[4mm]\hfill\text{if n is even.}\end{cases}$$

The integrals above may be evaluated by the calculus of residues, although the initial form of the answer obtained might be different from that stated by

Ramanujan. See also Gradshteyn and Ryzhik's tables [1, p. 414, formula 3.738, No. 2], where again the evaluation is given in a different formulation and a bracket { is misplaced. Since a similar calculation is performed in the proofs of Entries 33(i), (ii), we suppress the details.

Entry 30(i). *If n is a nonnegative integer, then*

$$\int_0^\infty \frac{\sin^{2n+1} x}{x}\, dx = \int_0^\infty \frac{\sin^{2n+2} x}{x^2}\, dx = \frac{\sqrt{\pi}\,\Gamma(n + \tfrac{1}{2})}{2n!}.$$

A proof of Entry 30(i) may be found in Fichtenholz's text [1, p. 656]. These integrals actually are special cases of Entries 16(i), (ii) in Chapter 14. For further references to Entries 30(i), (ii), see a problem of Wang [1].

Entry 30(ii). *If p > 2 and n − p + 1 > 0, then*

$$(p - 1)(p - 2)\varphi(n, p) = n(n - 1)\varphi(n - 2, p - 2) - n^2 \varphi(n, p - 2),$$

where

$$\varphi(n, p) = \int_0^\infty \frac{\sin^n x}{x^p}\, dx. \tag{30.1}$$

PROOF. Integrating by parts twice, we find that

$$\varphi(n, p) = \frac{n}{(p - 1)(p - 2)} \int_0^\infty \frac{(n - 1)\sin^{n-2} x \cos^2 x - \sin^n x}{x^{p-2}}\, dx,$$

which is easily seen to be equivalent to the proposed recursion formula. □

Corollary (i). *If n is a nonnegative integer, then*

$$\varphi(2n + 3, 3) = \frac{\sqrt{\pi}(n + \tfrac{3}{2})\Gamma(n + \tfrac{1}{2})}{4(n + 1)!},$$

where φ is defined by (30.1).

PROOF. By Entries 30(ii) and (i), respectively,

$$\varphi(2n + 3, 3) = \tfrac{1}{2}(2n + 3)(2n + 2)\varphi(2n + 1, 1) - \tfrac{1}{2}(2n + 3)^2 \varphi(2n + 3, 1)$$

$$= (2n + 3)(n + 1)\frac{\sqrt{\pi}\,\Gamma(n + \tfrac{1}{2})}{2n!} - (2n + 3)^2 \frac{\sqrt{\pi}\,\Gamma(n + \tfrac{3}{2})}{4(n + 1)!},$$

which, upon simplification, yields the desired result. □

Corollary (ii). *If n is a nonnegative integer and φ is defined by (30.1), then*

$$\varphi(2n + 4, 4) = \frac{\sqrt{\pi}(n + 2)\Gamma(n + \tfrac{1}{2})}{6(n + 1)!}.$$

PROOF. The proof is like that of Corollary (i); simply apply Entries 30(ii) and (i) and then simplify.   □

Example (i). *If* $0 < p < n + 1$ *and* φ *is given by* (30.1), *then*

$$\varphi(n, p) = \frac{1}{\Gamma(p)} \int_0^\infty \sin^n x \int_0^\infty e^{-tx} t^{p-1} \, dt \, dx.$$

PROOF. From the definition of the gamma function,

$$\frac{1}{x^p} = \frac{1}{\Gamma(p)} \int_0^\infty e^{-tx} t^{p-1} \, dt, \qquad x, p > 0.$$

The desired result now follows from (30.1).   □

Examples (ii), (iii). *If* $a > 0$ *and* n *is a nonnegative integer, then*

$$\int_0^\infty e^{-ax} \sin^{2n+1} x \, dx = \frac{(2n + 1)!}{(a^2 + 1^2)(a^2 + 3^2) \cdots (a^2 + (2n + 1)^2)}$$

and

$$\int_0^\infty e^{-ax} \sin^{2n} x \, dx = \frac{(2n)!}{a(a^2 + 2^2)(a^2 + 4^2) \cdots (a^2 + (2n)^2)}.$$

These formulas are classical (Gradshteyn and Ryzhik [1, p. 478]) and follow readily by induction.

In the sequel, a prime (′) on a summation sign, $\sum'_{a \le k \le b} f(k)$, indicates that if a and/or b is an integer, then only $\frac{1}{2} f(a)$ and/or $\frac{1}{2} f(b)$, respectively, is counted.

Entry 31(i). *Let* $h, \alpha, \beta > 0$ *with* $\alpha\beta = 2\pi$. *Let* φ *be a continuous function of bounded variation on* $[0, h]$. *Define*

$$\psi(r) = \int_0^h \varphi(x) \cos(rx) \, dx.$$

Then, if n *is real,*

$$\alpha \sum_{0 \le k \le h/\alpha}' \varphi(\alpha k) \cos(\alpha n k) = \psi(n) + \sum_{k=1}^\infty \{\psi(\beta k + n) + \psi(\beta k - n)\}. \quad (31.1)$$

PROOF. We shall employ the Poisson summation formula in the form

$$\sum_{a \le k \le b}' f(k) = \int_a^b f(x) \, dx + 2 \sum_{k=1}^\infty \int_a^b f(x) \cos(2\pi k x) \, dx, \quad (31.2)$$

where f is a continuous function of bounded variation on $[a, b]$. In (31.2), let $a = 0$, $b = h/\alpha$, and $f(x) = \varphi(\alpha x) \cos(\alpha n x)$. Thus, putting $u = \alpha x$, we find that

$$\sideset{}{'}\sum_{0 \le k \le h/\alpha} \varphi(\alpha k) \cos(\alpha n k)$$

$$= \frac{1}{\alpha}\psi(n) + \frac{2}{\alpha}\sum_{k=1}^{\infty}\int_0^h \varphi(u)\cos(nu)\cos(\beta k u)\,du$$

$$= \frac{1}{\alpha}\psi(n) + \frac{1}{\alpha}\sum_{k=1}^{\infty}\int_0^h \varphi(u)\{\cos(\beta k + n)u + \cos(\beta k - n)u\}\,du.$$

Upon using the definition of ψ, we complete the proof of (31.1). ☐

Entry 31 (ii). *Let φ and ψ be defined as in Entry 31(i). Let $h > 0$, and assume that n is an integer. Then*

$$\int_0^h \frac{\sin(nx)}{\sin x}\varphi(x)\,dx = \pi \sideset{}{'}\sum_{0 \le k \le h/\pi}(-1)^k \varphi(k\pi)\cos(kn\pi)$$

$$- 2\sum_{k=0}^{\infty}\psi(n + 2k + 1). \qquad (31.3)$$

PROOF. We shall induct on n. First, in (31.1), put $\alpha = \pi$, $\beta = 2$, and $n = 1$ to obtain

$$\pi \sideset{}{'}\sum_{0 \le k \le h/\pi}(-1)^k \varphi(k\pi) = 2\sum_{k=0}^{\infty}\psi(2k + 1).$$

But this equality is precisely (31.3) in the case $n = 0$.
   Second, let $\alpha = \pi$, $\beta = 2$, and $n = 0$ in (31.1) to find that

$$\pi \sum_{0 \le k \le h/\pi}\varphi(k\pi) = \psi(0) + 2\sum_{k=1}^{\infty}\psi(2k).$$

This equality is easily seen to be equivalent to (31.3) in the case $n = 1$.
   Now assume that (31.3) holds up to a fixed integer n. Then, by induction,

$$\int_0^h \frac{\sin(n \pm 2)x}{\sin x}\varphi(x)\,dx = \int_0^h \frac{\sin(nx)}{\sin x}\varphi(x)\,dx \pm 2\int_0^h \cos(n \pm 1)x\,\varphi(x)\,dx$$

$$= \pi \sideset{}{'}\sum_{0 \le k \le h/\pi}(-1)^k \varphi(k\pi)\cos(kn\pi)$$

$$- 2\sum_{k=0}^{\infty}\psi(n + 2k + 1) \pm 2\psi(n \pm 1)$$

$$= \pi \sideset{}{'}\sum_{0 \le k \le h/\pi}(-1)^k \varphi(k\pi)\cos(k(n \pm 2)\pi)$$

$$- 2\sum_{k=0}^{\infty}\psi(n \pm 2 + 2k + 1),$$

which is (31.3) with n replaced by $n \pm 2$. ☐

Corollary (i). *Let* α, $\beta > 0$ *with* $\alpha\beta = 2\pi$. *Let* φ *be a continuous function of bounded variation on* $(0, \infty)$. *Suppose that* φ *is integrable over* $(0, \infty)$. *Put*

$$\psi(r) = \int_0^\infty \varphi(x) \cos(rx)\, dx.$$

Then

$$\alpha \sum_{k=0}^{\infty}{}' \varphi(\alpha k) = \psi(0) + 2 \sum_{k=1}^{\infty} \psi(\beta k).$$

PROOF. In (31.1), let $n = 0$ and let h tend to ∞. (To justify this, see Titchmarsh's book [2, pp. 61, 62].) □

Corollary (ii). *Under the assumptions of Entry* 31(ii),

$$\lim_{n\to\infty} \left(\int_0^h \frac{\sin(nx)}{\sin x} \varphi(x)\, dx - \pi \sum_{0 \le k \le h/\pi}{}' (-1)^k \varphi(k\pi) \cos(kn\pi) \right) = 0.$$

PROOF. The desired result is an immediate consequence of (31.3), since clearly

$$\sum_{k=1}^{\infty} \psi(k)$$

converges. □

Entry 32(i). *Let* h, α, $\beta > 0$ *with* $\alpha\beta = 2\pi$. *Let* φ *be a continuous function of bounded variation on* $[0, h]$. *Define*

$$\psi(r) = \int_0^h \varphi(x) \sin(rx)\, dx.$$

Then, if n *is real,*

$$\alpha \sum_{0 \le k \le h/\alpha}{}' \varphi(\alpha k) \sin(\alpha n k) = \psi(n) + \sum_{k=1}^{\infty} \{\psi(\beta k + n) - \psi(\beta k - n)\}.$$

PROOF. In the Poisson summation formula (31.2), put $a = 0$, $b = h/\alpha$, and $f(x) = \varphi(\alpha x) \sin(\alpha n x)$. Then

$$\sum_{0 \le k \le h/\alpha}{}' \varphi(\alpha k) \sin(\alpha n k)$$

$$= \frac{1}{\alpha}\psi(n) + \frac{1}{\alpha} \sum_{k=1}^{\infty} \int_0^h \varphi(u) \{\sin(n + \beta k)u + \sin(n - \beta k)u\}\, du,$$

from which the proposed formula follows. □

Entry 32(ii) is another version of the Euler–Maclaurin summation formula (10.5).

Corollary. *Let* α, $\beta > 0$ *with* $\alpha\beta = \pi/2$. *Let* $\varphi(x)$ *be continuous on* $(0, \infty)$, *integrable over* $(0, \delta)$, *of bounded variation on* (δ, ∞), *and tend to* 0 *as* x *tends to* ∞, *where* $0 < \delta < \pi/2$. *Define*

$$\psi(r) = \int_0^\infty \varphi(x) \sin(rx)\, dx.$$

Then

$$\alpha \sum_{k=0}^\infty (-1)^k \varphi((2k+1)\alpha) = \sum_{k=0}^\infty (-1)^k \psi((2k+1)\beta).$$

This corollary gives the Poisson summation formula for Fourier sine transforms (Titchmarsh [2, p. 66]).

Ramanujan concludes Section 32 by remarking that integrals such as

$$\int_0^h \frac{\cos(nx)}{\cos x} \varphi(x)\, dx, \quad \int_0^h \frac{\sin(nx)}{\cos x} \varphi(x)\, dx, \quad \text{and} \quad \int_0^h \frac{\cos(nx)}{\sin x} \varphi(x)\, dx$$

may be determined. Ramanujan is evidently indicating that analogues of Entry 31(ii) exist.

Entries 33(i), (ii). *Let* n *and* l *denote nonnegative integers with* $n > l$. *Let* $p > 0$. *Then*

$$I := \int_0^\infty \left\{ \frac{(-x^2)^l}{1 - x^{2n}} + \frac{(-1)^l}{n(x^2 - 1)} \right\} \cos(px)\, dx$$

$$= \begin{cases} \dfrac{\pi}{2n} e^{-p} + \dfrac{\pi}{n} \displaystyle\sum_{k=1}^{n/2-1} e^{-p\cos(\pi k/n)} \cos\left(\dfrac{\pi(2l+1)k}{n} - p\sin\dfrac{\pi k}{n} \right), & \text{if } n \text{ is even,} \\[4mm] \dfrac{\pi}{n} \displaystyle\sum_{k=1}^{(n-1)/2} e^{-p\cos((2k-1)\pi/2n)} \cos\left(\dfrac{\pi(2l+1)(2k-1)}{2n} - p\sin\dfrac{(2k-1)\pi}{2n} \right), & \\ & \text{if } n \text{ is odd.} \end{cases}$$

PROOF. First observe that

$$(-1)^l I = \int_0^\infty \frac{nx^{2l} - x^{2n-2} - \cdots - x^2 - 1}{n(1 - x^{2n})} \cos(px)\, dx.$$

Let $R(z_0)$ denote the residue of

$$f(z) := \frac{nz^{2l} - z^{2n-2} - \cdots - z^2 - 1}{n(1 - z^{2n})} e^{ipz}$$

at a pole z_0. In the upper half-plane, $f(z)$ has simple poles at $z = \exp(\pi i k/n)$, $1 \le k \le n - 1$. Hence, by a familiar argument from the calculus of residues,

$$(-1)^l I = \pi i \sum_{k=1}^{n-1} R(e^{\pi i k/n})$$

$$= -\frac{\pi i}{2n} \sum_{k=1}^{n-1} e^{\pi i(2l+1)k/n} \exp(ipe^{\pi i k/n})$$

$$= \frac{\pi}{2n} \sum_{k=1}^{n-1} e^{-p \sin(\pi k/n)} \sin\left(\frac{\pi(2l+1)k}{n} + p \cos\frac{\pi k}{n}\right).$$

Observe that the terms with indices k and $n - k$ are equal.

First, suppose that n is even. Singling out the term with $k = n/2$, we then find that

$$(-1)^l I = \frac{\pi}{2n}(-1)^l e^{-p} + \frac{\pi}{n} \sum_{k=1}^{n/2-1} e^{-p \sin(\pi k/n)} \sin\left(\frac{\pi(2l+1)k}{n} + p \cos\frac{\pi k}{n}\right)$$

$$= \frac{\pi}{2n}(-1)^l e^{-p} + (-1)^l \frac{\pi}{n} \sum_{k=1}^{n/2-1} e^{-p \cos(\pi k/n)} \cos\left(\frac{\pi(2l+1)k}{n} - p \sin\frac{\pi k}{n}\right),$$

where we have replaced k by $n/2 - k$ in the former sum.

Second, suppose that n is odd. Then

$$(-1)^l I = \frac{\pi}{n} \sum_{k=1}^{(n-1)/2} e^{-p \sin(\pi k/n)} \sin\left(\frac{\pi(2l+1)k}{n} + p \cos\frac{\pi k}{n}\right)$$

$$= (-1)^l \frac{\pi}{n} \sum_{k=1}^{(n-1)/2} e^{-p \cos((2k-1)\pi/2n)}$$

$$\times \cos\left(\frac{\pi(2l+1)(2k-1)}{2n} - p \sin\frac{(2k-1)\pi}{2n}\right),$$

where we have replaced k by $(n+1)/2 - k$ in the former sum.   □

Entry 34. *If x is arbitrary, then*

(i) $$\frac{\pi \cos(\theta x)}{x \sin(\pi x)} = \frac{1}{x^2} + 2 \sum_{k=1}^{\infty} \frac{(-1)^{k+1} \cos(k\theta)}{k^2 - x^2}, \qquad |\theta| \le \pi,$$

and

(ii) $$\frac{\pi \sin(\theta x)}{4x \cos(\frac{1}{2}\pi x)} = \sum_{k=0}^{\infty} \frac{(-1)^k \sin((2k+1)\theta)}{(2k+1)^2 - x^2}, \qquad |\theta| \le \pi/2.$$

Corollary. *If x is arbitrary, then*

(i) $$\frac{\pi \cosh(\theta x)}{x \sinh(\pi x)} = \frac{1}{x^2} + 2 \sum_{k=1}^{\infty} \frac{(-1)^k \cos(k\theta)}{k^2 + x^2}, \qquad |\theta| \le \pi,$$

and

(ii) $$\frac{\pi \sinh(\theta x)}{4x \cosh(\frac{1}{2}\pi x)} = \sum_{k=0}^{\infty} \frac{(-1)^k \sin((2k+1)\theta)}{(2k+1)^2 + x^2}, \qquad |\theta| \le \pi/2.$$

PROOFS OF ENTRY 34 AND COROLLARY. First, Corollary (i) is proved in Bromwich's text [1, p. 368, Eq. (4.1)].

Entry 34(i) is easily obtained from Corollary (i) by replacing x by ix.

We next prove Entry 34(ii). Recall that the set of functions $\sin\{(2k + 1)\theta\}$, $0 \le k < \infty$, is a complete orthogonal set on $-\pi/2 \le \theta \le \pi/2$. Calculating the Fourier series of $\sin(\theta x)$, when x is real, with respect to this orthogonal set, we readily deduce Entry 34(ii) for real x. By analytic continuation, Entry 34(ii) holds for complex x as well.

Lastly, Corollary (ii) follows from Entry 34(ii) by replacing x by ix.   □

Entry 35. *Let n denote a nonnegative integer, and let $\alpha, \beta > 0$ with $\alpha\beta = \pi$. Then*

$$\sqrt{\alpha}\left\{1 + 2 \sum_{k=1}^{\infty} \frac{1}{(1 + \alpha^2 k^2)^{n+1}}\right\} = \sqrt{\beta}\,\frac{\Gamma(n + \tfrac{1}{2})}{\Gamma(n + 1)}\left\{1 + 2 \sum_{k=1}^{\infty} e^{-2\beta k}\varphi(4\beta k)\right\},$$

where

$$\varphi(t) = \frac{n!}{(2n)!} \sum_{k=0}^{n} \frac{(n + k)!\,t^{n-k}}{(n - k)!\,k!}.$$

PROOF. In the Poisson summation formula (31.2), set $a = 0$, $b = \infty$, and $f(x) = 2(1 + \alpha^2 x^2)^{-n-1}$. Thus,

$$\sum_{k=0}^{\infty}{}' \frac{2}{(1 + \alpha^2 k^2)^{n+1}} = 2 \int_0^{\infty} \frac{dx}{(1 + \alpha^2 x^2)^{n+1}} + 4 \sum_{k=1}^{\infty} \int_0^{\infty} \frac{\cos(2\pi kx)}{(1 + \alpha^2 x^2)^{n+1}}\,dx. \quad (35.1)$$

By Entry 26(i),

$$4 \int_0^{\infty} \frac{\cos(2\pi kx)}{(1 + \alpha^2 x^2)^{n+1}}\,dx = \frac{4\pi(\pi k/\alpha)^n e^{-2\pi k/\alpha}}{2\alpha n!} \sum_{j=0}^{n} \frac{(n + j)!}{(4\pi k/\alpha)^j (n - j)!\,j!}$$

$$= \frac{\beta e^{-2\beta k}}{2^{2n-1} n!} \sum_{j=0}^{n} \frac{(n + j)!(4\beta k)^{n-j}}{(n - j)!\,j!}$$

$$= \sqrt{\frac{\beta}{\alpha}}\,\frac{\Gamma(n + \tfrac{1}{2})}{\Gamma(n + 1)}\,\frac{2e^{-2\beta k} n!}{(2n)!} \sum_{j=0}^{n} \frac{(n + j)!(4\beta k)^{n-j}}{(n - j)!\,j!}$$

$$= \sqrt{\frac{\beta}{\alpha}}\,\frac{\Gamma(n + \tfrac{1}{2})}{\Gamma(n + 1)}\,2e^{-2\beta k}\varphi(4\beta k). \quad (35.2)$$

Substituting (35.2) into (35.1), we complete the proof.   □

Entry 36. *Let N be any positive integer. As $m^2 + n^2$ tends to ∞,*

$$m \sum_{k=0}^{\infty}{}' \frac{1}{m^2 + (n + k)^2} = \tan^{-1}(m/n) + \sum_{k=1}^{N} \frac{B_{2k}}{2k}\,\frac{\sin(2k \tan^{-1}(m/n))}{(m^2 + n^2)^k}$$

$$+ O((m^2 + n^2)^{-N-1}),$$

where B_j, $0 \le j < \infty$, denotes the jth Bernoulli number.

PROOF. Letting $a = 0$, $b = \infty$, and $f(x) = \{m^2 + (n + x)^2\}^{-1}$ in the Euler–Maclaurin summation formula (10.5), we find that

$$\sum_{k=0}^{\infty} f(k) = \int_0^{\infty} f(x)\, dx + \tfrac{1}{2}f(0) - \sum_{k=1}^{N} \frac{B_{2k}}{(2k)!} f^{(2k-1)}(0) + R_{N+1}. \quad (36.1)$$

First,

$$\int_0^{\infty} f(x)\, dx = \int_n^{\infty} \frac{du}{m^2 + u^2} = \frac{1}{m}\left(\frac{\pi}{2} - \tan^{-1}\frac{n}{m}\right) = \frac{1}{m}\tan^{-1}\frac{m}{n}. \quad (36.2)$$

Next, a straightforward calculation shows that

$$f^{(2k-1)}(x) = -\frac{(2k-1)!}{2m}\left\{\frac{i^{2k-1}}{(m + (n + x)i)^{2k}} + \frac{(-i)^{2k-1}}{(m - (n + x)i)^{2k}}\right\}, \quad k \geq 1.$$

Thus, for $k \geq 1$,

$$\begin{aligned}
f^{(2k-1)}(0) &= -\frac{(-1)^k(2k-1)!}{2mi}\left\{\frac{(m - ni)^{2k} - (m + ni)^{2k}}{(m^2 + n^2)^{2k}}\right\} \\
&= -\frac{(-1)^k(2k-1)!}{2mi(m^2 + n^2)^k}\left(e^{-2ik\tan^{-1}(n/m)} - e^{2ik\tan^{-1}(n/m)}\right) \\
&= -\frac{(2k-1)!}{m(m^2 + n^2)^k}\sin(2k\tan^{-1}(m/n)). \quad (36.3)
\end{aligned}$$

Hence, using (36.2) and (36.3) in (36.1), we deduce that

$$\begin{aligned}
\sum_{k=0}^{\infty}\frac{1}{m^2 + (n + k)^2} &= \frac{1}{m}\tan^{-1}(m/n) + \frac{1}{2(m^2 + n^2)} \\
&\quad + \frac{1}{m}\sum_{k=1}^{N}\frac{B_{2k}}{2k}\frac{\sin(2k\tan^{-1}(m/n))}{(m^2 + n^2)^k} + R_{N+1}.
\end{aligned}$$

The remainder R_{N+1} is easily estimated, and the desired result readily follows.   □

Corollary. *As n tends to ∞,*

$$n\sum_{k=0}^{\infty}{}'\frac{1}{n^2 + (n + k)^2} \sim \frac{\pi}{4} + \sum_{k=0}^{\infty}\frac{(-1)^k B_{4k+2}}{(2k + 1)2^{2k+2}n^{4k+2}}.$$

PROOF. Let $m = n$ in Entry 36.   □

CHAPTER 14

Infinite Series

Since Ramanujan's death in 1920, there have perhaps been more published papers establishing results in Chapter 14 than in any of the remaining 20 chapters. In many cases, the authors were unaware that their discoveries are found in Ramanujan's notebooks. In [6] and [7], the author showed that several results in Chapter 14, as well as many others as well, arise from a general transformation formula for a large class of analytic Eisenstein series. It should be emphasized, however, that Chapter 14 also contains many other types of results.

Chapter 14 is primarily concerned with identities involving infinite series. In Ramanujan's *Collected Papers* [16, p. xxv], Hardy remarked: "There is always more in one of Ramanujan's formulae than meets the eye, as anyone who sets to work to verify those which look the easiest will soon discover. In some the interest lies very deep, in others comparatively near the surface; but there is not one which is not curious and entertaining." There could not be a more apt comment about Chapter 14 than this last sentence of Hardy. Some of the formulas are fairly easy to prove; others require considerable effort. As previously indicated, many of the formulas in Chapter 14 have their genesis in elliptic modular functions. A large number of formulas arise from partial fraction decompositions. Some formulas are instances of the Poisson summation formula. Six formulas lie in the realm of hypergeometric series. There are also a few integral evaluations.

In the sequel, $R(f, z_0) = R(z_0)$ denotes the residue of f at a pole z_0. Also, $\chi(n)$ always denotes the primitive character of modulus 4; that is,

$$\chi(n) = \begin{cases} 0, & \text{if } n \equiv 0 \ (\text{mod } 2), \\ 1, & \text{if } n \equiv 1 \ (\text{mod } 4), \\ -1, & \text{if } n \equiv 3 \ (\text{mod } 4). \end{cases} \tag{0.1}$$

Entry 1. *For $z^2 \neq -n(n + 1)/2$, where n is a nonnegative integer, we have*

$$z^{-2} \prod_{n=1}^{\infty} \left(1 + \frac{2z^2}{n(n + 1)}\right)^{-1} = \sum_{n=0}^{\infty} \frac{(-1)^n(2n + 1)}{z^2 + n(n + 1)/2}. \tag{1.1}$$

PROOF. From the partial fraction decomposition (Whittaker and Watson [1, p. 136])

$$\operatorname{sech} x = 4\pi \sum_{n=0}^{\infty} \frac{(-1)^n(2n + 1)}{(2n + 1)^2\pi^2 + 4x^2}, \tag{1.2}$$

we obtain, after some simplification,

$$2\pi \operatorname{sech}(\pi\sqrt{2z^2 - \tfrac{1}{4}}) = \sum_{n=0}^{\infty} \frac{(-1)^n(2n + 1)}{z^2 + n(n + 1)/2}.$$

From the product expansion (Gradshteyn and Ryzhik [1, p. 37])

$$\cosh z = \prod_{n=0}^{\infty} \left(1 + \frac{4z^2}{(2n + 1)^2\pi^2}\right)$$

and Wallis's product (Gradshteyn and Ryzhik [1, p. 12])

$$\frac{\pi}{4} = \prod_{n=1}^{\infty} \frac{4n(n + 1)}{(2n + 1)^2},$$

we find that

$$2\pi \operatorname{sech}(\pi\sqrt{2z^2 - \tfrac{1}{4}}) = \frac{\pi}{4}z^{-2} \prod_{n=1}^{\infty} \left(1 + \frac{8z^2 - 1}{(2n + 1)^2}\right)^{-1}$$

$$= z^{-2} \prod_{n=1}^{\infty} \left\{\frac{(2n + 1)^2}{4n(n + 1)}\left(1 + \frac{8z^2 - 1}{(2n + 1)^2}\right)\right\}^{-1}$$

$$= z^{-2} \prod_{n=1}^{\infty} \left(1 + \frac{2z^2}{n(n + 1)}\right)^{-1}.$$

The result now follows. ☐

Corollary. *For $\operatorname{Re} z > 0$,*

$$\sum_{n=1}^{\infty} \frac{(-1)^{n+1}(2n + 1)}{\sqrt{n(n + 1)}(e^{2\pi z\sqrt{n(n+1)}} - 1)} + \frac{1}{z}\sum_{n=1}^{\infty} \operatorname{sech}\left(\frac{\pi}{z}\sqrt{n^2 - z^2/4}\right)$$

$$= \frac{1}{2\pi z} + \frac{\pi z}{6} - C, \tag{1.3}$$

where

$$C = \frac{1}{2} + \frac{1}{2}\sum_{n=1}^{\infty}{}^* \frac{(-1)^{n+1}(2n + 1)}{\sqrt{n(n + 1)}}, \tag{1.4}$$

where the asterisk (*) on the summation sign above indicates that the terms must be added in successive pairs in order for the series to converge.

We first show that the series defining C converges. We have

$$\frac{2n+1}{\sqrt{n(n+1)}} - \frac{2n+3}{\sqrt{(n+1)(n+2)}}$$

$$= \frac{1}{\sqrt{n+1}}\left\{\frac{2n+1}{\sqrt{n}} - \frac{2n+3}{\sqrt{n}}\left(1 - \frac{1}{n} + \frac{3}{2n^2} + O\left(\frac{1}{n^3}\right)\right)\right\}$$

$$= \frac{1}{\sqrt{n+1}}\left\{O\left(\frac{1}{n^{5/2}}\right)\right\}$$

$$= O(n^{-3}),$$

and so C is well defined.

Formula (1.3) does not agree with the corresponding entry in the notebooks in that Ramanujan claims that C should be replaced by

$$\frac{1}{2} + \sum_{n=1}^{\infty} \frac{(-1)^{n+1}}{2n+1+2\sqrt{n(n+1)}}$$

$$= 1 - \frac{\pi}{8} + \frac{1}{2}\sum_{n=1}^{\infty} \frac{(-1)^{n+1}}{(2n+1)\{2n+1+2\sqrt{n(n+1)}\}^2}. \qquad (1.5)$$

It is not difficult to prove the foregoing equality. Indeed, let C' denote the left side of (1.5). Using Gregory's series for $\pi/4$, we find that

$$C' = 1 - \frac{\pi}{8} + \sum_{n=1}^{\infty}\left\{\frac{(-1)^n}{2(2n+1)} + \frac{(-1)^{n+1}}{2n+1+2\sqrt{n(n+1)}}\right\}$$

$$= 1 - \frac{\pi}{8} + \sum_{n=1}^{\infty} (-1)^n \frac{\{2\sqrt{n(n+1)} - (2n+1)\}}{2(2n+1)\{2n+1+2\sqrt{n(n+1)}\}}$$

$$= 1 - \frac{\pi}{8} + \sum_{n=1}^{\infty} (-1)^n \frac{\{4n(n+1) - (2n+1)^2\}}{2(2n+1)\{2n+1+2\sqrt{n(n+1)}\}^2},$$

and (1.5) easily follows.

Calculations of J. Hill first demonstrated that the constant given by Ramanujan is incorrect. In fact, $C' = 0.61144169\cdots$, while $C = 0.54661949\cdots$. The formula for C given in (1.4) can be transformed into another formula that exhibits Ramanujan's error. Letting $a_n = (2n+1)/\sqrt{n(n+1)}$, we have

$$C = \frac{1}{2} + \frac{1}{2}\sum_{\substack{n=1 \\ n\text{ odd}}}^{\infty} (a_n - a_{n+1})$$

$$= \frac{1}{2} + \frac{1}{2}\sum_{\substack{n=1 \\ n\text{ odd}}}^{\infty} \{(a_n - 1) - (a_{n+1} - 1)\}$$

$$= \frac{1}{2} + \sum_{n=1}^{\infty} (-1)^{n+1}\left\{\frac{n + \frac{1}{2}}{\sqrt{n(n+1)}} - 1\right\}$$

$$= \frac{1}{2} + \sum_{n=1}^{\infty} (-1)^{n+1} \frac{2n + 1 - 2\sqrt{n(n+1)}}{2\sqrt{n(n+1)}}$$

$$= \frac{1}{2} + \sum_{n=1}^{\infty} \frac{(-1)^{n+1}}{2\sqrt{n(n+1)}(2n + 1 + 2\sqrt{n(n+1)})}.$$

Comparing the formula above with (1.5), we find that Ramanujan neglected a factor of $2\sqrt{n(n+1)}$ in the denominators of the summands on the left side of (1.5). The correct formula for C was first conjectured by R. Lamphere who verified it numerically.

After stating the corollary of Entry 1, Ramanujan declares: "Similarly any function whose denominator is in the form of a product can be expressed as the sum of partial fractions and many other theorems may be deduced from the result." But nonetheless, we have been unable to prove that (1.3) is a corollary of (1.1). The following proof of (1.3) is due to R. J. Evans.

PROOF OF COROLLARY TO ENTRY 1. We prove the result for $z = x > 0$; the more general result will then hold by analytic continuation.

For $n \geq 1$, let a_n be as defined above and put

$$f_n(x) = \frac{1}{e^{2\pi x\sqrt{n(n+1)}} - 1} - \frac{1}{2\pi x\sqrt{n(n+1)}}.$$

Thus,

$$\sum_{n=1}^{\infty} (-1)^{n+1} a_n f_n(x) = \sum_{n=1}^{\infty} (-1)^{n+1} \frac{2n + 1}{\sqrt{n(n+1)}(e^{2\pi x\sqrt{n(n+1)}} - 1)}$$

$$- \frac{1}{2\pi x} \sum_{n=1}^{\infty} \frac{(-1)^{n+1}(2n + 1)}{n(n+1)}. \tag{1.6}$$

By combining successive terms, we find after an elementary calculation that

$$\sum_{n=1}^{\infty} \frac{(-1)^{n+1}(2n + 1)}{n(n+1)} = \sum_{\substack{n=1 \\ n \text{ odd}}}^{\infty} \left(\frac{1}{n} - \frac{1}{n + 2}\right) = 1. \tag{1.7}$$

Putting (1.7) into (1.6) and comparing the resulting equality with (1.3), we find that we must show that

$$\sum_{n=1}^{\infty} (-1)^{n+1} a_n f_n(x) + \frac{1}{x} \sum_{n=1}^{\infty} \text{sech}\left(\frac{\pi}{x}\sqrt{n^2 - x^2/4}\right) - \frac{\pi x}{6} = -C, \tag{1.8}$$

where

$$C = \frac{1}{2} + \frac{1}{2} \sum_{\substack{n=1 \\ n \text{ odd}}}^{\infty} (a_n - a_{n+1}).$$

From Whittaker and Watson's text [1, p. 136],

$$\frac{1}{e^x - 1} = \frac{1}{2} \coth\left(\frac{x}{2}\right) - \frac{1}{2} = \frac{1}{x} - \frac{1}{2} + \sum_{m=1}^{\infty} \frac{2x}{x^2 + 4\pi^2 m^2}. \tag{1.9}$$

Using (1.2) and (1.9) in (1.8) and then simplifying, we find that

$$\sum_{n=1}^{\infty} \left\{ \frac{1}{2}(-1)^n a_n + \frac{x}{\pi} \sum_{m=1}^{\infty} \frac{(-1)^{n+1}(2n+1)}{n(n+1)x^2 + m^2} \right\}$$

$$= -C + \frac{\pi x}{6} - \frac{x}{\pi} \sum_{n=1}^{\infty} \sum_{m=0}^{\infty} \frac{(-1)^m(2m+1)}{m(m+1)x^2 + n^2}$$

$$= -C + \frac{x}{\pi} \sum_{n=1}^{\infty} \sum_{m=1}^{\infty} \frac{(-1)^{m+1}(2m+1)}{m(m+1)x^2 + n^2}. \tag{1.10}$$

Letting

$$B(m, n) = \frac{2n+1}{n(n+1)x^2 + m^2},$$

we see that (1.10) may be written as

$$\frac{x}{\pi} \sum_{\substack{n=1 \\ n \text{ even}}}^{\infty} \sum_{m=1}^{\infty} \{B(m, n-1) - B(m, n)\}$$

$$= -\frac{1}{2} + \frac{x}{\pi} \sum_{m=1}^{\infty} \sum_{\substack{n=1 \\ n \text{ even}}}^{\infty} \{B(m, n-1) - B(m, n)\}. \tag{1.11}$$

A brief calculation gives

$$B(m, n-1) - B(m, n) = \frac{2n^2x^2 - 2m^2}{(m^2 + n^2x^2 - nx^2)(m^2 + n^2x^2 + nx^2)}.$$

Replacing x by $x/2$, we see then that (1.11) is equivalent to

$$\sum_{n=1}^{\infty} \sum_{m=1}^{\infty} \frac{n^2x^2 - m^2}{(m^2 + n^2x^2)^2 - n^2x^4/4} = -\frac{\pi}{2x} + \sum_{m=1}^{\infty} \sum_{n=1}^{\infty} \frac{n^2x^2 - m^2}{(m^2 + n^2x^2)^2 - n^2x^4/4}. \tag{1.12}$$

By a brief calculation,

$$\sum_{n=1}^{\infty} \sum_{m=1}^{\infty} \left\{ \frac{n^2x^2 - m^2}{(m^2 + n^2x^2)^2 - n^2x^4/4} - \frac{n^2x^2 - m^2}{(m^2 + n^2x^2)^2} \right\}$$

is seen to be an absolutely convergent double series, and so an inversion in order of summation is justified. Thus, (1.12) is seen to be equivalent to

$$\sum_{n=1}^{\infty} \sum_{m=1}^{\infty} \frac{n^2x^2 - m^2}{(m^2 + n^2x^2)^2} = -\frac{\pi}{2x} + \sum_{m=1}^{\infty} \sum_{n=1}^{\infty} \frac{n^2x^2 - m^2}{(m^2 + n^2x^2)^2}$$

$$= -\frac{\pi}{2x} - x^{-2} \sum_{n=1}^{\infty} \sum_{m=1}^{\infty} \frac{n^2x^2 - m^2}{(m^2 + n^2x^{-2})^2}, \tag{1.13}$$

where on the right side we have replaced the indices m and n by n and m, respectively. Let the left side of (1.13) be denoted by $F(x)$. Thus, (1.13) may be rewritten as

$$F(x) + x^{-2}F(1/x) = -\pi/(2x). \tag{1.14}$$

Now return to (1.9). Replace x by $2\pi nx$ and differentiate the extremal sides with respect to x. After some simplification, we find that

$$\frac{2\pi^2}{(e^{\pi nx} - e^{-\pi nx})^2} - \frac{1}{2n^2x^2} = \sum_{m=1}^{\infty} \frac{2n^2x^2}{(n^2x^2 + m^2)^2} - \sum_{m=1}^{\infty} \frac{1}{n^2x^2 + m^2}$$

$$= \sum_{m=1}^{\infty} \frac{n^2x^2 - m^2}{(n^2x^2 + m^2)^2}. \tag{1.15}$$

Summing both sides of (1.15) on n, $1 \le n < \infty$, we deduce that

$$\frac{1}{2}\pi^2 \sum_{n=1}^{\infty} \operatorname{csch}^2(\pi nx) - \frac{\pi^2}{12x^2} = F(x).$$

Thus, (1.14) is seen to be equivalent to

$$\pi x \sum_{n=1}^{\infty} \operatorname{csch}^2(\pi nx) + \frac{\pi}{x} \sum_{n=1}^{\infty} \operatorname{csch}^2(\pi n/x) = -1 + \frac{\pi}{6}\left(x + \frac{1}{x}\right).$$

If we put $\alpha = \pi x$ and $\beta = \pi/x$, we find that for α, $\beta > 0$ and $\alpha\beta = \pi^2$,

$$\alpha \sum_{n=1}^{\infty} \operatorname{csch}^2(\alpha n) + \beta \sum_{n=1}^{\infty} \operatorname{csch}^2(\beta n) = -1 + (\alpha + \beta)/6. \tag{1.16}$$

In summary, we have shown that (1.3) is equivalent to (1.16). But the author [6, Proposition 2.25] has previously proved (1.16), and hence the proof is complete. $\square$

Observe that (1.13) provides a beautiful example of a nonabsolutely convergent double series whose order of summation cannot be inverted.

Entry 2. *Let m, n, x, and y be complex numbers. Suppose that $\Gamma(1 + xz)$ and $\Gamma(1 + yz)$ have no coincident poles and that $z = 1$ is not a pole of either. Then if $\operatorname{Re}(m + n) > 0$,*

$$\sum_{k=1}^{\infty} \frac{(-1)^{k+1}\Gamma(1 - ky/x)}{\Gamma(m - k + 1)\Gamma(n + 1 - ky/x)\Gamma(k)(x + k)}$$

$$+ \sum_{k=1}^{\infty} \frac{(-1)^{k+1}\Gamma(1 - kx/y)}{\Gamma(n - k + 1)\Gamma(m + 1 - kx/y)\Gamma(k)(y + k)}$$

$$= \frac{\Gamma(x + 1)\Gamma(y + 1)}{\Gamma(x + m + 1)\Gamma(y + n + 1)}. \tag{2.1}$$

PROOF. Let

$$f(z) = \frac{\Gamma(1 + xz)\Gamma(1 + yz)}{\Gamma(m + xz + 1)\Gamma(n + yz + 1)(z - 1)}.$$

Then f has poles at $z = 1$, $-j/x$, and $-k/y$, where $1 \le j, k < \infty$, and all poles are simple by hypothesis. Routine calculations yield

$$R(1) = \frac{\Gamma(x + 1)\Gamma(y + 1)}{\Gamma(m + x + 1)\Gamma(n + y + 1)},$$

$$R(-j/x) = \frac{(-1)^j \Gamma(1 - jy/x)}{\Gamma(m - j + 1)\Gamma(n - jy/x + 1)(j + x)\Gamma(j)},$$

and

$$R(-k/y) = \frac{(-1)^k \Gamma(1 - kx/y)}{\Gamma(m - kx/y + 1)\Gamma(n - k + 1)(k + y)\Gamma(k)}.$$

Let C_N be a positively oriented square centered at the origin and with vertical and horizontal sides of length $2N$. We shall let N tend to ∞ on some countable subset of the positive real numbers chosen so that the sides of C_N never get closer than some fixed positive distance from the set of poles of f. Using Stirling's formula, we find that

$$f(z) = O(|z|^{-\text{Re}(m+n)-1}),$$

as $|z|$ tends to ∞. Hence, if $\text{Re}(m + n) > 0$, we deduce that

$$\int_{C_N} f(z)\, dz = o(1), \tag{2.2}$$

as N tends to ∞.

Now integrate f over C_N and apply the residue theorem. Let N tend to ∞ and use (2.2). We then deduce (2.1) immediately.   □

Corollary 1. *Let m, n, and x be complex numbers such that x is not an integer and that $\text{Re}(m + n) > -1$. Then*

$$\sum_{k=-\infty}^{\infty} \frac{(-1)^k}{(x + k)\Gamma(m + 1 - k)\Gamma(n + 1 + k)}$$

$$= \frac{\pi}{\sin(\pi x)\Gamma(m + x + 1)\Gamma(n - x + 1)}. \tag{2.3}$$

Corollary 2. *Let α and β be complex numbers with $\text{Re}(\alpha + \beta) > 0$. Then*

$$\sum_{k=0}^{\infty} \frac{(-1)^k}{(2k + 1)\Gamma(\alpha - k)\Gamma(\beta + k + 1)} + \sum_{k=0}^{\infty} \frac{(-1)^k}{(2k + 1)\Gamma(\beta - k)\Gamma(\alpha + k + 1)}$$

$$= \frac{\pi}{2\Gamma(\alpha + \frac{1}{2})\Gamma(\beta + \frac{1}{2})}. \tag{2.4}$$

Corollaries 1 and 2 are not really corollaries of Entry 2. Ramanujan evidently means to imply that the *proofs* of the present results are very much like the proof of the preceding theorem.

PROOF OF COROLLARY 1. Let

$$f(z) = \frac{\pi}{\sin(\pi z)(z + x)\Gamma(m + 1 - z)\Gamma(n + 1 + z)}.$$

Observe that f has a simple pole at $z = -x$ and at each integer k. Routine calculations give

$$R(-x) = -\frac{\pi}{\sin(\pi x)\Gamma(m + 1 + x)\Gamma(n + 1 - x)}$$

and

$$R(k) = \frac{(-1)^k}{(k + x)\Gamma(m + 1 - k)\Gamma(n + 1 + k)}.$$

Let C_N be the positively oriented square centered at the origin with vertical and horizontal sides passing through $\pm(N + \frac{1}{2})$ and $\pm(N + \frac{1}{2})i$, respectively, where N is a positive integer. By Stirling's formula,

$$f(z) = O(|z|^{-\text{Re}(m+n)-2}),$$

as $|z|$ tends to ∞. Hence, for $\text{Re}(m + n) > -1$,

$$\int_{C_N} f(z)\, dz = o(1), \tag{2.5}$$

as N tends to ∞. Apply the residue theorem to the integral of f over C_N. Let N tend to ∞. Using (2.5), we deduce (2.3) at once. □

PROOF OF COROLLARY 2. Integrate

$$\frac{\pi}{\sin(\pi z)(z - \frac{1}{2})\Gamma(\alpha + z)\Gamma(\beta - z + 1)}$$

over the same square as in the foregoing proof. The present proof follows along precisely the same lines, and we omit it.

A second proof can be given as follows. Let the left side of (2.4) be denoted by $g(\alpha, \beta)$. After a little manipulation, we see that $g(\alpha, \beta)$ may be written as

$$g(\alpha, \beta) = \frac{\sin(\pi\alpha)}{2\pi} \sum_{k=-\infty}^{\infty} \frac{\Gamma(\frac{1}{2} + k)\Gamma(1 - \alpha + k)}{\Gamma(\frac{3}{2} + k)\Gamma(1 + \beta + k)}, \tag{2.6}$$

which converges absolutely for $\text{Re}(\alpha + \beta) > 0$ by Stirling's formula. Now apply Dougall's formula (Henrici [2, p. 52]) to the right side of (2.6) to obtain

$$g(\alpha, \beta) = \frac{\pi}{2\Gamma(\alpha + \frac{1}{2})\Gamma(\beta + \frac{1}{2})}. □$$

Corollary 1 may also be proved with the aid of Dougall's theorem. However, Dougall's theorem is not applicable to Entry 2. The next entry is also an instance of Dougall's theorem.

Entry 3. *Let α, β, γ, and δ be complex numbers such that $\mathrm{Re}(\alpha + \beta + \gamma + \delta) >$*
-1. Then

$$\sum_{k=0}^{\infty} \frac{1}{\Gamma(\alpha - k + 1)\Gamma(\beta - k + 1)\Gamma(\gamma + k + 1)\Gamma(\delta + k + 1)}$$

$$+ \sum_{k=1}^{\infty} \frac{1}{\Gamma(\alpha + k + 1)\Gamma(\beta + k + 1)\Gamma(\gamma - k + 1)\Gamma(\delta - k + 1)}$$

$$= \frac{\Gamma(\alpha + \beta + \gamma + \delta + 1)}{\Gamma(\alpha + \gamma + 1)(\beta + \gamma + 1)\Gamma(\alpha + \delta + 1)\Gamma(\beta + \delta + 1)}. \tag{3.1}$$

PROOF. The left side of (3.1) may be written as

$$\frac{\sin(\pi\alpha)\sin(\pi\beta)}{\pi^2} \sum_{k=-\infty}^{\infty} \frac{\Gamma(k - \alpha)\Gamma(k - \beta)}{\Gamma(\gamma + k + 1)\Gamma(\delta + k + 1)},$$

which converges absolutely for $\mathrm{Re}(\alpha + \beta + \gamma + \delta) > -1$ by Stirling's formula. A straightforward application of Dougall's theorem (Henrici [2, p. 52]) yields (3.1) immediately.   □

Entry 4. *If $z \neq me^{\pm\pi i/3}$, where m is a nonzero integer, then*

$$\sum_{n=1}^{\infty} \frac{1}{n^2 + z^2 + z^4/n^2} = \frac{\pi}{2z\sqrt{3}} \frac{\sinh(\pi z\sqrt{3}) - \sqrt{3}\sin(\pi z)}{\cosh(\pi z\sqrt{3}) - \cos(\pi z)}. \tag{4.1}$$

PROOF. Let $f(z)$ denote the right side of (4.1). We expand f into partial fractions. Since

$$\cosh(\pi z\sqrt{3}) - \cos(\pi z) = 2\sin(\pi z e^{\pi i/3})\sin(\pi z e^{-\pi i/3}),$$

f has simple poles at $z = ne^{\pm\pi i/3}$ for each nonzero integer n. Now

$$R(ne^{-\pi i/3}) = \frac{\sinh(\pi n e^{-\pi i/3}\sqrt{3}) - \sqrt{3}\sin(\pi n e^{-\pi i/3})}{4n(-1)^n\sqrt{3}\sin(\pi n e^{-2\pi i/3})}.$$

Note that $R(-ne^{-\pi i/3}) = -R(ne^{-\pi i/3})$. The residues of the poles at $\pm ne^{\pi i/3}$ are obtained by replacing $e^{-\pi i/3}$ by $e^{\pi i/3}$ above. For each positive integer n, the sum of the principal parts for the four poles $\pm ne^{\pm\pi i/3}$ is then

$$\frac{(-1)^n}{2\sqrt{3}} \left\{ \frac{e^{-\pi i/3}\{\sinh(\pi n e^{-\pi i/3}\sqrt{3}) - \sqrt{3}\sin(\pi n e^{-\pi i/3})\}}{\sin(\pi n e^{-2\pi i/3})(z^2 - n^2 e^{-2\pi i/3})} \right.$$

$$\left. + \frac{e^{\pi i/3}\{\sinh(\pi n e^{\pi i/3}\sqrt{3}) - \sqrt{3}\sin(\pi n e^{\pi i/3})\}}{\sin(\pi n e^{2\pi i/3})(z^2 - n^2 e^{2\pi i/3})} \right\}. \tag{4.2}$$

Elementary calculations give

$$\sin(\pi n e^{\pm 2\pi i/3}) = \begin{cases} \pm i(-1)^m \sinh(\pi m\sqrt{3}), & \text{if } n = 2m, \\ (-1)^{m+1}\cosh(\pi n\sqrt{3}/2), & \text{if } n = 2m + 1, \end{cases}$$

and

$$e^{\pm\pi i/3}\{\sinh(\pi n e^{\pm\pi i/3}\sqrt{3}) - \sqrt{3}\sin(\pi n e^{\pm\pi i/3})\}$$

$$= \begin{cases} 2(-1)^m \sinh(\pi m\sqrt{3}), & \text{if } n = 2m, \\ \pm 2i(-1)^{m+1}\cosh(\pi n\sqrt{3}/2), & \text{if } n = 2m+1. \end{cases}$$

Using the calculations above, we find that (4.2) simplifies to $n^2/(z^4 + n^2 z^2 + n^4)$ for both n even and n odd. Hence,

$$f(z) = \sum_{n=1}^{\infty} \frac{n^2}{z^4 + n^2 z^2 + n^4} + g(z),$$

where g is entire. However, as $|z|$ tends to ∞, we clearly see that $g(z)$ tends to 0. Thus, g is a bounded entire function. By Liouville's theorem, $g(z)$ is constant, and this constant is obviously 0. Hence, the proof is complete. □

Corollary. *For each nonzero integer* n,

$$\sum_{k=1}^{\infty} \frac{1}{k^2 + (2n)^2 + (2n)^4/k^2} = \frac{1}{12n^2} + \frac{1}{2}\sum_{k=1}^{\infty}\frac{1}{k^2 + 3n^2}.$$

PROOF. In the derivation below, we shall employ (1.9) and the decomposition (Whittaker and Watson [1, p. 136])

$$\operatorname{csch}(\pi z) = \frac{1}{\pi z} + \frac{2z}{\pi}\sum_{k=1}^{\infty}\frac{(-1)^k}{z^2 + k^2}.$$

In Entry 4 let $z = 2n$ to get

$$\sum_{k=1}^{\infty}\frac{1}{k^2 + (2n)^2 + (2n)^4/k^2} = \frac{\pi}{4n\sqrt{3}}\{\coth(2\pi n\sqrt{3}) + \operatorname{csch}(2\pi n\sqrt{3})\}$$

$$= \frac{1}{12n^2} + \sum_{k=1}^{\infty}\frac{1}{12n^2 + k^2} + \sum_{k=1}^{\infty}\frac{(-1)^k}{12n^2 + k^2},$$

and the result follows. □

Entry 5(i). *Let* $0 < x < \pi/(n + \frac{1}{2})$, *where* n *is a positive integer. Then*

$$\sum_{k=1}^{\infty}\frac{\sin^{2n+1}(kx)}{k} = \frac{\sqrt{\pi}}{2}\frac{\Gamma(n + \frac{1}{2})}{\Gamma(n + 1)}.$$

PROOF. Since (Gradshteyn and Ryzhik [1, p. 25])

$$\sin^{2n+1}x = 2^{-2n}\sum_{j=0}^{n}(-1)^{n+j}\binom{2n+1}{j}\sin\{(2n+1-2j)x\},$$

we have

$$\sum_{k=1}^{\infty}\frac{\sin^{2n+1}(kx)}{k} = \frac{(-1)^n}{2^{2n}}\sum_{j=0}^{n}(-1)^j\binom{2n+1}{j}\sum_{k=1}^{\infty}\frac{\sin\{(2n+1-2j)kx\}}{k}. \tag{5.1}$$

Using the familiar result (Gradshteyn and Ryzhik [1, p. 38])

$$\sum_{k=1}^{\infty} \frac{\sin(kx)}{k} = \frac{\pi - x}{2}, \qquad 0 < x < 2\pi, \tag{5.2}$$

we find that, for $0 < x < \pi/(n + \frac{1}{2})$,

$$\sum_{k=1}^{\infty} \frac{\sin^{2n+1}(kx)}{k} = \frac{(-1)^n}{2^{2n}} \sum_{j=0}^{n} (-1)^j \binom{2n+1}{j} \frac{\pi - (2n + 1 - 2j)x}{2}$$

$$= \frac{\pi}{2^{2n+1}} \binom{2n}{n} - \frac{(-1)^n}{2^{2n+1}} \sum_{j=0}^{n} (-1)^j \binom{2n+1}{j} (2n + 1 - 2j)x, \tag{5.3}$$

where we used the evaluation (Gradshteyn and Ryzhik [1, p. 3])

$$\sum_{j=0}^{m} (-1)^j \binom{k}{j} = (-1)^m \binom{k-1}{m}, \tag{5.4}$$

with $m = n$ and $k = 2n + 1$.

We next show that the sum on the far right side of (5.3) vanishes. We have

$$2 \sum_{j=0}^{n} (-1)^j \binom{2n+1}{j} (2n + 1 - 2j) = \sum_{j=0}^{2n+1} (-1)^j \binom{2n+1}{j} (2n + 1 - 2j)$$

$$= (2n + 1) \sum_{j=0}^{2n+1} (-1)^j \binom{2n+1}{j}$$

$$- 2 \sum_{j=0}^{2n+1} (-1)^j \binom{2n+1}{j} j$$

$$= 0,$$

where we have used (5.4) and (Gradshteyn and Ryzhik [1, p. 4])

$$\sum_{j=0}^{n} (-1)^j j \binom{n}{j} = 0.$$

Hence, from (5.3),

$$\sum_{k=1}^{\infty} \frac{\sin^{2n+1}(kx)}{k} = \frac{\pi}{2^{2n+1}} \binom{2n}{n},$$

which is easily seen to be equivalent to the desired result by the Legendre duplication formula. $\qquad\square$

Entry 5(ii). *Let $0 \leq x \leq \pi/(n + 1)$, where n is a positive integer. Then*

$$\sum_{k=1}^{\infty} \frac{\sin^{2n+2}(kx)}{k^2} = \frac{\sqrt{\pi}}{2} \frac{\Gamma(n + \frac{1}{2})}{\Gamma(n + 1)} x. \tag{5.5}$$

PROOF. Let $f(x)$ denote the left side of (5.5). Since (Gradshteyn and Ryzhik [1, p. 25])

$$\sin^{2n+2} x$$

$$= 2^{-2n-2}\left\{\sum_{j=0}^{n}(-1)^{n+1+j}2\binom{2n+2}{j}\cos\{2(n+1-j)x\}+\binom{2n+2}{n+1}\right\},$$

we find that

$$f(x) = \frac{(-1)^{n+1}}{2^{2n+1}}\sum_{j=0}^{n}(-1)^{j}\binom{2n+2}{j}\sum_{k=1}^{\infty}\frac{\cos\{2(n+1-j)kx\}}{k^2}$$

$$+\frac{1}{2^{2n+2}}\binom{2n+2}{n+1}\frac{\pi^2}{6}. \tag{5.6}$$

Now (Gradshteyn and Ryzhik [1, p. 39])

$$\sum_{k=1}^{\infty}\frac{\cos(kx)}{k^2}=\frac{\pi^2}{6}-\frac{\pi x}{2}+\frac{x^2}{4}, \qquad 0 \le x \le 2\pi.$$

Employing the formula above and (5.4) with $m = n$ and $k = 2n + 2$, we find that (5.6) becomes

$$f(x) = \frac{(-1)^{n+1}}{2^{2n+1}}\sum_{j=0}^{n}(-1)^{j}\binom{2n+2}{j}\{(n+1-j)^2x^2-\pi(n+1-j)x\}$$

$$-\frac{1}{2^{2n+1}}\binom{2n+1}{n}\frac{\pi^2}{6}+\frac{1}{2^{2n+2}}\binom{2n+2}{n+1}\frac{\pi^2}{6}, \tag{5.7}$$

where $0 \le x \le \pi/(n + 1)$. First,

$$2\sum_{j=0}^{n}(-1)^{j}\binom{2n+2}{j}(n+1-j)^2 = \sum_{j=0}^{2n+2}(-1)^{j}\binom{2n+2}{j}(n+1-j)^2$$

$$= \sum_{j=0}^{2n+2}(-1)^{j}\binom{2n+2}{j}j^2$$

$$= 0. \tag{5.8}$$

Next, from two applications of (5.4), we find that

$$\sum_{j=0}^{n}(-1)^{j}\binom{2n+2}{j}(n+1-j)$$

$$= (2n+2)\sum_{j=0}^{n}(-1)^{j}\binom{2n+1}{j}-(n+1)\sum_{j=0}^{n}(-1)^{j}\binom{2n+2}{j}$$

$$= (2n+2)(-1)^{n}\binom{2n}{n}-(n+1)(-1)^{n}\binom{2n+1}{n}$$

$$= (-1)^{n}\binom{2n}{n}. \tag{5.9}$$

Substituting (5.8) and (5.9) into (5.7), we find that

$$f(x) = \frac{\pi x}{2^{2n+1}}\binom{2n}{n},$$

which is again equivalent to the desired result by the Legendre duplication
formula.   □

Entry 6. *For $n > 0$, let*

$$\varphi(\beta) = \prod_{k=0}^{\infty} \left\{ 1 + \left(\frac{\beta}{n+k} \right)^2 \right\}^{-1}.$$

Let $\alpha, \beta > 0$ with $\alpha\beta = \pi$. Then

$$\sqrt{\alpha} \left\{ \frac{1}{2} + \sum_{k=1}^{\infty} \operatorname{sech}^{2n}(\alpha k) \right\} = \frac{\Gamma(n)}{\Gamma(n+\frac{1}{2})} \sqrt{\beta} \left\{ \frac{1}{2} + \sum_{k=1}^{\infty} \varphi(\beta k) \right\}.$$

PROOF. Recall the Poisson summation formula. If f is a continuous function
of bounded variation on $[a, b]$, then

$$\sum_{a \le k \le b}' f(k) = \int_a^b f(x)\,dx + 2 \sum_{k=1}^{\infty} \int_a^b f(x) \cos(2\pi k x)\,dx, \qquad (6.1)$$

where the prime on the summation sign at the left indicates that if a or b is
an integer, then only $\frac{1}{2}f(a)$ or $\frac{1}{2}f(b)$, respectively, is counted.

Now $\varphi(x)$ was studied by Ramanujan in [8], [16, pp. 53–58]. On page 54
of [16] Ramanujan remarks that

$$\varphi(x) = \frac{\Gamma(n+ix)\Gamma(n-ix)}{\Gamma^2(n)}.$$

This is not too difficult to prove; use the Weierstrass product formula for
the quotient of Γ-functions above, and after considerable simplification, the
desired equality follows. We shall apply (6.1) with $f(x) = \varphi(\beta x)$, $a = 0$, and
$b = \infty$. By using Stirling's formula for $|\Gamma(n+ix)\Gamma(n-ix)|$, as x tends to ∞,
we easily justify letting b tend to ∞. Furthermore, for m real and $n > 0$, by
Entry 15 in Chapter 13,

$$\int_0^{\infty} \varphi(x) \cos(2mx)\,dx = \frac{\sqrt{\pi}}{2} \frac{\Gamma(n+\frac{1}{2})}{\Gamma(n)} \operatorname{sech}^{2n} m.$$

Hence, since $\varphi(0) = 1$, (6.1) yields

$$\frac{1}{2} + \sum_{k=1}^{\infty} \varphi(\beta k) = \frac{1}{\beta} \int_0^{\infty} \varphi(x)\,dx + \frac{2}{\beta} \sum_{k=1}^{\infty} \int_0^{\infty} \varphi(x) \cos(2\pi k x/\beta)\,dx$$

$$= \sqrt{\frac{\alpha}{\beta}} \frac{\Gamma(n+\frac{1}{2})}{\Gamma(n)} \left\{ \frac{1}{2} + \sum_{k=1}^{\infty} \operatorname{sech}^{2n}(\pi k/\beta) \right\},$$

which is easily seen to be equivalent to the desired result.   □

Entry 7. *Let $\alpha, \beta > 0$ with $\alpha\beta = \pi$ and let z be an arbitrary complex number.
Then*

$$e^{z^2/4}\sqrt{\alpha}\left\{\frac{1}{2} + \sum_{k=1}^{\infty} e^{-\alpha^2 k^2} \cos(\alpha zk)\right\} = \sqrt{\beta}\left\{\frac{1}{2} + \sum_{k=1}^{\infty} e^{-\beta^2 k^2} \cosh(\beta zk)\right\}.$$

PROOF. Apply the Poisson formula (6.1) with $f(x) = \exp(-\alpha^2 x^2) \cos(\alpha zx)$, $a = 0$, and $b = \infty$. Now (Gradshteyn and Ryzhik [1, p. 480]),

$$\int_0^{\infty} e^{-cx^2} \cos(rx) \, dx = \frac{1}{2}\sqrt{\frac{\pi}{c}} e^{-r^2/(4c)}, \tag{7.1}$$

where Re $c > 0$ and r is arbitrary. With the use of the above evaluation, all the calculations are quite routine, and the desired formula follows with no difficulty.   □

Corollary. *Let* $\alpha, \beta > 0$ *with* $\alpha\beta = \pi$. *Then*

$$\sqrt{\alpha}\left\{\frac{1}{2} + \sum_{k=1}^{\infty} e^{-\alpha^2 k^2}\right\} = \sqrt{\beta}\left\{\frac{1}{2} + \sum_{k=1}^{\infty} e^{-\beta^2 k^2}\right\}.$$

PROOF. Let $z = 0$ in Entry 7.   □

Note that the formula above is simply the functional equation for the classical theta-function.

Entry 8(i). *Let* $\alpha, \beta, n > 0$ *with* $\alpha\beta = \pi$ *and* $0 < \beta n < \pi$. *Then*

$$\alpha \sum_{k=1}^{\infty} \frac{\sinh(2\alpha nk)}{e^{2\alpha^2 k} - 1} + \beta \sum_{k=1}^{\infty} \frac{\sin(2\beta nk)}{e^{2\beta^2 k} - 1} = \tfrac{1}{4}\alpha \coth(\alpha n) - \tfrac{1}{4}\beta \cot(\beta n) - \tfrac{1}{2}n.$$

Entry 8(i) arises from the transformation formulas of a function akin to the logarithm of the Dedekind eta-function. The first proof of Entry 8(i) preceded that by Ramanujan and was found by Schlömilch [1], [2, p. 156]. Later proofs have been given by Rao and Ayyar [1], J. Lagrange [1], and the author [6, Eq. (3.31)], [2, Eq. (11.21)].

Entry 8(ii). *Let* $\alpha, \beta, n > 0$ *with* $\alpha\beta = \pi$ *and* $0 < \alpha n < \pi$. *Then*

$$2 \sum_{k=1}^{\infty} \frac{\cos(2\alpha nk)}{k(e^{2\alpha^2 k} - 1)} - 2 \sum_{k=1}^{\infty} \frac{\cosh(2\beta nk)}{k(e^{2\beta^2 k} - 1)} = n^2 - \tfrac{1}{6}(\alpha^2 - \beta^2) + \text{Log}\left\{\frac{\sin(\alpha n)}{\sinh(\beta n)}\right\}.$$

Entry 8(ii) arises from the transformation formulas of a function that generalizes the logarithm of the Dedekind eta-function. Proofs have been given by J. Lagrange [1] and the author [6, Proposition 3.4].

Entry 8(iii). *Let* $\alpha, \beta, n, r, t > 0$ *with* $\alpha\beta = \pi$, $r = n\beta$, *and* $t = \pi/\beta^2$. *Let* C *be the positively oriented parallelogram with vertices* $\pm i$ *and* $\pm t$. *Let* $\varphi(z)$ *be entire. Let* m *be a positive integer and put* $M = m + \tfrac{1}{2}$. *Define*

$$f_m(z) = \frac{\varphi(rMz)}{z(e^{-2\pi Mz} - 1)(e^{2\pi iMz/t} - 1)},$$

and assume that $f_m(z)$ tends to 0 boundedly on $C' = C - \{\pm i, \pm t\}$ as m tends to ∞. Then

$$\sum_{k=1}^{\infty} \frac{\varphi(\alpha nk) + \varphi(-\alpha nk)}{k(e^{2\alpha^2 k} - 1)} + \sum_{k=1}^{\infty} \frac{\varphi(\alpha nk)}{k} - \sum_{k=1}^{\infty} \frac{\varphi(\beta nki) + \varphi(-\beta nki)}{k(e^{2\beta^2 k} - 1)} - \sum_{k=1}^{\infty} \frac{\varphi(\beta nki)}{k}$$

$$= \frac{\pi i \varphi(0)}{2} - \frac{\alpha^2 \varphi(0)}{6} + \frac{\beta^2 \varphi(0)}{6} - \frac{\alpha n \varphi'(0)}{2} + \frac{\beta n i \varphi'(0)}{2} - \frac{n^2 \varphi''(0)}{4}, \tag{8.1}$$

provided that all series above converge.

The obviously very restrictive hypotheses on φ are of a technical nature. We could state these hypotheses more specifically, but an even lengthier statement of the theorem would be necessary.

PROOF. We integrate $f_m(z)$ over C. On the interior of C, f_m has simple poles at $z = \pm ik/M$ and at $z = \pm kt/M$, $1 \le k \le m$. Also, there is a triple pole at $z = 0$. Straightforward calculations give

$$R(ik/M) = \frac{\varphi(rki)}{2\pi ik} \left\{ \frac{1}{e^{2\pi k/t} - 1} + 1 \right\},$$

$$R(-ik/M) = \frac{\varphi(-rki)}{2\pi ik(e^{2\pi k/t} - 1)},$$

$$R(kt/M) = -\frac{\varphi(rkt)}{2\pi ik} \left\{ \frac{1}{e^{2\pi kt} - 1} + 1 \right\},$$

and

$$R(-kt/M) = -\frac{\varphi(-rkt)}{2\pi ik(e^{2\pi kt} - 1)},$$

where $1 \le k \le m$. Now,

$$f_m(z) = \frac{it}{(2\pi M)^2 z^3} \{ \varphi(0) + \varphi'(0)rMz + \tfrac{1}{2}\varphi''(0)(rMz)^2 + \cdots \}$$

$$\times \{ 1 + \pi Mz + \tfrac{1}{3}(\pi Mz)^2 + \cdots \} \left\{ 1 - \frac{\pi iMz}{t} - \frac{1}{3}\left(\frac{\pi Mz}{t}\right)^2 + \cdots \right\},$$

and so

$$R(0) = \frac{\varphi(0)}{4} + \frac{it\varphi(0)}{12} - \frac{i\varphi(0)}{12t} + \frac{irt\varphi'(0)}{4\pi} + \frac{r\varphi'(0)}{4\pi} + \frac{ir^2 t\varphi''(0)}{8\pi^2}.$$

Applying the residue theorem and letting M tend to ∞, we find that

$$\lim_{m \to \infty} \int_C f_m(z)\, dz = \sum_{k=1}^{\infty} \frac{\varphi(rki) + \varphi(-rki)}{k(e^{2\pi k/t} - 1)} + \sum_{k=1}^{\infty} \frac{\varphi(rki)}{k} - \sum_{k=1}^{\infty} \frac{\varphi(rkt) + \varphi(-rkt)}{k(e^{2\pi kt} - 1)}$$

$$- \sum_{k=1}^{\infty} \frac{\varphi(rkt)}{k} + \frac{\pi i \varphi(0)}{2} - \frac{\pi t \varphi(0)}{6} + \frac{\pi \varphi(0)}{6t}$$

$$- \frac{rt\varphi'(0)}{2} + \frac{ir\varphi'(0)}{2} - \frac{r^2 t \varphi''(0)}{4\pi}. \tag{8.2}$$

By our hypotheses and the bounded convergence theorem, the limit on the left side of (8.2) is 0. Substituting $r = n\beta$ and $t = \pi/\beta^2$ in (8.2) and rearranging, we deduce (8.1). ∎

We next show that Entry 8(ii) is a special instance of Entry 8(iii).

Let $\varphi(z) = \exp(2iz)$. Thus, $\varphi(\alpha nk) + \varphi(-\alpha nk) = 2\cos(2\alpha nk)$ and $\varphi(\beta nki) + \varphi(-\beta nki) = 2\cosh(2\beta nk)$. Since $0 < \alpha n < \pi$, by a standard result found in Gradshteyn and Ryzhik's book [1, p. 38] and (5.2), we have

$$\sum_{k=1}^{\infty} \frac{\varphi(\alpha nk)}{k} = \sum_{k=1}^{\infty} \frac{e^{2\alpha nki}}{k} = -\text{Log}\{2\sin(\alpha n)\} + i\frac{\pi - 2\alpha n}{2}.$$

Second, an elementary calculation gives

$$\sum_{k=1}^{\infty} \frac{\varphi(\beta nki)}{k} = \sum_{k=1}^{\infty} \frac{e^{-2\beta nk}}{k} = \beta n - \text{Log}\{2\sinh(\beta n)\}.$$

Thus,

$$\sum_{k=1}^{\infty} \frac{\varphi(\beta nki)}{k} - \sum_{k=1}^{\infty} \frac{\varphi(\alpha nk)}{k} + \frac{\pi i \varphi(0)}{2} - \frac{\alpha^2 \varphi(0)}{6} + \frac{\beta^2 \varphi(0)}{6}$$

$$- \frac{\alpha n \varphi'(0)}{2} + \frac{\beta ni \varphi'(0)}{2} - \frac{n^2 \varphi''(0)}{4}$$

$$= \text{Log}\left\{\frac{\sin(\alpha n)}{\sinh(\beta n)}\right\} + n^2 - \tfrac{1}{6}\alpha^2 + \tfrac{1}{6}\beta^2.$$

Hence, formally, Entry 8(ii) follows readily from Entry 8(iii).

It remains to check the hypotheses concerning the parallelogram C. This is easily done by parameterizing each side of C. In the first quadrant, $f_m(z)$ trivially tends to 0 boundedly on C'. The same is true on C' in the second quadrant, but the hypothesis $r > 0$ is needed. Since $0 < \alpha n < \pi$, $f_m(z)$ tends to 0 boundedly on that part of C' in the lower half-plane.

Corollary (i). *Let $\alpha, \beta > 0$ with $\alpha\beta = \pi^2$. Then*

$$\alpha \sum_{k=1}^{\infty} \frac{k}{e^{2\alpha k} - 1} + \beta \sum_{k=1}^{\infty} \frac{k}{e^{2\beta k} - 1} = \frac{\alpha + \beta}{24} - \frac{1}{4}. \tag{8.3}$$

This entry is really not a corollary of Entry 8(iii); however, a proof can be given along somewhat the same lines.

Formula (8.3) was first established by Schlömilch [1], [2, p. 157]. Other proofs have been given by Malurkar [1], Rao and Ayyar [1], J. Lagrange [1], Grosswald [2], Sitaramachandrarao [2], and the author [6, Proposition 2.11], [2, Eq. (11.7)]. In essence, (8.3) was also established by Hurwitz [1], [2] and Guinand [1], although neither author explicitly states the formula.

Corollary (ii). *Let* $\alpha, \beta > 0$ *with* $\alpha\beta = \pi^2$. *Then*

$$e^{(\alpha-\beta)/12} = \left(\frac{\alpha}{\beta}\right)^{1/4} \prod_{k=1}^{\infty} \frac{1 - e^{-2\alpha k}}{1 - e^{-2\beta k}}.$$

PROOF. Let $u, v > 0$ with $uv = \pi^2$. Write Corollary (i) in the form

$$\sum_{k=1}^{\infty} \frac{ke^{-2uk}}{1 - e^{-2uk}} + \frac{v}{u} \sum_{k=1}^{\infty} \frac{ke^{-2vk}}{1 - e^{-2vk}} = \frac{1}{24} + \frac{v/u}{24} - \frac{1}{4u}.$$

Integrate both sides of the equality above with respect to u over the interval $[\pi, \alpha]$ to obtain

$$\frac{1}{2} \sum_{k=1}^{\infty} \text{Log} \frac{1 - e^{-2\alpha k}}{1 - e^{-2\pi k}} + \sum_{k=1}^{\infty} k \int_{\pi}^{\alpha} \frac{e^{-2vk}(v/u)}{1 - e^{-2vk}} du$$

$$= \frac{\alpha - \pi}{24} + \frac{1}{24} \int_{\pi}^{\alpha} (v/u)\, du + \tfrac{1}{4} \text{Log}(\pi/\alpha).$$

In the integrals that remain, make the change of variable $u = \pi^2/v$. By the hypothesis, the limits π and α are transformed into π and β, respectively. Thus, the last equality becomes

$$\frac{1}{2} \sum_{k=1}^{\infty} \text{Log} \frac{1 - e^{-2\alpha k}}{1 - e^{-2\beta k}} = \frac{\alpha - \beta}{24} + \tfrac{1}{8} \text{Log}(\beta/\alpha).$$

Multiplying both sides by 2 and then exponentiating both sides yields the desired result. □

Example. *We have*

$$\sum_{k=1}^{\infty} \frac{k}{e^{2\pi k} - 1} = \frac{1}{24} - \frac{1}{8\pi}. \tag{8.4}$$

This example is obtained from Corollary (i) by setting $\alpha = \beta = \pi$. Ramanujan stated (8.4) as a problem in [3], [16, p. 326]. He later gave a proof of (8.4) in [7, p. 361], [16, p. 34] by using some formulas from the theory of elliptic functions. But, as already indicated, (8.4) was first established by Schlömilch [1], [2, p. 157]. Proofs of (8.4) have also been given by Krishnamachari [1], Watson [1], Sandham [1], Lewittes [1], [2], and Ling [3], in addition to the authors listed after Corollary (i).

Entry 9(i). *Let $\alpha, \beta > 0$ with $\alpha\beta = \pi/2$. Let $h > 0$ be chosen so that $h/\alpha > 1$ and h/α is not an odd integer. Let m be the greatest odd integer that is less than h/α. Let n be an arbitrary real number. Let $\varphi(x)$ be continuous and of bounded variation on $[0, h]$ and define*

$$\psi(t) = \int_0^h \varphi(x) \cos(tx) \, dx. \tag{9.1}$$

If χ is defined by (0.1), then

$$\alpha \sum_{k=1}^m \chi(k) \sin(\alpha nk)\varphi(\alpha k) = \frac{1}{2} \sum_{k=1}^\infty \chi(k)\{\psi(\beta k - n) - \psi(\beta k + n)\}.$$

PROOF. Let f be a continuous function of bounded variation on $[a, b]$. Then the Poisson formula for sine transforms (Titchmarsh [2, p. 66])

$$\sideset{}{'}\sum_{a \le k \le b} \chi(k)f(k) = \sum_{k=1}^\infty \chi(k) \int_a^b f(x) \sin(\pi kx/2) \, dx \tag{9.2}$$

is valid, where the prime on the summation sign on the left side has the same meaning as in (6.1). Let $f(x) = \sin(\alpha nx)\varphi(\alpha x)$, $a = 0$, and $b = h/\alpha$. Then

$$\sum_{k=1}^m \chi(k) \sin(\alpha nk)\varphi(\alpha k) = \sum_{k=1}^\infty \chi(k) \int_0^{h/\alpha} \sin(\alpha nx)\varphi(\alpha x) \sin(\pi kx/2) \, dx. \tag{9.3}$$

The integrals on the right side of (9.3) are easily calculated by (9.1) to complete the proof. $\qquad\square$

Entry 9(ii). *Let $\alpha, \beta, h, m, n,$ and φ satisfy the same hypotheses as in Entry 9(i). Define*

$$\psi(t) = \int_0^h \varphi(x) \sin(tx) \, dx.$$

Then

$$\alpha \sum_{k=1}^m \chi(k) \cos(\alpha nk)\varphi(\alpha k) = \frac{1}{2} \sum_{k=1}^\infty \chi(k)\{\psi(\beta k - n) + \psi(\beta k + n)\}.$$

PROOF. The proof is completely analogous to that for Entry 9(i). $\qquad\square$

Ramanujan stated Entries 9(i) and 9(ii) with the extra condition $|n| < \beta$, but this hypothesis does not seem necessary. Entries 9(i) and 9(ii) are analogues of Entries 31(i) and 32(ii) in Chapter 13, respectively.

Entry 10. *Let $\alpha, \beta > 0$ with $\alpha\beta = \pi/4$, and let z be an arbitrary complex number. Then*

$$e^{z^2/4}\sqrt{\alpha} \sum_{k=1}^\infty \chi(k)e^{-\alpha^2 k^2} \sin(\alpha zk) = \sqrt{\beta} \sum_{k=1}^\infty \chi(k)e^{-\beta^2 k^2} \sinh(\beta zk).$$

Entry 10 should be compared with Entry 7.

PROOF. Apply (9.2) with $f(x) = e^{-\alpha^2 x^2} \sin(\alpha z x)$, $a = 0$, and $b = \infty$. By (7.1),

$$\int_0^\infty e^{-\alpha^2 x^2} \sin(\alpha z x) \sin(\pi k x/2)\, dx = \frac{1}{4\alpha}\sqrt{\pi}\left(e^{-(\alpha z + \pi k/2)^2/(4\alpha^2)} - e^{(\alpha z + \pi k/2)^2/(4\alpha^2)}\right)$$

$$= \sqrt{\beta/\alpha}\, e^{-k^2\beta^2 - z^2/4} \sinh(\beta z k).$$

The entry now readily follows. $\qquad\qquad\qquad\qquad\qquad\qquad\qquad\qquad\qquad\qquad$ □

Entry 11. *Let $\alpha, \beta > 0$ with $\alpha\beta = \pi$, and let n be real with $|n| < \beta/2$. Then*

$$\alpha\left\{\tfrac{1}{4}\sec(\alpha n) + \sum_{k=1}^\infty \chi(k)\frac{\cos(\alpha n k)}{e^{\alpha^2 k} - 1}\right\} = \beta\left\{\frac{1}{4} + \frac{1}{2}\sum_{k=1}^\infty \frac{\cosh(2\beta n k)}{\cosh(\beta^2 k)}\right\}. \qquad (11.1)$$

PROOF. We shall use a transformation formula, Theorem 3(i), from the author's paper [4]. Because the statements of the relevant theorem and notation from our paper [4] would require considerable space, we kindly ask the reader to refer to [4].

Let $V(z) = -1/z$, $r_1 = 0$, and $-1 < r_2 = r < 0$. Then $R_1 = r$, $R_2 = 0$, and $\rho = 0$. Also, let $s = -N = 1$. By [4, Eq. (4.5)], we find that

$$f^*(z, 1; 0, r; 1, \mu) = 2\pi i\left\{\frac{1}{8z} + B_1\left(\frac{\mu - r}{4}\right)\right\},$$

where $B_1(x)$ denotes the first Bernoulli polynomial. It follows that

$$\sum_{\mu=0}^3 \chi(\mu) f^*(z, 1; 0, r; 1, \mu) = -\pi i. \qquad (11.2)$$

We next calculate, for $\mathrm{Im}\, z > 0$,

$$H_2(z, 1; \chi; 0, r) = \sum_{m=1}^\infty \sum_{k=1}^\infty \chi(k) e^{\pi i k(mz + r)/2} + \sum_{m=1}^\infty \sum_{k=1}^\infty \chi(k) e^{\pi i k(mz - r)/2}$$

$$= 2\sum_{k=1}^\infty \chi(k)\frac{\cos(\pi k r/2)}{e^{-\pi i k z/2} - 1}.$$

Hence,

$$z^{-1}(-2\pi i/4)G(\chi)H_2(-1/z, 1; \chi; 0, r) = \frac{2\pi}{z}\sum_{k=1}^\infty \chi(k)\frac{\cos(\pi k r/2)}{e^{\pi i k/(2z)} - 1}. \qquad (11.3)$$

Next, we calculate, for $\mathrm{Im}\, z > 0$,

$$-2\pi i H_1(z, 1; \chi; r, 0)$$

$$= -2\pi i \sum_{m=1}^\infty \sum_{k=1}^\infty \chi(m) e^{2\pi i k(m+r)z} - 2\pi i \sum_{m=1}^\infty \sum_{k=1}^\infty \chi(m) e^{2\pi i k(m-r)z}$$

$$= -4\pi i \sum_{k=1}^\infty \cos(2\pi k r z) \sum_{j=0}^3 \chi(j) \sum_{m=0}^\infty e^{2\pi i k(4m+j)z}$$

$$= -2\pi i \sum_{k=1}^\infty \frac{\cos(2\pi k r z)}{\cosh(2\pi i k z)}. \qquad (11.4)$$

Lastly, we need to calculate $\mathscr{L}_+(1, \chi, r)$, where

$$\mathscr{L}_+(s, \chi, r) = L(s, \chi, r) - e^{\pi is}L(s, \chi, -r)$$

and where, for Re $s > 0$ and a real,

$$L(s, \chi, a) = \sum_{k > -a} \chi(k)(k + a)^{-s}.$$

Also define, for a, x real and Re $s > 1$,

$$L(s, x, a, \chi) = \sum_{k=0}^{\infty}{}' e^{\pi ikx/2}\chi(k)(k + a)^{-s},$$

where the prime on the summation sign indicates that the possible term $k = -a$ is omitted from the summation. The functions $L(s, \chi, a)$, $\mathscr{L}_+(s, \chi, a)$, and $L(s, x, a, \chi)$ possess analytic continuations into the entire complex s-plane. Now apply the functional equation for $L(s, x, a, \chi)$ (Berndt [3, Theorem 5.1]) to get, for all s,

$$L(1 - s, -r, 0, \chi) = \Gamma(s)(2/\pi)^s(i/2)e^{-\pi is/2}\mathscr{L}_+(s, \chi, r).$$

Hence,

$$\mathscr{L}_+(1, \chi, r) = \pi L(0, -r, 0, \chi). \tag{11.5}$$

Now, for Re $s > 0$,

$$L(s, -r, 0, \chi) = \sum_{k=1}^{\infty} e^{-\pi ikr/2}\chi(k)k^{-s}$$

$$= e^{-\pi ir/2} \sum_{k=0}^{\infty} e^{-2\pi ikr}(4k + 1)^{-s} - e^{-3\pi ir/2} \sum_{k=0}^{\infty} e^{-2\pi ikr}(4k + 3)^{-s}$$

$$= e^{-\pi ir/2}4^{-s}\varphi(-r, \tfrac{1}{4}, s) - e^{-3\pi ir/2}4^{-s}\varphi(-r, \tfrac{3}{4}, s), \tag{11.6}$$

where, for x, a real and Re $s > 1$,

$$\varphi(x, a, s) = \sum_{k=0}^{\infty}{}' e^{2\pi ikx}(k + a)^{-s}$$

denotes Lerch's zeta-function. By analytic continuation, the extreme left and right sides of (11.6) are equal for all s. Now from Apostol's paper [2, p. 164],

$$\varphi(x, a, 0) = \frac{i}{2}\cot(\pi x) + \frac{1}{2}. \tag{11.7}$$

Hence, from (11.5)–(11.7),

$$z^{-1}\mathscr{L}_+(1, \chi, r) = \frac{\pi}{z}\left(e^{-\pi ir/2}\left\{\frac{1}{2} - \frac{i}{2}\cot(\pi r)\right\} - e^{-3\pi ir/2}\left\{\frac{1}{2} - \frac{i}{2}\cot(\pi r)\right\}\right)$$

$$= \frac{\pi}{z}e^{-\pi ir}\sin(\pi r/2)\{\cot(\pi r) + i\}$$

$$= \frac{\pi}{2z}\sec(\pi r/2). \tag{11.8}$$

Substitute (11.2), (11.3), (11.4), and (11.8) into Eq. (4.6) of our paper [4] to get

$$\frac{2\pi}{z} \sum_{k=1}^{\infty} \chi(k) \frac{\cos(\pi k r/2)}{e^{\pi i k/(2z)} - 1} + \frac{\pi}{2z} \sec(\pi r/2) = -2\pi i \sum_{k=1}^{\infty} \frac{\cos(2\pi k r z)}{\cosh(2\pi i k z)} - \pi i,$$

where Im $z > 0$ and $0 < -r < 1$. Now let $z = i\pi/(2\alpha^2)$ and $r = 2n/\beta$, where $\alpha\beta = \pi$. Thus, $0 < -n < \beta/2$. Hence,

$$-4\alpha^2 i \sum_{k=1}^{\infty} \chi(k) \frac{\cos(\alpha n k)}{e^{\alpha^2 k} - 1} - \alpha^2 i \sec(\alpha n) = -2\pi i \sum_{k=1}^{\infty} \frac{\cosh(2\beta n k)}{\cosh(\beta^2 k)} - \pi i.$$

Multiplying the last formula by $i/(4\alpha)$ yields (11.1). Now note that both sides of (11.1) are even functions of n. Thus, (11.1) is valid for $0 < |n| < \beta/2$ and, hence, by continuity, for $|n| < \beta/2$.   □

We remark that the differentiation of (11.1) with respect to n yields the last formula in our paper [2] after suitable redefinitions of the parameters. However, it appears to be difficult to deduce (11.1) from the latter formula.

Entry 12. *Let* $\alpha, \beta > 0$ *with* $\alpha\beta = \pi/2$, *and let* $0 < n < \pi/(2\alpha)$. *Then*

$$\alpha \sum_{k=1}^{\infty} \chi(k) \frac{\sin(\alpha n k)}{\cosh(\alpha^2 k)} = \beta \sum_{k=1}^{\infty} \chi(k) \frac{\sinh(\beta n k)}{\cosh(\beta^2 k)}. \tag{12.1}$$

PROOF. In our paper [6, Eq. (4.23)] we showed that if $0 < r < 1$ and $\alpha, \beta > 0$ with $\alpha\beta = \pi^2/16$, then

$$\sqrt{\alpha} \sum_{k=1}^{\infty} \chi(k) \frac{\sin(\pi r k/2)}{\cosh(2\alpha k)} = \sqrt{\beta} \sum_{k=1}^{\infty} \chi(k) \frac{\sinh(2\beta r k)}{\cosh(2\beta k)}. \tag{12.2}$$

Replace α by $\alpha^2/2$ and β by $\beta^2/2$; hence, in the new notation $\alpha\beta = \pi/2$. Let $r = 2\alpha n/\pi$. Thus, we need $0 < n < \pi/(2\alpha)$. With these substitutions, we easily find that (12.2) is transformed into (12.1).   □

Corollary. *Let* $\alpha, \beta, t > 0$ *with* $\alpha\beta = \pi/2$ *and* $t = \alpha/\beta$. *Let* C *be the positively oriented parallelogram with vertices* $\pm i$ *and* $\pm t$. *Let* $\varphi(z)$ *be entire. For each positive integer* N, *define*

$$f_N(z) = \frac{\varphi(4\beta N z)}{\cosh(2\pi N z)\cosh(2\pi i N z/t)},$$

and assume that $N f_N(z)$ *tends to* 0 *boundedly on* C *as* N *tends to* ∞. *Then*

$$\alpha \sum_{k=1}^{\infty} \chi(k) \frac{\{\varphi(\alpha k) - \varphi(-\alpha k)\}}{\cosh(\alpha^2 k)} + i\beta \sum_{k=1}^{\infty} \chi(k) \frac{\{\varphi(i\beta k) - \varphi(-i\beta k)\}}{\cosh(\beta^2 k)} = 0. \tag{12.3}$$

The above entry is not a corollary of Entry 12. In fact, as we shall see later, the converse is true. As with Entry 8(iii), at the expense of brevity, the hypotheses on $f_N(z)$ can be made more explicit.

PROOF. We integrate $f_N(z)$ over C. On the interior of C, $f_N(z)$ has simple poles at $z = i(2k + 1)/(4N)$ and at $z = (2k + 1)t/(4N)$, $-2N \leq k < 2N$. Straightforward calculations give

$$R(i(2k + 1)/(4N)) = \frac{(-1)^k \varphi(i\beta(2k + 1))}{2\pi i N \cosh\{(2k + 1)\pi/(2t)\}}$$

and

$$R((2k + 1)t/(4N)) = -\frac{(-1)^k t \varphi(2k + 1))}{2\pi N \cosh\{(2k + 1)\pi t/2\}}.$$

Applying the residue theorem and letting N tend to ∞, we find that

$$\lim_{N \to \infty} N \int_C f_N(z)\, dz = \sum_{k=0}^{\infty} \frac{(-1)^k \{\varphi(i\beta(2k + 1)) - \varphi(-i\beta(2k + 1))\}}{\cosh\{(2k + 1)\pi t/2\}}$$
$$- it \sum_{k=0}^{\infty} \frac{(-1)^k \{\varphi(\beta t(2k + 1)) - \varphi(-\beta t(2k + 1))\}}{\cosh\{(2k + 1)\pi t/2\}}.$$
$$(12.4)$$

Putting $t = \alpha/\beta$ in (12.4) and using the fact that $Nf_N(z)$ tends to 0 boundedly on C, we readily deduce (12.3). $\qquad\square$

Next, we show that Entry 12 is a corollary of the preceding entry. Let $\varphi(z) = e^{inz}$, where $n > 0$. We see at once that (12.3) then reduces to (12.1). It is easily seen that the hypotheses on $f_N(z)$ are satisfied on the two sides of C in the upper half-plane. In the lower half-plane on C, $Nf_N(z)$ tends to 0 boundedly if and only if $n < \pi/(2\alpha)$, which is precisely a hypothesis of Entry 12.

Entry 13. *Let $\alpha, \beta > 0$ with $\alpha\beta = \pi^2$, and let n be an integer greater than* 1. *Then*

$$\alpha^n \sum_{k=1}^{\infty} \frac{k^{2n-1}}{e^{2\alpha k} - 1} - (-\beta)^n \sum_{k=1}^{\infty} \frac{k^{2n-1}}{e^{2\beta k} - 1} = \{\alpha^n - (-\beta)^n\} \frac{B_{2n}}{4n}.$$

Entry 13 is stated without proof by Ramanujan in [13, p. 269], [16, p. 190]. The first published proof known to the author is by Rao and Ayyar [1]. Malurkar [1] and Hardy [3], [7, pp. 537–539] gave proofs shortly afterward. Later proofs were found by Nanjundiah [1], J. Lagrange [1], Grosswald [2], Sitaramachandrarao [2], and the author [2, Eq. (11.10)], [6, Proposition 2.6].

Corollary (i). $\displaystyle \sum_{k=1}^{\infty} \frac{k^5}{e^{2\pi k} - 1} = \frac{1}{504}.$

Corollary (ii). $\displaystyle \sum_{k=1}^{\infty} \frac{k^9}{e^{2\pi k} - 1} = \frac{1}{264}.$

Corollary (iii). $\displaystyle \sum_{k=1}^{\infty} \frac{k^{13}}{e^{2\pi k} - 1} = \frac{1}{24}.$

Corollary (iv). *If* n *is a positive integer, then*

$$\sum_{k=1}^{\infty} \frac{k^{4n+1}}{e^{2\pi k} - 1} = \frac{B_{4n+2}}{8n + 4}. \tag{13.1}$$

If $\alpha = \beta = \pi$ and n is odd, then Entry 13 reduces to (13.1) if n is replaced by $2n + 1$. Corollaries (i)–(iii) are special instances of Corollary (iv). Corollary (iii) was communicated by Ramanujan in a letter to Hardy [16, p. xxvi]. Sandham [1] also proved this special case. M. V. Aiyar [1] and Ling [3] established Corollaries (i)–(iii). The more general Corollary (iv) was actually first proved earlier by Glaisher [3] in 1889. In addition to the authors who have proved Entry 13, Corollary (iv) has also been established by Krishnamachari [1], Watson [1], Sandham [2], and Zucker [1].

As usual, let $\sigma_\nu(n) = \sum_{d|n} d^\nu$. It is easy to show that

$$\sum_{k=1}^{\infty} \sigma_\nu(k) e^{-ky} = \sum_{d=1}^{\infty} \frac{d^\nu}{e^{dy} - 1}, \tag{13.2}$$

where $y > 0$. Thus, Entry 13 may be rewritten in terms of the left side of (13.2). In this form, Entry 13 was established in Hurwitz's thesis [1], [2] in 1881 and may be even older than 1881. Later proofs were found by Koshliakov [1], Guinand [1], and Chandrasekharan and Narasimhan [1].

Entry 14. *Let* $\alpha, \beta > 0$ *with* $\alpha\beta = \pi^2$, *and let* n *be a positive integer. Then*

$$\alpha^n \sum_{k=1}^{\infty} \chi(k) \frac{k^{2n-1}}{\cosh(\alpha k/2)} + (-\beta)^n \sum_{k=1}^{\infty} \chi(k) \frac{k^{2n-1}}{\cosh(\beta k/2)} = 0. \tag{14.1}$$

Entry 14 has been established by Malurkar [1], Nanjundiah [1], and the author [6, Proposition 4.7].

Corollary of Entry 14. *If* n *is a positive integer, then*

$$\sum_{k=0}^{\infty} \frac{(-1)^k (2k + 1)^{4n-1}}{\cosh\{(2k + 1)\pi/2\}} = 0. \tag{14.2}$$

If $\alpha = \beta = \pi$ and n is even in (14.1), then (14.1) reduces to (14.2) upon the replacement of n by $2n$.

This corollary was, in fact, first established by Cauchy [1, pp. 313, 362]. Ramanujan stated (14.2) as a problem in [2]. In addition to the authors who have proved Entry 14, (14.2) has been established by Rao and Ayyar [2], Chowla [1], Sandham [2], Riesel [1], and Ling [3].

Entry 15. *Let* $\alpha, \beta > 0$ *with* $\alpha\beta = \pi^2/4$. *Then*

$$2 \sum_{n=1}^{\infty} \chi(n) \tan^{-1}(e^{-\alpha n}) + 2 \sum_{n=1}^{\infty} \chi(n) \tan^{-1}(e^{-\beta n})$$

$$= \sum_{n=1}^{\infty} \chi(n) \frac{\operatorname{sech}(\alpha n)}{n} + \sum_{n=1}^{\infty} \chi(n) \frac{\operatorname{sech}(\beta n)}{n} = \frac{\pi}{4}. \tag{15.1}$$

PROOF. A proof of the rightmost equality in (15.1) has been given by Malurkar [1], Nanjundiah [1], and the author [6, Proposition 4.5].

The leftmost equality in (15.1) follows from

$$\sum_{n=1}^{\infty} \chi(n) \frac{\text{sech}(ny)}{n} = 2 \sum_{n=1}^{\infty} \frac{\chi(n)}{n} e^{-ny} \sum_{k=0}^{\infty} (-1)^k e^{-2nky}$$

$$= 2 \sum_{k=1}^{\infty} \sum_{n=1}^{\infty} \frac{\chi(nk)}{n} e^{-nky}$$

$$= 2 \sum_{k=1}^{\infty} \chi(k) \tan^{-1}(e^{-ky}),$$

where $y > 0$. □

Corollary. *We have*

$$\sum_{n=1}^{\infty} \chi(n) \tan^{-1}(e^{-\pi n/2}) = \pi/16.$$

The corollary follows trivially from (15.1) upon setting $\alpha = \beta = \pi/2$. Rao and Ayyar [2] have also established this result. Chowla [1] has proved some formulas similar in appearance to (15.1).

Entry 16(i). *Let m and n be nonnegative integers. Then*

$$\int_0^{\infty} \frac{\sin^{2n+1}x}{x} \cos^{2m} x \, dx = \frac{\Gamma(m + \frac{1}{2})\Gamma(n + \frac{1}{2})}{2\Gamma(m + n + 1)}$$

$$= \int_0^{\infty} \frac{\sin^{2n+2} x}{x^2} \cos^{2m} x \, dx.$$

Readers should compare the formulas for $m = 0$ with Entries 5(i), (ii).

PROOF. The first equality can be found in Gradshteyn and Ryzhik's tables [1, p. 457], but since the second is not in [1], we give a brief proof. (A proof of the first equality can, in fact, be given along the same lines.) Let the integral on the right side above be denoted by $I(m, n)$. We induct on m. For $m = 0$,

$$I(0, n) = \frac{\Gamma(\frac{1}{2})\Gamma(n + \frac{1}{2})}{2\Gamma(n + 1)},$$

by the tables of Gradshteyn and Ryzhik [1, p. 446]. Proceeding by induction, we have

$$I(m, n) = I(m - 1, n) - I(m - 1, n + 1)$$

$$= \frac{\Gamma(m - \frac{1}{2})\Gamma(n + \frac{1}{2})}{2\Gamma(m + n)} - \frac{\Gamma(m - \frac{1}{2})\Gamma(n + \frac{3}{2})}{2\Gamma(m + n + 1)}$$

$$= \frac{\Gamma(m + \frac{1}{2})\Gamma(n + \frac{1}{2})}{2\Gamma(m + n + 1)},$$

and the proof is complete. □

The equalities below with $p = 0$ should be compared with Entries 5(i), (ii).

Entry 16(ii). *Let n and p be nonnegative integers. Then*

$$\int_0^\infty \frac{\sin^{2n+1} x}{x} \cos(2px)\, dx = (-1)^p \frac{\sqrt{\pi}}{2} \frac{\Gamma(n+1)\Gamma(n+\tfrac{1}{2})}{\Gamma(n-p+1)\Gamma(n+p+1)}$$

$$= \int_0^\infty \frac{\sin^{2n+2} x}{x^2} \cos(2px)\, dx.$$

PROOF. We prove the first equality; the proof of the second is virtually the same. Let $I(n, p)$ denote the integral on the left side above. For $p = 0$, the proposed formula is true by Entry 16(i). Thus, we assume that $p > 0$ for the remainder of the proof. We induct on n. For $n = 0$, it is easy to show that $I(0, p) = 0$ (Gradshteyn and Ryzhik [1, p. 414]), which agrees with the proposed result. Using the identities $2 \sin^2 x = 1 - \cos(2x)$ and $2 \cos(2x) \cos(2px) = \cos\{2(p + 1)x\} + \cos\{2(p - 1)x\}$, we find that, by the induction hypothesis,

$$I(n, p) = \tfrac{1}{2} I(n - 1, p) - \tfrac{1}{4} I(n - 1, p + 1) - \tfrac{1}{4} I(n - 1, p - 1)$$

$$= \frac{(-1)^p \sqrt{\pi}}{4} \Gamma(n)\Gamma(n - \tfrac{1}{2}) \left\{ \frac{1}{\Gamma(n - p)\Gamma(n + p)} \right.$$

$$+ \frac{1}{2\Gamma(n - p - 1)\Gamma(n + p + 1)} + \frac{1}{2\Gamma(n - p + 1)\Gamma(n + p - 1)} \right\}$$

$$= \frac{(-1)^p \sqrt{\pi}\, \Gamma(n + 1)\Gamma(n + \tfrac{1}{2})}{2\Gamma(n - p + 1)\Gamma(n + p + 1)},$$

after several applications of the functional equation of $\Gamma(z)$.   □

Entry 17(i). *Let $\alpha, \beta, n > 0$ with $\alpha\beta = 2\pi$. Suppose that $\pi/(2\alpha)$ is not an integer, and let $m = [\pi/(2\alpha)]$. Let p be real. Then*

$$\alpha \left\{ \frac{1}{2} + \sum_{k=1}^m \cos^n(\alpha k) \cos(\alpha p k) \right\}$$

$$= \frac{\pi n!}{2^{n+1}} \left\{ \frac{1}{\Gamma\{\tfrac{1}{2}(n + p) + 1\}\Gamma\{\tfrac{1}{2}(n - p) + 1\}} \right.$$

$$+ \sum_{k=1}^\infty \left(\frac{1}{\Gamma\{\tfrac{1}{2}(n - p + \beta k) + 1\}\Gamma\{\tfrac{1}{2}(n + p - \beta k) + 1\}} \right.$$

$$+ \left. \left. \frac{1}{\Gamma\{\tfrac{1}{2}(n + p + \beta k) + 1\}\Gamma\{\tfrac{1}{2}(n - p - \beta k) + 1\}} \right) \right\}. \qquad (17.1)$$

PROOF. By Stirling's formula, the right side of (17.1) converges absolutely for $n > 0$.

Apply the Poisson formula (6.1) with $f(x) = \cos^n(\alpha x) \cos(\alpha p x)$, $a = 0$, and $b = \pi/(2\alpha)$. After a simple change of variable, we find that

$$\frac{1}{2} + \sum_{k=1}^{m} \cos^n(\alpha k) \cos(\alpha p k) = \frac{1}{\alpha} \int_0^{\pi/2} \cos^n t \cos(pt) \, dt$$

$$+ \frac{2}{\alpha} \sum_{k=1}^{\infty} \int_0^{\pi/2} \cos^n t \cos(pt) \cos(\beta k t) \, dt. \quad (17.2)$$

Now for $v > 0$ and arbitrary a (Gradshteyn and Ryzhik [1, p. 372]),

$$\int_0^{\pi/2} \cos^{v-1} x \cos(ax) \, dx = \frac{\pi \Gamma(v+1)}{2^v v \Gamma\{\frac{1}{2}(v+a+1)\} \Gamma\{\frac{1}{2}(v-a+1)\}}. \quad (17.3)$$

If we calculate all the integrals in (17.2) with the aid of (17.3), we arrive at (17.1) forthwith.   □

Entry 17(ii). *Let $\alpha, \beta, n > 0$ with $\alpha\beta = \pi/2$. Suppose that $\pi/(2\alpha)$ is not an odd integer, and let $m = [\pi/(2\alpha)]$. Let p be real. Then*

$$\alpha \sum_{k=1}^{m} \chi(k) \cos^n(\alpha k) \sin(\alpha p k)$$

$$= \frac{\pi n!}{2^{2n+2}} \sum_{k=1}^{\infty} \chi(k) \left\{ \frac{1}{\Gamma\{\frac{1}{2}(n-p+\beta k)+1\} \Gamma\{\frac{1}{2}(n+p-\beta k)+1\}} \right.$$

$$\left. - \frac{1}{\Gamma\{\frac{1}{2}(n+p+\beta k)+1\} \Gamma\{\frac{1}{2}(n-p-\beta k)+1\}} \right\}. \quad (17.4)$$

PROOF. As before, the series on the right side of (17.4) converges absolutely for $n > 0$.

Apply the Poisson formula for sine transforms (9.2) with $f(x) = \cos^n(\alpha x) \sin(\alpha p x)$, $a = 0$, and $b = \pi/(2\alpha)$. After a simple change of variable, we find that

$$\sum_{k=1}^{m} \chi(k) \cos^n(\alpha k) \sin(\alpha p k) = \frac{1}{\alpha} \sum_{k=1}^{\infty} \chi(k) \int_0^{\pi/2} \cos^n t \sin(pt) \sin(\beta k t) \, dt. \quad (17.5)$$

If we calculate the integrals in (17.5) with the use of (17.3), we deduce (17.4) immediately.   □

Corollary 1. *Let $\alpha = \pi/(n+j)$, where n and j are positive integers of opposite parity. Let $m = [\pi/(2\alpha)]$. Then*

$$\frac{1}{2} + \sum_{k=1}^{m} \cos^{2n}(\alpha k) = \frac{\sqrt{\pi} \Gamma(n+\frac{1}{2})}{2\alpha n!}. \quad (17.6)$$

PROOF. In Entry 17(i) replace n by $2n$ and let $p = 0$. Let $2f_n(\alpha)$ denote the infinite series on the right side of (17.1); that is,

$$f_n(\alpha) = \sum_{k=1}^{\infty} \frac{1}{\Gamma(n+1+k\pi/\alpha)\Gamma(n+1-k\pi/\alpha)}.$$

Since $f_n(\pi/(n + j)) = 0$, we see that (17.1) reduces to

$$\alpha\left\{\frac{1}{2} + \sum_{k=1}^{m} \cos^{2n}(\alpha k)\right\} = \frac{\pi}{2^{2n+1}}\binom{2n}{n},$$

which can be transformed into the desired result by the use of Legendre's duplication formula. $\square$

In fact, Ramanujan claimed that (17.6) is valid for $0 \le \alpha \le \pi/(n + 1)$, that is, $f_n(\alpha) \equiv 0$, $0 \le \alpha \le \pi/(n + 1)$, provided that $\pi/(2\alpha)$ is not an integer. (Of course, for $\alpha = 0$ the result is false.) In general, $f_n(\alpha)$ does not vanish for all α in $(0, \pi/(n + 1))$, as the following counterexample shows.

Let $n = 1$ and put $f(\alpha) \equiv f_1(\alpha)$. Let $\alpha = 2\pi/5 < \pi/2$. Then

$$f(2\pi/5) = \sum_{k=1}^{\infty} \frac{1}{\Gamma(2 + 5k/2)\Gamma(2 - 5k/2)},$$

$$= \sum_{k=1}^{\infty} \frac{1}{(1 - (5k/2)^2)(5k/2)\Gamma(5k/2)\Gamma(1 - 5k/2)}$$

$$= \frac{2}{5\pi}\sum_{k=1}^{\infty} \frac{\sin(5\pi k/2)}{(1 - (5k/2)^2)k}$$

$$= \frac{1}{5\pi}\sum_{k=-\infty}^{\infty} \frac{(-1)^k}{(1 - \{5(2k + 1)/2\}^2)(2k + 1)}.$$

The latter series can be evaluated by the residue theorem. Let

$$h(z) = \frac{\sec(\pi z)}{\{1 - (5z)^2\}z},$$

which has simple poles at $z = 0$, $\pm\frac{1}{5}$, and $(2k + 1)/2$, where k is an integer. Routine calculations give

$$R(0) = 1, \qquad R(\tfrac{1}{5}) = -\tfrac{1}{2}\sec(\pi/5) = R(-\tfrac{1}{5}),$$

and

$$R((2k + 1)/2) = \frac{2(-1)^{k+1}}{\pi(1 - \{5(2k + 1)/2\}^2)(2k + 1)}.$$

Integrate $h(z)$ over a positively oriented square C_n with center at the origin and horizontal and vertical sides of length $2n$, where n is a positive integer. As n tends to ∞,

$$\int_{C_n} h(z)\, dz = o(1).$$

Hence, applying the residue theorem and then letting n tend to ∞, we find that

$$f(2\pi/5) = \tfrac{1}{10}(1 - \sec(\pi/5)) \ne 0,$$

which disproves Ramanujan's claim.

Corollary 2. *Let* $\alpha = \pi/(n - j)$, *where n and j are integers of opposite parity such that* $n > 0$ *and* $0 \le j \le (n - 1)/2$. *Let* $m = [\pi/(2\alpha)]$. *Then*

$$\alpha\left\{\frac{1}{2} + \sum_{k=1}^{m} \cos^{2n}(\alpha k)\right\} = \frac{\sqrt{\pi}\,\Gamma(n + \tfrac{1}{2})}{2n!}\left\{1 + \frac{2(n!)^2}{\Gamma(n + 1 + \pi/\alpha)\Gamma(n + 1 - \pi/\alpha)}\right\}.$$
$$(17.7)$$

PROOF. In Entry 17(i) replace n by $2n$ and let $p = 0$. After some manipulation, we find that

$$\alpha\left\{\frac{1}{2} + \sum_{k=1}^{m} \cos^{2n}(\alpha k)\right\}$$

$$= \frac{\sqrt{\pi}\,\Gamma(n + \tfrac{1}{2})}{2n!}\left\{1 + \frac{2(n!)^2}{\Gamma(n + 1 + \pi/\alpha)\Gamma(n + 1 - \pi/\alpha)} + 2(n!)^2 g_n(\alpha)\right\},$$

where

$$g_n(\alpha) = \sum_{k=2}^{\infty} \frac{1}{\Gamma(n + 1 + k\pi/\alpha)\Gamma(n + 1 - k\pi/\alpha)}.$$

For $\alpha = \pi/(n - j)$, $0 \le j \le (n - 1)/2$, $g_n(\alpha) = 0$, and so the proof is complete. $\square$

Ramanujan, in fact, claimed that (17.7) is true for $\pi/n \le \alpha \le 2\pi/(n + 1)$, that is, $g_n(\alpha) \equiv 0$, $\pi/n \le \alpha \le 2\pi/(n + 1)$, provided that $\pi/(2\alpha)$ is not an integer. Again, this claim is false, in general, and we give a counterexample.

Let $n = 3$ and put $\alpha = 2\pi/5$; so $\pi/3 < \alpha < \pi/2$. Then

$$g_3(2\pi/5) = \sum_{k=2}^{\infty} \frac{1}{\Gamma(4 + 5k/2)\Gamma(4 - 5k/2)}$$

$$= \frac{2}{5\pi} \sum_{k=1}^{\infty} \frac{(-1)^k}{P((2k + 1)/2)(2k + 1)},$$

where $P(z) = (9 - 25z^2)(4 - 25z^2)(1 - 25z^2)$. This series can be evaluated by the same method as used in the previous counterexample. Accordingly, we find that

$$g_3(2\pi/5) = \frac{1}{10}\left\{\frac{4 - \sqrt{5}}{45} - \frac{4^4}{2079\pi}\right\} \neq 0,$$

which disproves Ramanujan's claim.

Entry 18. *Let* a_n, b_n, p_n, q_n, P_n, *and* Q_n *be complex numbers with* $a_n b_n \neq 0$. *Let x and y be complex variables with* $xy \neq 0$. *Let*

$$\varphi(x) = \sum_n \frac{P_n}{p_n - a_n x} \quad and \quad \psi(y) = \sum_n \frac{Q_n}{q_n - b_n y}.$$

Then

$$\varphi(x)\psi(y) = \sum_n \frac{P_n}{p_n - a_n x} \psi\left(\frac{p_n y}{a_n x}\right) + \sum_n \frac{Q_n}{q_n - b_n y} \varphi\left(\frac{q_n x}{b_n y}\right), \qquad (18.1)$$

where it is assumed that at least one of the two double series on the right side of (18.1) converges absolutely.

PROOF. Without loss of generality, assume that the latter double series on the right side of (18.1) converges absolutely. Inverting the order of summation below by absolute convergence, we have

$$\sum_n \frac{P_n}{p_n - a_n x} \psi\left(\frac{p_n y}{a_n x}\right) + \sum_k \frac{Q_k}{q_k - b_k y} \varphi\left(\frac{q_k x}{b_k y}\right)$$

$$= \sum_n \frac{P_n}{p_n - a_n x} \sum_k \frac{Q_k a_n x}{a_n q_k x - b_k p_n y} + \sum_k \frac{Q_k}{q_k - b_k y} \sum_n \frac{P_n b_k y}{b_k p_n y - a_n q_k x}$$

$$= \sum_n \frac{P_n}{p_n - a_n x} \sum_k \left\{ \frac{Q_k a_n x}{a_n q_k x - b_k p_n y} + \frac{Q_k b_k y(p_n - a_n x)}{(q_k - b_k y)(b_k p_n y - a_n q_k x)} \right\}$$

$$= \sum_n \frac{P_n}{p_n - a_n x} \sum_k \frac{Q_k}{q_k - b_k y}$$

$$= \varphi(x)\psi(y). \qquad \square$$

Despite the simplicity of the result above, Ramanujan found many interesting applications of it, as we shall see in the sequel. However, each of the following corollaries may be alternatively established by using partial fraction decompositions directly and not employing Entry 18. The following entries are valid except for obvious singularities which we shall not state.

Corollary 1. *Let θ and φ be real with $|\theta|, |\varphi| < \pi$. Then for n, x, and y complex, with x/y not purely imaginary,*

$$\pi^2 n^2 xy \frac{\cos(\theta nx) \cosh(\varphi ny)}{\sin(\pi nx) \sinh(\pi ny)} = 1 + 2\pi n^2 xy \sum_{k=1}^{\infty} \frac{(-1)^k k \cos(k\varphi) \cosh(k\theta x/y)}{(k^2 + n^2 y^2) \sinh(\pi kx/y)}$$

$$- 2\pi n^2 xy \sum_{k=1}^{\infty} \frac{(-1)^k k \cos(k\theta) \cosh(k\varphi y/x)}{(k^2 - n^2 x^2) \sinh(\pi ky/x)}.$$

PROOF. For $|\theta| \le \pi$ (Knopp [1, p. 377]),

$$\frac{\pi nx \cos(\theta nx)}{\sin(\pi nx)} = 1 + n^2 x^2 \sum_{\substack{k=-\infty \\ k \ne 0}}^{\infty} \frac{(-1)^k \cos(k\theta)}{k(nx - k)}.$$

Similarly, for $|\varphi| \le \pi$,

$$\frac{\pi ny \cosh(\varphi ny)}{\sinh(\pi ny)} = \frac{i\pi ny \cos(i\varphi ny)}{\sin(i\pi ny)} = 1 - n^2 y^2 \sum_{\substack{k=-\infty \\ k \ne 0}}^{\infty} \frac{(-1)^k \cos(k\varphi)}{k(iny - k)}.$$

Define the functions φ, f, ψ, and g by

$$\varphi(x) = \frac{\pi nx \, \cos(\theta nx)}{\sin(\pi nx)} - 1 = f(x) - 1$$

and

$$\psi(y) = \frac{\pi ny \, \cosh(\varphi ny)}{\sinh(\pi ny)} - 1 = g(y) - 1.$$

Thus, in the notation of Entry 18, $P_k = n^2 x^2 (-1)^k \cos(k\theta)$, $p_k = -k^2$, $a_k = -kn$, $Q_k = -n^2 y^2 (-1)^k \cos(k\varphi)$, $q_k = -k^2$, and $b_k = -ikn$. Applying Entry 18, we find that, for $|\theta|$, $|\varphi| < \pi$ and y/x not purely imaginary,

$$\varphi(x)\psi(y) = \frac{\pi^2 n^2 xy \, \cos(\theta nx) \, \cosh(\varphi ny)}{\sin(\pi nx) \sinh(\pi ny)} - f(x) - g(y) + 1$$

$$= -f(x) + 1 - g(y) + 1 + \pi n^2 xy \sum_{\substack{k=-\infty \\ k \neq 0}}^{\infty} \frac{(-1)^k k \, \cos(k\theta) \, \cosh(k\varphi y/x)}{(knx - k^2) \sinh(\pi ky/x)}$$

$$- \pi n^2 xy \sum_{\substack{k=-\infty \\ k \neq 0}}^{\infty} \frac{(-1)^k k \, \cos(k\varphi) \, \cosh(k\theta x/y)}{(kiny - k^2) \sinh(\pi kx/y)},$$

which yields the desired result after some simplification. □

Corollary 2. *Let θ and φ be real with $|\theta|$, $|\varphi| \leq \pi/2$. Let n, x, and y be complex with y/x not purely imaginary. Then*

$$\frac{\pi \, \sin(\theta nx) \, \sinh(\varphi ny)}{4n^2 \, \cos(\pi nx/2) \, \cosh(\pi ny/2)} = y^2 \sum_{k=1}^{\infty} \frac{\chi(k) \, \sin(k\varphi) \, \sinh(k\theta x/y)}{k(k^2 + n^2 y^2) \, \cosh(\pi kx/(2y))}$$

$$+ x^2 \sum_{k=1}^{\infty} \frac{\chi(k) \, \sin(k\theta) \, \sinh(k\varphi y/x)}{k(k^2 - n^2 x^2) \, \cosh(\pi ky/(2x))}. \quad (18.2)$$

PROOF. The set of functions $\sin\{(2k+1)\theta\}$, $0 \leq k < \infty$, is orthogonal and complete on $[-\pi/2, \pi/2]$. An elementary calculation gives the Fourier series of $\sin(\theta nx)$ with respect to this orthogonal set. Accordingly, we find that, for $|\theta| < \pi/2$,

$$\varphi(x) := \frac{\sin(\theta nx)}{x \, \cos(\pi nx/2)} = \frac{2}{\pi x} \sum_{k=-\infty}^{\infty} \frac{(-1)^{k+1} \sin\{(2k+1)\theta\}}{nx + 2k + 1}.$$

Similarly, for $|\varphi| < \pi/2$,

$$\varphi(y) := \frac{\sinh(\varphi ny)}{y \, \cosh(\pi ny/2)} = \frac{\sin(i\varphi ny)}{iy \, \cos(i\pi ny/2)} = \frac{2i}{\pi y} \sum_{k=-\infty}^{\infty} \frac{(-1)^k \sin\{(2k+1)\varphi\}}{iny + 2k + 1}.$$

Apply Entry 18 to $\varphi(x)$ and $\psi(y)$ as defined above. Then $P_k = (2/(\pi x)) \times (-1)^{k+1} \sin\{(2k+1)\theta\}$, $p_k = 2k+1$, $a_k = -n$, $Q_k = (2i/(\pi y))(-1)^k \times \sin\{(2k+1)\varphi\}$, $q_k = 2k+1$, and $b_k = -in$. A straightforward application of Entry 18 yields (18.2) for $|\theta|$, $|\varphi| < \pi/2$. By continuity, (18.2) holds for $|\theta|$, $|\varphi| \leq \pi/2$. □

Corollary 3. *Let θ and φ be real with $|\theta|, |\varphi| \leq \pi/2$. Let n, x, and y be complex with y/x not purely imaginary. Then*

$$\frac{\pi \cos(\theta nx) \sinh(\varphi ny)}{4 \sin(\pi nx/2) \cosh(\pi ny/2)}$$

$$= \frac{\varphi y}{2x} + n^2 y^2 \sum_{k=0}^{\infty} \frac{(-1)^{k+1} \sin\{(2k+1)\varphi\} \cosh\{(2k+1)\theta x/y\}}{(2k+1)\{(2k+1)^2 + n^2 y^2\} \sinh\{(2k+1)\pi x/(2y)\}}$$

$$+ n^2 x^2 \sum_{k=1}^{\infty} \frac{(-1)^{k+1} \cos(2k\theta) \sinh(2k\varphi y/x)}{2k\{(2k)^2 - n^2 x^2\} \cosh(\pi k y/x)}. \tag{18.3}$$

PROOF. We first calculate the Fourier series of $\cos(\theta nx)$ with respect to the complete orthogonal set $\cos(2k\theta)$, $0 \leq k < \infty$, on $[-\pi/2, \pi/2]$. Accordingly, we find that

$$\frac{\cos(\theta nx)}{x \sin(\pi nx/2)} = \frac{2}{\pi x} \sum_{k=-\infty}^{\infty} \frac{(-1)^k \cos(2k\theta)}{nx + 2k}.$$

Define $\varphi(x) = \cos(\theta nx)/(x \sin(\pi nx/2)) - g(x)$, where $g(x) = 2/(\pi nx^2)$. Thus, in the notation of Entry 18, $P_k = (2/(\pi x))(-1)^k \cos(2k\theta)$, $p_k = 2k$, and $a_k = -n$, where $k \neq 0$. Let $\psi(y)$ be as in the previous corollary. Thus, by Entry 18, for $|\theta|, |\varphi| < \pi/2$ and y/x not purely imaginary,

$$\frac{\cos(\theta nx) \sinh(\varphi ny)}{xy \sin(\pi nx/2) \cosh(\pi ny/2)} - \psi(y)g(x)$$

$$= f(x, y) + \frac{n}{\pi y} \sum_{\substack{k=-\infty \\ k \neq 0}}^{\infty} \frac{(-1)^k \cos(2k\theta) \sinh(2k\varphi y/x)}{k(nx + 2k) \cosh(\pi k y/x)}$$

$$+ \frac{2in}{\pi x} \sum_{k=-\infty}^{\infty} \frac{(-1)^{k+1} \sin\{(2k+1)\varphi\} \cosh\{(2k+1)\theta x/y\}}{(2k+1)(iny + 2k + 1) \sinh\{(2k+1)\pi x/(2y)\}}, \tag{18.4}$$

where

$$f(x, y) = \frac{4iny}{\pi^2 x^2} \sum_{k=-\infty}^{\infty} \frac{(-1)^k \sin\{(2k+1)\varphi\}}{(iny + 2k + 1)(2k+1)^2}$$

$$= -g(x)\psi(y) + \frac{4iny}{\pi^2 x^2} \sum_{k=-\infty}^{\infty} \frac{(-1)^k \sin\{(2k+1)\varphi\}}{iny + 2k + 1} \left\{ \frac{1}{(2k+1)^2} + \frac{1}{n^2 y^2} \right\}$$

$$= -g(x)\psi(y) + \frac{8}{\pi^2 x^2} \sum_{k=0}^{\infty} \frac{(-1)^k \sin\{(2k+1)\varphi\}}{(2k+1)^2}$$

$$= -g(x)\psi(y) + \frac{2\varphi}{\pi x^2}. \tag{18.5}$$

In this last step, we have used the Fourier series of φ with respect to the complete orthogonal set $\sin\{(2k+1)\varphi\}$, $0 \leq k < \infty$, on $[-\pi/2, \pi/2]$. If we substitute (18.5) into (18.4), we obtain (18.3) for $|\theta|, |\varphi| < \pi/2$, after some simplification. By continuity, (18.3) is valid for $|\theta|, |\varphi| \leq \pi/2$.   □

Entry 19(i). *We have*

$$\pi^2 xy \cot(\pi x) \coth(\pi y)$$

$$= 1 + 2\pi xy \sum_{n=1}^{\infty} \frac{n \coth(\pi n x/y)}{n^2 + y^2} - 2\pi xy \sum_{n=1}^{\infty} \frac{n \coth(\pi n y/x)}{n^2 - x^2}. \tag{19.1}$$

We have stated Entry 19(i) with no hypotheses because, in general, the two series on the right side of (19.1) do not converge. Ramanujan evidently used Entry 18 to derive Entry 19(i), and so we formally derive Entry 19(i) in this way. From (1.9), we have

$$\pi x \cot(\pi x) = 1 + x^2 \sum_{\substack{n=-\infty \\ n \neq 0}}^{\infty} \frac{1}{nx - n^2} \tag{19.2}$$

and

$$\pi y \coth(\pi y) = 1 + y^2 \sum_{\substack{n=-\infty \\ n \neq 0}}^{\infty} \frac{1}{n^2 - iny}.$$

Apply Entry 18 to $\varphi(x) = \pi x \cot(\pi x) - 1$ and $\psi(y) = \pi y \coth(\pi y) - 1$. Ignoring the fact that the resulting two series on the right side of (18.1) diverge, we arrive at (19.1) quite easily.

R. Sitaramachandrarao [1], [2] has found a corrected version of Entry 19(i), namely,

$$\pi^2 xy \cot(\pi x) \coth(\pi y) = 1 + \frac{\pi^2}{3}(y^2 - x^2) - 2\pi xy^3 \sum_{n=1}^{\infty} \frac{\coth(\pi n x/y)}{n(n^2 + y^2)}$$

$$- 2\pi x^3 y \sum_{n=1}^{\infty} \frac{\coth(\pi n y/x)}{n(n^2 - x^2)}. \tag{19.3}$$

We give Sitaramachandrarao's proof. From (1.9),

$$\pi^2 xy \cot(\pi x) \coth(\pi y)$$

$$= \left(1 + 2x^2 \sum_{n=1}^{\infty} \frac{1}{x^2 - n^2}\right)\left(1 + 2y^2 \sum_{m=1}^{\infty} \frac{1}{y^2 + m^2}\right)$$

$$= 1 + 2 \sum_{n=1}^{\infty} \left(\frac{x^2}{x^2 - n^2} + \frac{y^2}{y^2 + n^2}\right)$$

$$+ 4x^2 y^2 \sum_{m,n=1}^{\infty} \frac{1}{m^2 x^2 + n^2 y^2}\left(\frac{x^2}{x^2 - n^2} - \frac{y^2}{y^2 + m^2}\right)$$

$$= 1 + 2 \sum_{n=1}^{\infty} \left(\frac{x^2}{x^2 - n^2} + \frac{y^2}{y^2 + n^2}\right)$$

$$+ 4y^2 \sum_{n=1}^{\infty} \frac{x^2}{x^2 - n^2}\left(\frac{\pi \coth(\pi n y/x)}{2ny/x} - \frac{1}{2n^2 y^2/x^2}\right)$$

$$- 4x^2 \sum_{m=1}^{\infty} \frac{y^2}{y^2 + m^2}\left(\frac{\pi \coth(\pi m x/y)}{2mx/y} - \frac{1}{2m^2 x^2/y^2}\right)$$

$$= 1 + 2 \sum_{n=1}^{\infty} \left(\frac{x^2}{x^2 - n^2} - \frac{x^4}{n^2(x^2 - n^2)} + \frac{y^2}{y^2 + n^2} + \frac{y^4}{n^2(y^2 + n^2)} \right)$$

$$+ 2\pi x^3 y \sum_{n=1}^{\infty} \frac{\coth(\pi n y/x)}{n(x^2 - n^2)} - 2\pi x y^3 \sum_{n=1}^{\infty} \frac{\coth(\pi n x/y)}{n(y^2 + n^2)}$$

$$= 1 + \frac{\pi^2}{3}(y^2 - x^2) - 2\pi x y^3 \sum_{n=1}^{\infty} \frac{\coth(\pi n x/y)}{n(n^2 + y^2)} - 2\pi x^3 y \sum_{n=1}^{\infty} \frac{\coth(\pi n y/x)}{n(n^2 - x^2)},$$

which completes the proof of (19.3).

Entry 19(ii). *Let x and y be complex numbers such that x/y is not purely imaginary. Then*

$$\pi^2 x y \csc(\pi x) \operatorname{csch}(\pi y)$$

$$= 1 + 2\pi x y \sum_{n=1}^{\infty} \frac{(-1)^n n \operatorname{csch}(\pi n x/y)}{n^2 + y^2} - 2\pi x y \sum_{n=1}^{\infty} \frac{(-1)^n n \operatorname{csch}(\pi n y/x)}{n^2 - x^2}.$$

PROOF. From Whittaker and Watson's text [1, p. 136],

$$\varphi(x) := \pi x \csc(\pi x) - 1 = x^2 \sum_{\substack{n=-\infty \\ n \neq 0}}^{\infty} \frac{(-1)^n}{nx - n^2}$$

and

$$\psi(y) := \pi y \operatorname{csch}(\pi y) - 1 = y^2 \sum_{\substack{n=-\infty \\ n \neq 0}}^{\infty} \frac{(-1)^n}{n^2 - iny}.$$

Apply Entry 18 with $\varphi(x)$ and $\psi(y)$ as defined above. Thus, $P_n = (-1)^n x^2$, $p_n = -n^2$, $a_n = -n$, $Q_n = (-1)^n y^2$, $q_n = n^2$, and $b_n = in$. Hence,

$$\varphi(x)\psi(y) = x^2 \sum_{\substack{n=-\infty \\ n \neq 0}}^{\infty} \frac{(-1)^n}{nx - n^2} \left\{ \frac{\pi n y}{x} \operatorname{csch}\left(\frac{\pi n y}{x} \right) - 1 \right\}$$

$$+ y^2 \sum_{\substack{n=-\infty \\ n \neq 0}}^{\infty} \frac{(-1)^n}{n^2 - iny} \left\{ \frac{\pi i n x}{y} \csc\left(\frac{\pi i n x}{y} \right) - 1 \right\}.$$

The completion of the proof is straightforward, and we omit it. □

Entry 19(iii). *Let x and y be complex numbers such that y/x is not purely imaginary. Then*

$$\frac{\pi}{4} \tan(\pi x/2) \tanh(\pi y/2)$$

$$= y^2 \sum_{n=0}^{\infty} \frac{\tanh\{(2n + 1)\pi x/(2y)\}}{(2n + 1)\{(2n + 1)^2 + y^2\}} + x^2 \sum_{n=0}^{\infty} \frac{\tanh\{(2n + 1)\pi y/(2x)\}}{(2n + 1)\{(2n + 1)^2 - x^2\}}.$$

PROOF. From Gradshteyn and Ryzhik's tables [1, p. 36],

$$\varphi(x) := \frac{1}{x} \tan(\pi x/2) = -\frac{2}{\pi x} \sum_{n=-\infty}^{\infty}{}' \frac{1}{2n + 1 + x}$$

and

$$\psi(y) := \frac{1}{y} \tanh(\pi y/2) = \frac{2i}{\pi y} \sum_{n=-\infty}^{\infty}{}' \frac{1}{2n + 1 + iy},$$

where the prime on the summation sign on each right side above indicates that the sum is to be interpreted as $\lim_{N \to \infty} \sum_{n=-N}^{N}$. Apply Entry 18 to $\varphi(x)$ and $\psi(y)$ as defined above. Thus, $P_n = -2/(\pi x)$, $p_n = 2n + 1$, $a_n = -1$, $Q_n = 2i/(\pi y)$, $q_n = 2n + 1$, and $b_n = -i$. Hence,

$$\varphi(x)\psi(y) = -\frac{2}{\pi y} \sum_{n=-\infty}^{\infty} \frac{\tanh\{(2n + 1)\pi y/(2x)\}}{(2n + 1)(2n + 1 + x)}$$

$$+ \frac{2i}{\pi x} \sum_{n=-\infty}^{\infty} \frac{\tanh\{(2n + 1)\pi x/(2y)\}}{(2n + 1)(2n + 1 + iy)},$$

and, after a little simplification, the desired result follows. □

Entry 19(iv). *Let x and y be complex numbers such that y/x is not purely imaginary. Then*

$$\frac{\pi}{4} \sec(\pi x/2) \operatorname{sech}(\pi y/2)$$

$$= \sum_{n=1}^{\infty} \frac{\chi(n)n \operatorname{sech}\{\pi n x/(2y)\}}{n^2 + y^2} + \sum_{n=1}^{\infty} \frac{\chi(n)n \operatorname{sech}\{\pi n y/(2x)\}}{n^2 - x^2}.$$

PROOF. From (1.2),

$$\varphi(x) := \sec(\pi x/2) = \frac{2}{\pi} \sum_{n=-\infty}^{\infty} \frac{(-1)^n}{2n + 1 + x}$$

and

$$\psi(y) := \operatorname{sech}(\pi y/2) = \frac{2}{\pi} \sum_{n=-\infty}^{\infty} \frac{(-1)^n}{2n + 1 + iy}. \tag{19.4}$$

Apply Entry 18 with φ and ψ defined as above, and we readily obtain the desired result. □

Entry 19(v). *Let x and y be complex numbers such that y/x is not purely imaginary. Then*

$$\frac{\pi}{4} \cot(\pi x/2) \operatorname{sech}(\pi y/2)$$

$$= \frac{1}{2x} - y \sum_{n=1}^{\infty} \frac{\chi(n) \coth\{\pi n x/(2y)\}}{n^2 + y^2} - x \sum_{n=1}^{\infty} \frac{\operatorname{sech}(\pi n y/x)}{(2n)^2 - x^2}.$$

PROOF. From (19.2),

$$\varphi(x) := \cot(\pi x/2) - \frac{2}{\pi x} = \frac{x}{2\pi} \sum_{\substack{n=-\infty \\ n \neq 0}}^{\infty} \frac{1}{nx/2 - n^2}.$$

Apply Entry 18 to $\varphi(x)$ given above and to $\psi(y)$ given by (19.4). Hence,

$$\varphi(x)\psi(y) = \frac{x}{2\pi} \sum_{\substack{n=-\infty \\ n \neq 0}}^{\infty} \frac{\operatorname{sech}(\pi n y/x)}{nx/2 - n^2}$$

$$+ \frac{2}{\pi} \sum_{n=-\infty}^{\infty} \frac{(-1)^n}{2n + 1 + iy} \left\{ \cot\left(\frac{\pi i(2n + 1)x}{2y}\right) + \frac{2iy}{(2n + 1)\pi x} \right\}$$

$$= -\frac{4x}{\pi} \sum_{n=1}^{\infty} \frac{\operatorname{sech}(\pi n y/x)}{(2n)^2 - x^2} - \frac{2}{\pi x} \operatorname{sech}(\pi y/2)$$

$$- \frac{4y}{\pi} \sum_{n=0}^{\infty} \frac{(-1)^n \coth\{(2n + 1)\pi x/(2y)\}}{(2n + 1)^2 + y^2}$$

$$+ \frac{4}{\pi^2 x} \sum_{n=-\infty}^{\infty} \frac{(-1)^n}{2n + 1 + iy} \left\{ 1 + \frac{iy}{2n + 1} \right\}.$$

The last series above reduces to twice Gregory's series for $\pi/4$. Hence, after a little simplification, the formula above reduces to the desired result. $\square$

After Entry 19(v), Ramanujan remarks that similar formulas can be derived for $\tan(\pi x/2) \coth(\pi y/2)$ and $\sec(\pi x/2) \coth(\pi y/2)$.

Entry 20(i). *We have*

$$\pi^2 z^2 \cot(\pi z) \coth(\pi z) = 1 - 4\pi z^4 \sum_{n=1}^{\infty} \frac{n \coth(\pi n)}{n^4 - z^4}.$$

Note that if we set $x = y = z$ in (19.1), we obtain the equality above. However, as previously observed, the two series on the right side of (19.1) do not converge for $x = y$. A correct proof of Entry 20(i) is obtained from setting $x = y = z$ in (19.3).

Corollary. *We have*

$$\pi^2 z^2 \frac{\cosh(\pi z\sqrt{2}) + \cos(\pi z\sqrt{2})}{\cosh(\pi z\sqrt{2}) - \cos(\pi z\sqrt{2})} = 1 + 4\pi z^4 \sum_{n=1}^{\infty} \frac{n \coth(\pi n)}{n^4 + z^4}.$$

PROOF. In Entry 20(i) replace z by $e^{\pi i/4}z$. We see that we must calculate

$$i \cot(\pi e^{\pi i/4}z) \coth(\pi e^{\pi i/4}z) = \frac{\cosh(\pi z(1 - i)/\sqrt{2}) \cosh(\pi z(1 + i)/\sqrt{2})}{\sinh(\pi z(1 - i)/\sqrt{2}) \sinh(\pi z(1 + i)/\sqrt{2})}$$

$$= \frac{\cosh(\pi z\sqrt{2}) + \cos(\pi z\sqrt{2})}{\cosh(\pi z\sqrt{2}) - \cos(\pi z\sqrt{2})}.$$

The desired equality now follows. $\square$

Entry 20(ii). *We have*

$$\pi^2 z^2 \csc(\pi z) \operatorname{csch}(\pi z) = 1 - 4\pi z^4 \sum_{n=1}^{\infty} \frac{(-1)^n n \operatorname{csch}(\pi n)}{n^4 - z^4}.$$

PROOF. Let $x = y = z$ in Entry 19(ii), and the result follows. □

Corollary. *We have*

$$\frac{2\pi^2 z^2}{\cosh(\pi z \sqrt{2}) - \cos(\pi z \sqrt{2})} = 1 + 4\pi z^4 \sum_{n=1}^{\infty} \frac{(-1)^n n \operatorname{csch}(\pi n)}{n^4 + z^4}.$$

PROOF. In Entry 20(ii) replace z be $e^{\pi i/4} z$. Use part of the calculation in the proof of the corollary of Entry 20(i), and the desired result easily follows. □

Entry 20(iii). *We have*

$$\frac{\pi}{8z^2} \tan(\pi z/2) \tanh(\pi z/2) = \sum_{n=0}^{\infty} \frac{(2n+1) \tanh\{(2n+1)\pi/2\}}{(2n+1)^4 - z^4}.$$

PROOF. Put $x = y = z$ in Entry 19(iii), and the result readily follows. □

Corollary. *We have*

$$\frac{\pi}{8z^2} \frac{\cosh(\pi z/\sqrt{2}) - \cos(\pi z/\sqrt{2})}{\cosh(\pi z/\sqrt{2}) + \cos(\pi z/\sqrt{2})} = \sum_{n=0}^{\infty} \frac{(2n+1) \tanh\{(2n+1)\pi/2\}}{(2n+1)^4 + z^4}.$$

PROOF. Replace z by $e^{\pi i/4} z$ in Entry 20(iii). The calculation that is needed is precisely of the same type as that given in the proof of the corollary of Entry 20(i). □

Entry 20(iv). *We have*

$$\frac{\pi}{8} \sec(\pi z/2) \operatorname{sech}(\pi z/2) = \sum_{n=1}^{\infty} \chi(n) \frac{n^3 \operatorname{sech}(\pi n/2)}{n^4 - z^4}.$$

PROOF. Let $x = y = z$ in Entry 19(iv), and the result follows forthwith. □

Corollary. *We have*

$$\frac{\pi/4}{\cosh(\pi z/\sqrt{2}) + \cos(\pi z/\sqrt{2})} = \sum_{n=1}^{\infty} \chi(n) \frac{n^3 \operatorname{sech}(\pi n/2)}{n^4 + z^4}.$$

PROOF. The corollary follows from Entry 20(iv) upon the replacement of z by $e^{\pi i/4} z$ and from the calculation in the proof of Entry 20(i). □

Entry 21(i). *Let $\alpha, \beta > 0$ with $\alpha\beta = \pi^2$, and let n be any nonzero integer. Then*

$$\alpha^{-n}\left\{\tfrac{1}{2}\zeta(2n+1) + \sum_{k=1}^{\infty} \frac{k^{-2n-1}}{e^{2\alpha k}-1}\right\}$$

$$= (-\beta)^{-n}\left\{\tfrac{1}{2}\zeta(2n+1) + \sum_{k=1}^{\infty} \frac{k^{-2n-1}}{e^{2\beta k}-1}\right\}$$

$$- 2^{2n}\sum_{k=0}^{n+1} (-1)^k \frac{B_{2k}}{(2k)!}\frac{B_{2n+2-2k}}{(2n+2-2k)!} \alpha^{n+1-k}\beta^k,$$

where B_j denotes the jth Bernoulli number.

Entry 21(i) is perhaps the most well-known result in Chapter 14. For $\alpha = \beta = \pi$ and n odd and positive, the theorem is first due to Lerch [1]. A proof of the more general Entry 21(i) was first given by Malurkar [1]. Other proofs of the aforementioned special case or of the full result have been given by Grosswald [1], [2], Smart [1], Katayama [1], [4], Riesel [1], S. N. Rao [1], N. Zhang [1] (see also the paper of N. Zhang and S. Zhang [1]), Sitaramachandrarao [2], and the author [5], [6]. Several other authors have established transformation formulas from which Entry 21(i) readily follows. Thus, although Entry 21(i) was not explicitly stated by them, Guinand [1], [2], Apostol [1], Mikolás [1], Iseki [1], Chandrasekharan and Narasimhan [1], Glaeske [1], [2], Bodendiek [1], and Bodendiek and Halbritter [1] have essentially proved Entry 21(i). For a more detailed discussion of this formula, see the author's expository paper [1]. Lastly, note that for $n < -1$, Entry 21(i) yields Entry 13 (with n replaced by $-n$).

Many generalizations of Ramanujan's formula for $\zeta(2n+1)$ have been given. First, analogues have been established for L-functions by Berndt [4], Katayama [2], [3], and Toyoizumi [2], [3]. A special case is Entry 21(iii) below. Other generalizations have been found by Katayama [3], [4], Goldstein and Razar [1], and Nagasaka [1]. Some related formulas have been derived by Terras [1].

Matsuoka instigated a series of papers by himself and Toyoizumi in a different direction. Each [1], [1] first established formulas for $\zeta(s)$ at half-integral arguments. Matsuoka [2] generalized his result for rational arguments. Toyoizumi [2], [3], [4] found some analogous results for L-functions and Dedekind zeta functions attached to imaginary quadratic fields.

Interesting applications of Entry 21(i) and some of its corollaries have been made by P. Kirschenhofer and H. Prodinger [1] to the analysis of special data structures and algorithms.

Entry 21(ii). *Let $\alpha, \beta > 0$ with $\alpha\beta = \pi^2/4$. Let n be any integer. Then*

$$\alpha^{-n}\sum_{k=1}^{\infty} \chi(k)\frac{\operatorname{sech}(\alpha k)}{k^{2n+1}} + (-\beta)^{-n}\sum_{k=1}^{\infty} \chi(k)\frac{\operatorname{sech}(\beta k)}{k^{2n+1}}$$

$$= \frac{\pi}{4}\sum_{k=0}^{n} (-1)^k \frac{E_{2k}}{(2k)!}\frac{E_{2n-2k}}{(2n-2k)!} \alpha^{n-k}\beta^k,$$

where E_j denotes the jth Euler number.

Note that the latter equality in Entry 15 is the case $n = 0$ of Entry 21(ii). Also observe that Entry 21(ii) reduces to Entry 14 when $n < 0$. (The parameters n, α, and β must be replaced by $-n$, $\alpha/2$, and $\beta/2$, respectively, to obtain Entry 14.)

Proofs of Entry 21(ii) have been given first by Malurkar [1] and then by Nanjundiah [1] and the author [6, Proposition 4.5].

For Re $s > 0$, let

$$L(s) = \sum_{n=1}^{\infty} \chi(n) n^{-s}. \tag{21.1}$$

Note that $L(s)$ is the Dirichlet L-function associated with the primitive character χ and so can be analytically continued to an entire function.

Entry 21(iii). *Let α, $\beta > 0$ with $\alpha\beta = \pi^2$, and let n be any integer. Then*

$$\alpha^{-n+1/2}\left\{\tfrac{1}{2}L(2n) + \sum_{k=1}^{\infty} \frac{\chi(k)}{k^{2n}(e^{\alpha k} - 1)}\right\}$$

$$= \frac{(-1)^n \beta^{-n+1/2}}{2^{2n+1}} \sum_{k=1}^{\infty} \frac{1}{k^{2n}\cosh(\beta k)}$$

$$+ \frac{1}{4}\sum_{k=0}^{n} \frac{(-1)^k}{2^{2k}} \frac{E_{2k}}{(2k)!} \frac{B_{2n-2k}}{(2n-2k)!} \alpha^{n-k}\beta^{k+1/2}.$$

The first published proof of Entry 21(iii) was given by Chowla [1, Eq. (1.2)]. The author [6, Eq. (3.20)] has also given a proof. (Unfortunately, formula (3.20) contains an error; replace $(\beta/8)^k$ by $\beta^{k+1/2}2^{-4k}$ at the end of (3.20).) Entry 21(iii) also follows from results of Katayama [2], [3].

Entry 22(i). *Let x and y be complex numbers with y/x not purely imaginary. Then*

$$\pi^2 xy \frac{\cosh\{\pi(x+y)\sqrt{2}\} + \cos\{\pi(x-y)\sqrt{2}\} - \cosh\{\pi(x-y)\sqrt{2}\} - \cos\{\pi(x+y)\sqrt{2}\}}{\{\cosh(\pi x\sqrt{2}) - \cos(\pi x\sqrt{2})\}\{\cosh(\pi y\sqrt{2}) - \cos(\pi y\sqrt{2})\}}$$

$$= 2 + 4\pi xy^3 \sum_{n=1}^{\infty} \frac{n\coth(\pi nx/y)}{n^4 + y^4} + 4\pi x^3 y \sum_{n=1}^{\infty} \frac{n\coth(\pi ny/x)}{n^4 + x^4}. \tag{22.1}$$

PROOF. Let

$$zf(z) = \pi^2 \cot(\pi zx)\coth(\pi zy) \quad \text{and} \quad zg(z) = \pi^2 \cot(\pi zy)\coth(\pi zx).$$

If we expand $f(z)$ and $g(z)$ into partial fractions, we obtain

$$xy\{f(z) + g(z)\} = \frac{2}{z^3} + 4\pi x^3 yz \sum_{n=1}^{\infty} \frac{n\coth(\pi ny/x)}{x^4 z^4 - n^4}$$

$$+ 4\pi xy^3 z \sum_{n=1}^{\infty} \frac{n\coth(\pi nx/y)}{y^4 z^4 - n^4}.$$

If $z = 1$, the equality above becomes

$$\pi^2 xy \{\cot(\pi x) \coth(\pi y) + \cot(\pi y) \coth(\pi x)\}$$

$$= 2 - 4\pi x^3 y \sum_{n=1}^{\infty} \frac{n \coth(\pi n y/x)}{n^4 - x^4} - 4\pi xy^3 \sum_{n=1}^{\infty} \frac{n \coth(\pi n x/y)}{n^4 - y^4}. \quad (22.2)$$

Replace x by $e^{\pi i/4}x$ and y by $e^{\pi i/4}y$ in the formula above. The right side of (22.2) then becomes the right side of (22.1). On the left side of (22.2) we have

$$\pi^2 xy \left\{ \frac{\cosh(a - ia)\cosh(b + ib)}{\sinh(a - ia)\sinh(b + ib)} + \frac{\cosh(b - ib)\cosh(a + ia)}{\sinh(b - ib)\sinh(a + ia)} \right\}$$

$$= \frac{\pi^2 xy \{F(a, b) + F(b, a)\}}{G(a, b)}, \quad (22.3)$$

where $a = \pi x/\sqrt{2}$, $b = \pi y/\sqrt{2}$,

$$F(a, b) = \cosh(a - ia)\sinh(a + ia)\cosh(b + ib)\sinh(b - ib),$$

and

$$G(a, b) = \sinh(a - ia)\sinh(a + ia)\sinh(b - ib)\sinh(b + ib).$$

Now,

$$F(a, b) = \tfrac{1}{4}\{\sinh(2a) + i\sin(2a)\}\{\sinh(2b) - i\sin(2b)\},$$

and so

$$F(a, b) + F(b, a) = \tfrac{1}{2}\{\sinh(2a)\sinh(2b) + \sin(2a)\sin(2b)\}$$

$$= \tfrac{1}{4}(\cosh\{2(a + b)\} - \cosh\{2(a - b)\}$$

$$+ \cos\{2(a - b)\} - \cos\{2(a + b)\}). \quad (22.4)$$

Also,

$$G(a, b) = \tfrac{1}{4}\{\cosh(2a) - \cos(2a)\}\{\cosh(2b) - \cos(2b)\}. \quad (22.5)$$

If we substitute (22.4) and (22.5) into (22.3), we find that (22.3) is transformed into the left side of (22.1). This completes the proof. □

Entry 22(i) in the second notebook is slightly in error. Ramanujan has replaced the numerator on the left side of (22.1) by

$$\cosh\{\pi(x + y)\sqrt{2}\} \cos\{\pi(x - y)\sqrt{2}\}$$

$$- \cosh\{\pi(x - y)\sqrt{2}\} \cos\{\pi(x + y)\sqrt{2}\}.$$

It also may be remarked that *formally* (22.2) can be derived from Entry 19(i).

Entry 22(ii). *Let $n \geq 0$. Then*

$$\int_0^{\infty} \frac{\cos(2nx)\, dx}{\cosh(\pi\sqrt{x}) + \cos(\pi\sqrt{x})} = \sum_{k=1}^{\infty} \frac{\chi(k)k}{\cosh(\pi k/2)} e^{-nk^2}. \quad (22.6)$$

PROOF. Let

$$f(z) = \frac{1}{\cosh(\pi z) + \cos(\pi z)}.$$

We expand f into its partial fraction decomposition. There are simple poles at $z = (2k + 1)(\pm 1 + i)/2, -\infty < k < \infty$. Since

$$R((2k + 1)(1 + i)/2) = -\frac{(-1)^k(1 + i)}{2\pi \cosh\{(2k + 1)\pi/2\}}$$

and

$$R((2k + 1)(-1 + i)/2) = \frac{(-1)^k(1 - i)}{2\pi \cosh\{(2k + 1)\pi/2\}},$$

we readily find that

$$f(z) = \frac{1}{\pi} \sum_{k=0}^{\infty} \frac{(-1)^k(2k + 1)^3}{\cosh\{(2k + 1)\pi/2\}(z^4 + (2k + 1)^4/4)} + g(z), \quad (22.7)$$

where $g(z)$ is entire. By the same argument as that used in the proof of Entry 4, $g(z) \equiv 0$.

Letting $z = \sqrt{x}$, we multiply both sides of (22.7) by $\cos(2nx)$ and integrate with respect to x over $[0, \infty)$. Inverting the order of integration and summation by absolute convergence and using a result from Ramanujan's quarterly reports (Part I [9, p. 322]),

$$\int_0^{\infty} \frac{\cos(ax)\, dx}{x^2 + b^2} = \frac{\pi}{2b} e^{-ab}, \quad a \geq 0, \quad b > 0,$$

we find that

$$\int_0^{\infty} f(x) \cos(2nx)\, dx = \frac{1}{\pi} \sum_{k=1}^{\infty} \frac{\chi(k)k^3}{\cosh(\pi k/2)} \int_0^{\infty} \frac{\cos(2nx)\, dx}{x^2 + k^4/4}$$

$$= \sum_{k=1}^{\infty} \frac{\chi(k)k}{\cosh(\pi k/2)} e^{-nk^2},$$

which completes the proof of (22.6). □

Ramanujan claimed that the next entry is a corollary of Entry 22(ii). We cannot show this and so proceed from scratch.

Corollary. *Let $\alpha, \beta > 0$ with $\alpha\beta = \pi^3/4$. Then*

$$\sum_{n=1}^{\infty} \frac{\chi(n)}{n\{\cosh\sqrt{\alpha n} + \cos\sqrt{\alpha n}\}} + \sum_{n=1}^{\infty} \frac{\chi(n)}{n \cosh(\pi n/2) \cosh^2(\beta n^2)} = \frac{\pi}{8}. \quad (22.8)$$

PROOF. Let N be an even positive integer. We shall let N tend to ∞, but we shall further restrict N by requiring that N^2 remain at a bounded distance from the numbers $(2n + 1)\alpha/\pi^2$, where n is a positive integer. Let

$$f_N(z) = \frac{1}{z\{\cosh(\pi Nz) + \cos(\pi Nz)\} \cos(2\beta N^2 z^2)}.$$

Elementary considerations show that $f(z)$ has simple poles at $z = 0$, at $z = (2n + 1)(\pm 1 + i)/(2N)$, where n is an integer, and at $z = \pm\sqrt{(2k + 1)\alpha/(N\pi)}$, where k is an integer. Straightforward calculations yield $R(0) = \frac{1}{2}$,

$$R((2n + 1)(\pm 1 + i)/(2N)) = \frac{(-1)^{n+1}}{\pi(2n + 1)\cosh\{(2n + 1)\pi/2\}\cosh\{(2n + 1)^2\beta\}},$$

and

$$R(\pm\sqrt{(2k + 1)\alpha}/(N\pi)) = \frac{(-1)^{k+1}}{\pi(2k + 1)\{\cosh\sqrt{(2k + 1)\alpha} + \cos\sqrt{(2k + 1)\alpha}\}}.$$

Let C denote the positively oriented rhombus with vertices ± 1 and $\pm i$. Hence, employing the residue theorem and letting N tend to ∞, we find that

$$\lim_{N\to\infty} \frac{1}{2\pi i} \int_C f_N(z)\, dz$$

$$= \frac{1}{2} + \frac{2}{\pi} \sum_{n=-\infty}^{\infty} \frac{(-1)^{n+1}}{(2n + 1)\cosh\{(2n + 1)\pi/2\}\cosh\{(2n + 1)^2\beta\}}$$

$$+ \frac{2}{\pi} \sum_{k=-\infty}^{\infty} \frac{(-1)^{k+1}}{(2k + 1)\{\cosh\sqrt{(2k + 1)\alpha} + \cos\sqrt{(2k + 1)\alpha}\}}. \quad (22.9)$$

By the definition of f_N and the choice of N, it is easily seen that the limit on the left side of (22.9) is zero. A slight rearrangement of (22.9) yields (22.8), and we are done. $\square$

Entry 22(iii). *Let α, $\beta > 0$ with $\alpha\beta = 4\pi^3$, and let γ denote Euler's constant. Then*

$$\frac{7\alpha}{720} + \frac{1}{2} \sum_{n=1}^{\infty} \frac{\cos\sqrt{\alpha n}}{n(\cosh\sqrt{\alpha n} - \cos\sqrt{\alpha n})}$$

$$= \frac{\gamma + \text{Log}(2\pi/\beta)}{4} + \frac{\beta}{48\pi} + \sum_{n=1}^{\infty} \frac{1}{n(e^{2\pi n} - 1)} + \sum_{n=1}^{\infty} \frac{\coth(\pi n)}{n(e^{\beta n^2} - 1)}. \quad (22.10)$$

Furthermore,

$$\sum_{n=1}^{\infty} \frac{1}{n(e^{2\pi n} - 1)} = \frac{1}{4}\text{Log}(4/\pi) - \frac{\pi}{12} + \text{Log }\Gamma(\tfrac{3}{4}). \quad (22.11)$$

In the notebooks, formula (22.11) contains a misprint; Log $\Gamma(\tfrac{3}{4})$ is replaced by $\frac{1}{4}$ Log $\Gamma(\tfrac{3}{4})$.

Proof. We first prove (22.11). A direct calculation gives

$$\sum_{n=1}^{\infty} \frac{1}{n(e^{2\pi n} - 1)} = -\text{Log} \prod_{n=1}^{\infty} (1 - q^{2n}), \qquad (22.12)$$

where $q = e^{-\pi}$. Now from Whittaker and Watson's text [1, p. 488, problem 10],

$$\prod_{n=1}^{\infty} (1 - q^{2n})^6 = \frac{2kk'K^3}{\pi^3 q^{1/2}}, \qquad (22.13)$$

where k, k', and K have their standard meanings in the theory of elliptic functions. Here, $k = k' = 1/\sqrt{2}$ and $K = \pi^{3/2}/(2\Gamma^2(\frac{3}{4}))$. (See Zucker's paper [1], for example.) Thus, (22.12) and (22.13) yield

$$\sum_{n=1}^{\infty} \frac{1}{n(e^{2\pi n} - 1)} = -\frac{1}{6} \text{Log} \left\{ \frac{\pi^{3/2}}{2^3 \Gamma^6(\frac{3}{4})} \right\} - \frac{\pi}{12}$$

$$= \tfrac{1}{4} \text{Log}(4/\pi) + \text{Log } \Gamma(\tfrac{3}{4}) - \pi/12,$$

as desired.

We now prove (22.10). Let $N = n + \frac{1}{2}$, where n is a positive integer. We shall let N tend to ∞ through a sequence such that $N^2\pi^2/\alpha$ remains at a bounded distance away from the positive integers. Let

$$f_N(z) = \frac{\coth(\pi Nz) \cot(\pi Nz)}{z(e^{\beta N^2 z^2} - 1)}.$$

The function $f_N(z)$ has simple poles at $z = \pm\sqrt{\alpha}k(1 + i)/(2\pi N)$, at $z = ik/N$, and at $z = ik/N$, where k is a nonzero integer. In addition, $f_N(z)$ has a quintuple pole at $z = 0$. Using elementary trigonometric identities, we find, after some calculation, that

$$R(\pm\sqrt{\alpha}k(1 + i)/(2\pi N)) = R(\pm\sqrt{-\alpha}k(1 + i)/(2\pi N))$$

$$= -\frac{1}{4\pi k} \left\{ \frac{2 \cos \sqrt{\alpha}k}{\cosh \sqrt{\alpha}k - \cos \sqrt{\alpha}k} + 1 \right\}. \qquad (22.14)$$

Easier calculations yield

$$R(\pm ik/N) = -\frac{\coth(\pi k)}{\pi k(e^{-\beta k^2} - 1)}$$

and

$$R(\pm k/N) = \frac{\coth(\pi k)}{\pi k(e^{\beta k^2} - 1)}.$$

Observe that

$$R(\pm ik/N) + R(\pm k/N) = \frac{2 \coth(\pi k)}{\pi k(e^{\beta k^2} - 1)} + \frac{\coth(\pi k)}{\pi k}. \qquad (22.15)$$

To calculate the residue at $z = 0$, write

$$f_N(z) = \frac{1}{z}\left\{\frac{1}{\pi N z} + \frac{\pi N z}{3} - \frac{(\pi N z)^3}{45} + \cdots\right\}\left\{\frac{1}{\pi N z} - \frac{\pi N z}{3} - \frac{(\pi N z)^3}{45} + \cdots\right\}$$

$$\times \frac{1}{\beta N^2 z^2}\left\{1 - \frac{\beta N^2 z^2}{2} + \frac{(\beta N^2 z^2)^2}{12} + \cdots\right\}.$$

After some simplification, we find that

$$R(0) = \frac{\beta}{12\pi^2} - \frac{7\alpha}{180\pi}. \tag{22.16}$$

Let C denote the positively oriented rhombus with vertices ± 1 and $\pm i$. By our choice of N, there are no poles of f_N on C. Applying the residue theorem and employing (22.14)–(22.16), we find that

$$\frac{1}{2\pi i}\int_C f_N(z)\,dz = -\frac{2}{\pi}\sum_{1 \le k \le \pi^2 N^2/\alpha}\frac{\cos\sqrt{\alpha}k}{k(\cosh\sqrt{\alpha}k - \cos\sqrt{\alpha}k)}$$

$$-\frac{1}{\pi}\sum_{1 \le k \le \pi^2 N^2/\alpha}\frac{1}{k} + \frac{4}{\pi}\sum_{1 \le k \le N}\frac{\coth(\pi k)}{k(e^{\beta k^2} - 1)}$$

$$+\frac{2}{\pi}\sum_{1 \le k \le N}\frac{\coth(\pi k)}{k} + \frac{\beta}{12\pi^2} - \frac{7\alpha}{180\pi}. \tag{22.17}$$

Next, we calculate directly the integral on the left side of (22.17). Let C_j denote that part of C in the jth quadrant, $1 \le j \le 4$. On C_1 set $z = 1 - x + ix$, $0 \le x \le 1$, and on C_3 set $z = x - 1 - ix$, $0 \le x \le 1$. Then in either case,

$$\lim_{N \to \infty} f_N(z) = \begin{cases} 0, & 0 < x < \frac{1}{2}, \\ i/z, & \frac{1}{2} < x < 1. \end{cases} \tag{22.18}$$

On C_2 set $z = -x + (1 - x)i$, $0 \le x \le 1$, and on C_4 set $z = x + (x - 1)i$, $0 \le x \le 1$. Then in either case,

$$\lim_{N \to \infty} f_N(z) = \begin{cases} -i/z, & 0 < x < \frac{1}{2}, \\ 0, & \frac{1}{2} < x < 1. \end{cases} \tag{22.19}$$

By the choice of N, the convergence in (22.18) and (22.19) is bounded on C as N tends to ∞. Hence, by the bounded convergence theorem,

$$\lim_{N \to \infty}\frac{1}{2\pi i}\int_C f_N(z)\,dz = \frac{1}{2\pi}\left\{\int_{(1+i)/2}^{i} - \int_i^{(-1+i)/2} + \int_{(-1-i)/2}^{-i} - \int_{-i}^{(1-i)/2}\right\}\frac{dz}{z}$$

$$= \frac{1}{\pi}\operatorname{Log} 2. \tag{22.20}$$

Returning to (22.17), we examine

$$2\sum_{1 \le k \le N}\frac{\coth(\pi k)}{k} - \sum_{1 \le k \le \pi^2 N^2/\alpha}\frac{1}{k}$$

$$= 4\sum_{1 \le k \le N}\frac{1}{k(e^{2\pi k} - 1)} + 2\sum_{1 \le k \le N}\frac{1}{k} - \sum_{1 \le k \le \pi^2 N^2/\alpha}\frac{1}{k}. \tag{22.21}$$

Now from Ayoub's text [1, p. 43],

$$2 \sum_{1 \le k \le N} \frac{1}{k} - \sum_{1 \le k \le \pi^2 N^2/\alpha} \frac{1}{k}$$

$$= 2\{\text{Log } N + \gamma + O(1/N)\} - \{\text{Log}(\pi^2 N^2/\alpha) + \gamma + O(1/N^2)\}$$

$$= \gamma - 2 \text{ Log } \pi + \text{Log } \alpha + O(1/N). \tag{22.22}$$

Thus, letting N tend to ∞ in (22.17), using (22.20)–(22.22), and multiplying both sides by π, we deduce that

$$\text{Log } 2 = -2 \sum_{k=1}^{\infty} \frac{\cos \sqrt{\alpha k}}{k(\cosh \sqrt{\alpha k} - \cos \sqrt{\alpha k})} + 4 \sum_{k=1}^{\infty} \frac{\coth(\pi k)}{k(e^{\beta k^2} - 1)}$$

$$+ 4 \sum_{k=1}^{\infty} \frac{1}{k(e^{2\pi k} - 1)} + \gamma - 2 \text{ Log } \pi + \text{Log } \alpha + \frac{\beta}{12\pi} - \frac{7\alpha}{180},$$

which is equivalent to (22.10) after some elementary manipulation. □

Entry 23(i). *We have*

$$\frac{\varphi(0)}{4\pi} + \sum_{k=1}^{\infty} \sum_{n=0}^{\infty} k \coth(\pi k)(-1)^n (kx)^{4n} \varphi(4n) = \frac{\pi}{2x^2} \{\tfrac{1}{2}\varphi(-2) + h\}, \tag{23.1}$$

where the error h is nearly equal to

$$\varphi(-2) + \sum_{n=0}^{\infty} \frac{(2\pi)^{2n+1} \cos\{3(2n+1)\pi/4\} \varphi(-2n-3)}{x^{2n+1}(2n+1)!}, \tag{23.2}$$

if x is small. (It is not clear whether the entry reads $\varphi(2)$ or $\varphi(-2)$ on the right side of (23.1).)

It is not clear what interpretation should be given to Entry 23(i). It is surprising that a power series in x is to be approximated near $x = 0$ by a power series in $1/x$. Perhaps (23.2) is an asymptotic series for h. It seems quite certain that Ramanujan derived Entry 23(i) in a purely formal manner. We shall show that *perhaps* Ramanujan made a mistake, because a formal argument seems to produce a slightly different formula. For most of the discussion which follows, we are very grateful to D. Zagier.

For each integer n, set $\varphi(n) = \psi(n + 2)$. Define

$$F(x) = \sum_{n=0}^{\infty} (-1)^n \psi(4n + 2)x^{4n}. \tag{23.3}$$

(Ramanujan seems to tacitly assume that F is entire.) Thus, Entry 23(i) may be rewritten in the form

$$\frac{\psi(2)}{4\pi} + \sum_{k=1}^{\infty} k \coth(\pi k)F(kx) = \frac{\pi}{2x^2} \{\tfrac{1}{2}\psi(0) + h\}, \tag{23.4}$$

where

$$h \approx \psi(0) + \sum_{\substack{n=1 \\ n \text{ odd}}}^{\infty} \frac{\cos(3\pi n/4)\psi(-n)}{n!}\left(\frac{2\pi}{x}\right)^n. \tag{23.5}$$

In his theory of integral transforms, discussed by Hardy [9, pp. 188–193], [4], [8, pp. 280–289] and the author [9], Ramanujan often writes

$$\int_0^\infty x^{n-1}G(x)\,dx := \int_0^\infty x^{n-1}\sum_{k=0}^\infty \frac{\varphi(k)(-x)^k}{k!}\,dx.$$

It is quite clear that Ramanujan is not assuming that G is an entire function; he is simply indicating the form of the Taylor series of G about $x = 0$. Likewise, in the setting at hand, Ramanujan is undoubtedly assuming that F has the expansion given by (23.3), only for x sufficiently small.

As an example, let $\psi(s) = \lambda^{-s}$, where $\lambda > 0$. Then $F(x) = \lambda^{-2}(1 + x^4/\lambda^4)^{-1}$. Letting $f(x)$ denote the left side of (23.4), we deduce that

$$f(x) = \frac{\lambda^{-2}}{4\pi}\left(1 + 4\pi\sum_{k=1}^\infty \frac{k\coth(\pi k)}{1 + k^4 x^4/\lambda^4}\right).$$

The sum on the right side may be evaluated by letting $z = e^{\pi i/4}\lambda/x$ in Entry 20(i). Temporarily putting $u = \pi\lambda/x$, we then find that

$$f(x) = \frac{\pi i}{4x^2}\cot\left(\frac{u}{\sqrt{2}}(1 + i)\right)\coth\left(\frac{u}{\sqrt{2}}(1 + i)\right)$$

$$= \frac{\pi}{4x^2}\frac{\cos^2(u/\sqrt{2})\cosh^2(u/\sqrt{2}) + \sin^2(u/\sqrt{2})\sinh^2(u/\sqrt{2})}{\sin^2(u/\sqrt{2})\cosh^2(u/\sqrt{2}) + \cos^2(u/\sqrt{2})\sinh^2(u/\sqrt{2})}$$

$$= \frac{\pi}{4x^2}\frac{\cosh(\sqrt{2}u) + \cos(\sqrt{2}u)}{\cosh(\sqrt{2}u) - \cos(\sqrt{2}u)}.$$

Thus, in the notation (23.4), as x tends to 0,

$$2h = \frac{\cosh(\sqrt{2}u) + \cos(\sqrt{2}u)}{\cosh(\sqrt{2}u) - \cos(\sqrt{2}u)} - 1$$

$$= \frac{2\cos(\sqrt{2}u)}{\cosh(\sqrt{2}u)} + O(e^{-2\sqrt{2}u})$$

$$= 4e^{-\sqrt{2}u}\cos(\sqrt{2}u) + O(e^{-2\sqrt{2}u})$$

$$= 2(e^{-2ue^{-\pi i/4}} + e^{-2ue^{\pi i/4}}) + O(e^{-2\sqrt{2}u})$$

$$= 2\sum_{n=0}^\infty \frac{(-1)^n(2u)^n}{n!}(e^{-\pi in/4} + e^{\pi in/4}) + O(e^{-2\sqrt{2}u})$$

$$= 4\sum_{n=0}^\infty \frac{(-1)^n(2u)^{4n}}{(4n)!} - 2\sqrt{2}\sum_{n=0}^\infty \frac{(-1)^n(2u)^{4n+1}}{(4n+1)!}$$

$$+ 2\sqrt{2}\sum_{n=0}^\infty \frac{(-1)^n(2u)^{4n+3}}{(4n+3)!} + O(e^{-2\sqrt{2}u}).$$

Thus,

$$h \approx 2 - \sqrt{2} \sum_{n=0}^{\infty} \frac{(-1)^n (2u)^{4n+1}}{(4n+1)!} + \sqrt{2} \sum_{n=0}^{\infty} \frac{(-1)^n (2u)^{4n+3}}{(4n+3)!} + 2 \sum_{n=1}^{\infty} \frac{(-1)^n (2u)^{4n}}{(4n)!}.$$
(23.6)

According to (23.5), Ramanujan claims that

$$h \approx 1 + \sum_{\substack{n=1 \\ n \text{ odd}}}^{\infty} \frac{\cos(3\pi n/4)(2u)^n}{n!}$$

$$= 1 - \frac{1}{\sqrt{2}} \sum_{n=1}^{\infty} \frac{(-1)^n (2u)^{4n+1}}{(4n+1)!} + \frac{1}{\sqrt{2}} \sum_{n=1}^{\infty} \frac{(-1)^n (2u)^{4n+3}}{(4n+3)!}.$$
(23.7)

A comparison of (23.6) and (23.7) indicates that apparently the error h is twice what Ramanujan claims. Furthermore, (23.6) contains an extra power series

$$2 \sum_{n=1}^{\infty} \frac{(-1)^n (2u)^{4n}}{(4n)!}.$$

Observe that $G(t) := e^{-\lambda t}$ is the inverse Mellin transform of $\Gamma(s)\lambda^{-s}$. Our calculations above have shown that

$$h \approx G\left(\frac{2\pi}{x} e^{-\pi i/4}\right) + G\left(\frac{2\pi}{x} e^{\pi i/4}\right).$$

Because of the close proximity of Entry 23(i) to Entry 20(i), we conjecture that Ramanujan probably proceeded as we have above and then more generally considered those φ having the shape

$$\varphi(s) = \sum_{j=0}^{\infty} c_j \lambda_j^{-s}.$$

In regard to Entry 23(i), some series transformations of S. N. Aiyar [1], published in 1913, might be mentioned. Recall that S. N. Aiyar was the manager of the Madras Port Trust office when Ramanujan worked there as a clerk for about 15 months during 1912–1913.

Entry 23(ii). *We have*

$$\sum_{k=1}^{\infty} \sum_{n=0}^{\infty} k^{-1} \chi(k) \operatorname{sech}(\pi k/2)(-1)^n (kx)^{4n} \varphi(4n) = \frac{\pi}{8} \varphi(0) - \frac{\pi}{2} h, \quad (23.8)$$

where h is very nearly equal to

$$\sum_{n=0}^{\infty} \frac{(-1)^n (\pi/\sqrt{2})^n \varphi(-n)}{x^n n!},$$

if x is small.

Comments similar to those made after Entry 23(i) can be made about this mysterious formula as well. However, as we shall shortly see, if we assume that the double series in (23.8) converges absolutely, then, in fact, (23.8) is indeed true with $h \equiv 0$. Of course, we are unable to make this hypothesis about the double series in (23.1).

PROOF. Assume that the double series in (23.8) converges absolutely. Then inverting the order of summation and employing the corollary of Entry 14 and Entry 15, we find that

$$\sum_{k=1}^{\infty} \sum_{n=0}^{\infty} k^{-1}\chi(k) \operatorname{sech}(\pi k/2)(-1)^n (kx)^{4n} \varphi(4n)$$

$$= \sum_{n=0}^{\infty} (-1)^n x^{4n} \varphi(4n) \sum_{k=1}^{\infty} k^{4n-1}\chi(k) \operatorname{sech}(\pi k/2)$$

$$= \frac{\pi}{8} \varphi(0),$$

which establishes (23.8) with $h \equiv 0$.   □

We are indebted to D. Zagier for the following very perceptive remarks on Entry 23(ii).

Let $G(t)$ be analytic at $t = 0$, and suppose that $G(t) = O(t^{-c})$ as t tends to ∞ for every $c > 0$. Define, for Re $s > 0$,

$$\varphi(s) = \frac{1}{\Gamma(s)} \int_0^{\infty} G(t)t^{s-1} \, dt.$$

Then φ is entire. Also, formally,

$$G(t) = \sum_{n=0}^{\infty} \frac{(-1)^n \varphi(-n)}{n!} t^n.$$

In view of Ramanujan's work on Mellin transforms in his quarterly reports (see Part I [9, p. 298]), we have determined the coefficients of $G(t)$ from the converse of Ramanujan's Master Theorem.

For x sufficiently small, suppose that

$$F(x) = \sum_{n=0}^{\infty} (-1)^n \varphi(4n) x^{4n}.$$

As in Entry 23(i), on the surface, it appears from (23.8) that Ramanujan is assuming that F is entire, but this is not the case. With F and G defined above, (23.8) can be rewritten in the form

$$\sum_{k=1}^{\infty} \frac{\chi(k)F(kx)}{k \cosh(\pi k/2)} = \frac{\pi}{8} \varphi(0) - \frac{\pi}{2} h, \tag{23.9}$$

where

$$h \approx G\left(\frac{\pi}{x\sqrt{2}}\right), \tag{23.10}$$

as x tends to 0.

We now discuss certain cases.

Case 1. Suppose that $G(t)$ is continuous for $t \geq 0$ and that $G(t) = 0$ for $t \geq t_0$; that is, $G(t)$ has compact support on $[0, t_0]$. It follows immediately from the definition of φ that, for all $s \geq 1$,

$$\Gamma(s)\varphi(s) \ll t_0^s.$$

Then

$$F(x) \ll \sum_{n=1}^{\infty} \frac{(xt_0)^{4n}}{(4n-1)!} \ll e^{xt_0}.$$

Hence, the left side of (23.9) converges absolutely as a double series for $xt_0 < \pi/2$. By our proof above, $h \equiv 0$. Now

$$\frac{\pi}{x\sqrt{2}} > \frac{\pi}{2x} > t_0.$$

Thus, $G(\pi/(x\sqrt{2})) = 0$. Hence, our findings are consistent with Ramanujan's claim (23.10).

On the other hand, suppose that the left side of (23.9) converges absolutely as a double series for $0 \leq x < x_0$. It follows that

$$\sum_{n=1}^{\infty} \sum_{k=1}^{\infty} |\varphi(4n)| k^{4n-1} e^{-\pi k/2} x^{4n} < \infty, \qquad x < x_0.$$

Comparing the sum on k with the integral of $t^{4n-1} e^{-\pi t/2}$ over $0 \leq t < \infty$, we deduce that

$$\sum_{n=1}^{\infty} |\varphi(4n)| \frac{(4n-1)!}{(\pi/2)^{4n}} x^{4n} < \infty, \qquad x < x_0.$$

Hence,

$$|\Gamma(4n)\varphi(4n)| < \left(\frac{\pi}{2x_0} + o(1)\right)^{4n},$$

as n tends to ∞. Moreover, if $\varphi(s)$ is reasonably smooth,

$$|\Gamma(s)\varphi(s)| < \left(\frac{\pi}{2x_0} + o(1)\right)^{s},$$

as s tends to ∞. By examining the inverse Mellin transform of $\Gamma(s)\varphi(s)$ and moving the line of integration to the right, we deduce that $G(t) = 0$ for $t > \pi/(2x_0) =: t_0$. Again, this is consistent with Ramanujan's claim (23.10), since

$$\frac{\pi}{x\sqrt{2}} > \frac{\pi}{x_0\sqrt{2}} = t_0\sqrt{2} > t_0,$$

and so $G(\pi/(x\sqrt{2})) = 0$.

Case 2. Suppose that $\varphi(s) = \lambda^{-s}$, where $\lambda > 0$. Then $G(t) = e^{-\lambda t}$ and $F(x) = (1 + x^4/\lambda^4)^{-1}$. If $f(x)$ denotes the left side of (23.9), we then find that

$$f(x) = \sum_{k=1}^{\infty} \frac{\chi(k) \, \text{sech}(\pi k/2)}{k(1 + k^4 x^4/\lambda^4)}$$

$$= \sum_{k=1}^{\infty} \frac{\chi(k)}{k \cosh(\pi k/2)} - \sum_{k=1}^{\infty} \frac{\chi(k)k^3 \, \text{sech}(\pi k/2)}{k^4 + \lambda^4/x^4}.$$

Applying Entry 25(vii) (or Entry 15) and the corollary to Entry 20(iv), and letting $\pi\lambda/x = u$, we deduce that

$$f(x) = \frac{\pi}{8} - \frac{\pi/4}{\cosh(u/\sqrt{2}) + \cos(u/\sqrt{2})}.$$

Comparing this with (23.9), we see that

$$h = \frac{2}{\cosh(u/\sqrt{2}) + \cos(u/\sqrt{2})} = e^{-u/\sqrt{2}} + O(e^{-u\sqrt{2}})$$

$$= G\left(\frac{\pi}{x\sqrt{2}}\right) + O(e^{-u\sqrt{2}}), \qquad (23.11)$$

as x tends to 0. Hence, (23.10) is established.

We have therefore shown that Entry 23(ii) is valid when $\varphi(s) = \lambda^{-s}$.

Observe, from (23.11), that we may write h in the form

$$h = \sum_{j=1}^{\infty} (-1)^{j-1} \sum_{\substack{|k| \leq j-1 \\ k \equiv j-1 (\text{mod } 2)}} e^{-\pi j\lambda/(x\sqrt{2}) - \pi i k\lambda/(x\sqrt{2})}$$

$$= \sum_{j=1}^{\infty} (-1)^{j-1} \sum_{\substack{|k| \leq j-1 \\ k \equiv j-1 (\text{mod } 2)}} G\left(\frac{(j + ik)\pi}{x\sqrt{2}}\right),$$

where $G(t) = e^{-\lambda t}$. This suggests that, for more general functions φ and G,

$$h = \sum_{j=1}^{\infty} (-1)^{j-1} \sum_{\substack{|k| \leq j-1 \\ k \equiv j-1 (\text{mod } 2)}} G\left(\frac{(j + ik)\pi}{x\sqrt{2}}\right), \qquad (23.12)$$

under suitable hypotheses. We now establish such a theorem.

It is clear that we now need to define $G(z)$ in the quadrant $Q := \{z: |\arg z| \leq \pi/4\}$, instead of on just $[0, \infty)$. Thus, suppose that G is analytic on Q and that $G(z) = O(z^{-c})$ as z tends to ∞ in Q, for every constant $c > 0$. Define, as before, for $\text{Re } s > 0$,

$$\varphi(s) = \frac{1}{\Gamma(s)} \int_0^{\infty} G(t)t^{s-1} \, dt.$$

Hence, if $\omega = \exp(\pi i/4)$,

$$F(x) = \varphi(0) - x \sum_{n=1}^{\infty} \frac{(-1)^{n-1}x^{4n-1}}{(4n-1)!} \int_0^{\infty} G(t)t^{4n-1}\, dt$$

$$= \varphi(0) - \tfrac{1}{2}x \int_0^{\infty} G(t)\{\omega \sin(\omega xt) + \omega^{-1} \sin(\omega^{-1}xt)\}\, dt$$

$$= \varphi(0) - \tfrac{1}{2}\omega xH(\omega x) - \tfrac{1}{2}\omega^{-1}xH(\omega^{-1}x),$$

by absolute convergence, where

$$H(u) = \int_0^{\infty} G(t)\sin(tu)\, dt. \qquad (23.13)$$

Using Entry 25(vii) and the last expression for $F(x)$, we find that

$$\frac{\pi}{8}\varphi(0) - \sum_{k=1}^{\infty} \frac{\chi(k)F(kx)}{k\cosh(\pi k/2)}$$

$$= x \sum_{m=1}^{\infty} \frac{\chi(m)}{2\cosh(\pi m/2)}\{\omega H(\omega mx) + \omega^{-1}H(\omega^{-1}mx)\}. \qquad (23.14)$$

Since

$$\frac{1}{2\cosh y} = \sum_{n=1}^{\infty} \chi(n)e^{-ny}, \qquad y > 0,$$

the right side of (23.14) may be written in the form

$$x \sum_{n=1}^{\infty} \chi(n)\left\{\omega \sum_{m=1}^{\infty} \chi(m)e^{-\pi mn/2}H(\omega mx) + \omega^{-1} \sum_{m=1}^{\infty} \chi(m)e^{-\pi mn/2}H(\omega^{-1}mx)\right\}.$$
$$(23.15)$$

We now assume that the real and imaginary parts of $H(\omega xu)e^{-\pi nu/2}$ and $H(\omega^{-1}xu)e^{-\pi nu/2}$, for each positive integer n, are integrable over $(0, \delta)$ for some $\delta, 0 < \delta < \pi/2$, are of bounded variation over (δ, ∞), and tend to 0 as u tends to ∞. Then by Poisson's summation formula for Fourier sine transforms (Titchmarsh [2, p. 66]) (see also (9.2) above), the expression within curly brackets in (23.15) equals

$$\sum_{m=1}^{\infty} \chi(m)\left\{\omega \int_0^{\infty} e^{-\pi nu/2}H(\omega xu)\sin(\pi mu/2)\, du\right.$$

$$\left. + \omega^{-1} \int_0^{\infty} e^{-\pi nu/2}H(\omega^{-1}xu)\sin(\pi mu/2)\, du\right\}. \qquad (23.16)$$

Next, replace u by $\omega^{-1}u$ and ωu, respectively, in the two integrals above. Assume that $H(z)$ decays sufficiently rapidly in Q so that we may apply Cauchy's theorem to replace the paths $(0, \omega^{-1}\infty)$ and $(0, \omega\infty)$, respectively, by $(0, \infty)$. Collecting together the calculations from (23.14)–(23.16), we deduce that

$$\frac{\pi}{8}\varphi(0) - \sum_{k=1}^{\infty} \frac{\chi(k)F(kx)}{k\cosh(\pi k/2)}$$

$$= x\sum_{n=1}^{\infty}\sum_{m=1}^{\infty}\chi(mn)\int_0^{\infty} H(xu)\{e^{-\pi n\omega^{-1}u/2}\sin(\pi m\omega^{-1}u/2)$$

$$+ e^{-\pi n\omega u/2}\sin(\pi m\omega u/2)\}\,du. \tag{23.17}$$

Assume that the iterated sum above is equal to its double sum. Since the coefficient $\chi(mn)$ is symmetric in m and n, we may interchange the roles of m and n in the second expression above. If we also employ the identity

$$e^{-\pi n\omega^{-1}u/2}\sin(\pi m\omega^{-1}u/2) + e^{-\pi m\omega u/2}\sin(\pi n\omega u/2) = \sin\{\pi(\omega n + \omega^{-1}m)u/2\},$$

we find, from (23.9) and (23.17), that

$$\frac{\pi}{2}h = \sum_{m,n=1}^{\infty}\chi(mn)\int_0^{\infty} H(u)\sin\{\pi(\omega n + \omega^{-1}m)u/(2x)\}\,du,$$

where $x > 0$. By the Fourier sine inversion of (23.13),

$$G(t) = \frac{2}{\pi}\int_0^{\infty} H(u)\sin(tu)\,du.$$

Hence, for $x > 0$,

$$h = \sum_{m,n=1}^{\infty}\chi(mn)G\left(\frac{\pi(\omega n + \omega^{-1}m)}{2x}\right).$$

Setting $j = (n + m)/2$ and $k = (n - m)/2$, we find that the conditions m, n odd and positive are transformed into the conditions j, k integral, $j \geq |k| + 1$, and $j \equiv k + 1 \pmod 2$. Also, $\chi(mn) = (-1)^{j-1}$ and $\omega n + \omega^{-1}m = (j + ik)\sqrt{2}$. Thus, for $x > 0$, we deduce (23.12).

As an example, let

$$G(z) = ze^{-cz^2},$$

where $c > 0$. Then, initially for Re $s > -1$,

$$\varphi(s) = \frac{1}{\Gamma(s)}\int_0^{\infty} e^{-ct^2}t^s\,dt = \frac{c^{-(s+1)/2}\Gamma\left(\dfrac{s+1}{2}\right)}{2\Gamma(s)} = \frac{c^{-(s+1)/2}\sqrt{\pi}}{2^s\Gamma\left(\dfrac{s}{2}\right)},$$

where lastly s is any complex number, by analytic continuation. Also,

$$H(u) = \int_0^{\infty} te^{-ct^2}\sin(tu)\,dt = \frac{\sqrt{\pi}u}{4c^{3/2}}e^{-u^2/(4c)},$$

which can be obtained from differentiating formula 3.896, No. 4, p. 480, in Gradshteyn and Ryzhik's tables [1]. Lastly,

$$F(x) = \varphi(0) - \tfrac{1}{2}\omega x H(\omega x) - \tfrac{1}{2}\omega^{-1} x H(\omega^{-1} x)$$

$$= -\frac{ix^2\sqrt{\pi}}{8c^{3/2}} e^{-ix^2/(4c)} + \frac{ix^2\sqrt{\pi}}{8c^{3/2}} e^{ix^2/(4c)}$$

$$= -\frac{x^2\sqrt{\pi}}{4c^{3/2}} \sin\left(\frac{x^2}{4c}\right).$$

Thus, we have shown that, for $x > 0$,

$$\frac{x^3}{c^{3/2}} \sum_{k=1}^{\infty} \frac{\chi(k)k \sin(k^2 x^2/(4c))}{\cosh(\pi k/2)}$$

$$= (2\pi^3)^{1/2} \sum_{j=1}^{\infty} (-1)^{j-1} \sum_{\substack{|k| \le j-1 \\ k \equiv j-1 (\mathrm{mod}\ 2)}} (j+ik) \exp\left(-\frac{c(j+ik)^2\pi^2}{2x^2}\right).$$

The analysis above can be strengthened by beginning with the Fourier sine transform (23.13), imposing conditions on $H(u)$, and then defining F and G in terms of H. Furthermore, an analogue of (23.12) undoubtedly holds for Entry 23(i) as well. However, in view of the limited applications that any more rigorous and/or stronger versions of Entries 23(i), (ii) might have, it seems best here to end our discussion of these entries.

Entry 24. *For z complex,*

$$\frac{\pi e^{-2\pi z}}{2z\{\cosh(2\pi z) - \cos(2\pi z)\}} = \frac{1}{8\pi z^3} - \frac{1}{4z^2} + \frac{\pi}{4z} - \sum_{n=1}^{\infty} \frac{1}{z^2 + (z+n)^2}$$

$$+ 4z \sum_{n=1}^{\infty} \frac{n}{(e^{2\pi n} - 1)(4z^4 + n^4)}.$$

PROOF. Let $f(z)$ denote the left side above. We shall expand f by partial fractions. The function f has a triple pole at $z = 0$ and simple poles at $z = \pm n(1 \pm i)/2$, where n is a positive integer. By division of power series, it is easily calculated that the principal part of f about $z = 0$ is

$$\frac{1}{8\pi z^3} - \frac{1}{4z^2} + \frac{\pi}{4z}. \tag{24.1}$$

Straightforward calculations show that

$$R(n(1+i)/2) = \frac{1}{2in(e^{2\pi n} - 1)} = -R(n(1-i)/2).$$

Replacing n by $-n$ above and manipulating slightly, we find that

$$R(-n(1+i)/2) = \frac{1}{2in(e^{2\pi n} - 1)} + \frac{1}{2in} = -R(-n(1-i)/2).$$

Now,

$$\frac{1}{2in}\left\{\frac{1}{z + n(1 + i)/2} - \frac{1}{z + n(1 - i)/2}\right\} = -\frac{1}{z^2 + (z + n)^2}. \quad (24.2)$$

After much, but routine, simplification, we get

$$\frac{1}{2in(e^{2\pi n} - 1)}\left\{\frac{1}{z - n(1 + i)/2} - \frac{1}{z - n(1 - i)/2}\right.$$

$$\left. + \frac{1}{z + n(1 + i)/2} - \frac{1}{z + n(1 - i)/2}\right\} = \frac{4nz}{(e^{2\pi n} - 1)(4z^4 + n^4)}. \quad (24.3)$$

Using the principal parts in (24.1)–(24.3), we easily deduce the desired result after employing an argument like that at the end of the proof of Entry 4. □

Entry 24(i). *For complex z we have*

$$\frac{1}{2z^2} + \sum_{n=1}^{\infty} \frac{1}{z^2 + n^2} = \frac{\pi}{2z} + \frac{\pi}{z(e^{2\pi z} - 1)}.$$

This result is just a reformulation of (1.9).

Entry 24(ii). *Let z be complex. Then*

$$\frac{1}{z(e^{\pi z} + 1)} = \frac{1}{2z} - \frac{2}{\pi}\sum_{n=0}^{\infty} \frac{1}{z^2 + (2n + 1)^2}.$$

A proof of Entry 24(ii) is easily obtained by expanding the function on the left side above into partial fractions.

The next entry is complementary to Entry 24.

Entry 25. *Let z be complex. Then*

$$\frac{\pi e^{-\pi z}}{4z\{\cosh(\pi z) + \cos(\pi z)\}} = \frac{\pi}{8z} - \sum_{n=0}^{\infty} \frac{1}{z^2 + (z + 2n + 1)^2}$$

$$- 4z \sum_{n=0}^{\infty} \frac{2n + 1}{(e^{(2n+1)\pi} + 1)(4z^4 + (2n + 1)^4)}.$$

PROOF. Let $f(z)$ denote the left side above. We expand f into partial fractions. The function f has a simple pole at $z = 0$, and the principal part about 0 is easily seen to be $\pi/(8z)$. Also, f has simple poles at $z = \pm(2n + 1)(1 \pm i)/2$, where n is a nonnegative integer. Routine calculations give

$$R((2n + 1)(1 + i)/2) = \frac{i}{2(2n + 1)(e^{(2n+1)\pi} + 1)}$$

$$= -R((2n + 1)(1 - i)/2)$$

and

$$R(-(2n + 1)(1 + i)/2) = \frac{i}{2(2n + 1)(e^{(2n+1)\pi} + 1)} - \frac{i}{2(2n + 1)}$$

$$= -R(-(2n + 1)(1 - i)/2).$$

The sum of the principal parts for the four poles $\pm(2n + 1)(1 \pm i)/2$ is thus found to be

$$-\frac{1}{z^2 + (z + 2n + 1)^2} - \frac{4z(2n + 1)}{(e^{(2n+1)\pi} + 1)(4z^4 + (2n + 1)^4)}.$$

The theorem now readily follows. □

Entries 25(i), (ii). *We have*

$$\sum_{k=1}^{\infty} \frac{\coth(\pi k)}{k^3} = \frac{7\pi^3}{180} \tag{25.1}$$

and

$$\sum_{k=1}^{\infty} \frac{\coth(\pi k)}{k^7} = \frac{19\pi^7}{56,700}. \tag{25.2}$$

Both (25.1) and (25.2) are special cases of the more general formula

$$\sum_{k=1}^{\infty} \frac{\coth(\pi k)}{k^{2n+1}} = 2^{2n}\pi^{2n+1} \sum_{k=0}^{n+1} (-1)^{k+1} \frac{B_{2k}}{(2k)!} \frac{B_{2n+2-2k}}{(2n + 2 - 2k)!}, \tag{25.3}$$

where n is an odd positive integer and B_j denotes the jth Bernoulli number. Ramanujan does not state the general formula (25.3) in his notebooks. However, it does follow quite easily from Entry 21(i). (See our paper [6, p. 155].) Formula (25.2) was communicated by Ramanujan in one of his letters to Hardy [16, p. xxvi]. Entry 25(i), in fact, was long ago established by Cauchy [1, pp. 320, 361]. Cauchy does not state the general formula (25.3), but he does give a general method for evaluating the series on the left side of (25.3). Preece [3] has established (25.1) and Sandham [1] has proved (25.2). The first statement of (25.3) known to the author is by Lerch [1]. Later proofs of (25.3) have been given by Watson [1], Sandham [2], Smart [1], Sayer [1], Sitaramachandrarao [2], and the author [6, p. 155], [5].

Entries 25(iii), (iv). *We have*

$$\sum_{k=0}^{\infty} \frac{\tanh\{(2k + 1)\pi/2\}}{(2k + 1)^3} = \frac{\pi^3}{32}$$

and

$$\sum_{k=0}^{\infty} \frac{\tanh\{(2k + 1)\pi/2\}}{(2k + 1)^7} = \frac{7\pi^7}{23,040}.$$

Both entries follow from the more general formula

$$\sum_{k=0}^{\infty} \frac{\tanh\{(2k+1)\pi/2\}}{(2k+1)^{4n+3}} = \frac{\pi^{4n+3}}{8} \sum_{k=0}^{\infty} (-1)^k \frac{E_{2k+1}(0)}{(2k+1)!} \frac{E_{4n+1-2k}(0)}{(4n+1-2k)!}, \quad (25.4)$$

where n is a nonnegative integer and $E_j(x)$ denotes the jth Euler polynomial. Formula (25.4) cannot be found in the notebooks. The first proof of (25.4) was given by Phillips [1]. Later proofs have been given by Nanjundiah [1], Sandham [2], Smart [1], Sayer [1], and the author [7, Corollary 4.10].

Formula (25.4) is, in fact, a special case of a more general formula. Let $\alpha, \beta > 0$ with $\alpha\beta = \pi^2$. Then

$$\alpha^{-n} \sum_{k=0}^{\infty} \frac{\tanh\left\{(2k+1)\dfrac{\alpha}{2}\right\}}{(2k+1)^{2n+1}} = (-\beta)^{-n} \sum_{k=0}^{\infty} \frac{\tanh\left\{(2k+1)\dfrac{\beta}{2}\right\}}{(2k+1)^{2n+1}}$$

$$+ \frac{1}{4} \sum_{k=0}^{n-1} (-1)^k \frac{E_{2k+1}(0)}{(2k+1)!} \frac{E_{2n-2k-1}(0)}{(2n-2k-1)!} \alpha^{n-k} \beta^{k+1},$$

where n is a positive integer. The first proof of this formula appears to be by Nanjundiah [1]. The author [7, Corollary 4.9] has also given a proof. The case $n = 1$ was established by Grosjean [3]. The case $n = 2$ was proved by de Saint-Venant [1] in 1856 and occurs in the determination of the torsional rigidity of a beam of rectangular cross section. This motivated a problem by Boersma [1] who asked for an asymptotic expansion of the series on the left side as α tends to 0. The identity above easily yields such a result.

The author [7, Theorem 4.11] has evaluated

$$\sum_{k=0}^{\infty} \frac{\tan\left\{(2k+1)\dfrac{\theta}{2}\right\}}{(2k+1)^{2n+1}}$$

for a very general class of real quadratic irrationalities θ.

Entries 25(v), (vi). *We have*

$$\sum_{k=1}^{\infty} \frac{(-1)^{k+1} \operatorname{csch}(\pi k)}{k^3} = \frac{\pi^3}{360}$$

and

$$\sum_{k=1}^{\infty} \frac{(-1)^{k+1} \operatorname{csch}(\pi k)}{k^7} = \frac{13\pi^7}{453,600}.$$

Both entries follow from the more general formula

$$\sum_{k=1}^{\infty} \frac{(-1)^{k+1} \operatorname{csch}(\pi k)}{k^{4n+3}} = \frac{1}{2}(2\pi)^{4n+3} \sum_{k=0}^{2n+2} (-1)^k \frac{B_{2k}(\frac{1}{4})}{(2k)!} \frac{B_{4n+4-2k}(\frac{1}{4})}{(4n+4-2k)!}, \quad (25.5)$$

where n is an integer and $B_j(x)$ denotes the jth Bernoulli polynomial. Formula (25.5) is essentially due to Cauchy [1, pp. 311, 361] who gave a somewhat less explicit formulation. Otherwise, (25.5) was first established by Mellin [1]. Later proofs have been given by Malurkar [1], Phillips [1], Nanjundiah [1], Sandham [2], Riesel [1], Sayer [1], and the author [7, Corollary 3.2]. The general formula (25.5) does not appear in the notebooks.

Formula (25.5) is an immediate consequence of the following more general result. Let $\alpha, \beta > 0$ with $\alpha\beta = \pi^2$. Then, for any integer n,

$$\alpha^{-n} \sum_{k=1}^{\infty} (-1)^{k+1} k^{-2n-1} \operatorname{csch}(\alpha k)$$

$$= (-\beta)^{-n} \sum_{k=1}^{\infty} (-1)^{k+1} k^{-2n-1} \operatorname{csch}(\beta k)$$

$$+ 2^{2n+1} \sum_{k=0}^{n+1} (-1)^k \frac{B_{2k}(\frac{1}{2})}{(2k)!} \frac{B_{2n+2-2k}(\frac{1}{2})}{(2n+2-2k)!} \alpha^{n+1-k}\beta^k,$$

which has been proved by Mellin [1], Malurkar [1], Nanjundiah [1], and the author [7, Theorem 3.1].

Entries 25(vii), (viii), (ix). *We have*

$$\sum_{k=1}^{\infty} \chi(k) \frac{\operatorname{sech}(\pi k/2)}{k} = \frac{\pi}{8},$$

$$\sum_{k=1}^{\infty} \chi(k) \frac{\operatorname{sech}(\pi k/2)}{k^5} = \frac{\pi^5}{768},$$

and

$$\sum_{k=1}^{\infty} \chi(k) \frac{\operatorname{sech}(\pi k/2)}{k^9} = \frac{23\pi^9}{1,720,320}.$$

All three entries follow from the general formula

$$\sum_{k=1}^{\infty} \chi(k) \frac{\operatorname{sech}(\pi k/2)}{k^{4n+1}} = \frac{1}{4}\left(\frac{\pi}{2}\right)^{4n+1} \sum_{k=0}^{2n} (-1)^k \frac{E_{2k}}{(2k)!} \frac{E_{4n-2k}}{(4n-2k)!}, \tag{25.6}$$

which can be easily deduced from Entry 21(ii). Here n is any integer. Entry 25(vii) is a simple consequence of Entry 15 and was proved by Preece [3]. Zucker [2] has established Entry 25(vii) as well as some related results. Entry 25(vii) was also submitted as a problem to the *Mathematical Gazette* [1], where several solutions are indicated and considerable discussion is found. Entry 25(viii) appeared in one of Ramanujan's letters to Hardy [16, p. xxvi]. In addition to the proofs mentioned after Entry 21(ii), proofs of (25.6) have been given by Watson [1], Sandham [2], Riesel [1], and Sayer [1].

Entry 25(x). *We have*

$$\sum_{k=1}^{\infty} \frac{\chi(k)}{k^2(e^{\pi k} - 1)} + \frac{1}{8} \sum_{k=1}^{\infty} \frac{1}{k^2 \cosh(\pi k)} = \frac{5\pi^2}{96} - \frac{1}{2} \int_0^1 \frac{\tan^{-1}x}{x} dx. \tag{25.7}$$

PROOF. Let

$$f(z) = \frac{1}{z^2(e^{2\pi z} - 1)\cos(\pi z)}.$$

We shall integrate f over the positively oriented rectangle C_N whose horizontal sides pass through $\pm(N + \frac{1}{2})i$ and whose vertical sides pass through $\pm N$, where N is a positive integer. The function f has a triple pole at $z = 0$ and simple poles at $z = \pm(2k + 1)/2$, where k is a nonnegative integer, and at $z = \pm ki$, where k is a positive integer. Routine calculations yield $R(0) = 5\pi/12$,

$$R((2k + 1)/2) = \frac{4(-1)^{k+1}}{(2k + 1)^2 \pi (e^{(2k+1)\pi} - 1)},$$

$$R(-(2k + 1)/2) = R((2k + 1)/2) + \frac{4(-1)^{k+1}}{(2k + 1)^2 \pi},$$

and

$$R(ki) = -\frac{1}{2\pi k^2 \cosh(\pi k)} = R(-ki).$$

Hence, applying the residue theorem and letting N tend to ∞, we find that

$$0 = \lim_{N \to \infty} \frac{1}{2\pi i} \int_{C_N} f(z)\, dz$$

$$= -\frac{8}{\pi} \sum_{k=1}^{\infty} \frac{\chi(k)}{k^2(e^{\pi k} - 1)} - \frac{4}{\pi} L(2) - \frac{1}{\pi} \sum_{k=1}^{\infty} \frac{1}{k^2 \cosh(\pi k)} + \frac{5\pi}{12}, \quad (25.8)$$

where $L(s)$ is defined by (21.1). A comparison of (25.8) with (25.7) indicates that it remains to show that

$$\int_0^1 \frac{\tan^{-1} x}{x}\, dx = L(2). \quad (25.9)$$

Integrating termwise the Maclaurin expansion

$$\frac{\tan^{-1} x}{x} = \sum_{k=0}^{\infty} \frac{(-1)^k x^{2k}}{2k + 1},$$

we readily deduce (25.9), and the proof is complete.   □

Entry 25(xi). *We have*

$$\sum_{k=1}^{\infty} \frac{1}{\{k^2 + (k + 1)^2\}(\cosh\{(2k + 1)\pi\} - \cosh \pi)}$$

$$= \frac{1}{2 \sinh \pi} \left\{ \frac{1}{\pi} + \coth \pi - \frac{\pi}{2} \tanh^2(\pi/2) \right\}. \quad (25.10)$$

Entry 25(xi) is in error in the notebooks, for Ramanujan has written $\sinh\{(2k+1)\pi\} - \sinh \pi$ instead of $\cosh\{(2k+1)\pi\} - \cosh \pi$ on the left side of (25.10). Ramanujan communicated (25.10), with the same error, in one of his letters to Hardy [16, p. 349]. Watson [1] established (25.10) by contour integration. Because Watson's proof contains a few errors, we briefly sketch another proof by contour integration below. The calculations in both proofs are extremely laborious. Sitaramachandrarao [1] has found another proof based on Entry 20(i). Since his proof is very elegant and is likely the one which Ramanujan had, we shall give this proof as well.

FIRST PROOF. Let

$$f(z) = \frac{\pi \sinh \pi}{z\{\cosh(\pi z) + \cosh \pi\}\{\cos(\pi z) + \cosh \pi\}},$$

which has a simple pole at $z = 0$ and poles at $z = i(2k+1) \pm 1$, if k is an integer, and at $z = 2n + 1 \pm i$, if n is an integer. These poles are simple except when $k = 0, -1$ and $n = 0, -1$ when the two sets coalesce to give double poles. Very lengthy calculations yield

$$R(0) = \frac{\pi \tanh^2(\pi/2)}{\sinh \pi},$$

$$R(i(2k+1) \pm 1) = \frac{\pm 1}{\{i(2k+1) \pm 1\}(\cosh\{(2k+1)\pi\} - \cosh \pi)}, \quad k \neq 0, -1,$$

$$R(2n+1 \pm i) = \frac{\pm i}{(2n+1 \pm i)(\cosh\{(2n+1)\pi\} - \cosh \pi)}, \quad n \neq 0, -1,$$

and

$$R(\pm 1 \pm i) = -\frac{\coth \pi}{2 \sinh \pi} - \frac{1}{2\pi \sinh \pi}.$$

Integrate f over a square with vertical and horizontal sides passing through $\pm 2N$ and $\pm 2Ni$, respectively, where N is an integer. Apply the residue theorem and let N tend to ∞ to deduce (25.10). □

SECOND PROOF. Let S denote the left side of (25.10). Since an elementary calculation shows that

$$\frac{\coth(k\pi) - \coth\{(k+1)\pi\}}{2 \sinh \pi} = \frac{1}{\cosh\{(2k+1)\pi\} - \cosh \pi},$$

we readily deduce that

$$S = \frac{1}{2 \sinh \pi} \sum_{k=1}^{\infty} \frac{\coth(k\pi) - \coth\{(k+1)\pi\}}{k^2 + (k+1)^2}.$$

Transforming the right side by partial summation, we deduce that

$$2(\sinh \pi)S = \frac{\coth \pi}{5} - \sum_{k=2}^{\infty} \coth(\pi k) \left\{ \frac{1}{2k^2 - 2k + 1} - \frac{1}{2k^2 + 2k + 1} \right\}$$

$$= \frac{\coth \pi}{5} - 4 \sum_{k=2}^{\infty} \frac{k \coth(\pi k)}{4k^4 + 1}$$

$$= \coth \pi - \sum_{k=1}^{\infty} \frac{k \coth(\pi k)}{k^4 + \frac{1}{4}}.$$

Setting $z = (1 + i)/2$ in Entry 20(i), we easily find that

$$\sum_{k=1}^{\infty} \frac{k \coth(\pi k)}{k^4 + \frac{1}{4}} = -\frac{1}{\pi} + \frac{\pi}{2} \tanh^2(\pi/2).$$

Using this in the foregoing equality, we complete the second proof of Entry 25(xi).   □

Entry 25(xii). *We have*

$$\sum_{k=0}^{\infty} \frac{2k + 1}{\{25 + (2k + 1)^4/100\}(e^{(2k+1)\pi} + 1)} = \frac{4689}{11,890} - \frac{\pi}{8} \coth^2(5\pi/2). \quad (25.11)$$

This entry again was communicated by Ramanujan in one of his letters to Hardy [16, p. 349]. The right side of (25.11), however, had the wrong sign on both terms. This error is also made in the notebooks. Furthermore, the left side of (25.11) is replaced by only the first three terms of the series in the notebooks, and the second term contains another misprint. It may be of interest to determine how well the first three terms on the left side of (25.11) approximate the right side. We note that

$$\frac{1}{25.01(e^\pi + 1)} + \frac{3}{25.81(e^{3\pi} + 1)} + \frac{5}{31.25(e^{5\pi} + 1)} = 0.001665694154\cdots,$$

while on the other hand,

$$\frac{4689}{11,890} - \frac{\pi}{8} \coth^2(5\pi/2) = 0.001665694195\cdots.$$

Watson [1] has given a proof of (25.11) by contour integration. It will be shown below that Entry 25(xii) is a corollary of Entry 25; hence, this is probably the method used by Ramanujan to establish (25.11).

PROOF. In Entry 25 put $z = 5i$. After some simplification and rearrangement, we find that

$$\sum_{k=0}^{\infty} \frac{2k + 1}{(e^{(2k+1)\pi} + 1)\{25 + (2k + 1)^4/100\}} - 5i \sum_{k=0}^{\infty} \frac{1}{(2k + 1)^2 + 10(2k + 1)i - 50}$$

$$= -\frac{\pi \coth^2(5\pi/2)}{8}. \quad (25.12)$$

A comparison of (25.12) with (25.11) indicates that it remains to show that

$$5i \sum_{k=0}^{\infty} \frac{1}{(2k+1)^2 + 10(2k+1)i - 50} = \frac{4689}{11,890}, \tag{25.13}$$

or equivalently that

$$50 \sum_{k=0}^{\infty} \frac{2k+1}{(2k+1)^4 + 2500} = \frac{4689}{11,890}, \tag{25.14}$$

since (25.12) obviously implies that the imaginary part of the left side of (25.13) is zero. To show (25.14), write

$$50 \sum_{k=0}^{\infty} \frac{2k+1}{(2k+1)^4 + 2500}$$

$$= \frac{5}{2} \sum_{k=0}^{\infty} \left\{ \frac{1}{(2k+1)^2 - 10(2k+1) + 50} - \frac{1}{(2k+1)^2 + 10(2k+1) + 50} \right\}$$

$$= \frac{5}{2} \left\{ \sum_{k=0}^{\infty} \frac{1}{(2k+1)^2 - 10(2k+1) + 50} \right.$$

$$\left. - \sum_{k=5}^{\infty} \frac{1}{(2k+1-10)^2 + 10(2k+1-10) + 50)} \right\}$$

$$= \frac{5}{2} \sum_{k=0}^{4} \frac{1}{(2k+1)^2 - 10(2k+1) + 50}$$

$$= \frac{4689}{11,890},$$

and the proof of (25.14), and hence (25.11), is complete. □

Infinite series involving the hyperbolic functions have attracted the attention of many authors. Our papers [6], [7] contain many such results as well as numerous references. Readers may also wish to consult papers by Cauchy [1], Zucker [1], [2], Ling [1], [2], [3], Forrester [1], and Bruckman [1] for additional results not examined in the aforementioned papers. The papers of Berndt [8] and Klusch [1] offer some hyperbolic series of different types.

CHAPTER 15

Asymptotic Expansions and Modular Forms

The title of Chapter 15 does not entirely reflect its contents, because this chapter contains several diverse topics. Of the 21 chapters in the second notebook, Chapter 15 contains more disparate topics than the remaining chapters. Ramanujan appears to have collected here several "odds and ends." While much of the material is fascinating, a few parts have little substance.

The first seven sections are devoted primarily to asymptotic expansions of series. For example, Ramanujan derives asymptotic series, as x tends to $0+$, for

$$\sum_{k=1}^{\infty} e^{-xk^p} k^{m-1}, \quad \sum_{k=1}^{\infty} e^{-kx} \operatorname{Log} k, \quad \text{and} \quad \sum_{k=1}^{\infty} \frac{k^{m-1}}{(1 + xk^p)^l}.$$

Frequently, theorems about such series are established by us in greater generality than indicated by Ramanujan. These generalizations not explicitly stated by Ramanujan are labeled as "Theorems" in the sequel, in contrast to our usual designations by "Entries" in relating Ramanujan's results. This has necessitated some reordering of Ramanujan's findings in our description of Sections 2–7 below.

Ramanujan's discourse in Section 1 seems to indicate that he used the Euler–Maclaurin summation formula to establish his asymptotic expansions. However, his comments are so cryptic and obscure that we have been unable to find a proper interpretation for them. At any rate, it appears that Ramanujan's use of the Euler–Maclaurin formula was formal and non-rigorous. Despite the nature of his methods, Ramanujan's results are correct, except for some minor errors. Our original proofs were likewise based on the Euler–Maclaurin summation formula. These proofs were rather lengthy and involved. We are extremely grateful to P. Flajolet for suggesting that considerably shorter proofs could be achieved via the use of Mellin transforms.

In the second half of Chapter 15, modular forms, in particular, Eisenstein series, are at center stage. However, as our summary below indicates, there are many themes.

One of the most interesting theorems in Chapter 15 is found in (8.3) below. This undoubtedly new result gives an inversion formula for a certain modified theta-function. It may be surprising that an exact formula of this type exists.

Entry 11 is a beautiful and new reciprocity formula reminiscent of some of the formulas in Chapter 14.

Section 12 contains several results found in Ramanujan's famous paper [11], [16, pp. 136–162]. We mention, in particular, Entry 12(x) which is equivalent to the very interesting identity

$$\sum_{k=0}^{n} \sigma_1(2k + 1)\sigma_3(n - k) = \frac{1}{240}\sigma_5(2n + 1), \qquad n \ge 0,$$

where $\sigma_\nu(m) = \sum_{d|m} d^\nu$, $m \neq 0$, and $\sigma_3(0) = \frac{1}{240}$. Ramanujan states this identity without proof in [11], [16, p. 146] and indicates that he has two proofs, one of which is elementary. We have not been able to find an elementary proof in the literature nor can we produce one ourselves. All the results in Section 13 can also be found in [11].

Entry 14 offers a new recursion formula for Eisenstein series. It is quite distinct from the most well-known recursion formula for Eisenstein series which was discovered by Ramanujan in [11], [16, p. 140].

In Section 10, Ramanujan defines some terminology in the theory of infinite series. His definitions are rather vague and do not seem to be important.

The motivation for most of the material in the last two sections of Chapter 15 is unclear. However, some of the work gains meaning when one realizes that it is precursory to Ramanujan's profound work on modular equations in Chapters 19–21. This will be described in Part III [11].

Most of Ramanujan's results in Section 2–7 are expressed in terms of Bernoulli numbers B_n. Recall that, for example, from Titchmarsh's book [3, p. 19], for each positive integer n,

$$\zeta(1 - n) = (-1)^{n+1}B_n/n, \tag{0.1}$$

where $\zeta(s)$ denotes the Riemann zeta-function. Since the values $\zeta(1 - n)$, rather than Bernoulli numbers, arise naturally in our proofs, and since the former notation is more economical, we shall generally express Ramanujan's results in terms of $\zeta(s)$.

We quote precisely Ramanujan's cryptic formulation of Entry 1 and its corollary.

Entry 1.

$$\text{``}h \sum_{k=1}^{\infty} \varphi(kh) = \int_0^{\infty} \varphi(x)\, dx + F(h),$$

where $F(h)$ can be found by expanding the left and writing the constant instead of a series and $F(0) = 0$."

If we formally apply the Euler–Maclaurin formula, (10.5) of Chapter 13, to $f(x) = \varphi(hx)$ on $[0, \infty)$, we find that

$$F(h) = h \sum_{k=1}^{\infty} \varphi(kh) - \int_0^{\infty} \varphi(x) \, dx$$

$$= \sum_{k=1}^{m} \frac{(-1)^{k+1} B_k}{k!} \varphi^{(k-1)}(0) h^k + hR_m, \qquad (1.1)$$

where here

$$R_m = \frac{(-1)^{m-1}}{m!} \int_0^{\infty} B_m(t - [t]) f^{(m)}(t) \, dt$$

and where we have assumed that $\varphi^{(k-1)}(\infty) = 0$ for $1 \le k \le m$. If φ is such that (1.1) is valid and if furthermore hR_m tends to 0 as h tends to 0, then $F(0) = 0$ as stated in Entry 1.

Corollary. "*If*

$$h\varphi(h) = ah^p + bh^q + ch^r + dh^s + \cdots,$$

then

$$h \sum_{k=1}^{\infty} \varphi(kh) = \int_0^{\infty} \varphi(x) \, dx - \frac{aB_p h^p}{p} - \frac{bB_q h^q}{q} - \cdots ."$$

Apparently, Ramanujan assumes that $p, q, \ldots$ are integers with $2 \le p < q < \cdots$ in his application of (1.1). If (1.1) holds for φ and $R_m = O(h^m)$ for each $m \ge 1$, then this corollary yields a valid asymptotic formula as h tends to 0.

Ramanujan next observes that "if the expansion of $\varphi(h)$ be an infinite series, then that of $F(h)$ also will be an infinite series; but if most of the numbers p, q, r, s, t, etc., be odd integers $F(h)$ appears to terminate. In this case the hidden part of $F(h)$ can't be expanded in ascending powers of h and is very rapidly diminishing when h is slowly diminishing and consequently can be neglected for practical purposes when h is small."

The first part of this observation refers to the fact that the coefficients B_p, $B_q, \ldots$ in the corollary vanish when $p, q, \ldots$ are odd integers greater than 1. The latter part about "the hidden part of $F(h)$" refers to situations as in the following Example 1 (where $\varphi(h) = (1 + h^2)^{-1}$) and Example 2 (where $\varphi(h) = \exp(-h^2)$), in which (1.1) takes the form $F(h) = -h/2 + hR_m$ for any positive integer m. Ramanujan's claim is that $F(h) + h/2 = hR_m$ tends rapidly to 0 as h tends to 0. Indeed, in the corrected versions of his Examples 1 and 2 which we are about to present, it will be seen that $F(h) + h/2 \sim \exp(-2\pi/h)$ and $F(h) + h/2 \sim \exp(-\pi^2/h^2)$, respectively. It would be interesting to obtain asymptotic estimates of $F(h)$ for other classes of even meromorphic functions $\varphi(h)$ as h tends to 0 (see Example (iv) of Section 3 and (8.3)).

Example 1. *If* $\varphi(h) = 1/(1 + h^2)$, *then*

$$F(h) = \frac{\pi}{e^{2\pi/h} - 1} - \frac{h}{2},$$

and

$$F(\tfrac{1}{10}) = \pi/(e^{20\pi} - 1) - \tfrac{1}{20}.$$

PROOF. By Entry 1,

$$F(h) = h \sum_{k=1}^{\infty} \frac{1}{1 + h^2 k^2} - \int_0^{\infty} \frac{dx}{1 + x^2}$$

$$= \frac{\pi}{2} \coth \frac{\pi}{h} - \frac{h}{2} - \frac{\pi}{2}$$

$$= \frac{\pi}{e^{2\pi/h} - 1} - \frac{h}{2}. \qquad \square$$

Ramanujan (p. 181) claims that $F(h) = 2\pi/(e^{2\pi/h} - 1)$, and so his value of $F(\tfrac{1}{10})$ is also incorrect.

Example 2. *If* $\varphi(h) = \exp(-h^2)$, *then*

$$F(\tfrac{1}{10}) \approx -\tfrac{1}{20} + \sqrt{\pi} e^{-100\pi^2}.$$

PROOF. By Entry 1 and the well-known transformation formula for the classical theta-function $\theta(z)$, found in the corollary to Entry 7 of Chapter 14,

$$F(h) = h \sum_{k=1}^{\infty} e^{-h^2 k^2} - \int_0^{\infty} e^{-x^2} \, dx$$

$$= -\frac{h}{2} + \frac{\sqrt{\pi}}{2} \sum_{k=-\infty}^{\infty} e^{-\pi^2 k^2/h^2} - \frac{\sqrt{\pi}}{2}$$

$$= -\frac{h}{2} + \sqrt{\pi} \sum_{k=1}^{\infty} e^{-\pi^2 k^2/h^2}. \tag{1.2}$$

The proposed approximation for $F(\tfrac{1}{10})$ readily follows. $\qquad \square$

In contrast, Ramanujan asserts that "$F(\tfrac{1}{10})$ is very nearly $10\sqrt{\pi} e^{-100\pi^2}$." Entry 2 is the special case $p = 1$ of Entry 3 below.

Example (i). *As x tends to $0+$,*

$$\sum_{k=1}^{\infty} e^{-kx} \operatorname{Log} k \sim \frac{-\gamma - \operatorname{Log} x}{x} + \tfrac{1}{2} \operatorname{Log}(2\pi),$$

where γ denotes Euler's constant.

PROOF. This follows from the case $p = m = 1$ in Theorem 3.2, since $\zeta'(0) = -\frac{1}{2} \text{Log}(2\pi)$ (Titchmarsh [3, p. 20]).   □

Example (ii). *Let $d(n)$ denote the number of positive integral divisors of the positive integer n. Then as x tends to ∞,*

$$\sum_{n \leq x} d(n) \sim x \text{ Log } x + (2\gamma - 1)x, \tag{2.1}$$

where γ denotes Euler's constant.

This asymptotic formula is a well-known result in elementary number theory. Let $\Delta(x)$ denote the difference of the left and right sides in (2.1). By elementary methods, $\Delta(x) = O(\sqrt{x})$, as x tends to ∞. (See, for example, Hardy and Wright's book [1, p. 264].) At present, the best O-estimate that we have is

$$\Delta(x) = O(x^{7/22 + \varepsilon}),$$

for each $\varepsilon > 0$, which is due to Iwaniec and Mozzochi [1]. On the other hand, Hafner [1] has shown that, for some constant $c > 0$,

$$\Delta(x) = \Omega_+((x \text{ Log } x)^{1/4}(\text{Log Log } x)^{(3 + 2 \text{ Log } 2)/4} \exp(-c(\text{Log Log Log } x)^{1/2})),$$

which is the best Ω theorem at present. The problem of determining the order of $\Delta(x)$ is known as the "divisor problem" and is one of the most difficult and famous problems in the analytic theory of numbers. It is conjectured that $\Delta(x) = O(x^{1/4 + \varepsilon})$ for every $\varepsilon > 0$. For a fuller discussion of the divisor problem along with historical references, consult the books of Ivić [1, Chapter 3] and Graham and Kolesnik [1].

Example (iii). *If p_k denotes the kth prime, then*

$$\sum_{k=1}^{\infty} e^{-kx} p_k \sim -\frac{\text{Log } x}{x^2},$$

as x tends to $0+$.

PROOF. From Landau's treatise [1, p. 215],

$$p_k = k \text{ Log } k + O(k \text{ Log Log } k),$$

as k tends to ∞. Thus, as t tends to ∞,

$$F(t) := \sum_{k \leq t} p_k = \frac{1}{2}t^2 \text{ Log } t + O(t^2 \text{ Log Log } t).$$

Therefore, by partial summation,

$$\sum_{k=1}^{\infty} e^{-kx} p_k = \int_0^{\infty} F(t)xe^{-tx} \, dt$$

$$= x \int_0^{\infty} \frac{1}{2}t^2(\text{Log } t + O(\text{Log Log } t))e^{-tx} \, dt$$

$$= \frac{1}{2x^2} \int_0^\infty e^{-u}u^2 \operatorname{Log}(u/x)\,du + O\left(x^{-2} \int_{3x}^\infty e^{-u}u^2 \operatorname{Log} \operatorname{Log}(u/x)\,du\right)$$

$$\sim \frac{-\operatorname{Log} x}{2x^2} \int_0^\infty e^{-u}u^2\,du = \frac{-\operatorname{Log} x}{x^2},$$

as x tends to $0+$. $\qquad\square$

Example (iv). *Write*

$$\left(\sum_{k=-\infty}^\infty (-x)^{k^2}\right)^{-1} = \sum_{n=0}^\infty I(n)x^n, \qquad |x| < 1.$$

Then "$I(n)$ is of the order

$$\frac{1}{4n}\left\{\cosh \pi\sqrt{n} - \frac{\sinh \pi\sqrt{n}}{\pi\sqrt{n}}\right\}.$$"

This declaration essentially appears in Ramanujan's first two letters to Hardy [16, pp. xxvii, 352] and is not correctly stated. (This was pointed out by Watson [2].) However, a corrected version can be found in the famous paper of Hardy and Ramanujan [1] (Hardy [6, p. 334], Ramanujan [16, p. 304]), wherein their asymptotic formula for the partition function $p(n)$ is established. Example (iv) is an analogue of this theorem in that the generating function for $p(n)$, the Dedekind eta-function, is essentially replaced by the classical theta-function $\theta(\tau)$, where $x = -e^{-\pi i\tau}$. Littlewood [1] (see also Andrews' book [2, pp. 68–69]) has written that in the collaboration with Hardy on $p(n)$, Ramanujan kept insisting that a highly accurate formula for $p(n)$ existed. This persistence especially pushed Hardy to the discovery of their amazingly precise asymptotic formula. Example (iv) shows that the foundation for Ramanujan's confidence originated in India several years earlier.

In 1937, Rademacher [1] discovered an exact formula for $p(n)$, which yields, of course, Hardy and Ramanujan's asymptotic formula as a corollary. (See also Ayoub's text [1, Chap. 3] for a proof of Rademacher's theorem.) Zuckerman [1] shortly thereafter found an exact formula for $I(n)$ in Example (iv) as well as for the Fourier coefficients of the reciprocals of other modular forms including all the classical theta-functions. Simpler formulas for the Fourier coefficients of the reciprocals of the classical theta-functions, and simpler proofs, have been derived by Goldberg [1].

Riesel [1] examined Entries 3–5 below and asserted that they were incorrect, because he interpreted them as exact formulas. As asymptotic formulas, Entries 3–5 are, indeed, correct.

We begin Section 3 by stating Entry 3 and its corollary, as Ramanujan probably intended them. Throughout, γ denotes Euler's constant and B_n is the nth Bernoulli number.

Entry 3. *Suppose that m and p denote positive integers. Then as x approaches $0+$,*

$$\sum_{k=1}^{\infty} e^{-k^p x} k^{m-1} \sim \frac{\Gamma(m/p)}{px^{m/p}} - \sum_{k=0}^{\infty} (-1)^{m+pk} \frac{B_{m+pk}}{m+pk} \frac{(-x)^k}{k!}.$$

Corollary. *Suppose that p is a positive integer. Then as x approaches $0+$,*

$$\sum_{k=1}^{\infty} \frac{e^{-k^p x}}{k} \sim \frac{-\text{Log } x}{p} + \gamma(1 - 1/p) - \sum_{k=1}^{\infty} (-1)^{pk} \frac{B_{pk}}{pk} \frac{(-x)^k}{k!}.$$

By (0.1), Entry 3 and its corollary follow from our Theorem 3.1 below.

Throughout the sequel, Z^- denotes the set of nonpositive integers and $\sigma = \text{Re } s$. The symbol $\sum_k^*$ indicates that those values of k yielding undefined summands are excluded from the summation. The residue of a meromorphic function f at a pole z_0 is denoted by $R(z_0)$. (The identification of f will always be clear.)

Theorem 3.1. *Let $p > 0$, let m denote a complex number, and define*

$$f(x) = \sum_{k=1}^{\infty} e^{-xk^p} k^{m-1}.$$

Then as x tends to $0+$,

$$f(x) \sim \frac{\Gamma(m/p)}{px^{m/p}} + \sum_{k=0}^{\infty} \zeta(1 - m - pk) \frac{(-x)^k}{k!}, \tag{3.1}$$

if $m/p \notin Z^-$, while if $m/p = -r \in Z^-$,

$$f(x) \sim \left\{ \frac{1}{p}(H_r - \gamma) + \gamma - \frac{1}{p} \text{Log } x \right\} \frac{(-x)^r}{r!}$$

$$+ \sum_{k=0}^{\infty}{}^* \zeta(1 - m - pk) \frac{(-x)^k}{k!}, \tag{3.2}$$

where γ denotes Euler's constant and

$$H_r = \sum_{k=1}^{r} \frac{1}{k}.$$

PROOF. We shall assume that m is real; the more general result can be established by similar lines of reasoning.

Using the definition of f and inverting the order of summation and integration by absolute convergence, we easily find that

$$\int_0^{\infty} f(x) x^{s-1} \, dx = \Gamma(s) \zeta(1 - m + ps),$$

provided that $\sigma > \sup\{0, m/p\}$. By Mellin's inversion formula (Titchmarsh [3, p. 33]),

$$f(x) = \frac{1}{2\pi i} \int_{a-i\infty}^{a+i\infty} \Gamma(s) \zeta(1 - m + ps) x^{-s} \, ds, \tag{3.3}$$

where $a > \sup\{0, m/p\}$.

Consider now

$$I_{M,T} = \frac{1}{2\pi i} \int_{C_{M,T}} \Gamma(s)\zeta(1 - m + ps)x^{-s}\,ds,$$

where $C_{M,T}$ is the positively oriented rectangle with vertices $a \pm iT$ and $-M \pm iT$, where $T > 0$ and $M = N + \frac{1}{2}$. Here N is an integer chosen sufficiently large to ensure that $N > |m|/p$. The integrand has simple poles at $s = m/p$ and $s = 0, -1, -2, \ldots, -N$ on the interior of $C_{M,T}$, unless $m/p = -r \in Z^-$, in which case there exists a double pole at $s = -r$. By the residue theorem, if $m/p \notin Z^-$,

$$I_{M,T} = \frac{\Gamma(m/p)}{px^{m/p}} + \sum_{k=0}^{N} \zeta(1 - m - pk)\frac{(-x)^k}{k!}, \qquad (3.4)$$

while if $m/p = -r \in Z^-$,

$$I_{M,T} = R(-r) + \sum_{k=0}^{N}{}^* \zeta(1 - m - pk)\frac{(-x)^k}{k!}. \qquad (3.5)$$

Now,

$$\Gamma(s) = \frac{(-1)^r}{r!(s+r)} + \frac{(-1)^r}{r!}(H_r - \gamma) + \cdots, \qquad 0 < |s + r| < 1,$$

$$\zeta(1 - m + ps) = \frac{1}{p(s+r)} + \gamma + \cdots, \qquad (3.6)$$

and

$$x^{-s} = x^r - x^r(\text{Log } x)(s + r) + \cdots. \qquad (3.7)$$

Hence, if $m/p = -r \in Z^-$,

$$R(-r) = \left\{\frac{1}{p}(H_r - \gamma) + \gamma - \frac{1}{p}\text{Log } x\right\}\frac{(-x)^r}{r!}. \qquad (3.8)$$

Putting (3.8) into (3.5), we see from (3.3)–(3.5) that, in order to establish (3.1) and (3.2), it suffices to show that

$$\int_{-M}^{a} \Gamma(\sigma + iT)\zeta(1 - m + p(\sigma \pm iT))^{-\sigma \mp iT}\,d\sigma = o(1) \qquad (3.9)$$

as T tends to ∞, and then that

$$\int_{-\infty}^{\infty} \Gamma(-M + it)\zeta(1 - m + p(-M + it))x^{M-it}\,dt \ll x^M, \qquad (3.10)$$

as x approaches $0+$.

Recall (Copson [2, p. 224]) the following form of Stirling's formula. Uniformly for σ in any finite interval, as $|t|$ tends to ∞,

$$|\Gamma(s)| \sim (2\pi)^{1/2}e^{-\pi|t|/2}|t|^{\sigma - 1/2}. \qquad (3.11)$$

Also (Titchmarsh [3, p. 81]), uniformly for $\sigma \geq \sigma_0$, there exists a constant

$k = k(\sigma_0) > 0$, such that

$$\zeta(s) = O(|t|^k), \tag{3.12}$$

as $|t|$ tends to ∞. Estimates (3.9) and (3.10) clearly follow from (3.11) and (3.12), and the proof of Theorem 3.1 is complete.   □

The proofs that follow are similar to that of Theorem 3.1, and so we shall not provide all the details. In particular, estimates analogous to (3.9) and (3.10) are always needed, and they are always obtained in a manner very much like that described above.

The case $m = p = 1$ of Theorem 3.2 below yields Example (i) of Section 2. Observe that Theorem 3.2 follows formally from Theorem 3.1 via differentiation with respect to m. This (nonrigorous) procedure may be what Ramanujan used to deduce Example (i) of Section 2.

Theorem 3.2. *Let $p > 0$ and define*

$$g(x) = \sum_{k=1}^{\infty} e^{-xk^p} k^{m-1} \operatorname{Log} k.$$

Then, if $m/p \notin Z^-$, as x tends to $0+$,

$$g(x) \sim \frac{\Gamma'(m/p) - \Gamma(m/p) \operatorname{Log} x}{p^2 x^{m/p}} - \sum_{k=0}^{\infty} \zeta'(1 - m - pk) \frac{(-x)^k}{k!}.$$

PROOF. As before, we may, without loss of generality, assume that m is real.

Inverting the order of summation and integration by absolute convergence, we readily deduce that

$$\int_0^{\infty} g(x) x^{s-1} \, dx = -\Gamma(s)\zeta'(1 - m + ps),$$

if $\sigma > \sup\{0, m/p\}$. From Mellin's inversion formula,

$$g(x) = -\frac{1}{2\pi i} \int_{a-i\infty}^{a+i\infty} \Gamma(s)\zeta'(1 - m + ps)x^{-s} \, ds,$$

where $a > \sup\{0, m/p\}$.

Consider next

$$I_{M,T} = -\frac{1}{2\pi i} \int_{C_{M,T}} \Gamma(s)\zeta'(1 - m + ps)x^{-s} \, ds,$$

where $C_{M,T}$ denotes the same rectangular contour as in the proof of Theorem 3.1. Since $m/p \notin Z^-$, by the residue theorem,

$$I_{M,T} = \frac{\Gamma'(m/p) - \Gamma(m/p) \operatorname{Log} x}{p^2 x^{m/p}} - \sum_{k=0}^{N} \zeta'(1 - m - pk) \frac{(-x)^k}{k!}.$$

The remainder of the proof is similar to that of Theorem 3.1, but an estimate just like (3.12) is needed for $\zeta'(s)$ in place of $\zeta(s)$. This can be obtained by differentiating the formula for $\zeta(s)$ obtained from the Euler–Maclaurin formula, (10.5) in Chapter 13, with $f(t) = t^{-s}, a = 1, b = \infty$, and m sufficiently large.   □

Ramanujan now records several examples to illustrate his results. Example (i) is the case $p = 2$ of the corollary following Entry 3. Example (ii) is the case $p = 4, m = 2$ of Entry 3. Example (iii) is the case $p = 3, m = 2$ of Entry 3.

Example (iv). *As x tends to $0+$,*

$$\sum_{k=1}^{\infty} \frac{e^{-k^2 x}}{k^2} = \frac{\pi^2}{6} + \frac{x}{2} - \sqrt{\pi x} + O(\sqrt{x} e^{-\pi^2/x}).$$

PROOF. From (1.2), as t tends to $0+$,

$$\sum_{k=1}^{\infty} e^{-k^2 t} = -\frac{1}{2} + \frac{1}{2}\sqrt{\frac{\pi}{t}} + O\left(\frac{e^{-\pi^2/t}}{\sqrt{t}}\right).$$

Integrating this equality over $[0, x]$, we find that

$$\sum_{k=1}^{\infty} \frac{(1 - e^{-k^2 x})}{k^2} = -\frac{x}{2} + \sqrt{\pi x} + O(\sqrt{x} e^{-\pi^2/x}),$$

as x tends to $0+$. Since $\zeta(2) = \pi^2/6$, the proposed result follows.   □

Example (v) records the case $m = 3, p = 6$ of Entry 3.

Entry 4. *Let $p > 0$ and let m and d be complex numbers with $\text{Re}(pd - m) > 0$. Define*

$$h(x) = \sum_{k=1}^{\infty} \frac{k^{m-1}}{(1 + xk^p)^d}.$$

Then, if $m/p \notin Z^-$, as x tends to $0+$,

$$h(x) \sim \frac{\Gamma(m/p)\Gamma(d - m/p)}{p\Gamma(d)x^{m/p}} + \sum_{k=0}^{\infty} (d)_k \zeta(1 - m - pk)\frac{(-x)^k}{k!},$$

where, as usual, $(d)_k = \Gamma(d + k)/\Gamma(d)$.

PROOF. As in previous proofs, we may assume without loss of generality that m and d are real.

Inverting the order of summation and integration by absolute convergence, we deduce that

$$\int_0^{\infty} h(x)x^{s-1}\, dx = \sum_{k=1}^{\infty} k^{m-1-ps} \int_0^{\infty} \frac{u^{s-1}}{(1 + u)^d}\, du$$

$$= \zeta(1 - m + ps)\frac{\Gamma(s)\Gamma(d - s)}{\Gamma(d)},$$

provided that $m/p < \sigma < d$ and $\sigma > 0$. By Mellin's inversion formula,

$$h(x) = \frac{1}{2\pi i} \int_{a-i\infty}^{a+i\infty} \frac{\Gamma(s)\Gamma(d-s)}{\Gamma(d)} \zeta(1-m+ps)x^{-s}\, ds,$$

where $m/p < a < d$ and $a > 0$.

With the same oriented rectangle $C_{M,T}$ as in previous proofs, we find that

$$\frac{1}{2\pi i} \int_{C_{M,T}} \frac{\Gamma(s)\Gamma(d-s)}{\Gamma(d)} \zeta(1-m+ps)x^{-s}\, ds$$

$$= \frac{\Gamma(m/p)\Gamma(d-m/p)}{p\Gamma(d)x^{m/p}} + \sum_{k=0}^{N} (d)_k \zeta(1-m-pk)\frac{(-x)^k}{k!},$$

by the residue theorem. The remainder of the proof now parallels that of
Theorem 3.1.   □

Ramanujan concludes Section 4 with an example, which is the case $p = 8$,
$m = 2$, $d = \frac{1}{2}$ of Entry 4.

Entry 5 is the case $n = q = 1$, $m \neq p$ of Theorem 6.1 in Section 6. The
corollary of Entry 5 is the case $n = q = 1$, $m = p$ of Theorem 6.1.

In the case $m/p \neq n/q$, Theorem 6.1 is a somewhat more general version of
Entry 6, and in the case $m/p = n/q$, Theorem 6.1 generalizes a result marked
by "N.B." immediately following Entry 6.

Theorem 6.1. *Let $p, q > 0$ and let m and n be complex numbers. Define*

$$f_{m,n}(x) = \sum_{j,k=1}^{\infty} e^{-xk^p j^q} k^{m-1} j^{n-1}.$$

Then, if $m/p, n/q \notin Z^-$, as x tends to $0+$,

$$f_{m,n}(x) \sim \frac{\Gamma(m/p)}{px^{m/p}} \zeta(1-n+qm/p) + \frac{\Gamma(n/q)}{qx^{n/q}} \zeta(1-m+pn/q)$$

$$+ \sum_{k=0}^{\infty} \zeta(1-m-pk)\zeta(1-n-qk)\frac{(-x)^k}{k!}, \quad \text{if } pn \neq qm,$$

and

$$f_{m,n}(x) \sim \frac{\Gamma(m/p)}{pqx^{m/p}}\left\{(p+q)\gamma + \frac{\Gamma'}{\Gamma}\left(\frac{m}{p}\right) - \text{Log } x\right\}$$

$$+ \sum_{k=0}^{\infty} \zeta(1-m-pk)\zeta(1-n-qk)\frac{(-x)^k}{k!}, \quad \text{if } pn = qm.$$

In the case $p = q = 1$,

$$f_{1,1}(x) = \sum_{r=1}^{\infty} e^{-rx}d(r),$$

where $d(r)$ denotes the number of positive divisors of r. Titchmarsh [3, p. 140] used the asymptotic expansion for $f_{1,1}(x)$, which was first proved by Wigert [1], and which is a special case of Theorem 6.1, in obtaining mean value theorems for $\zeta(s)$.

PROOF. Without loss of generality, assume that m and n are real.

Inverting the order of summation and integration, we readily see that

$$\int_0^\infty f_{m,n}(x)x^{s-1}\,dx = \Gamma(s)\zeta(1-m+ps)\zeta(1-n+qs),$$

under the assumption that $\sigma > \sup\{0, m/p, n/q\}$. By Mellin's inversion formula,

$$f_{m,n}(x) = \frac{1}{2\pi i}\int_{a-i\infty}^{a+i\infty}\Gamma(s)\zeta(1-m+ps)\zeta(1-n+qs)x^{-s}\,ds,$$

provided that $a > \sup\{0, m/p, n/q\}$.

Let $C_{M,T}$ denote the rectangular contour given in the proof of Theorem 3.1. By the residue theorem,

$$\frac{1}{2\pi i}\int_{C_{M,T}}\Gamma(s)\zeta(1-m+ps)\zeta(1-n+qs)x^{-s}\,ds$$

$$= \delta_{m,n}(R(m/p) + R(n/q)) + \sum_{k=0}^{N}\zeta(1-m-pk)\zeta(1-n-qk)\frac{(-x)^k}{k!},$$

where $\delta_{m,n} = 1$ if $pn \neq qm$, and $\delta_{m,n} = \frac{1}{2}$ if $pn = qm$. If $pn \neq qm$, the residues $R(m/p)$ and $R(n/q)$ are routinely calculated. In the case that $pn = qm$, the integrand has a double pole at $m/p = n/q$ instead of simple poles at m/p and n/q. With the use of (3.6) (and its analogue with q and n in place of p and m, respectively), (3.7), and the Taylor expansion of $\Gamma(s)$ about $s = m/p$, we may easily calculate $R(m/p)$ for this double pole. The remainder of the proof is almost identical to that of Theorem 3.1.   □

Theorems 6.2 and 6.3 below supplement Theorem 6.1 by providing asymptotic expansions when m or n or both are equal to 0.

Theorem 6.2. *Let $p, q > 0$ and let n be complex. Then, if $n/q \notin Z^-$,*

$$f_{0,n}(x) \sim \frac{\Gamma(n/q)\zeta(1+pn/q)}{qx^{n/q}} - \frac{\zeta(1-n)\,\mathrm{Log}\,x}{p} + A_n$$

$$+ \sum_{k=1}^{\infty}\zeta(1-pk)\zeta(1-n-qk)\frac{(-x)^k}{k!},$$

as x tends to $0+$, where

$$A_n = \frac{\zeta'(1-n)q}{p} + \gamma\left(1 - \frac{1}{p}\right)\zeta(1-n).$$

PROOF. The proof is identical to that of Theorem 6.1 except in one respect. The former simple poles of the integrand at $s = 0$ and $s = m/p$ now coalesce to form a double pole at $s = 0$. An elementary calculation yields

$$R(0) = A_n - \frac{\zeta(1 - n) \, \text{Log} \, x}{p},$$

and the desired result follows.   □

Theorem 6.3. *Let* $p, q > 0$. *Let*

$$A_0 = \frac{\pi^2}{12pq} + \gamma^2 \left(1 + \frac{1}{2pq} - \frac{1}{p} - \frac{1}{q} \right) - c_1 \left(\frac{p}{q} + \frac{q}{p} \right),$$

where

$$c_1 = \lim_{n \to \infty} \left(\sum_{r=1}^{n} \frac{\text{Log} \, r}{r} - \frac{\text{Log}^2 \, n}{2} \right).$$

Then as x *tends to* $0+$,

$$f_{0,0}(x) \sim \frac{\text{Log}^2 \, x}{2pq} - \frac{(p + q - 1)\gamma \, \text{Log} \, x}{pq} + A_0$$

$$+ \sum_{k=1}^{\infty} \zeta(1 - pk)\zeta(1 - qk) \frac{(-x)^k}{k!}.$$

PROOF. The proof is the same as for Theorem 6.1, except that the former simple poles at $s = 0$, m/p, and n/q now coalesce to yield a triple pole at $s = 0$. To calculate $R(0)$, we require the Laurent expansions

$$\Gamma(s) = \frac{1}{s} - \gamma + \left(\frac{1}{2} \gamma^2 + \frac{\pi^2}{12} \right) s + \cdots, \qquad 0 < |s| < 1,$$

$$\zeta(1 + ps) = \frac{1}{ps} + \gamma - c_1 ps + \cdots,$$

$$\zeta(1 + qs) = \frac{1}{qs} + \gamma - c_1 qs + \cdots,$$

and

$$x^{-s} = 1 - s \, \text{Log} \, x + \tfrac{1}{2} s^2 \, \text{Log}^2 \, x + \cdots.$$

The Laurent expansion for $\Gamma(s)$ about $s = 0$ may be calculated from the Weierstrass product representation for $\Gamma(s)$, while the Laurent expansion of $\zeta(s)$ about $s = 1$ is found in Entry 13 of Chapter 7 (Part I [9]). It transpires that

$$R(0) = \frac{\text{Log}^2 \, x}{2pq} - \frac{(p + q - 1)\gamma \, \text{Log} \, x}{pq} + A_0.$$

The desired result now follows as in the proof of Theorem 6.1.   □

Note that A_n does not approach A_0 as n tends to 0.

Ramanujan concludes Section 6 with two examples. Example (i) is the case $p = 2, n = q = 1$ of Theorem 6.2, and Example (ii) is the case $m = 3, p = q = n = 1$ of Theorem 6.1.

As customary, put $\sigma_r(s) = \sum d^r$, where the sum is over the positive integers d which divide s.

We next offer a corrected version of Ramanujan's Entry 7. (Ramanujan mistakenly indicates $\sigma_{n-1}(s)$ instead of $\sigma_{n-m}(s)$ below.)

Entry 7. *As x tends to* $0+$,

$$\sum_{s=1}^{\infty} \frac{s^{m-1}\sigma_{n-m}(s)}{e^{sx}-1} \sim \frac{\Gamma(m)}{x^m}\zeta(m)\zeta(1+m-n)$$

$$+ \frac{\Gamma(n)}{x^n}\zeta(n)\zeta(1+n-m) + \frac{\zeta(2-m)\zeta(2-n)}{x}$$

$$+ \sum_{k=0}^{\infty} \zeta(1-m-k)\zeta(1-n-k)\zeta(-k)\frac{(-x)^k}{k!}, \tag{7.1}$$

provided that $m \neq n$, $m \neq 1$, $n \neq 1$, *and* $m, n \notin Z^-$.

Setting $s = jk$ below, we see that the sum on the left side of (7.1) equals

$$\sum_{s=1}^{\infty} s^{m-1}\sigma_{n-m}(s) \sum_{i=1}^{\infty} e^{-isx} = \sum_{i,j,k=1}^{\infty} e^{-ijkx}(jk)^{m-1}j^{n-m};$$

that is, the left side of (7.1) equals

$$f_{m,n,l}(x) := \sum_{i,j,k=1}^{\infty} e^{-xk^pj^qi^r}k^{m-1}j^{n-1}i^{l-1} \tag{7.2}$$

for the special choices $p = q = r = l = 1$. In Theorem 7.1 below, we give an asymptotic formula, as x tends to $0+$, for the triple sum in (7.2), under the general conditions m/p, n/q, $l/r \notin Z^-$, $qm \neq np$, $rm \neq pl$, and $rn \neq ql$. Ramanujan's formula (7.1) then follows from Theorem 7.1 upon setting $p = q = r = l = 1$.

Theorem 7.1. *Let* $p, q, r > 0$ *and let* $m, n,$ *and* l *denote complex numbers. Let* $f_{m,n,l}(x)$ *be defined by* (7.2). *Suppose that* m/p, n/q, $l/r \notin Z^-$, $qm \neq np$, $rm \neq pl$, *and* $rn \neq ql$. *Then as* x *tends to* $0+$,

$$f_{m,n,l}(x) \sim \frac{\Gamma(m/p)}{px^{m/p}}\zeta(1-n+qm/p)\zeta(1-l+rm/p)$$

$$+ \frac{\Gamma(n/q)}{qx^{n/q}}\zeta(1-m+pn/q)\zeta(1-l+rn/q)$$

$$+ \frac{\Gamma(l/r)}{rx^{l/r}}\zeta(1-m+pl/r)\zeta(1-n+ql/r)$$

$$+ \sum_{k=0}^{\infty} \zeta(1-m-pk)\zeta(1-n-qk)\zeta(1-l-rk)\frac{(-x)^k}{k!}.$$

PROOF. The proof follows precisely along the same lines as the proofs of Theorems 3.1 and 6.1.   □

Ramanujan concludes Section 7 with an example, which is the case $m = 3$, $n = 5$ of (7.1). Again he mistakenly indicates $\sigma_{n-1}(s)$ in place of $\sigma_{n-m}(s)$ in the example, and, moreover, he inadvertently omits the term $-1/(1440x)$ in the asymptotic expansion.

Clearly, the theorems that we have proved can be generalized and extended even further. In particular, restrictions imposed on the parameters can be lifted. The computation of the residues would then be somewhat more difficult.

At the beginning of Section 8, Ramanujan remarks that "if $F(h)$ in XV 1 terminates we do not know how far the result is true. But from the following and similar ways we can calculate the error in such cases." To illustrate these cryptic remarks, Ramanujan indicates a method for calculating the error in the asymptotic expansion

$$\sum_{k=1}^{\infty} \frac{1}{e^{k^2 x} - 1} = \frac{\pi^2}{6x} + \frac{1}{2}\sqrt{\frac{\pi}{x}}\zeta(\tfrac{1}{2}) + \tfrac{1}{4} + o(1), \tag{8.1}$$

as x tends to $0+$, which is the case $p = 2$, $q = m = n = 1$ in Theorem 6.1. In fact, he indicates that the equalities

$$\int_0^{\infty} \sum_{k=1}^{\infty} e^{-k^2 x} \cos(ax)\, dx = \sum_{k=1}^{\infty} \frac{k^2}{a^2 + k^4}$$

$$= \frac{\pi}{2\sqrt{2a}} \frac{\sinh(\pi\sqrt{2a}) - \sin(\pi\sqrt{2a})}{\cosh(\pi\sqrt{2a}) - \cos(\pi\sqrt{2a})} \tag{8.2}$$

can be used to deduce the following exact formula extending (8.1).

Entry 8. If $x > 0$, then

$$\sum_{k=1}^{\infty} \frac{1}{e^{k^2 x} - 1} = \frac{\pi^2}{6x} + \frac{1}{2}\sqrt{\frac{\pi}{x}}\zeta(\tfrac{1}{2}) + \tfrac{1}{4}$$

$$+ \sqrt{\frac{\pi}{2x}} \sum_{k=1}^{\infty} \frac{1}{\sqrt{k}} \left\{ \frac{\cos\left(\frac{\pi}{4} + 2\pi\sqrt{\frac{\pi k}{x}}\right) - e^{-2\pi\sqrt{\pi k/x}}\cos\left(\frac{\pi}{4}\right)}{\cosh\left(2\pi\sqrt{\frac{\pi k}{x}}\right) - \cos\left(2\pi\sqrt{\frac{\pi k}{x}}\right)} \right\}. \tag{8.3}$$

This is truly a remarkable formula. The left side can be construed as a modification of the theta-function

$$\theta(x) = 1 + 2\sum_{k=1}^{\infty} \frac{1}{e^{k^2 x}}.$$

Thus, Entry 8 is an analogue of the inversion formula for $\theta(x)$.

Before proving Entry 8, we first establish (8.2).

The first equality easily follows from inverting the order of summation and integration on the left side and using a well-known integral evaluation (Gradshteyn and Ryzhik [1, p. 477]).

To prove the second equality, we shall expand the right side into partial fractions. An elementary calculation shows that the nonzero zeros of $\cosh(\pi\sqrt{2a}) - \cos(\pi\sqrt{2a})$ are at $a = \pm k^2 i$, $1 \le k < \infty$, and that they are simple. Thus, if $R(z_0)$ denotes the residue of the function on the far right side of (8.2) at a simple pole z_0, we find that

$$R(\pm k^2 i) = \mp \frac{i}{2}.$$

Thus, for some entire function $g(a)$,

$$\frac{\pi}{2\sqrt{2a}} \frac{\sinh(\pi\sqrt{2a}) - \sin(\pi\sqrt{2a})}{\cosh(\pi\sqrt{2a}) - \cos(\pi\sqrt{2a})} = \frac{i}{2} \sum_{k=1}^{\infty} \left\{ -\frac{1}{a - k^2 i} + \frac{1}{a + k^2 i} \right\} + g(a)$$

$$= \sum_{k=1}^{\infty} \frac{k^2}{a^2 + k^4} + g(a).$$

Letting a tend to ∞ on both sides above, we find that $g(a)$ tends to 0. Hence, $g(a)$ is a bounded entire function, and so by Liouville's theorem $g(a)$ is constant. Clearly, this constant is zero. Hence, the proof of the second equality in (8.2) is complete.

A different proof of the second equality in (8.2) may be found in a paper of Glaisher [1].

PROOF OF ENTRY 8. Setting $x = \pi y$, we restate (8.3) in the form

$$\sum_{k=1}^{\infty} \frac{1}{e^{\pi k^2 y} - 1} - \left(\frac{\pi}{6y} + \frac{\zeta(\frac{1}{2})}{2\sqrt{y}} + \frac{1}{4} \right) = R, \tag{8.4}$$

where

$$R = \frac{1}{2} \sum_{k=1}^{\infty} \frac{1}{\sqrt{ky}} \left\{ \frac{\cos(2\pi\sqrt{k/y}) - \sin(2\pi\sqrt{k/y}) - e^{-2\pi\sqrt{k/y}}}{\cosh(2\pi\sqrt{k/y}) - \cos(2\pi\sqrt{k/y})} \right\}$$

$$= \frac{2}{\pi y} \sum_{k=1}^{\infty} \frac{\pi}{2\sqrt{2a}} \left\{ \frac{\sinh(\pi\sqrt{2a}) - \sin(\pi\sqrt{2a})}{\cosh(\pi\sqrt{2a}) - \cos(\pi\sqrt{2a})} - 1 \right\}, \tag{8.5}$$

where $a := a_k := 2k/y$.

For brevity, set

$$\psi(u) = \sum_{k=1}^{\infty} e^{-\pi k^2 u}, \qquad u > 0.$$

Thus, by (8.2) and (8.5), with $a = 2k/y$,

$$R = 2 \sum_{k=1}^{\infty} \left\{ \int_0^{\infty} \psi(u) \cos(2\pi k u/y) \frac{du}{y} - \frac{1}{4\sqrt{ky}} \right\}$$

$$= 2 \sum_{k=1}^{\infty} \left\{ \int_0^{\infty} \psi(uy) \cos(2\pi k u) \, du - \frac{1}{4\sqrt{ky}} \right\}. \tag{8.6}$$

Now, for $y > 0$,

$$\sum_{k=1}^{\infty} \psi(ky) = \sum_{k=1}^{\infty} \sum_{j=1}^{\infty} e^{-\pi j^2 ky}$$

$$= \sum_{j=1}^{\infty} \sum_{k=1}^{\infty} e^{-\pi j^2 ky}$$

$$= \sum_{j=1}^{\infty} \frac{1}{e^{\pi j^2 y} - 1}. \tag{8.7}$$

By (8.6) and (8.7), the proposed formula (8.4) now becomes

$$\sum_{k=1}^{\infty} \psi(ky) - \frac{\pi}{6y} - \frac{\zeta(\frac{1}{2})}{2\sqrt{y}} - \frac{1}{4} = 2 \sum_{k=1}^{\infty} \left\{ \int_0^{\infty} \psi(uy) \cos(2\pi ku) \, du - \frac{1}{4\sqrt{ky}} \right\}. \tag{8.8}$$

Let $0 < \varepsilon < 1$. Applying the Poisson summation formula, (6.1) of Chapter 14, we deduce that

$$\sum_{k=1}^{\infty} \psi(ky) = \int_{\varepsilon}^{\infty} \psi(uy) \, du + 2 \sum_{k=1}^{\infty} \int_{\varepsilon}^{\infty} \psi(uy) \cos(2\pi ku) \, du. \tag{8.9}$$

From (8.2),

$$\int_0^{\infty} \psi(uy) \, du = \frac{\pi}{6y}.$$

Hence, by (8.9),

$$\sum_{k=1}^{\infty} \psi(ky) - \frac{\pi}{6y} = \lim_{\varepsilon \to 0+} 2 \sum_{k=1}^{\infty} \int_{\varepsilon}^{\infty} \psi(uy) \cos(2\pi ku) \, du. \tag{8.10}$$

By (8.8) and (8.10), it remains to prove that

$$-\frac{1}{4} - \frac{\zeta(\frac{1}{2})}{2\sqrt{y}} = \lim_{\varepsilon \to 0+} 2 \sum_{k=1}^{\infty} \left\{ \int_0^{\varepsilon} \psi(uy) \cos(2\pi ku) \, du - \frac{1}{4\sqrt{ky}} \right\}. \tag{8.11}$$

From the corollary to Entry 7 of Chapter 14, for $u > 0$,

$$\psi(uy) = -\frac{1}{2} + \frac{1}{2\sqrt{uy}} + \frac{1}{\sqrt{uy}} \psi\left(\frac{1}{uy}\right).$$

Therefore,

$$\int_0^{\varepsilon} \psi(uy) \cos(2\pi ku) \, du = -\frac{1}{2} \int_0^{\varepsilon} \cos(2\pi ku) \, du + \frac{1}{2} \int_0^{\varepsilon} \frac{\cos(2\pi ku)}{\sqrt{uy}} \, du$$

$$+ \int_0^{\varepsilon} \frac{\cos(2\pi ku)}{\sqrt{uy}} \psi\left(\frac{1}{uy}\right) du. \tag{8.12}$$

The first term on the right side of (8.12) gives to (8.11) the contribution

$$\lim_{\varepsilon \to 0+} -2 \sum_{k=1}^{\infty} \frac{\sin(2\pi k\varepsilon)}{4\pi k} = \lim_{\varepsilon \to 0+} \tfrac{1}{2}(\varepsilon - [\varepsilon] - \tfrac{1}{2}) = -\tfrac{1}{4}, \qquad (8.13)$$

where we have used (5.2) of Chapter 14. The third expression on the right side of (8.12) contributes to (8.11)

$$\lim_{\varepsilon \to 0+} 2 \sum_{k=1}^{\infty} \int_0^\varepsilon \frac{\cos(2\pi ku)}{\sqrt{uy}} \psi\left(\frac{1}{uy}\right) du = 0, \qquad (8.14)$$

which can be seen after two integrations by parts. By (8.11)–(8.14), it remains to prove that

$$-\frac{\zeta(\tfrac{1}{2})}{2} = \lim_{\varepsilon \to 0+} \sum_{k=1}^{\infty} \left\{ \int_0^\varepsilon \frac{\cos(2\pi ku)}{\sqrt{u}} du - \frac{1}{2\sqrt{k}} \right\}. \qquad (8.15)$$

Now (Gradshteyn and Ryzhik [1, p. 395]),

$$\int_0^\infty \frac{\cos(2\pi ku)}{\sqrt{u}} du = \sqrt{\frac{2}{\pi k}} \int_0^\infty \cos u^2 \, du = \frac{1}{2\sqrt{k}}.$$

Using this in (8.15), we find that (8.15) becomes

$$\zeta(\tfrac{1}{2}) = \lim_{\varepsilon \to 0+} 2 \sum_{k=1}^{\infty} \int_\varepsilon^\infty \frac{\cos(2\pi ku)}{\sqrt{u}} du. \qquad (8.16)$$

We shall again apply the Poisson summation formula. Let $0 < \varepsilon < 1 < N$ and suppose N is *not* an integer. Then

$$\sum_{\varepsilon < k < N} \frac{1}{\sqrt{k}} - \int_\varepsilon^N \frac{du}{\sqrt{u}} = 2 \sum_{k=1}^{\infty} \int_\varepsilon^N \frac{\cos(2\pi ku)}{\sqrt{u}} du. \qquad (8.17)$$

The left side of (8.17) may be written as

$$\int_\varepsilon^N \frac{d([u] - u)}{\sqrt{u}} = \frac{[u] - u}{\sqrt{u}}\Big|_\varepsilon^N + \frac{1}{2} \int_\varepsilon^N \frac{[u] - u}{u^{3/2}} du.$$

Using this in (8.17) and letting N tend to ∞, we deduce that

$$\sqrt{\varepsilon} + \frac{1}{2} \int_\varepsilon^\infty \frac{[u] - u}{u^{3/2}} du = 2 \sum_{k=1}^{\infty} \int_\varepsilon^\infty \frac{\cos(2\pi ku)}{\sqrt{u}} du, \qquad (8.18)$$

where letting N tend to ∞ inside the summation sign is justified by two integrations by parts. Combining (8.16) and (8.18), we see that we must show that

$$\zeta(\tfrac{1}{2}) = \frac{1}{2} \int_0^\infty \frac{[u] - u}{u^{3/2}} du.$$

But this last formula follows immediately from a well-known representation for $\zeta(s)$ found in Titchmarsh's treatise [3, p. 14, Eq. (2.1.5)]. Hence, the proof of (8.3) is complete. $\qquad \square$

In the sequel, we shall set

$$F_{m,n}(x) = \sum_{j,k=1}^{\infty} j^m k^n e^{-jkx}, \tag{9.1}$$

where $x > 0$ and m and n are nonnegative integers. Without loss of generality, assume that $m \geq n$. In Theorem 6.1, an asymptotic expansion is given for $F_{m,n}(x)$ as x tends to $0+$. Ramanujan begins Section 9 with the special case $p = q = 1$, $m \neq n$ of Theorem 6.1. He then defines, for $|q| < 1$,

$$L = 1 - 24 \sum_{k=1}^{\infty} \frac{kq^k}{1 - q^k},$$

$$M = 1 + 240 \sum_{k=1}^{\infty} \frac{k^3 q^k}{1 - q^k},$$

and

$$N = 1 - 504 \sum_{k=1}^{\infty} \frac{k^5 q^k}{1 - q^k}.$$

The functions L, M, and N were thoroughly studied in a famous paper [11], [16, pp. 136–162] by Ramanujan, where L, M, and N are denoted by P, Q, and R, respectively. We now show that L, M, and N are essentially the Eisenstein series of weights 2, 4, and 6, respectively, on the full modular group $\Gamma(1)$. To see this, first let $q = \exp(2\pi i\tau)$, where τ is in the upper half-plane $\mathcal{H}$, and write

$$\Phi_\nu(q) := \sum_{k=1}^{\infty} \frac{k^\nu q^k}{1 - q^k} = \sum_{k=1}^{\infty} k^\nu \sum_{j=1}^{\infty} e^{2\pi ijk\tau} = \sum_{r=1}^{\infty} \sigma_\nu(r) e^{2\pi ir\tau}, \tag{9.2}$$

where we put $jk = r$ and where $\sigma_\nu(r) = \sum_{k|r} k^\nu$. Next recall that the Fourier expansions of the Eisenstein series $E_n(\tau)$, where n is an even positive integer, are given by (Rankin [2, p. 194])

$$E_2(\tau) = 1 - 24 \sum_{k=1}^{\infty} \sigma_1(k) e^{2\pi ik\tau} - \frac{3}{\pi y}$$

$$= 1 - 24\Phi_1(q) - \frac{3}{\pi y} \tag{9.3}$$

and

$$E_n(\tau) = 1 - \frac{2n}{B_n} \sum_{k=1}^{\infty} \sigma_{n-1}(k) e^{2\pi ik\tau}$$

$$= 1 - \frac{2n}{B_n} \Phi_{n-1}(q), \qquad n > 2, \tag{9.4}$$

where $y = \text{Im } \tau > 0$ and where B_n denotes the nth Bernoulli number. Hence, $L = E_2(\tau) + 3/(\pi y)$, $M = E_4(\tau)$, and $N = E_6(\tau)$.

Ramanujan next claims that if $m + n$ is an odd, positive integer, then the function $F_{m,n}(x)$ of (9.1) can be evaluated exactly in terms of L, M, and N. First, observe that, by setting $jk = r$ in the definition of $F_{m,n}(x)$, we obtain

$$F_{m,n}(x) = \sum_{r=1}^{\infty} r^n \sigma_{m-n}(r) e^{-rx}.$$

Thus, with $x = -2\pi i \tau$, $F_{m,n}(x)$ is essentially an n-fold derivative of an Eisenstein series of even weight if $m - n$ is odd. If $m - n = 1$ and $n \geq 1$, then $F_{m,n}(x)$ is clearly a multiple of an n-fold derivative of L. Suppose now that $m - n$ is odd and > 1. By a theorem in Rankin's text [2, p. 199], each modular form of even positive weight can be expressed as a polynomial in $E_4(\tau)$ and $E_6(\tau)$. Thus,

$$\sum_{r=1}^{\infty} \sigma_{m-n}(r) e^{-rx}$$

can be so expressed, and since $F_{m,n}(x)$ is, up to a factor of ± 1, an n-fold derivative of the function above, then $F_{m,n}(x)$ can be represented as a polynomial in M, N, and their derivatives.

For further remarks and discussion, see Venkatachaliengar's monograph [1, pp. 30, 31].

Entry 10(i) (First Part). *For each positive integer $n \geq 2$,*

$$-\frac{B_{2n}}{4n} E_{2n}(\tau) = -\frac{B_{2n}}{4n} + \sum_{k=1}^{\infty} \sigma_{2n-1}(k) e^{2\pi i k \tau}$$

can be expressed as a polynomial in M and N.

This statement was verified in Section 9 where we appealed to Rankin's book [2, p. 199]. See also (14.2) and Entry 14 below.

Entry 10(i) (Second Part). *For each positive integer n,*

$$f_n(x) := \sum_{k=1}^{\infty} \frac{k^{2n} q^k}{(1 - q^k)^2} - \delta_n \frac{nL}{6} \left\{ -\frac{B_{2n}}{4n} + \sum_{k=1}^{\infty} \frac{k^{2n-1} q^k}{1 - q^k} \right\}$$

can be expressed as a polynomial in M and N. Here $\delta_1 = \frac{1}{2}$ and $\delta_n = 1$ if $n \geq 2$.

PROOF. By (9.3) and (9.4),

$$\sum_{k=1}^{\infty} \frac{k^{2n} q^k}{(1 - q^k)^2} = \frac{1}{2\pi i} \frac{d}{d\tau} \left(-\frac{B_{2n}}{4n} E_{2n}^*(\tau) \right),$$

where

$$E_{2n}^*(\tau) = \begin{cases} E_2(\tau) + \dfrac{3}{\pi y} = L, & \text{if } n = 1, \\[2ex] E_{2n}(\tau), & \text{if } n > 1. \end{cases}$$

Thus, for $n \geq 1$,

$$f_n(x) = F_n(\tau) := \frac{1}{2\pi i} \frac{d}{d\tau} \left(-\frac{B_{2n}}{4n} E_{2n}^*(\tau) \right) + \frac{\delta_n n E_2^*(\tau)}{6} \frac{B_{2n}}{4n} E_{2n}^*(\tau).$$

By the aforementioned theorem in Rankin's treatise [2, p. 199], it suffices to prove that $F_n(\tau)$ is a modular form on $\Gamma(1)$ of weight $2n + 2$. We must therefore show that (Serre [1, Eq. (5), p. 80])

$$F_n(-1/\tau) = \tau^{2n+2} F_n(\tau), \qquad \tau \in \mathscr{H}. \tag{10.1}$$

Recall that for $V\tau = (a\tau + b)/(c\tau + d) \in \Gamma(1)$ (Schoeneberg [1, pp. 50, 68])

$$E_{2n}^*(V\tau) = \begin{cases} (c\tau + d)^2 E_2^*(\tau) - 6\pi^{-1} ic(c\tau + d), & \text{if } n = 1, \\ (c\tau + d)^{2n} E_{2n}(\tau), & \text{if } n > 1. \end{cases} \tag{10.2}$$

By (10.2), if $n > 1$,

$$\frac{4n}{B_{2n}} F_n(-1/\tau) = -\frac{\tau^2}{2\pi i} (2n\tau^{2n-1} E_{2n}(\tau) + \tau^{2n} E_{2n}'(\tau))$$

$$+ \frac{n}{6} \left(\tau^2 E_2^*(\tau) - \frac{6i\tau}{\pi} \right) \tau^{2n} E_{2n}(\tau)$$

$$= \tau^{2n+2} \left(-\frac{1}{2\pi i} E_{2n}'(\tau) + \frac{n}{6} E_2^*(\tau) E_{2n}(\tau) \right)$$

$$= \tau^{2n+2} \frac{4n}{B_{2n}} F_n(\tau).$$

This proves (10.1) for $n > 1$. A similar argument can be used for the case $n = 1$. □

Alternatively, for $n > 1$, (10.1) follows from the theorem in Ogg's survey [1, pp. 16, 17] that if $f(\tau)$ is a modular form of weight k, then $f'(\tau) - (2\pi i k/12) E_2^*(\tau) f(\tau)$ is a modular form of weight $k + 2$.

Ramanujan did not consider the case $n = 1$ in Entry 10(i).

In the remainder of this long section, Ramanujan makes several definitions and offers many examples to illustrate his definitions, which, for the most part, are imprecise. For each definition, we quote from the notebooks (pp. 186, 187).

Entry 10(ii). "*The degree of a series is the sum of the highest powers of the nth terms together with unity if the series contains all the powers of x or if the powers of x be in A.P. (arithmetic progression).*

If the coefficient of each nth term is homogeneous the series is said to be pure *and in other cases* mixed.

The theory of indices holds good in terms of degrees of series.

If F(h) in XV 1. terminates the series is said to be perfect. *If not it is said to be* imperfect.

If F(h) = 0 the series is said to be complete *in other cases* incomplete.

A series is said to be absolutely complete *when it remains complete when transformed or split up.*

A linear series can only be expressed by linear, double by double, treble by treble, pure by pure, perfect by perfect, imperfect by imperfect, and absolutely complete by absolutely complete adhering to the laws of indices in all cases. But a mixed series can be split up into a number of pure series of different degrees."

M. E. H. Ismail has suggested that the degree of a series is more properly defined in terms of the order of a singularity on the boundary of convergence of the series. Of course, this definition is possibly ambiguous if there is more than one singularity on the boundary. However, for some of Ramanujan's examples, Ismail's definition is more viable than Ramanujan's definition.

We do not know what is meant by "the theory of indices." The definition of $F(h)$ is given in Entry 1.

Example 1. Let

$$f_1(x) = \sum_{k=1}^{\infty} k^n x^k, \qquad |x| < 1,$$

where n is a nonnegative integer. First, f_1 has degree $n + 1$ because the degree of k^n is n and x^k contributes 1 to the degree. Since f_1 has a pole of order $n + 1$ at $x = 1$, f_1 has degree $n + 1$ by Ismail's definition as well. It is easily seen that k^n is homogeneous; that is, if $g(k) = k^n$, then $g(jk) = (jk)^n = j^n g(k)$. Thus, f_1 is pure. Here $\varphi(t) = t^n x^t$. Since $\varphi^{(2k-1)}(0)$ is not necessarily equal to 0 for each k sufficiently large, $F(h)$ does not terminate, and so f_1 is imperfect. It trivially follows that f_1 is incomplete. It is uncertain what Ramanujan means by "linear." But if he means that the series is not a multiple series, then it is clear that f_1 is linear.

Example 2. For x real, let

$$f_2(x) = \sum_{k=1}^{\infty} \frac{\sin(kx)}{k}.$$

Now $\sin(kx)$ probably has degree 1 in Ramanujan's definition. Since $1/k$ has degree -1, Ramanujan concludes that f_2 has degree 0. The singularities at $x = 2n\pi$, where n is an integer, are "jump" discontinuities, and so it is reasonable to say that they are of order 0. Hence, f_2 has degree 0 by this interpretation as well. The coefficients are equal to $1/k$, and so f_2 is pure. It is clear that f_2 is linear by the interpretation of "linear" given in Example 1. Now $F(h) \not\equiv 0$, but $\varphi(t) = \sin(tx)/t$ is an even function of t, and so $\varphi^{(2k-1)}(0) = 0, k \geq 1$. Hence, f_2 is perfect and incomplete.

Example 3. Consider $F_{m,n}(x)$, defined by (9.1). Clearly, $F_{m,n}$ is pure and is a double series. Now j^m, k^n, and e^{-x} are of degrees m, n, and 1, respectively, and so $F_{m,n}(x)$ has degree $m + n + 1$. By Theorem 6.1, the order of the singularity at $x = 0$ is not equal to $m + n + 1$, however. Also by Theorem 6.1, $F_{m,n}$ is incomplete. The series on the right side of Theorem 6.1 consists of terms of the form

$$\zeta(-m-k)\zeta(-n-k)\frac{(-x)^k}{k!} = \frac{B_{m+k+1}B_{n+k+1}(-x)^k}{(m+k+1)(n+k+1)k!},$$

by (0.1), if $k \geq 1$. If $m + n$ is even, these terms will not be equal to 0 when m and k are of opposite parity. Thus, $F_{m,n}$ is imperfect if $m + n$ is even. However, if $m + n$ is odd, then either B_{m+k+1} or B_{n+k+1} is equal to 0. Hence, if $m + n$ is odd, $F_{m,n}$ is perfect.

Example 4. Let m and n denote positive integers with $m \neq n$. Let

$$g_{m,n}(x) = \sum_{i,j,k=1}^{\infty} e^{-ijkx} j^m k^n.$$

Thus, in the notation of Section 7, $g_{m,n}(x) = f_{m+1,n+1,1}$. Ramanujan asserts that $g_{m,n}$ is a treble, pure series of degree $m + n + 1$, which is clear. Note that, by Theorem 7.1, the alternate definition of degree fails here. Also, by Theorem 7.1, $g_{m,n}$ is incomplete. A typical term in the asymptotic expansion for $g_{m,n}(x)$, by Theorem 7.1 and (0.1), equals

$$(-1)^{m+n}\frac{B_{m+k+1}B_{n+k+1}B_{k+1}x^k}{(m+k+1)(n+k+1)(k+1)k!}.$$

Thus, if both m and n are even, we see that the asymptotic series does not terminate, and so $g_{m,n}$ is imperfect. But if either m or n is odd, the expansion does terminate, and so $g_{m,n}$ is perfect in these cases.

Example 5. Let m, n, and x denote real numbers with $n > 0$. Put

$$h_{m,n}(x) = \sum_{k=1}^{\infty} \frac{k^m}{(e^{kx} + e^{-kx})^n}.$$

Ramanujan claims that $h_{m,n}(x)$ is a double series, so that he evidently writes $h_{m,n}(x)$ in the form

$$h_{m,n}(x) = \sum_{k=1}^{\infty} \sum_{j=0}^{\infty} \frac{(-1)^j(n)_j}{j!} k^m e^{-kx(n+2j)}, \qquad x > 0.$$

The coefficients are not homogeneous, and so the series is mixed. Ramanujan claims that $h_{m,n}$ has degree $m + n$, but it seems to us that the degree is equal to $m + n + 1$, since k^m, e^{-kxn}, and e^{-2jkx} have degrees m, n, and 1, respectively. Note that $h_{m,n}$ has an essential singularity at $x = 0$. It is easy to see that $h_{m,n}$ is incomplete.

Example 6. Consider the theta-function

$$f(x) = \frac{1}{2} + \sum_{k=1}^{\infty} x^{k^2} = \frac{1}{2} \sum_{k=-\infty}^{\infty} x^{k^2}, \qquad |x| < 1.$$

Clearly, f is pure. Since x^{t^2} is an even function of t, it is trivial that $F(h)$ terminates, and so f is perfect. Ramanujan also claims that f is a pure, double series. This is enigmatic, for if we expand x^{k^2} in a power series in k, f is no longer pure. However, possibly, in this instance, Ramanujan intends "double" to mean "bilateral," in which case, Ramanujan's assertion is correct. Lastly, he asserts that f has degree $\frac{1}{2}$. We are unable to justify this claim by using Ramanujan's definition of degree. Now f is analytic at $x = 0$. However, if we set $x = e^{\pi i \tau}$, $\tau \in \mathcal{H}$, then, by the theta-transformation formula (1.2), it may be loosely construed that $f(e^{\pi i \tau})$ has a "singularity of order $\frac{1}{2}$ at $\tau = 0$." Of course, this is not really the case, since the real axis is a natural boundary for $f(e^{\pi i \tau})$. Thus, a fuzzy interpretation of Ismail's definition has a modicum of viability.

Example 7. Ramanujan remarks that L, M, and N are perfect, pure double series of degrees 2, 4, and 6, respectively. By expanding $(1 - q^k)^{-1}$ in a geometric series, we readily see that L, M, and N are pure double series of degrees 2, 4, and 6, respectively, since q^{jk}, $1 \le j, k < \infty$, is of degree 1. Now apply Theorem 6.1 with $m = p = q = 1$, $e^{-x} = q$, and $n = 2$, 4, and 6, respectively. Since $B_{1+k}B_{n+k} = 0$, $k \ge 1$, L, M, and N are perfect. Lastly, Ramanujan asserts that M and N are complete, but L is incomplete. It appears to us, however, that all three series are incomplete, for in Theorem 6.1,

$$\frac{\zeta(2 - n)}{x} + \frac{\Gamma(n)\zeta(n)}{x^n} \ne 0.$$

Entry 11. If $\alpha, \beta > 0$ and $\alpha\beta = \pi^2$, then

$$\frac{1}{4} \sum_{k=1}^{\infty} \frac{1}{k^2 \sinh^2(\alpha k)} + \frac{1}{4} \sum_{k=1}^{\infty} \frac{1}{k^2 \sinh^2(\beta k)}$$

$$- 2\alpha \sum_{k=1}^{\infty} k^2 \, \mathrm{Log}(1 - e^{-2\alpha k}) - 2\beta \sum_{k=1}^{\infty} k^2 \, \mathrm{Log}(1 - e^{-2\beta k})$$

$$= \frac{\alpha^2 + \beta^2}{120} - \frac{\alpha\beta}{72}.$$

PROOF. By an elementary calculation,

$$\sum_{k=1}^{\infty} k^2 \, \mathrm{Log}(1 - e^{-2\alpha k}) = - \sum_{j=1}^{\infty} \frac{1}{j} \sum_{k=1}^{\infty} k^2 e^{-2\alpha jk}$$

$$= - \sum_{j=1}^{\infty} \frac{1}{j} \left\{ \frac{e^{-2\alpha j}}{(1 - e^{-2\alpha j})^2} + \frac{2e^{-4\alpha j}}{(1 - e^{-2\alpha j})^3} \right\}$$

$$= - \frac{1}{4} \sum_{j=1}^{\infty} \frac{\cosh(\alpha j)}{j \sinh^3(\alpha j)}. \tag{11.1}$$

With (11.1) as motivation, we define

$$f(z) = \pi \cot(\pi z)\left(\frac{1}{z^2 \sinh^2(\alpha z)} + \frac{2\alpha \cosh(\alpha z)}{z \sinh^3(\alpha z)}\right).$$

We shall integrate f over a suitable rectangle, to be described later, and apply the residue theorem. We let $R(z)$ denote the residue of a specified function at z.

First, f has simple poles at each nonzero integer k with

$$R(k) = \frac{1}{k^2 \sinh^2(\alpha k)} + \frac{2\alpha \cosh(\alpha k)}{k \sinh^3(\alpha k)}.$$

By (11.1), the sum of all such residues is equal to

$$2 \sum_{k=1}^{\infty} \frac{1}{k^2 \sinh^2(\alpha k)} - 16\alpha \sum_{k=1}^{\infty} k^2 \operatorname{Log}(1 - e^{-2\alpha k}). \tag{11.2}$$

Second, let $f_1(z) = p(z)/q(z)$, where $p(z) = \pi \cot(\pi z)$ and $q(z) = z^2 \sinh^2(\alpha z)$. The function $f_1(z)$ has double poles at $z = ik\pi/\alpha$, for each nonzero integer k. To calculate the residue at $ik\pi/\alpha$, we shall use a formula from Churchill's text [1, p. 160] for the residue of a double pole. Accordingly,

$$R(i\pi k/\alpha) = \frac{2p'(i\pi k/\alpha)}{q''(i\pi k/\alpha)} - \frac{2p(i\pi k/\alpha)q'''(i\pi k/\alpha)}{3\{q''(i\pi k/\alpha)\}^2}. \tag{11.3}$$

Elementary calculations yield

$$p(i\pi k/\alpha) = \pi \cot(\beta ki), \quad p'(i\pi k/\alpha) = -\pi^2 \csc^2(\beta ki),$$

$$q''(i\pi k/\alpha) = -2\pi^2 k^2, \quad \text{and} \quad q'''(i\pi k/\alpha) = 12\alpha\pi ki.$$

Using these values in (11.3), we find that

$$R(i\pi k/\alpha) = -\frac{1}{k^2 \sinh^2(\beta k)} - \frac{2 \coth(\beta k)}{\beta k^3}.$$

Thus, the sum of all such residues is

$$-2 \sum_{k=1}^{\infty} \frac{1}{k^2 \sinh^2(\beta k)} - \frac{4}{\beta} \sum_{k=1}^{\infty} \frac{\coth(\beta k)}{k^3}. \tag{11.4}$$

Consider a function $F(z) = p(z)/q(z)$, where p and q are analytic at z_0, $p(z_0) \neq 0$, and q has a zero of order 3 at z_0. Then a somewhat lengthy, but routine, exercise shows that

$$R(z_0) = \frac{3p''(z_0)}{q'''(z_0)} - \frac{3p'(z_0)q^{(4)}(z_0)}{2\{q'''(z_0)\}^2}$$

$$- \frac{3p(z_0)q^{(5)}(z_0)}{10\{q'''(z_0)\}^2} + \frac{3p(z_0)\{q^{(4)}(z_0)\}^2}{8\{q'''(z_0)\}^3}. \tag{11.5}$$

Now set $f_2(z) = p(z)/q(z)$, where $p(z) = 2\pi\alpha \cot(\pi z) \cosh(\alpha z)$ and $q(z) = z \sinh^3(\alpha z)$. The function $f_2(z)$ has triple poles at $z = i\pi k/\alpha$, for each nonzero integer k. Elementary calculations yield

$$p(i\pi k/\alpha) = -2(-1)^k \pi \alpha i \coth(\beta k),$$

$$p'(i\pi k/\alpha) = 2(-1)^k \pi^2 \alpha \operatorname{csch}^2(\beta k),$$

$$p''(i\pi k/\alpha) = 4(-1)^k \pi^3 \alpha i \operatorname{csch}^2(\beta k) \coth(\beta k) - 2(-1)^k \pi \alpha^3 i \coth(\beta k),$$

$$q'''(i\pi k/\alpha) = 6(-1)^k \pi \alpha^2 ki,$$

$$q^{(4)}(i\pi k/\alpha) = 24(-1)^k \alpha^3,$$

and

$$q^{(5)}(i\pi k/\alpha) = 60(-1)^k \pi \alpha^4 ki.$$

Using these values in (11.5), we find, after much simplification, that

$$R(i\pi k/\alpha) = \frac{2\beta}{k} \operatorname{csch}^2(\beta k) \coth(\beta k) + \frac{2}{k^2} \operatorname{csch}^2(\beta k) + \frac{2}{\beta k^3} \coth(\beta k).$$

Thus, the sum of all such residues, by (11.1), is equal to

$$-16\beta \sum_{k=1}^{\infty} k^2 \operatorname{Log}(1 - e^{-2\beta k}) + 4 \sum_{k=1}^{\infty} \frac{1}{k^2 \sinh^2(\beta k)} + \frac{4}{\beta} \sum_{k=1}^{\infty} \frac{\coth(\beta k)}{k^3}. \quad (11.6)$$

Lastly, f has a pole of order 5 at the origin. We have

$$f(z) = \pi \left(\frac{1}{\pi z} - \frac{\pi z}{3} - \frac{\pi^3 z^3}{45} + \cdots \right) \left\{ \frac{1}{z^2} \left(\frac{1}{\alpha z} - \frac{\alpha z}{6} + \frac{7\alpha^3 z^3}{360} + \cdots \right)^2 \right.$$

$$\left. + \frac{2\alpha}{z} \left(\frac{1}{\alpha z} + \frac{\alpha z}{3} - \frac{\alpha^3 z^3}{45} + \cdots \right) \left(\frac{1}{\alpha z} - \frac{\alpha z}{6} + \frac{7\alpha^3 z^3}{360} + \cdots \right)^2 \right\}$$

$$= \pi \left(\frac{1}{\pi z} - \frac{\pi z}{3} - \frac{\pi^3 z^3}{45} + \cdots \right) \left(\frac{3}{\alpha^2 z^4} - \frac{1}{3z^2} - \frac{\alpha^2}{15} + \cdots \right).$$

Hence,

$$R(0) = \pi \left(-\frac{\alpha^2}{15\pi} + \frac{\pi}{9} - \frac{\pi^3}{15\alpha^2} \right) = \frac{\alpha\beta}{9} - \frac{\alpha^2 + \beta^2}{15}. \quad (11.7)$$

Consider next

$$I_N := \frac{1}{2\pi i} \int_{C_N} f(z) \, dz,$$

where C_N is a positively oriented rectangle with sides parallel to the coordinate axes and passing through the points $\pm([\sqrt{N}] + \frac{1}{2})$ and $i\pi(N + \frac{1}{2})/\alpha$, where N is a positive integer. Note that C_N is free of poles of f. Estimating the integrand on the vertical and horizontal sides separately, we find that

$$I_N \ll \sqrt{N} e^{-2\alpha[\sqrt{N}]} + \frac{1}{\sqrt{N}} = o(1), \quad (11.8)$$

as N tends to ∞.

Apply the residue theorem to I_N and then let N tend to ∞. Using (11.2), (11.4), (11.6), (11.7), and (11.8), we deduce that

$$0 = 2 \sum_{k=1}^{\infty} \frac{1}{k^2 \sinh^2(\alpha k)} - 16\alpha \sum_{k=1}^{\infty} k^2 \, \mathrm{Log}(1 - e^{-2\alpha k})$$

$$+ 2 \sum_{k=1}^{\infty} \frac{1}{k^2 \sinh^2(\beta k)} - 16\beta \sum_{k=1}^{\infty} k^2 \, \mathrm{Log}(1 - e^{-2\beta k})$$

$$+ \frac{\alpha\beta}{9} - \frac{\alpha^2 + \beta^2}{15},$$

which is readily seen to be equivalent to the proposed identity.   □

Another proof of Entry 11 may be constructed from results in Berndt's paper [6, Theorems 2.2, 2.16] together with (11.1).

Entry 12. *Let L, M, and N be as defined in Section 9, and recall that $E_n(\tau)$, $n > 2$, and $\Phi_n(q)$ are defined by (9.4) and (9.2), respectively. Define the discriminant function $\Delta(\tau)$ by*

$$\Delta(\tau) = q \prod_{k=1}^{\infty} (1 - q^k)^{24}, \qquad q = e^{2\pi i \tau}, \quad \tau \in \mathcal{H}.$$

Then, for $|q| < 1$,

(i) $M^3 - N^2 = 1728\Delta(\tau)$,

(ii) $E_8(\tau) = M^2$,

(iii) $E_{10}(\tau) = MN$,

(iv) $E_{14}(\tau) = M^2 N$,

(v) $\displaystyle\sum_{k=1}^{\infty} \frac{k^2 q^k}{(1 - q^k)^2} = \frac{M - L^2}{288}$,

(vi) $\displaystyle\sum_{k=1}^{\infty} \frac{k^4 q^k}{(1 - q^k)^2} = \frac{LM - N}{720}$,

(vii) $\displaystyle\sum_{k=1}^{\infty} \frac{k^6 q^k}{(1 - q^k)^2} = \frac{M^2 - LN}{1008}$,

(viii) $\displaystyle\sum_{k=1}^{\infty} \frac{k^8 q^k}{(1 - q^k)^2} = \frac{LM^2 - MN}{720}$,

(ix) $\displaystyle L \sum_{k=0}^{\infty} (-1)^k (2k + 1) q^{k(k+1)/2} = \sum_{k=0}^{\infty} (-1)^k (2k + 1)^3 q^{k(k+1)/2}$,

(x) $\displaystyle M \sum_{k=1}^{\infty} \frac{(2k - 1)q^k}{1 - q^{2k-1}} = \sum_{k=1}^{\infty} \frac{(2k - 1)^5 q^k}{1 - q^{2k-1}}$.

PROOFS OF (i)–(viii). Formulas (i)–(iv) are very well known and are special cases of the general theorem in Rankin's book [2, p. 199] which we applied in Section 9. In particular, (i)–(iv) can be found in [2, pp. 195, 197, Eqs. (6.1.8), (6.1.9), and (6.1.14)]. These formulas were also derived by Ramanujan in [11], [16, p. 141].

Formulas (v)–(viii) are originally due to Ramanujan, and proofs can be found in his paper [11], [16, pp. 141, 142].   □

FIRST PROOF OF (ix). This formula is a special case of a general formula established by Ramanujan in Chapter 16. See Part III [11, Chap. 16, Entry 35(i)].   □

SECOND PROOF OF (ix). Rearranging in (ix), we find that

$$\sum_{j=1}^{\infty} \sigma(j)q^j \sum_{k=0}^{\infty} (-1)^k(2k+1)q^{k(k+1)/2}$$

$$= \frac{1}{24}\left\{\sum_{k=0}^{\infty} (-1)^k(2k+1)q^{k(k+1)/2} - \sum_{k=0}^{\infty} (-1)^k(2k+1)^3 q^{k(k+1)/2}\right\}.$$

Equating coefficients of q^n, $n \geq 0$, on both sides, we find that

$$\sigma(n) - 3\sigma(n-1) + 5\sigma(n-3) - 7\sigma(n-6) + \cdots = 0, \qquad (12.1)$$

if n is not a triangular number, while if $n = r(r+1)/2$ is a triangular number,

$$\sigma(n) - 3\sigma(n-1) + 5\sigma(n-3) - 7\sigma(n-6) + \cdots$$

$$= \frac{1}{24}\{(-1)^r(2r+1) - (-1)^r(2r+1)^3\}$$

$$= \frac{1}{6}(-1)^{r-1}r(r+1)(2r+1)$$

$$= (-1)^{r-1} \sum_{k=1}^{r} k^2. \qquad (12.2)$$

Thus, formula (ix) is equivalent to the arithmetic identities evinced in (12.1) and (12.2). These identities are due to Glaisher [2] in 1884, although they are really consequences of a formula proved seven years earlier by Halphen [1]. Hence, appealing to the theorem of Glaisher and Halphen, we have shown (ix).   □

For generalizations of Entry 12(ix), see two additional papers of Glaisher [4], [5]. For further references to the literature, consult Dickson's history [1, p. 289].

PROOF OF (x). If

$$f(\tau) = \sum_{n=0}^{\infty} a_n q^n, \qquad q = e^{2\pi i \tau},$$

define functions f_∞, f_0, and f_1 by

$$f_\infty(\tau) = f(2\tau) = \sum_{n=0}^{\infty} a_n q^{2n}, \qquad f_0(\tau) = f(\tau/2) = \sum_{n=0}^{\infty} a_n q^{n/2},$$

and

$$f_1(\tau) = f\left(\frac{\tau+1}{2}\right) = \sum_{n=0}^{\infty} a_n(-1)^n q^{n/2}.$$

Then Entry 12(x) may be rewritten in the form

$$N_1 - N_0 = 21M(L_1 - L_0). \tag{12.3}$$

If $w = (\tau + 1)/2$, observe that

$$\frac{1 - 1/\tau}{2} = \frac{w - 1}{2w - 1} \in \Gamma(1).$$

Thus, from (10.2), we readily find that

$$L_\infty(\tau + 1) = L_\infty(\tau), \quad L_0(\tau + 1) = L_1(\tau), \quad L_1(\tau + 1) = L_0(\tau),$$

$$L_\infty(-1/\tau) = \tfrac{1}{4}\tau^2 L_0(\tau) + \frac{3\tau}{\pi i}, \quad L_0(-1/\tau) = 4\tau^2 L_\infty(\tau) + \frac{12\tau}{\pi i},$$

$$L_1(-1/\tau) = \tau^2 L_1(\tau) + \frac{12\tau}{\pi i},$$

$$N_\infty(\tau + 1) = N_\infty(\tau), \quad N_0(\tau + 1) = N_1(\tau), \quad N_1(\tau + 1) = N_0(\tau),$$

$$N_\infty(-1/\tau) = \frac{1}{64}\tau^6 N_0(\tau), \quad N_0(-1/\tau) = 64\tau^6 N_\infty(\tau),$$

and

$$N_1(-1/\tau) = \tau^6 N_1(\tau).$$

Next, define

$$X_\infty = L_1 - L_0, \quad X_0 = 4L_\infty - L_1, \quad X_1 = L_0 - 4L_\infty,$$

$$Z_\infty = N_1 - N_0, \quad Z_0 = 64N_\infty - N_1, \quad Z_1 = N_0 - 64N_\infty.$$

Then the foregoing equalities readily imply that

$$\left.\begin{array}{l} X_\infty(\tau + 1) = -X_\infty(\tau), \quad X_0(\tau + 1) = -X_1(\tau), \quad X_1(\tau + 1) = -X_0(\tau), \\ X_\infty(-1/\tau) = -\tau^2 X_0(\tau), \quad X_0(-1/\tau) = -\tau^2 X_\infty(\tau), \quad X_1(-1/\tau) = -\tau^2 X_1(\tau), \end{array}\right\} \tag{12.4}$$

and

$$\left.\begin{array}{l} Z_\infty(\tau + 1) = -Z_\infty(\tau), \quad Z_0(\tau + 1) = -Z_1(\tau), \quad Z_1(\tau + 1) = -Z_0(\tau), \\ Z_\infty(-1/\tau) = -\tau^6 Z_0(\tau), \quad Z_0(-1/\tau) = -\tau^6 Z_\infty(\tau), \quad Z_1(-1/\tau) = -\tau^6 Z_1(\tau). \end{array}\right\} \tag{12.5}$$

Let M_k denote the space of modular forms of weight k on the modular subgroup $\Gamma(2)$. If $S(\tau) = \tau + 1$ and $T(\tau) = -1/\tau$, then, by a paper by Frasch [1, p. 245], generators of $\Gamma(2)$ are

$$S^2(\tau) \quad \text{and} \quad TS^2 T(\tau) = \frac{\tau}{-2\tau + 1}.$$

Using these generators and (12.4), we may easily verify that $X_0, X_\infty \in M_2$. Suppose that k is even. Then from Rankin's text [2, pp. 104, 105], dim $M_k =$

$1 + \frac{1}{2}k$. Moreover, since

$$X_0 = 3 - 24q^{1/2} + 72q + \cdots \tag{12.6}$$

and

$$X_\infty = 48q^{1/2} + 192q^{3/2} + \cdots \tag{12.7}$$

are obviously linearly independent in M_2, we conclude that $X_0^{k/2}$, $X_0^{k/2-1}X_\infty$, $\ldots$, $X_0 X_\infty^{k/2-1}$, $X_\infty^{k/2}$ form a basis for M_k. Now suppose that $f \in M_k$ and that $f(\tau) = o(q^{k/4})$, as q tends to 0. Then from (12.6) and (12.7), $f(\tau) \equiv 0$.

In our situation, we take $k = 6$. Clearly, $MX_\infty \in M_6$, and, from (12.5), we may verify that $Z_\infty \in M_6$. From the expansion

$$Z_\infty = 1008q^{1/2} + 245952q^{3/2} + \cdots,$$

(12.7), and the definition of M, we find that $21MX_\infty - Z_\infty = o(q^{3/2})$ as q tends to 0. Hence, $21MX_\infty - Z_\infty \equiv 0$, and (12.3) is proved. □

We are very grateful to D. W. Masser for supplying us with the proof above. Another proof of Entry 12(x) based on the theory of modular forms on $\Gamma_0(2)$ was constructed for us by A. O. L. Atkin.

Entry 12(x) was stated by Ramanujan in [11], [16, p. 146] without proof. Ramanujan indicated that he had two proofs, one of which was elementary, while the other used elliptic functions. However, he provided no hints to either proof. It is very unlikely that the proofs of Masser and Atkin are the same as either of Ramanujan's proofs. In her thesis, Ramamani [1, p. 59] has given a proof of Entry 12(x) that uses the theory of elliptic functions. Entry 12(x) is equivalent to the elegant identity

$$\sum_{k=0}^{n} \sigma_1(2k + 1)\sigma_3(n - k) = \frac{1}{240}\sigma_5(2n + 1), \qquad n \geq 0,$$

where $\sigma_3(0) = \frac{1}{240}$. It would be interesting to have an elementary proof of this identity and hence of Entry 12(x) as well.

In his paper [11], [16, pp. 136–162], Ramanujan studies

$$\Sigma_{r,s}(n) = \sum_{k=0}^{n} \sigma_r(k)\sigma_s(n - k),$$

where r and s are odd, positive integers and $\sigma_m(0) = \frac{1}{2}\zeta(-m)$. He establishes an asymptotic formula for $\Sigma_{r,s}(n)$ as n tends to ∞ with an error term. He, however, conjectured a better error term [11], [16, p. 136, Eq. (3)]. This conjecture remained unproved until 1978 when Levitt [1] proved Ramanujan's conjecture in his thesis. In some instances, Ramanujan showed that the error term is identically equal to 0. Levitt [1] established necessary and sufficient conditions for the vanishing of the error term and so showed that the instances of such found by Ramanujan are exhaustive. Such a theorem was also found by Grosjean [1], [2] who has made a systematic study of recursion formulas connected with $\Sigma_{r,s}(n)$.

An informative survey paper on convolutions involving $\sigma_k(n)$ has been written by Lehmer [1]. For other papers in this area, consult [1, Sect. A30], edited by LeVeque.

Entry 13. *Let $\Phi_n(q)$ be defined as in Entry 12. Then, for $|q| < 1$,*

(i) $691 + 65{,}520\,\Phi_{11}(q) = 441M^3 + 250N^2$,

(ii) $3617 + 16{,}320\,\Phi_{15}(q) = 1617M^4 + 2000MN^2$,

(iii) $43{,}867 - 28{,}728\,\Phi_{17}(q) = 38{,}367M^3N + 5500N^3$,

(iv) $174{,}611 + 13{,}200\,\Phi_{19}(q) = 53{,}361M^5 + 121{,}250M^2N^2$,

(v) $77{,}683 - 552\,\Phi_{21}(q) = 57{,}183M^4N + 20{,}500MN^3$,

(vi) $236{,}364{,}091 + 131{,}040\,\Phi_{23}(q) = 49{,}679{,}091M^6 + 176{,}400{,}000M^3N^2 + 10{,}285{,}000N^4$,

(vii) $657{,}931 - 24\,\Phi_{25}(q) = 392{,}931M^5N + 265{,}000M^2N^3$,

(viii) $3{,}392{,}780{,}147 + 6960\,\Phi_{27}(q) = 489{,}693{,}897M^7 + 2{,}507{,}636{,}250M^4N^2 + 395{,}450{,}000MN^4$,

(ix) $1{,}723{,}168{,}255{,}201 - 171{,}864\,\Phi_{29}(q) = 815{,}806{,}500{,}201M^6N + 881{,}340{,}705{,}000M^3N^3 + 26{,}021{,}050{,}000N^5$,

(x) $7{,}709{,}321{,}041{,}217 + 32{,}640\,\Phi_{31}(q) = 764{,}412{,}173{,}217M^8 + 5{,}323{,}905{,}468{,}000M^5N^2 + 1{,}621{,}003{,}400{,}000M^2N^4$.

Note.

$$q\frac{dL}{dq} = \frac{L^2 - M}{12}, \quad q\frac{dM}{dq} = \frac{LM - N}{3}, \quad \text{and} \quad q\frac{dN}{dq} = \frac{LN - M^2}{2}.$$

Examples. Define, for $|q| < 1$,

$$\Phi_{r,s}(q) = \sum_{j,k=1}^{\infty} j^r k^s q^{jk}.$$

(Thus, $\Phi_{0,s}(q) = \Phi_s(q)$.) Then

(i) $20{,}736\,\Phi_{4,5}(q) = 15LM^2 + 10L^3M - 20L^2N - 4MN - L^5$,

(ii) $1728\,\Phi_{2,7}(q) = 2LM^2 - MN - L^2N$,

(iii) $3456\,\Phi_{3,6}(q) = L^3M - 3L^2N + 3LM^2 - MN$.

All of the foregoing results may be found in Ramanujan's paper [11], [16, pp. 141, 142], where the method of proof is indicated.

Let ω_1 and ω_2 denote two complex numbers linearly independent over the real numbers. Put $\omega = m\omega_1 + n\omega_2$, where m and n are integers. Recall that the Weierstrass $\mathscr{P}$ function $\mathscr{P}(z)$ is defined by

$$\mathscr{P}(z) = \frac{1}{z^2} + \sum_{\omega \neq 0} \left(\frac{1}{(z - \omega)^2} - \frac{1}{\omega^2} \right),$$

where the sum is over all pairs of integers $(m, n) \neq (0, 0)$.

In order to prove Entry 14, we shall need the following facts about $\mathscr{P}(z)$ and Eisenstein series taken from Apostol's text [3, pp. 12, 13], as well as a lemma.

For $n \geq 1$, put

$$b(n) = 2(2n + 1)\zeta(2n + 2)E_{2n+2}(\tau), \tag{14.1}$$

where $E_n(\tau)$ is defined by (9.4). Then, for $n \geq 3$,

$$(2n + 3)(n - 2)b(n) = 3 \sum_{k=1}^{n-2} b(k)b(n - 1 - k). \tag{14.2}$$

(This is a more explicit version of the first part of Entry 10(i).) Furthermore, for $|z|$ sufficiently small,

$$\mathscr{P}(z) = \frac{1}{z^2} + \sum_{k=1}^{\infty} b(k)z^{2k}, \tag{14.3}$$

where $\omega_1 = 1$ and $\omega_2 = \tau$, with $\tau \in \mathscr{H}$. Lastly, $\mathscr{P}(z)$ satisfies the two differential equations

$$\{\mathscr{P}'(z)\}^2 = 4\mathscr{P}^3(z) - 20b(1)\mathscr{P}(z) - 28b(2) \tag{14.4}$$

and

$$\mathscr{P}''(z) = 6\mathscr{P}^2(z) - 10b(1). \tag{14.5}$$

In fact, (14.2) follows immediately from (14.5).

Lemma. *We have*

$$\mathscr{P}^{(4)}(z) = 30\{\mathscr{P}'(z)\}^2 + 240b(1)\mathscr{P}(z) + 504b(2).$$

PROOF. Differentiating (14.5) twice, we find that

$$\mathscr{P}^{(4)}(z) = 12\mathscr{P}'(z)^2 + 12\mathscr{P}(z)\mathscr{P}''(z). \tag{14.6}$$

Also, by (14.5),

$$12\mathscr{P}(z)\mathscr{P}''(z) = 72\mathscr{P}^3(z) - 120b(1)\mathscr{P}(z), \tag{14.7}$$

and by (14.4),

$$72\mathscr{P}^3(z) = 18\mathscr{P}'(z)^2 + 360b(1)\mathscr{P}(z) + 504b(2). \tag{14.8}$$

Substituting (14.8) into (14.7), we find that

$$12\mathscr{P}(z)\mathscr{P}''(z) = 18\mathscr{P}'(z)^2 + 240b(1)\mathscr{P}(z) + 504b(2). \tag{14.9}$$

Substituting (14.9) into (14.6), we complete the proof. $\qquad\square$

If n is an even positive integer, Ramanujan now defines

$$S_n = \frac{(-1)^{n/2-1}B_n}{2n} + (-1)^{n/2} \sum_{k=1}^{\infty} \frac{k^{n-1}q^k}{1 - q^k},$$

where $|q| < 1$ and B_n denotes the nth Bernoulli number. If $n > 1$ and $q = \exp(2\pi i\tau)$, with $\tau \in \mathscr{H}$, then, by (9.4),

$$S_{2n} = \frac{(-1)^{n-1}B_{2n}}{4n}E_{2n}(\tau).$$

Furthermore, from (14.1),

$$S_{2n+2} = \frac{(2n)!}{2(2\pi)^{2n+2}} b(n), \qquad n \geq 1. \tag{14.10}$$

In Entry 14, Ramanujan provides a recursion formula for S_{2n+2} which is different from (14.2). It should be remarked that in his paper [11], [16, p. 140, Eq. (22)], where a different definition of S_n is used, Ramanujan gives a very ingenious proof of (14.2). Rankin [1] has given an elementary proof of (14.2) as well as some other recursion formulas for S_{2n}. His paper also contains other references to the literature. However, the recursion formula of Entry 14, which is incompletely stated by Ramanujan in his notebooks (p. 191), does not appear to have been given elsewhere in the literature.

Entry 14. *If n is an even integer exceeding 4, then*

$$-\frac{(n+2)(n+3)}{2n(n-1)} S_{n+2} = -20 \binom{n-2}{2} S_4 S_{n-2}$$

$$+ \sum_{k=1}^{[(n-2)/4]} {}' \binom{n-2}{2k} \{(n+3-5k)(n-8-5k)$$

$$- 5(k-2)(k+3)\} S_{2k+2} S_{n-2k},$$

where the prime on the summation sign indicates that if $(n-2)/4$ is an integer, then the last term of the sum is to be multiplied by $\frac{1}{2}$.

PROOF. First, rewrite Entry 14 in the form

$$\frac{(n+2)(n+3)}{2} \frac{S_{n+2}}{n!} = 20 \frac{S_4}{2!} \frac{S_{n-2}}{(n-4)!} - \sum_{k=1}^{[(n-2)/4]} {}' \{(n+3-5k)(n-8-5k)$$

$$- 5(k-2)(k+3)\} \frac{S_{2k+2}}{(2k)!} \frac{S_{n-2k}}{(n-2k-2)!},$$

where n is even and at least 6. With $n = 2(m+1)$, where $m \geq 2$, the last equality may be rewritten as

$$(m+2)(2m+5)b(m+1) = 10b(1)b(m-1) + 10 \sum_{k=1}^{[m/2]} {}' k(m-k)b(k)b(m-k)$$

$$- (2m^2 - m) \sum_{k=1}^{[m/2]} {}' b(k)b(m-k), \tag{14.11}$$

where (14.10) has been employed. Now (14.2) can be written in the form

$$(2m+5)(m-1)b(m+1) = 6 \sum_{k=1}^{[m/2]} {}' b(k)b(m-k), \qquad m \geq 2, \tag{14.12}$$

where the prime on the summation sign indicates that if m is even, the last summand is to be multiplied by $\frac{1}{2}$. Using (14.12) in (14.11), we find that

$$(m + 2)(2m + 5)b(m + 1) = 10b(1)b(m - 1) + 10 \sum_{k=1}^{[m/2]}{}' k(m - k)b(k)b(m - k)$$

$$- \tfrac{1}{6}(2m^2 - m)(2m + 5)(m - 1)b(m + 1).$$

Thus, it remains to show that, for $m \geq 2$,

$$\tfrac{1}{30}(2m + 5)(m + 1)(2m^2 - 5m + 12)b(m + 1)$$

$$= 2b(1)b(m - 1) + \sum_{k=1}^{m-1} k(m - k)b(k)b(m - k). \tag{14.13}$$

Subtracting $2(m + 1)b(m + 1)$ from both sides of (14.13), we see that (14.13) is equivalent to

$$\tfrac{1}{30}m(m + 1)(2m - 1)(2m + 1)b(m + 1)$$

$$= 2b(1)b(m - 1) + \sum_{k=1}^{m-1} k(m - k)b(k)b(m - k) - 2(m + 1)b(m + 1), \tag{14.14}$$

for $m \geq 2$.

Now observe that the first expression $2b(1)b(m - 1)$ on the right side of (14.14) is the coefficient of z^{2m-2} in the power series for $2b(1)\mathscr{P}(z)$, by (14.3). Also, by (14.3), the latter two expressions on the right side of (14.14) constitute the coefficient of z^{2m-2} in the power series expansion for $\mathscr{P}'(z)^2/4$. Lastly, the left side of (14.14) is the coefficient of z^{2m-2} in the expansion of $\mathscr{P}^{(4)}(z)/120$. Thus, (14.14) follows from the lemma above, and this completes the proof. $\qquad\square$

Differentiating (14.5), we find that $\mathscr{P}'''(z) = 12\mathscr{P}(z)\mathscr{P}'(z)$, which yields another recursion formula for $b(n)$ midway in complexity between (14.2) and (14.14).

At first glance, the material in the next two sections appears uninteresting. However, it is a precursive introduction to Ramanujan's work in Chapters 18–21 on modular equations. The definition of "modular equation" given below is Ramanujan's personal one and is different from the standard definition which he used later and which can be found in Hardy's book [9, p. 214], for example. See the author's paper [10] for a discussion of the analogies between these two definitions.

With $F(x) = (1 - x)^{-1/2}$, Ramanujan begins Section 15(i) with the trivial identity

$$F\left(\frac{2t}{1 + t}\right) = (1 + t)F(t^2), \tag{15.1}$$

written in terms of binomial series. If we set $\alpha = 2t/(1 + t)$ and $\beta = \alpha^2/(2 - \alpha)^2$, then (15.1) may be written as

$$F(\alpha) = M_2(\alpha)F(\beta), \tag{15.2}$$

where $M_2(\alpha) = 2/(2 - \alpha)$. Ramanujan says that $\beta = \alpha^2/(2 - \alpha)^2$ is a modular equation of the second degree. The factor $M_2(\alpha)$ appearing in (15.2) is the "multiplier." Ramanujan also records the following representations for $M_2(\alpha)$:

$$M_2(\alpha) = 1 + \sqrt{\beta} = \sqrt{\frac{1-\beta}{1-\alpha}} = \sqrt{(1-\alpha)(1-\beta)} + 2\sqrt{\beta}.$$

Each of these formulas for $M_2(\alpha)$ is easily verified.

Consider now a more general equation

$$F(\alpha) = M_n(\alpha)F(\beta), \tag{15.3}$$

where $\beta = R_n(\alpha)$ is a function of "degree n" and $F(x)$ is not necessarily equal to $(1-x)^{-1/2}$. The factor $M_n(\alpha)$ is the "multiplier" of "degree n." The meaning of "degree" is not clear. In the sequel, modular equations and multipliers of degree 2^m will be obtained by iteration. We emphasize that in standard definitions of modular equations, the meaning of "degree" is precise.

Returning to the penultimate paragraph, we derive further modular equations by iteration. To obtain a modular equation for $n = 4$, iterate (15.2) to find that

$$F(\alpha) = \frac{2}{2-\alpha} \frac{2}{2-\alpha^2/(2-\alpha)^2} F\left(\frac{\{\alpha^2/(2-\alpha)^2\}^2}{\{2-\alpha^2/(2-\alpha)^2\}^2}\right)$$

$$= \frac{4(2-\alpha)}{\alpha^2 - 8\alpha + 8} F\left(\frac{\alpha^4}{(\alpha^2 - 8\alpha + 8)^2}\right).$$

Thus, $\beta = \alpha^4/(\alpha^2 - 8\alpha + 8)^2$ is a modular equation of degree 4. This procedure only yields modular equations when n is a positive power of 2. However, Ramanujan claims that the modular equation of degree n, *for any positive integer n*, is given by

$$\beta = \frac{4\alpha^n}{\{(1 + \sqrt{1-\alpha})^n + (1 - \sqrt{1-\alpha})^n\}^2}. \tag{15.4}$$

Possibly Ramanujan established (15.4) by induction when $n = 2^m$ and then "interpolated" to obtain a general formula for each positive integer n. Note that when $m = 0, 1, 2$, (15.4) is in agreement with our previous calculations. The inductive proof of (15.4) for $n = 2^m$ is straightforward, but rather tedious, and so we shall omit it.

We next calculate the function $M_n(\alpha)$ corresponding to (15.4). For brevity, set

$$P_n = (1 + \sqrt{1-\alpha})^n + (1 - \sqrt{1-\alpha})^n$$

and

$$Q_n = \frac{(1 + \sqrt{1-\alpha})^n - (1 - \sqrt{1-\alpha})^n}{\sqrt{1-\alpha}},$$

where $n \geq 1$. Then, by (15.3) and (15.4),

$$M_n(\alpha) = \frac{F(\alpha)}{F(\beta)} = \frac{F(\alpha)}{F(4\alpha^n/P_n^2)} = \frac{(P_n^2 - 4\alpha^n)^{1/2}}{P_n(1-\alpha)^{1/2}} = \frac{Q_n}{P_n},$$

after a straightforward calculation. Observe that $M_n(\alpha)$ is a rational function of α. In particular, if $n = 3$,

$$M_3(\alpha) = \frac{4-\alpha}{4-3\alpha}.$$

Ramanujan asserts that

$$M_3(\alpha) = 1 + 2\sqrt{\frac{\beta}{\alpha}} = \sqrt{\frac{1-\beta}{1-\alpha}},$$

and both equalities are readily verified.

In a corollary, Ramanujan claims that "if 2nd be $\alpha^2 + 2\alpha = \beta$, then the nth is $\beta = (\alpha + 1)^n - 1$." We have not been able to discern any connection between this statement and the original function F. It appears that Ramanujan is claiming that "modular equations" of degree 2^m can be obtained from the given "modular equation" of degree 2 by iteration. Since

$$\{(x+1)^k - 1\}^2 + 2\{(x+1)^k - 1\} = (x+1)^{2k} - 1,$$

for each positive integer k, Ramanujan's assertion is easily established when $n = 2^m$, $m \geq 0$. As above, Ramanujan evidently used an "interpolative" argument to establish his corollary for general n. It should be remarked that in his quarterly reports, Ramanujan defines the nth iterate of a function, for any *real* number n, by the same type of interpolative argument. (See Part I [9, pp. 324–326, 328–329].)

Ramanujan commences Section 15(ii) with the following theorem and corollary.

Entry 15(ii). "*If pth and qth be $\varphi(x)$ and $\psi(x)$ and rth be $f(x)$, then if pth and qth be $\varphi F(x)$ and $\psi F(x)$, then rth is $fF(x)$. And also if pth and qth be $F\varphi(x)$ and $F\psi(x)$ then rth is $Ff(x)$.*"

Corollary. "*Thus we may add or subtract any constant and multiply or divide by any constant to x in each function or to each function.*"

What can be said? It appears that Ramanujan is simply attempting to make some elementary remarks about the composition of functions.

Define, for $n = 2^m$, where m is any nonnegative integer,

$$F^{(n)}(x) = FF \cdots F(x), \tag{15.5}$$

where F occurs m times on the right side. In particular, $F^{(1)}(x) = x$.

Corollary (i). *If $f^{(1)}(x) = x$ and $f^{(2)}(x) = x^2 + 4x$, then*

$$f^{(n)}(x) = \left\{\left(\frac{\sqrt{x+4} + \sqrt{x}}{2}\right)^n - \left(\frac{\sqrt{x+4} - \sqrt{x}}{2}\right)^n\right\}^2.$$

Corollary (ii). *If $f^{(1)}(x) = x$ and $f^{(2)}(x) = x^2 - 2$, then*

$$f^{(n)}(x) = \left(\frac{x + \sqrt{x^2 - 4}}{2}\right)^n + \left(\frac{x - \sqrt{x^2 - 4}}{2}\right)^n.$$

As above, these two corollaries are statements about the iterates of functions when $n = 2^m$. Ramanujan then presumably is assuming that his formulas are valid for all positive integers n by interpolation. For both corollaries, the inductive proofs are completely straightforward.

Entry 15(iii). *"If $f(x)$ and $F(x)$ be of the pth and qth degree, find $\varphi(x)$ such that*

$$\sqrt[p]{\varphi f(x)} = \sqrt[q]{\varphi F(x)} = \chi(x) \tag{15.6}$$

suppose, then the function for the rth degree $= \varphi^{-1}\{\chi(x)\}^r$ *and the self-repeating series is* $\sqrt[n]{\varphi(x)/(\psi(x)\varphi'(x))}$, *where n is any quantity and $\psi(x)$ any suitable function. Supposing the series to be $S(x)$ we have*

$$\frac{SF(x)}{Sf(x)} = \sqrt[n]{\frac{p}{q} \frac{\psi f(x)}{\psi F(x)} \frac{F'(x)}{f'(x)}}. \text{"} \tag{15.7}$$

We have quoted Ramanujan (p. 192) for Entry 15(iii), which is very enigmatic indeed. There is no guarantee that the function φ exists. It also is not clear what a self-repeating series is.

We offer a proof under several assumptions.

PROOF. We shall assume that a function φ exists so that (15.6) holds. Without loss of generality, we assume that $p = 1$; thus $f(x) = x$. We furthermore suppose that q and r are nonnegative powers of 2 with $2 \le q \le 2r$. Since F is of "degree q", we put $F(x) = G^{(q)}(x)$, where $G^{(q)}(x)$ is defined by (15.5). With our assumptions, (15.6) now takes the form

$$\varphi^q(x) = \varphi(F(x)), \tag{15.8}$$

and we are required to prove that

$$G^{(r)}(x) = \varphi^{-1}(\varphi(x)^r). \tag{15.9}$$

We shall establish (15.9) by induction on r. For $r = 1$, (15.9) clearly holds. We shall now assume that (15.9) holds up to a fixed integer $r \ge 1$ and show that (15.9) is valid with r replaced by $2r$. Using (15.8), (15.9) with x replaced by $F(x)$, and (15.5), we deduce that

$$\varphi^{-1}(\varphi(x)^{2r}) = \varphi^{-1}(\{\varphi^q(x)\}^{2r/q})$$
$$= \varphi^{-1}(\varphi(F(x))^{2r/q})$$
$$= G^{(2r/q)}(F(x))$$
$$= G^{(2r/q)}(G^{(q)}(x))$$
$$= G^{(2r)}(x).$$

This concludes the proof of (15.9) and Ramanujan's first assertion in Entry 15(iii).

We next prove (15.7). There is now no need to make any restrictions on p and q, except that $pq \neq 0$. We do need to assume that f, F, and φ are differentiable.

Using the chain rule and (15.6), we find that

$$
\begin{aligned}
\frac{SF(x)}{Sf(x)} &= \left(\frac{\varphi F(x)F'(x)\psi f(x)d(\varphi f)/dx}{\psi F(x)d(\varphi F)/dx \; \varphi f(x)f'(x)} \right)^{1/n} \\
&= \left(\frac{\psi f(x)F'(x)}{\psi F(x)f'(x)} \; \frac{\varphi F(x)d(\varphi F)^{p/q}/dx}{\varphi f(x)d(\varphi F)/dx} \right)^{1/n} \\
&= \left(\frac{\psi f(x)F'(x)}{\psi F(x)f'(x)} \; \frac{\varphi F(x)\dfrac{p}{q}(\varphi F)^{p/q-1}}{\varphi f(x)} \right)^{1/n} \\
&= \left(\frac{\psi f(x)F'(x)}{\psi F(x)f'(x)} \frac{p}{q} \right)^{1/n},
\end{aligned}
$$

which completes the proof. $\qquad\square$

For the example below, which closes Section 15, we again quote Ramanujan (p. 192).

Example. "*If $I = x$ and $II = x^2 + 2nx$, then if x is great*

$$III = x^3 + 3nx^2 + \frac{3n(n+1)}{2}x - \frac{n(n-1)(n-2)x}{2x + 3(n+1)/2} \quad \text{nearly.}" \quad (15.10)$$

As in the examples above, we interpret this statement as an example in the iteration of functions. First, observe that, in the corollary in Section 15(i), the third function is equal to $x^3 + 3x^2 + 3x$, which agrees with (15.10) when $n = 1$. Second, by Corollary (i) in Section 15(ii), $f^{(3)}(x) = x^3 + 6x^2 + 9x$, which is in agreement with (15.10) in the case $n = 2$.

In accordance with our comments made earlier in Section 15, Ramanujan probably derived a representation for the rth iterate when $r = 2^m$ and then replaced r by an arbitrary positive integer. He then evidently derived a type of asymptotic formula for the third function and terminated the series to obtain the given approximation. Thus, for $r = 2^m$, $m \geq 0$, define a sequence of polynomials $P_r(x)$ by $P_1(x) = x$ and

$$P_{2r}(x) = P_r^2(x) + 2nP_r(x).$$

We can prove by induction that for $r = 2^m > 2$,

$$P_r(x) = x^r + rnx^{r-1} + \tfrac{1}{2}rn(1 + (r-2)n)x^{r-2}$$
$$+ \tfrac{1}{2}r(r-2)n^2(1 + n(r-4)/3)x^{r-3} + \cdots. \quad (15.11)$$

If we interpolate by setting $r = 3$ in (15.11), we find that

$$P_3(x) = x^3 + 3nx^2 + \tfrac{3}{2}n(n + 1)x + \tfrac{3}{2}n^2(1 - n/3) + \cdots. \qquad (15.12)$$

The first three terms on the right side of (15.12) agree with those in (15.10). However, the last term in (15.10) approaches, as x tends to ∞, $-n(n - 1) \times (n - 2)/2$, which differs from $\tfrac{3}{2}n^2(1 - n/3)$ in (15.12).

We do not know how to find a general closed formula for the coefficient of x^k in (15.11).

Entry 16. *If the modular equation of degree $n - 1$ is*

$$\sqrt[n]{\alpha\beta} + \sqrt[n]{(1 - \alpha)(1 - \beta)} = 1,$$

then the modular equation of degree $(n - 1)^2$ is

$$\{\sqrt[n]{\alpha(1 - \beta)} - \sqrt[n]{\beta(1 - \alpha)}\}^n = \{\sqrt[n]{\alpha} - \sqrt[n]{\beta}\}^n + \{\sqrt[n]{1 - \beta} - \sqrt[n]{1 - \alpha}\}^n.$$

PROOF. For brevity, set

$$A = \sqrt[n]{\alpha}, \quad B = \sqrt[n]{\beta}, \quad C = \sqrt[n]{\gamma}, \quad a = \sqrt[n]{1 - \alpha},$$
$$b = \sqrt[n]{1 - \beta}, \quad \text{and} \quad c = \sqrt[n]{1 - \gamma}.$$

The modular equation of degree $n - 1$ for γ as a function of α is

$$AC + ac = 1, \qquad (16.1)$$

and the modular equation of degree $n - 1$ for β as a function of γ is

$$BC + bc = 1. \qquad (16.2)$$

Thus, β is of degree $(n - 1)^2$ in α, and we can determine the modular equation of degree $(n - 1)^2$ by eliminating γ from (16.1) and (16.2). After subtracting (16.2) from (16.1), we readily find that

$$\frac{A - B}{b - a} = \left(\frac{1}{\gamma} - 1\right)^{1/n},$$

or

$$\gamma = \frac{1}{\left(\dfrac{A - B}{b - a}\right)^n + 1}.$$

Substituting in (16.1), we arrive at

$$A\left(\frac{1}{\left(\dfrac{A - B}{b - a}\right)^n + 1}\right)^{1/n} + a\,\frac{A - B}{b - a}\left(\frac{1}{\left(\dfrac{A - B}{b - a}\right)^n + 1}\right)^{1/n} = 1.$$

Multiplying both sides by $\{(A - B)^n + (b - a)^n\}^{1/n}$ and simplifying, we deduce that

$$Ab - aB = \{(A - B)^n + (b - a)^n\}^{1/n},$$

from which the identity that we sought follows.   □

References

Abramowitz, M. and Stegun, I. A., editors
[1] *Handbook of Mathematical Functions*, Dover, New York, 1965.
Achuthan, P. and Ponnuswamy, S.
[1] On an entry in Ramanujan's second notebook, to appear.
Aiyar, M. V.
[1] On some elliptic function series, *J. Indian Math. Soc.* **18**(1929/30), 46–50.
Aiyar, S. N.
[1] Some theorems in summation, *J. Indian Math. Soc.* **5**(1913), 183–186.
Allen, E. J.
[1] Continued radicals, *Math. Gaz.* **69**(1985), 261–263.
Andoyer, H.
[1] Sur une classe de fractions continues, *Bull. Sci. Math.* (2) **32**(1908), 207–221.
Andrews, G. E.
[1] Problems and prospects for basic hypergeometric functions, *Theory and Application of Special Functions*, R. A. Askey, editor, Academic Press, New York, 1975, pp. 191–224.
[2] *The Theory of Partitions*, Addison-Wesley, Reading, 1976.
[3] Notes on the Dyson conjecture, *SIAM J. Math. Anal.* **11**(1980), 787–792.
Andrews, G. E., Askey, R. A., Berndt, B. C., Ramanathan, K. G., and Rankin, R. A., editors
[1] *Ramanujan Revisited*, Academic Press, Boston, 1988.
Apéry, R.
[1] Interpolation de fractions continues et irrationalite de certaines constantes, *Bull. Section des Sci.*, Tome III, Bibliothéque Nationale, Paris, 1981, 37–63.
Apostol, T. M.
[1] Generalized Dedekind sums and transformation formulae of certain Lambert series, *Duke Math. J.* **17**(1950), 147–157.
[2] On the Lerch zeta function, *Pacific J. Math.* **1**(1951), 161–167.
[3] *Modular Functions and Dirichlet Series in Number Theory*, Springer-Verlag, New York, 1976.

Appell, P. E. and Kampé de Fériet, J.
[1]   *Fonctions Hypergéométriques et Hypersphériques*, Gauthier-Villars, Paris, 1926.
Appledorn, C. R.
[1]   Problem 87-6, *SIAM Rev.* **29**(1987), 297.
Askey, R.
[1]   A note on the history of series, MRC Technical Summary Report #1532, Madison, 1975.
[2]   Appendix to "Chapter 12 of Ramanujan's second notebook: Continued fractions," *Rocky Mt. J. Math.* **15**(1985), 311–318.
[3]   Ramanujan and important formulas, *Srinivasa Ramanujan* (1887–1920), Macmillan India Ltd., Madras, pp. 42–51.
Askey, R. and Wilson, J.
[1]   A set of hypergeometric orthogonal polynomials, *SIAM J. Math. Anal.* **13**(1982), 651–655.
Ayoub, R.
[1]   *An Introduction to the Analytic Theory of Numbers*, American Mathematical Society, Providence, 1963.
Bailey, W. N.
[1]   Products of generalized hypergeometric series, *Proc. London Math. Soc.* (2) **28**(1928), 242–254.
[2]   The partial sum of the coefficients of the hypergeometric series, *J. London Math. Soc.* **6**(1931), 40–41.
[3]   Some theorems concerning products of hypergeometric series, *Proc. London Math. Soc.* (2) **38**(1935), 377–384.
[4]   *Generalized Hypergeometric Series*, Stechert-Hafner, New York, 1964.
Barnes, E. W.
[1]   The asymptotic expansion of integral functions defined by generalized hypergeometric series, *Proc. London Math. Soc.* (2) **5**(1907), 59–116.
[2]   A new development of the theory of the hypergeometric functions, *Proc. London Math. Soc.* (2) **6**(1908), 141–177.
Batut, C. and Olivier, M.
[1]   Sur l'accélération de la convergence de certaines fractions continues, Sém. Théorie des Nombres, Bordeaux, 1979–1980, 25 pages.
Bauer, G.
[1]   Von den Coefficienten der Reihen von Kugelfunctionen einer Variabeln, *J. Reine Angew. Math.* **56**(1859), 101–121.
[2]   Von einem Kettenbruche Euler's und einem Theorem von Wallis, *Abh. Bayer. Akad. Wiss.* **11**(1872), 96–116.
Belevitch, V.
[1]   The Gauss hypergeometric ratio as a positive real function, *SIAM J. Math. Anal.* **13**(1982), 1024–1040.
Berndt, B. C.
[1]   Ramanujan's formula for $\zeta(2n + 1)$, *Professor Srinivasa Ramanujan Commemoration Volume*, Jupiter Press, Madras, 1974, pp. 2–9.
[2]   Periodic Bernoulli numbers, summation formulas and applications, *Theory and Application of Special Functions*, R. A. Askey, ed., Academic Press, New York, 1975, pp. 143–189.
[3]   Character analogues of the Poisson and Euler–Maclaurin summation formulas with applications, *J. Number Theory* **7**(1975), 413–445.
[4]   On Eisenstein series with characters and the values of Dirichlet *L*-functions, *Acta Arith.* **28**(1975), 299–320.
[5]   Dedekind sums and a paper of G. H. Hardy, *J. London Math. Soc.* (2) **13**(1976), 129–137.

[6] Modular transformations and generalizations of several formulae of Ramanujan, *Rocky Mt. J. Math.* **7**(1977), 147–189.

[7] Analytic Eisenstein series, theta-functions, and series relations in the spirit of Ramanujan, *J. Reine Angew. Math.* **303/304**(1978), 332–365.

[8] An arithmetic Poisson formula, *Pacific J. Math.* **103**(1982), 295–299.

[9] *Ramanujan's Notebooks, Part I*, Springer-Verlag, New York, 1985.

[10] Ramanujan's modular equations, *Ramanujan Revisited*, Academic Press, Boston, 1988, pp. 313–333.

[11] *Ramanujan's Notebooks, Part III*, Springer-Verlag, New York, 1991.

Blaum, M., Eisenberger, I., Lorden, G., and McEliece, R. J.

[1] More about the birthday surprise, preprint.

Boas, R. P. and Pollard, H.

[1] Continuous analogues of series, *Amer. Math. Monthly* **80**(1973), 18–25.

Bodendiek, R.

[1] *Über verschiedene Methoden zur Bestimmung der Transformationsformeln der achten Wurzeln der Integralmoduln $k^2(\tau)$ und $k'^2(\tau)$, ihrer Logarithmen sowie gewisser Lambertscher Reihen bei beliebigen Modulsubstitutionen*, Dissertation, der Universität Köln, 1968.

Bodendiek, R. and Halbritter, U.

[1] Über die Transformationsformel von log $\eta(\tau)$ und gewisser Lambertscher Reihen, *Abh. Math. Sem. Univ. Hamburg* **38**(1972), 147–167.

Boersma, J.

[1] Problem 559, Solutions by J. Boersma, H. Van Haeringen, M. T. Kosters, and S. W. Rienstra, *Nieuw Archief Wiss.* (3) **28**(1980), 286–289.

Bowman, K. O., Shenton, L. R., and Szekeres, G.

[1] A Poisson sum up to the mean and a Ramanujan problem, *J. Stat. Comput. Simul.* **20**(1984), 167–173.

Bromwich, T. J. I'A.

[1] *An Introduction to the Theory of Infinite Series*, 2nd ed., Macmillan, London, 1926.

Bruckman, P. S.

[1] On the evaluation of certain infinite series by elliptic functions, *Fibonacci Quart.* **15**(1977), 293–310.

Buckholtz, J. D.

[1] Concerning an approximation of Copson, *Proc. Amer. Math. Soc.* **14**(1963), 564–568.

Bühler, W. K.

[1] The hypergeometric function—a biographical sketch, *Math. Intell.* **7**(1985), No. 2, 35–40.

Bühring, W.

[1] The behavior at unit argument of the hypergeometric function $_3F_2$, *SIAM J. Math. Anal.* **18**(1987), 1227–1234.

Burkhardt, H., editor

[1] *Encyklopädie der Mathematischen Wissenschaften*, zweiter Band, *Analysis*, erster Teil, zweite Hälfte, B. G. Teubner, Leipzig, 1904.

Carlitz, L.

[1] The coefficients in an asymptotic expansion, *Proc. Amer. Math. Soc.* **16**(1965), 248–252.

Carlson, B. C.

[1] The hypergeometric function and the R-function near their branch points, International Conf. on Special Functions; Theory and Computation, Turin, 1984, *Rend. Sem. Mat. Univers. Politecn. Torino*, special issue, 1985, 63–89.

Carr, G. S.

     Formulas and Theorems in Pure Mathematics, 2nd ed., Chelsea, New York, 1970.

Cauchy, A.
[1]   *Oeuvres, Série II*, t. VII, Gauthier-Villars, Paris, 1889.
Chandrasekharan, K. and Narasimhan, R.
[1]   Hecke's functional equation and arithmetical identities, *Ann. Math.* **74**(1961), 1–23.
Chowla, S.
[1]   Some infinite series, definite integrals and asymptotic expansions, *J. Indian Math. Soc.* **17**(1927/28), 261–288.
Chrystal, G.
[1]   *Algebra*, Vol. 2, 7th ed., Chelsea, New York, 1964.
Churchill, R. V.
[1]   *Complex Variables and Applications*, 2nd ed., McGraw-Hill, New York, 1960.
Clausen, T.
[1]   Ueber die Fälle, wenn die Reihe von der Form $y = 1 + \frac{\alpha}{1} \cdot \frac{\beta}{\gamma} x + \frac{\alpha \cdot \alpha + 1}{1 \cdot 2} \cdot$ $\frac{\beta \cdot \beta + 1}{\gamma \cdot \gamma + 1} x^2 +$ etc. ein Quadrat von der Form $z = 1 + \frac{\alpha'}{1} \cdot \frac{\beta'}{\gamma'} \cdot \frac{\delta'}{\varepsilon'} x + \frac{\alpha' \cdot \alpha' + 1}{1 \cdot 2} \cdot$ $\frac{\beta' \cdot \beta' + 1}{\gamma' \cdot \gamma' + 1} \cdot \frac{\delta' \cdot \delta' + 1}{\varepsilon' \cdot \varepsilon' + 1} x^2 +$ etc. hat, *J. Reine Angew. Math.* **3**(1828), 89–91.
Coddington, E. A.
[1]   *An Introduction to Ordinary Differential Equations*, Prentice-Hall, Englewood Cliffs, NJ, 1961.
Cohen, H.
[1]   Acceleration de la convergence de certaines recurrences lineaires, Sém. Théorie des Nombres, Grenoble, 1980, 47 pp.
Copson, E. T.
[1]   An approximation connected with e^{-x}, *Proc. Edinburgh Math. Soc.* (2) **3**(1932), 201–206.
[2]   *Theory of Functions of a Complex Variable*, Clarendon Press, Oxford, 1935.
[3]   *Asymptotic Expansions*, University Press, Cambridge, 1965.
Darling, H. B. C.
[1]   On a proof of one of Ramanujan's theorems, *J. London Math. Soc.* **5**(1930), 8–9.
De Morgan, A.
[1]   On the reduction of a continued fraction to a series, *Philos. Mag.* (3) **24**(1844), 15–17.
de Saint-Venant, M.
[1]   Mémoire sur la torsion des prismes, *Mém. prés. par divers savants Acad. Sci. Inst. Impérial France Sci. Math. Phys.* **14**(1856), 233–560.
Dickson, L. E.
[1]   *History of the Theory of Numbers*, Vol. 1, Chelsea, New York, 1952.
Dixon, A. C.
[1]   Summation of a certain series, *Proc. London Math. Soc.* (1) **35**(1903), 285–289.
Dougall, J.
[1]   On Vandermonde's theorem and some more general expansions, *Proc. Edinburgh Math. Soc.* **25**(1907), 114–132.
Dutka, J.
[1]   Two results of Ramanujan, *SIAM J. Math. Anal.* **12**(1981), 471–476.
[2]   Wallis's product, Brouncker's continued fraction, and Leibniz's series, *Arch. History Exact Sci.* **26**(1982), 115–126.
[3]   The early history of the hypergeometric function, *Arch. History Exact Sci.* **31**(1984), 15–34.
Edwards, J.
[1]   *An Elementary Treatise on the Differential Calculus*, 2nd ed., Macmillan, London, 1892.

[2]   *A Treatise on the Integral Calculus*, Vol. 2, Macmillan, London, 1922.
Erdélyi, A., editor
[1]   *Higher Transcendental Functions*, Vol. 1, McGraw-Hill, New York, 1953.
Euler, L.
[1]   *Introductio in Analysin Infinitorum*, Vol. 1, Marcum-Michaelem Bousquet, Lausanne, 1748; *Opera Omnia*, Ser. I, Vol. 8, B. G. Teubner, Lipsiae, 1922.
[2]   *Institutiones Calculi Integralis*, Vol. 2, 3rd ed., Impensis Academiae Imperialis Scientiarum, Petropoli, 1827; *Opera Omnia*, Ser. 1, Vol. 12, B. G. Teubner, Lipsiae, 1914, pp. 1–413.
[3]   De transformationae serierum in fractiones continuas ubi simul haec theoria non mediocriter amplificatur, *Opera Omnia*, Ser. I, Vol. 15, B. G. Teubner, Lipsiae, 1927, pp. 661–700.
[4]   Methodus inveniendi formulas integrales, quae certis casibus datam inter se teneant rationem, ubi simul methodus traditur fractiones continuas summandi, *Opera Omnia*, Ser. I. Vol. 18, B. G. Teubner, Lipsiae, 1920, pp. 209–243.
[5]   De fractionibus continuis dissertatio, *Opera Omnia*, Ser. I, Vol. 14, B. G. Teubner, Lipsiae, 1925, pp. 187–215.
[6]   De fractionibus continuis observationes, *Opera Omnia*, Ser. I, Vol. 14, B. G. Teubner, Lipsiae, 1925, pp. 291–349.
[7]   Commentatio in fractionem continuam qua illustris La Grange potestates binomiales expressit, *Opera Omnia*, Ser. I, Vol. 16, B. G. Teubner, Lipsiae, 1935, pp. 232–240.
Evans, R. J.
[1]   Ramanujan's second notebook: Asymptotic expansions for hypergeometric series and related functions, *Ramanujan Revisited*, Academic Press, Boston, 1988, pp. 537–560.
Evans, R. J. and Stanton, D.
[1]   Asymptotic formulas for zero-balanced hypergeometric series, *SIAM J. Math. Anal.* **15**(1984), 1010–1020.
Fichtenholz, G. M.
[1]   *Differential-und Integralrechnung*, Band 2, VEB Deutscher Verlag der Wissenschaften, Berlin, 1966.
Fields, J. L.
[1]   Asymptotic expansions of a class of hypergeometric polynomials with respect to the order—III, *J. Math. Anal. Appl.* **12**(1965), 593–601.
Flajolet, P.
[1]   Combinatorial aspects of continued fractions, *Discrete Math.* **32**(1980), 125–161.
Fletcher, A., Miller, J. C. P., Rosenhead, L., and Comrie, L. J.
[1]   *An Index of Mathematical Tables*, Vol. I, 2nd ed., Addison-Wesley, Reading, 1962.
Forrester, P. J.
[1]   Extensions of several summation formulae of Ramanujan using the calculus of residues, *Rocky Mt. J. Math.* **13**(1983), 557–572.
Frank, E.
[1]   Corresponding type continued fractions, *Amer. J. Math.* **68**(1946), 89–108.
Fransén, A. and Wrigge, S.
[1]   High-precision values of the gamma function and of some related coefficients, *Math. Comp.* **34**(1980), 553–566.
Frasch, H.
[1]   Die Erzeugenden der Hauptkongruenzgruppen für Primzahlstufen, *Math. Ann.* **108**(1933), 229–252.
Gauss, C. F.
[1]   Disquisitiones generales circa seriem infinitam $1 + \dfrac{\alpha\beta}{1\cdot\gamma}x + \dfrac{\alpha(\alpha+1)\beta(\beta+1)}{1\cdot 2\cdot\gamma(\gamma+1)}xx$

$$+ \frac{\alpha(\alpha + 1)(\alpha + 2)\beta(\beta + 1)(\beta + 2)}{1 \cdot 2 \cdot 3 \cdot \gamma(\gamma + 1)(\gamma + 2)} x^3 + \text{etc., Pars prior, Comm. soc. regiae}$$

sci. Gottingensis rec. 2(1812); reprinted in *C. F. Gauss, Werke*, Band 3, Königlichen Gesellschaft der Wissenschaften, Göttingen, 1876, pp. 123–162.

Gessel, I. and Stanton, D.
[1] Strange evaluations of hypergeometric series, *SIAM J. Math. Anal.* 13(1982), 295–308.
[2] Short proofs of Saalschütz's and Dixon's theorems, *J. Comb. Theory Ser. A* 38(1985), 87–90.

Glaeske, H.-J.
[1] Eine einheitliche Herleitung einer gewissen Klasse von Transformationsformeln der analytischen Zahlentheorie (I), *Acta Arith.* 20(1972), 133–145.
[2] Eine einheitliche Herleitung einer gewissen Klasse von Transformationsformeln der analytischen Zahlentheorie (II), *Acta Arith.* 20(1972), 253–265.

Glaisher, J. W. L.
[1] On the summation by definite integrals of geometrical series of the second and higher orders, *Quart. J. Math. (Oxford)* 11(1871), 328–343.
[2] On the sum of the divisors of a number, *Proc. Cambridge Philos. Soc.* 5(1884), 108–120.
[3] On the series which represent the twelve elliptic and the four zeta functions, *Mess. Math.* 18(1889), 1–84.
[4] Recurring relations involving sums of powers of divisors, *Mess. Math.* 20(1890–91), 129–135.
[5] Recurring relations involving sums of powers of divisors (second paper), *Mess. Math.* 20(1890–91), 177–181.

Glasser, M. L.
[1] Evaluation of a class of definite integrals, *Univ. Beograd Publ. Elektrotehn. Fak. Ser. Mat. Fiz.*, No. 506 (1975), 49–50.

Goldberg, L. A.
[1] *Transformations of Theta-functions and Analogues of Dedekind Sums*, Ph.D. thesis, University of Illinois, Urbana–Champaign, 1981.

Goldstein, L. J. and Razar, M. J.
[1] Ramanujan type formulas for $\zeta(2k - 1)$, *J. Pure Appl. Algebra* 13(1978), 13–17.

Gould, H. W.
[1] *Combinatorial Identities*, Morgantown Printing and Binding Co., Morgantown, 1972.

Goulden, I. P. and Jackson, D. M.
[1] *Combinatorial Enumeration*, Wiley, New York, 1983.
[2] A combinatorial proof of a continued fraction expansion theorem from the Ramanujan notebooks, *Graph Theory and Combinatorics*, B. Bollabas, ed., Academic Press, London, 1984, pp. 161–169.

Gradshteyn, I. S. and Ryzhik, I. M.
[1] *Table of Integrals, Series, and Products*, 4th ed., Academic Press, New York, 1965.

Graham, S. and Kolesnik, G.
[1] *Van der Corput's Method of Exponential Sums*, Cambridge University Press, Cambridge, to appear.

Grosjean, C. C.
[1] An infinite set of recurrence formulae for the divisor sums, Part I, *Bull. Soc. Math. Belgique* 29(1977), 3–49.
[2] An infinite set of recurrence formulae for the divisor sums, Part II, *Bull. Soc. Math. Belgique* 29(1977), 95–138.
[3] Proof of a remarkable identity, *Simon Stevin* 58(1984), 219–241.

Grosswald, E.
[1] Die Werte der Riemannschen Zeta-funktion an ungeraden Argumentstellen, *Nachr. Akad. Wiss. Göttingen* (1970), 9–13.
[2] Comments on some formulae of Ramanujan, *Acta Arith.* **21**(1972), 25–34.
Guinand, A. P.
[1] Functional equations and self-reciprocal functions connected with Lambert series, *Quart. J. Math. (Oxford)* **15**(1944), 11–23.
[2] Some rapidly convergent series for the Riemann ζ-function, *Q. J. Math. (Oxford)* (2) **6**(1955), 156–160.
Hafner, J. L.
[1] New omega theorems for two classical lattice point problems, *Invent. Math.* **63**(1981), 181–186.
Halphen, M.
[1] Sur une formule récurrente concernant les sommes des diviseurs des nombres entiers, *Bull. Soc. Math. France* **5**(1877), 158–160.
Hansen, E. R.
[1] *A Table of Series and Products*, Prentice-Hall, Englewood-Cliffs, NJ, 1975.
Hardy, G. H.
[1] A chapter from Ramanujan's note-book, *Proc. Cambridge Philos. Soc.* **21**(1923), 492–503.
[2] Some formulae of Ramanujan, *Proc. London Math. Soc.* (2) **22**(1924), xii–xiii.
[3] A formula of Ramanujan, *J. London Math. Soc.* **3**(1928), 238–240.
[4] Ramanujan and the theory of Fourier transforms, *Quart. J. Math. (Oxford)* **8**(1937), 245–254.
[5] *Divergent Series*, Clarendon Press, Oxford, 1949.
[6] *Collected Papers*, Vol. 1, Clarendon Press, Oxford, 1966.
[7] *Collected Papers*, Vol. 4, Clarendon Press, Oxford, 1969.
[8] *Collected Papers*, Vol. 7, Clarendon Press, Oxford, 1979.
[9] *Ramanujan*, Chelsea, New York, 1978.
Hardy, G. H. and Ramanujan, S.
[1] Asymptotic formulae in combinatory analysis, *Proc. London Math. Soc.* (2) **17**(1918), 75–115.
Hardy, G. H. and Wright, E. M.
[1] *An Introduction to the Theory of Numbers*, 4th ed., Clarendon Press, Oxford, 1960.
Henrici, P.
[1] *Applied and Computational Complex Analysis*, Vol. 1, Wiley, New York, 1974.
[2] *Applied and Computational Complex Analysis*, Vol. 2, Wiley, New York, 1977.
Herschfeld, A.
[1] On infinite radicals, *Amer. Math. Monthly* **42**(1935), 419–429.
Hill, M. J. M.
[1] On a formula for the sum of a finite number of terms of the hypergeometric series when the fourth element is equal to unity, *Proc. London Math. Soc.* (2) **5**(1907), 335–341.
[2] On a formula for the sum of a finite number of terms of the hypergeometric series when the fourth element is unity (second communication), *Proc. London Math. Soc.* (2) **6**(1908), 339–348.
Hurwitz, A.
[1] Grundlagen einer independenten Theorie der elliptischen Modulfunktionen und Theorie der Multiplikator-Gleichungen erster Stufe, *Math. Ann.* **18**(1881), 528–592.
[2] *Mathematische Werke*, Band I, Emil Birkhäuser, Basel, 1932.
Iseki, S.
[1] The transformation formula for the Dedekind modular function and related functional equations, *Duke Math. J.* **24**(1957), 653–662.

Ivić, A.
    [1]   *The Riemann Zeta-function*, Wiley, New York, 1985.

Iwaniec, H. and Mozzochi, C. J.
    [1]   On the divisor and circle problems, *J. Number Theory* **29**(1988), 60–93.

Jackson, D. M.
    [1]   Some results on "product-weighted lead codes", *J. Comb. Theory Ser. A* **25**(1978), 181–187.

Jacobi, C. G. J.
    [1]   De seriebus ac differentiis observatiunculae, *J. Reine Angew. Math.* **36**(1847), 135–142.
    [2]   *Gesammelte Werke*, Band 6, Georg Reimer, Berlin, 1891, pp. 174–182.

Jacobsen, L.
    [1]   Composition of linear fractional transformations in terms of tail sequences, *Proc. Amer. Math. Soc.* **97**(1986), 97–104.
    [2]   General convergence of continued fractions, *Trans. Amer. Math. Soc.* **294**(1986), 477–485.
    [3]   Domains of validity for some of Ramanujan's continued fraction formulas, *J. Math. Anal. Appl.*, **143**(1989), 412–437.
    [4]   Compositions of contractions, J. Comput. Appl. Math. **32**(1990), 169–178.
    [5]   On the Bauer–Muir transformation for continued fractions and its applications, J. Math. Anal. Appl. **152**(1990), 496–514.

Jogdeo, K. and Samuels, S. M.
    [1]   Monotone convergence of binomial probabilities and a generalization of Ramanujan's equation, *Ann. Math. Stat.* **39**(1968), 1191–1195.

Jones, W. B. and Thron, W. J.
    [1]   *Continued Fractions: Analytic Theory and Applications*, Addison-Wesley, Reading, 1980.

Jordan, W. B.
    [1]   Problem 83-8, *SIAM Rev.* **26**(1984), 278–279.

Karlsson, P. W.
    [1]   On two hypergeometric summation formulas conjectured by Gosper, *Simon Stevin* **60**(1986), 329–337.

Katayama, K.
    [1]   On Ramanujan's formula for values of Riemann zeta-function at positive odd integers, *Acta Arith.* **22**(1973), 149–155.
    [2]   Ramanujan's formula for *L*-functions, *J. Math. Soc. Japan* **26**(1974), 234–240.
    [3]   Zeta-functions, Lambert series and arithmetic functions analogous to Ramanujan's τ-function. I, *J. Reine Angew. Math.* **268/269**(1974), 251–270.
    [4]   Zeta-functions, Lambert series and arithmetic functions analogous to Ramanujan's τ-function. II, *J. Reine Angew. Math.* **282**(1976), 11–34.

Khovanskii, A. N.
    [1]   *The Application of Continued Fractions and Their Generalizations to Problems in Approximation Theory*, trans. by P. Wynn, P. Noordhoff, Groningen, 1963.

Kirschenhofer, P. and Prodinger, H.
    [1]   On some applications of formulae of Ramanujan in the analysis of algorithms, Mathematika **38**(1991), 14–33.

Klamkin, M. S. and Newman, D. J.
    [1]   Extensions of the birthday surprise, *J. Comb. Theory* **3**(1967), 279–282.

Klein, F.
    [1]   *Vorlesungen über die hypergeometrische Funktion*, Springer-Verlag, Berlin, 1933.

Klusch, D.
    [1]   Mellin transforms and Fourier-Ramanujan expansions, *Math. Z.* **193**(1986), 515–526.

Knopp, K.

[1] *Theory and Application of Infinite Series*, Blackie & Son, Glasgow, 1951.

Knuth, D. E.

[1] *The Art of Computer Programming*, Vol. 1, 2nd ed., Addison-Wesley, Reading, 1973.

Koshliakov, N. S.

[1] Application of the theory of sumformulae to the investigation of a class of one-valued analytical functions in the theory of numbers, *Mess. Math.* **58**(1929), 1–23.

Krishnamachari, C.

[1] Certain definite integrals and series, *J. Indian Math. Soc.* **12**(1920), 14–31.

Krishnan, K. S.

[1] On the equivalence of certain infinite series and the corresponding integrals, *J. Indian Math. Soc.* **12**(1948), 79–88.

Kummer, E. E.

[1] Über die hypergeometrische Reihe $1 + \dfrac{\alpha \cdot \beta}{1 \cdot \gamma} x + \dfrac{\alpha(\alpha + 1)\beta(\beta + 1)}{1 \cdot 2 \cdot \gamma(\gamma + 1)} x^2 + \dfrac{\alpha(\alpha + 1)(\alpha + 2)\beta(\beta + 1)(\beta + 2)}{1 \cdot 2 \cdot 3 \cdot \gamma(\gamma + 1)(\gamma + 2)} x^3 + \cdots$, *J. Reine Angew. Math.* **15**(1836), 39–83, 127–172.

[2] *Collected Papers*, Vol. 2, Springer-Verlag, Berlin, 1975.

Lagrange, J.

[1] Une formule sommatoire et ses applications, *Bull. Sci. Math.* (2) **84**(1960), 105–110.

Lagrange, J. L.

[1] Sur l'usage des fractions continues dans le calcul intégral, Nouv. Mém. Acad. Roy. Sci. Berlin, 1776; *Oeuvres*, Tome 4, Gauthier-Villars, Paris, 1869, pp. 301–332.

Laguerre, E.

[1] Sur la fonction $\left(\dfrac{x + 1}{x - 1}\right)^{\omega}$, *Bull. Soc. Math. France* **8**(1879), 36–52.

Lambert, J. H.

[1] Mémoire sur quelques propriétés remarquables des quantitiés transcendantes, circulaires et logarithmiques, Hist. Acad. Roy. Sci. Berlin, 1761, 265–322; *Opera Mathematica*, Vol. 2, Orell Füssli, Zürich, 1948, pp. 112–159.

[2] Verwandlung der Brüche, Beyträge Gebrauche Math. Anwend., zweiter Teil, Berlin, 1770, 54–132; *Opera Mathematica*, Vol. 1, Orell Füssli, Zürich, 1946, pp. 133–188.

[3] Vorläufige Kenntnisse für die, so die Quadratur und Rectification des Circuls suchen, Beyträge Gebrauche Math. Anwend., zweiter Teil, Berlin, 1770, 140–169; *Opera Mathematica*, Vol. 1, Orell Füssli, Zürich, 1946, pp. 194–212.

Lamm, G. and Szabo, A.

[1] Atomic multipole polarizabilities in the extended Coulomb approximation, *J. Chem. Phys.* **67**(1977), 5942–5946.

[2] Analytic Coulomb approximations for dynamic multipole polarizabilities and dispersion forces, *J. Chem. Phys.* **72**(1980), 3354–3377.

Landau, E.

[1] *Handbuch der Lehre von der Verteilung der Primzahlen*, 2nd ed., Chelsea, New York, 1953.

Lavoie, J. L.

[1] Some evaluations for the generalized hypergeometric series, *Math. Comp.* **46**(1986), 215–218.

[2] Some summation formulas for the series $_3F_2(1)$, *Math. Comp.* **49**(1987), 269–274.

Lawden, D. F.

[1] Pseudo-random sequence loops, *Math. Gaz.* **68**(1984), 39–41.

Lebedev, N. N.

[1] *Special Functions and Their Applications*, Dover, New York, 1972.

Legendre, A. M.

[1] *Mémoires sur les Transcendantes Elliptiques*, C. DuPont, C. Firmin Didot, Paris, 1794.

Lehmer, D. H.

[1] Some functions of Ramanujan, *Math. Student* **27**(1959), 105–116.

Lerch, M.

[1] Sur la fonction $\zeta(s)$ pour valeurs impaires de l'argument, *J. Sci. Math. Astron.* pub. pelo Dr. F. Gomes Teixeira, Coimbra **14**(1901), 65–69.

LeVeque, W. J., editor

[1] *Reviews in Number Theory*, Vol. 1, American Mathematical Society, Providence, 1974.

Levitt, J.

[1] *On a Problem of Ramanujan*, M. Phil. thesis, University of Nottingham, 1978.

Lewittes, J.

[1] Analytic continuation of the series $\Sigma(m + nz)^{-s}$, *Trans. Amer. Math. Soc.* **159**(1971), 505–509.

[2] Analytic continuation of Eisenstein series, *Trans. Amer. Math. Soc.* **171**(1972), 469–490.

Lindelöf, E.

[1] *Le Calcul des Résidus*, Chelsea, New York, 1947.

Ling, C.-B.

[1] On summation of series of hyperbolic functions, *SIAM J. Math. Anal.* **5**(1974), 551–562.

[2] On summation of series of hyperbolic functions. II, *SIAM J. Math. Anal.* **6**(1975), 129–139.

[3] Generalization of certain summations due to Ramanujan, *SIAM J. Math. Anal.* **9**(1978), 34–48.

Littlewood, J. E.

[1] Review of Collected Papers of Srinivasa Ramanujan, *Math. Gaz.* **14**(1929), 427–428.

Luke, Y. L.

[1] *The Special Functions and Their Approximations*, Vol. 1, Academic Press, New York, 1969.

Macmahon, P. A.

[1] *Combinatory Analysis*, Vol. 1, University Press, Cambridge, 1915.

Madhava, K. B.

[1] Note on the continued fraction in Q. 713, *J. Indian Math. Soc.* **11**(1919), 230–234.

Malurkar, S. L.

[1] On the application of Herr Mellin's integrals to some series, *J. Indian Math. Soc.* **16**(1925/26), 130–138.

Marsaglia, J. C. W.

[1] The incomplete gamma function and Ramanujan's rational approximation to e^x, *J. Stat. Comput. Simul.* **24**(1986), 163–168.

Masson, D.

[1] Convergence and analytic continuation for a class of regular *C*-fractions, *Can. Math. Bull.* **28**(1985), 411–421.

[2] Difference equations, continued fractions, Jacobi matrices and orthogonal polynomials, in *Nonlinear Numerical Methods and Rational Approximation*, Reidel, Dordrecht-Boston, 1988, pp. 239–257.

[3]   Some continued fractions of Ramanujan and Meixner–Pollaczek polynomials, Canad. Math. Bull. **32**(1989), 177–181.

Matala-Aho, T. and Väänänen, K.

[1]   On the arithmetic properties of certain values of one Gauss hypergeometric function, Acta Univ. Ouluensis Ser. A. Sci. rerum Natur. No. 112 Math. No. 24, Oulu, 1981, 26 pp.

Mathematical Gazette

[1]   Solutions and comments on 67.E and 67.F, *Math. Gaz.* **68**(1984), 57–59.

[2]   Solutions and comments on 67.G and 67.H, *Math. Gaz.* **68**(1984), 139–141.

Matsuoka, Y.

[1]   On the values of the Riemann zeta function at half integers, *Tokyo J. Math.* **2**(1979), 371–377.

[2]   Generalizations of Ramanujan's formulae, *Acta Arith.* **41**(1982), 19–26.

McCabe, J. H.

[1]   A formal extension of the Padé table to include two point Padé quotients, *J. Inst. Math. Appl.* **15**(1975), 363–372.

McKay, J. H.

[1]   The William Lowell Putnam mathematical competition, *Amer. Math. Monthly* **74**(1967), 771–777.

Mellin, H.

[1]   Eine Formel für den Logarithmus transcendenter Funktionen von endlichem Geschlecht, *Acta Soc. Sci. Fennicae* **29**(1902), 49 pp.

Mikolás, M.

[1]   Über gewisse Lambertsche Reihen, I: Verallgemeinerung der Modulfunktion $\eta(\tau)$ und ihrer Dedekindschen Transformationsformel, *Math. Z.* **68**(1957), 100–110.

Morley, F.

[1]   On the series $1 + \left(\dfrac{p}{1}\right)^3 + \left\{\dfrac{p(p+1)}{1 \cdot 2}\right\}^3 + \cdots$, *Proc. London Math. Soc.* (1) **34**(1902), 397–402.

Muir, T.

[1]   New general formulae for the transformation of infinite series into continued fractions, *Trans. R. Soc. Edinburgh* **27**(1876), 467–471.

Nagasaka, C.

[1]   Eichler integrals and generalized Dedekind sums, *Mem. Fac. Sci. Kyushu Univ. Ser. A Math.* **37**(1983), 35–43.

Nanjundiah, T. S.

[1]   Certain summations due to Ramanujan, and their generalisations, *Proc. Indian Acad. Sci. Sec. A* **34**(1951), 215–228.

Nielsen, N.

[1]   *Handbuch der Theorie der Gammafunktion*, Chelsea, New York, 1965.

[2]   *Theorie des Integrallogarithmus und verwandter Transzendenten*, Chelsea, New York, 1965.

Nörlund, N. E.

[1]   Fractions continues et différences réciproques, *Acta Math.* **34**(1911), 1–108.

Ogg, A.

[1]   Survey of modular functions of one variable, *Modular Functions of One Variable I*, Lecture Notes in Math., No. 320, Springer-Verlag, Berlin, 1973, pp. 1–35.

Olver, F. W. J.

[1]   *Asymptotics and Special Functions*, Academic Press, New York, 1974.

Paris, R. B.

[1]   On a generalisation of a result of Ramanujan connected with the exponential

series, *Proc. Edinburgh Math. Soc.* **24**(1981), 179–195.

Perron, O.
[1] Über eine Formel von Ramanujan, *Sitz. Bayer. Akad. Wiss. München Math. Phys. Kl.* 1952, 197–213.
[2] Über die Preeceschen Kettenbrüche, *Sitz. Bayer, Akad. Wiss. München Math. Phys. Kl.* 1953, 21–56.
[3] *Die Lehre von den Kettenbrüchen*, Band 2, dritte Auf., B. G. Teubner, Stuttgart, 1957.

Pfaff, J. F.
[1] Observationes analyticae ad L. Euleri Institutiones Calculi Integralis, *Nova Acta Acad. Sci. Petropolitanae* **11**(1797), 38–57.

Phillips, E. G.
[1] Note on summation of series, *J. London Math. Soc.* **4**(1929), 114–116.

Pincherle, S.
[1] Delle Funzioni ipergeometriche e di varie questioni ad esse attinenti, *Giorn. Mat. Battaglini* **32**(1894), 209–291.

Pollak, H. O. and Shepp, L.
[1] Problem 64-1, An asymptotic expansion, solution by J. H. Van Lint, *SIAM Rev.* **8**(1966), 383–384.

Polya, G. and Szegö, G.
[1] *Problems and Theorems in Analysis*, Vol. 1, Springer-Verlag, Berlin, 1978.

Preece, C. T.
[1] Theorems stated by Ramanujan (VI): Theorems on continued fractions, *J. London Math. Soc.* **4**(1929), 34–39.
[2] Theorems stated by Ramanujan (X), *J. London Math. Soc.* **6**(1931), 22–32.
[3] Theorems stated by Ramanujan (XIII), *J. London Math. Soc.* **6**(1931), 95–99.

Rademacher, H.
[1] On the partition function $p(n)$, *Proc. London Math. Soc.* (2) **43**(1937), 241–254.

Ramamani, V.
[1] *On Some Identities Conjectured by Srinivasa Ramanujan Found in His Lithographed Notes Connected with Partition Theory and Elliptic Modular Functions —Their Proofs—Interconnection with Various Other Topics in the Theory of Numbers and Some Generalisations Thereon*, doctoral thesis, University of Mysore, 1970.

Ramanathan, K. G.
[1] Hypergeometric series and continued fractions, to appear.

Ramanujan, S.
[1] On question 330 of Prof. Sanjana, *J. Indian Math. Soc.* **4**(1912), 59–61.
[2] Question 358, *J. Indian Math. Soc.* **4**(1912), 78.
[3] Question 387, *J. Indian Math. Soc.* **4**(1912), 120.
[4] Question 294, *J. Indian Math. Soc.* **4**(1912), 151–152.
[5] Question 289, *J. Indian Math. Soc.* **4**(1912), 226.
[6] Question 295, *J. Indian Math. Soc.* **5**(1913), 65.
[7] Modular equations and approximations to π, *Q. J. Math.* **45**(1914), 350–372.
[8] Some definite integrals, *Mess. Math.* **44**(1915), 10–18.
[9] On the product $\prod_{n=0}^{n=\infty} \left[1 + \left(\dfrac{x}{a+nd} \right)^3 \right]$, *J. Indian Math. Soc.* **7**(1915), 209–211.
[10] Some definite integrals connected with Gauss's sums, *Mess. Math.* **44**(1915), 75–85.
[11] On certain arithmetical functions, *Trans. Cambridge Philos. Soc.* **22**(1916), 159–184.
[12] Question 769, solution by K. B. Madhava, M. K. Kewalramani, N. Durairajan, and S. V. Venkatachala Aiyar, *J. Indian Math. Soc.* **9**(1917), 120–121.

[13] On certain trigonometrical sums and their applications in the theory of numbers, *Trans. Cambridge Philos. Soc.* **22**(1918), 259–276.

[14] A class of definite integrals, *Q. J. Math.* **48**(1920), 294–310.

[15] *Notebooks* (2 volumes), Tata Institute of Fundamental Research, Bombay, 1957.

[16] *Collected Papers*, Chelsea, New York, 1962.

Rankin, R. A.

[1] Elementary proofs of relations between Eisenstein series, *Proc. R. Soc. Edinburgh* **76A**(1976), 107–117.

[2] *Modular Forms and Functions*, Cambridge University Press, Cambridge, 1977.

Rao, M. B. and Ayyar, M. V.

[1] On some infinite series and products, Part I, *J. Indian Math. Soc.* **15**(1923/24), 150–162.

[2] On some infinite products and series, Part II, *J. Indian Math. Soc.* **15**(1923/24), 233–247.

Rao, S. N.

[1] A proof of a generalized Ramanujan identity, *J. Mysore Univ. Sec. B* **28**(1981–82), 152–153.

Riesel, H.

[1] Some series related to infinite series given by Ramanujan, *BIT* **13**(1973), 97–113.

Riordan, J.

[1] *Combinatorial Identities*, Wiley, New York, 1968.

Rogers, L. J.

[1] Third memoir on the expansion of certain infinite products, *Proc. London Math. Soc.* **26**(1895), 15–32.

[2] On the representation of certain asymptotic series as convergent continued fractions, *Proc. London Math. Soc.* (2) **4**(1907), 72–89.

[3] Supplementary note on the representation of certain asymptotic series as convergent continued fractions, *Proc. London Math. Soc.* (2) **4**(1907), 393–395.

Roy, R.

[1] On a paper of Ramanujan on definite integrals, *Math. Student* **46**(1978), 130–132.

[2] Binomial identities and hypergeometric series, *Amer. Math. Monthly* **94**(1987), 36–46.

Saalschütz, L.

[1] Eine Summationsformel, *Z. Math. Phys.* **35**(1890), 186–188.

[2] Über einen Spezialfall der hypergeometrischen Reihe dritter Ordnung, *Z. Math. Phys.* **36**(1891), 278–295, 321–327.

Sandham, H. F.

[1] Three summations due to Ramanujan, *Q. J. Math.* (*Oxford*) (2) **1**(1950), 238–240.

[2] Some infinite series, *Proc. Amer. Math. Soc.* **5**(1954), 430–436.

Sayer, F. P.

[1] The sums of certain series containing hyperbolic functions, *Fibonacci Q.* **14**(1976), 215–223.

Schläfli, L.

[1] Einige Bemerkungen zu Herrn Neumann's Untersuchungen über die Bessel'schen Functionen, *Math. Ann.* **3**(1871), 134–149.

Schlömilch, O.

[1] Ueber einige unendliche Reihen, *Ber. Verh. K. Sachs. Gesell. Wiss. Leipzig* **29**(1877), 101–105.

[2] *Compendium der Höheren Analysis*, zweiter Band, 4th ed., Friedrich Vieweg und Sohn, Braunschweig, 1895.

Schoeneberg, B.

[1]   *Elliptic Modular Functions*, Springer-Verlag, New York, 1974.

Serre, J.-P.

[1]   *A Course in Arithmetic*, Springer-Verlag, New York, 1973.

Sitaramachandrarao, R.

[1]   Some formulae of S. Ramanujan III, *J. Madras Univ.*, to appear.

[2]   Ramanujan's formula for $\zeta(2n + 1)$, *Ramanujan Visiting Lecture Notes*, Madurai Kamaraj University, to appear.

Sizer, W. S.

[1]   Continued roots, *Math. Magazine* **59**(1986), 23–27.

Slater, L. J.

[1]   *Generalized Hypergeometric Functions*, University Press, Cambridge, 1966.

Smart, J. R.

[1]   On the values of the Epstein zeta function, *Glasgow Math. J.* **14**(1973), 1–12.

Srivastava, H. M.

[1]   Some formulas of Srinivasa Ramanujan involving products of hypergeometric functions, *Indian J. Math.* **29**(1987), 91–100.

Stanton, D.

[1]   Recent results for the q-Lagrange inversion formula, *Ramanujan Revisited*, Academic Press, Boston, 1988, pp. 525–536.

Stieltjes, T. J.

[1]   Sur quelques intégrales définies et leur développement en fractions continues, *Q. J. Math.* **24**(1890), 370–382.

[2]   Note sur quelques fractions continues, *Q. J. Math.* **25**(1891), 198–200.

[3]   Recherches sur les fractions continues, *Ann. Fac. Sci. Toulouse Sci. Math. et Sci. Phys.* **8**(1894), 1–122; **9**(1895), 1–47.

[4]   *Oeuvres Complètes*, Tome 2, P. Noordhoff, Groningen, 1918.

Szegö, G.

[1]   Über einige von S. Ramanujan gestellte Aufgaben, *J. London Math. Soc.* **3**(1928), 225–232.

[2]   *Collected Papers*, Vol. 2, Birkhäuser, Boston, 1982.

Terras, A.

[1]   Some formulas for the Riemann zeta function at odd integer argument resulting from the Fourier expansions of the Epstein zeta function, *Acta Arith.* **29**(1976), 181–189.

Thomae, J.

[1]   Ueber die Funktionen, welch durch Reihen von der Form dargestellt werden
$$1 + \frac{p}{1}\frac{p'}{q'}\frac{p''}{q''} + \frac{p}{1}\frac{p+1}{2}\frac{p'}{q'}\frac{p'+1}{q'+1}\frac{p''}{q''}\frac{p''+1}{q''+1} + \cdots, \quad J. \ Reine \ Angew. \ Math.$$
87(1879), 26–73.

Titchmarsh, E. C.

[1]   *The Theory of Functions*, 2nd ed., Clarendon Press, Oxford, 1939.

[2]   *Theory of Fourier Integrals*, 2nd ed., Clarendon Press, Oxford, 1948.

[3]   *The Theory of the Riemann Zeta-function*, Clarendon Press, Oxford, 1951.

Toyoizumi, M.

[1]   Formulae for the Riemann zeta function at half integers, *Tokyo J. Math.* **3**(1980), 177–186.

[2]   Formulae for the values of zeta and L-functions at half integers, *Tokyo J. Math.* **4**(1981), 193–201.

[3]   Ramanujan's formulae for certain Dirichlet series, *Comm. Math. Univ. Sancti Pauli* **30**(1981), 149–173.

[4]   On the values of the Dedekind zeta function of an imaginary quadratic field at $s = 1/3$, *Comm. Math. Univ. Sancti Pauli* **31**(1982), 159–161.

Tricomi, F. G. and Erdélyi, A.
[1]   The asymptotic expansion of a ratio of gamma functions, *Pacific J. Math.* 1(1951), 133–142.
Tschebyscheff, P.
[1]   Sur lé développement des fonctions à une seule variable, *Bull. Acad. Imp. Sci. St. Pétersbourg* 1(1860), 193–200; *Oeuvres*, Tome 1, l'Acad. Impériale des Sciences, St. Pétersbourg, 1899, pp. 501–508.
Venkatachaliengar, K.
[1]   Development of Elliptic Functions According to Ramanujan, Tech. Report 2, Madurai Kamaraj University, Madurai, 1988.
Waadeland, H.
[1]   Tales about tails, *Proc. Amer. Math. Soc.* 90(1984), 57–64.
Wall, H. S.
[1]   *Analytic Theory of Continued Fractions*, Van Nostrand, Toronto, 1948.
Wallis, J.
[1]   *Arithmetica Infinitorum*, originally published in 1655 and reprinted in *Opera Mathematica*, Band 1, Georg Olms Verlag, Hildesheim, 1972.
Wang, T. H.
[1]   Problem 1064, solutions by R. L. Young and T. M. Apostol, *Math. Magazine* 53(1980), 181–184.
Watson, G. N.
[1]   Theorems stated by Ramanujan II, *J. London Math. Soc.* 3(1928), 216–225.
[2]   Theorems stated by Ramanujan (IV): Theorems on approximate integration and summation of series, *J. London Math. Soc.* 3(1928), 282–289.
[3]   Theorems stated by Ramanujan (V): Approximations connected with e^x, *Proc. London Math. Soc.* (2) 29(1929), 293–308.
[4]   Theorems stated by Ramanujan (VIII): Theorems on divergent series, *J. London Math. Soc.* 4(1929), 82–86.
[5]   The constants of Landau and Lebesgue, *Q. J. Math. (Oxford)* 1(1930), 310–318.
[6]   Theorems stated by Ramanujan (XI), *J. London Math. Soc.* 6(1931), 59–65.
[7]   Ramanujan's note books, *J. London Math. Soc.* 6(1931), 137–153.
[8]   Ramanujan's continued fraction, *Proc. Cambridge Philos. Soc.* 31(1935), 7–17.
[9]   *A Treatise on the Theory of Bessel Functions*, 2nd ed., University Press, Cambridge, 1966.
Whipple, F. J. W.
[1]   On well-poised series, generalized hypergeometric series having parameters in pairs, each pair with the same sum, *Proc. London Math. Soc.* (2) 24(1926), 247–263.
Whittaker, E. T. and Watson, G. N.
[1]   *A Course of Modern Analysis*, 4th ed., University Press, Cambridge, 1962.
Wigert, S.
[1]   Sur la série de Lambert et son application à la théorie des nombres, *Acta Math.* 41(1916), 197–218.
Williamson, B.
[1]   *An Elementary Treatise on the Differential Calculus*, 4th ed., Longmans, Green, and Co., London, 1880.
Wilson, J. A.
[1]   Some hypergeometric orthogonal polynomials, *SIAM J. Math. Anal.* 11(1980), 690–701.
Zhang, N.
[1]   Ramanujan's formulas and the values of the Riemann zeta-function at odd positive integers (Chinese), *Adv. Math. Beijing* 12(1983), 61–71.

Zhang, N. and Zhang, S.
[1] Riemann zeta function, *analytic functions of one complex variable, Contemp. Math.* **48**(1985), 235–241.

Zippin, L.
[1] *Uses of Infinity*, Random House, New York, 1962.

Zucker, I. J.
[1] The summation of series of hyperbolic functions, *SIAM J. Math. Anal.* **10**(1979), 192–206.
[2] Some infinite series of exponential and hyperbolic functions, *SIAM J. Math. Anal.* **15**(1984), 406–413.

Zuckerman, H. S.
[1] On the coefficients of certain modular forms belonging to subgroups of the modular group, *Trans. Amer. Math. Soc.* **45**(1939), 298–321.

Index

A

Abel–Plana summation formula 221
Abel's formula 88
Absolutely complete series 320
Achuthan, P. 126, 130
Aiyar, M.V. *see* Ayyar
Aiyar, S.N. 285
Allen, E.J. 109
Andoyer, H. 137
Andrews, G.E. 1, 6, 15, 127, 305
Apéry, R. 155, 162
Apostol, T.M. 259, 276, 330
Appell, P.E. 8
Appledorn, C.R. 216
Askey, R.A. 6, 8–9, 36, 41, 86–87,
    137, 219, 225
Asymptotic expansions 5, 54–56, 168,
    181–184, 185–218, 238–239, 283–
    291, 300–314, 337
Atkin, A.O.L. 329
Ayoub, R. 73, 283, 305
Ayyar (Aiyar), M.V. 253, 256, 261–263

B

Bailey, W.N. 42, 48, 59
Barnes, E.W. 214
Batut, C. 162
Bauer, G. 24, 145

Bauer–Muir transformation 159, 162
Belevitch, V. 134
Berndt, B.C. 240, 245, 253, 256, 258–
    261, 263, 276–277, 293–295, 299,
    326, 333
Bernoulli number 36, 301
Bessel functions 51, 58–59, 133, 186,
    214
Binet, J. 206, 221
Binomial coefficient identities 69
Birthday surprise problem 182
Blaum, M. 182
Boas, R.P. 226
Bodendiek, R. 276
Boersma, J. 294
Bowman, K.O. 181
Bromwich, T.J.I'A. 238
Brouncker, Lord 145
Bruckman, P.S. 299
Buckholtz, J.D. 182
Bühler, W.K. 8
Bühring, W. 77
Burkhardt, H. 206

C

Carlitz, L. 182
Carlson, B.C. 15
Carr, G.S. 102, 134

Catalan's constant 40, 45, 151, 153
Cauchy, A. 187, 262, 293–294, 299
C-fraction 175
Chandrasekharan, K. 262, 276
Character (mod 4) 240
Chowla, S. 262–263, 277
Chrystal, G. 106
Churchill, R.V. 324
Clausen, T. 58
Coddington, E.A. 88
Cohen, H. 6, 151, 162, 175
Complete series 320
Continued fractions 4, 103–107, 112– 166, 168–172, 175–183
  of Bessel functions 133
  even part 121
  of gamma functions 132, 140–164
  general convergence 123
  of hypergeometric functions, see Hypergeometric functions
  kth numerator and denominator 105
  notation 104–105
  periodic 116
  tails 105, 107, 112, 115, 119, 120
Copson, E.T. 34, 66, 182, 203, 307
Coulomb approximation 40

D
Darling, H.B.C. 46–47
Dedekind eta-function 253, 281, 305
Dedekind zeta-function 276
Degree of a modular equation 334
Degree of a series 320–321
De Morgan, A. 127, 129
de Saint-Venant, M. 294
Dickson, L.E. 327
Discriminant function 326
Divisor problem 304
Divisor sums 301, 313–314, 319, 327, 329–330
Dixon, A.C. 4, 7, 15, 22, 24–25, 46, 60–62
Dougall, J. 4, 7–9, 11, 247–248
Dutka, J. 8, 40, 43, 47, 145
Dyson–Gunson–Wilson identity 15

E
Edwards, J. 102, 162, 192
Eisenstein series 5, 240, 301, 318–320, 326–333
Elliptic integral 38, 79–80, 281
Erdélyi, A. 44
Euler, L. 103, 106, 114, 129, 132–133, 137, 141, 168, 185
Euler–Maclaurin summation formula 208, 300–302
Evans, R.J. 5–6, 12, 40, 48, 65, 70, 77, 186, 202, 216, 243
Exponential integral 184
Exponential series 181

F
Fichtenholz, G.M. 232
Fields, J. 6, 12
Flajolet, P. 6, 130, 300
Fletcher, A. 96
Forrester, P.J. 226, 299
Fourier transform 223–224
Frank, E. 126
Fransén, A. 96
Frasch, H. 328
Frullani's theorem 162

G
Gamma function 5, 96, 172–175
  continued fractions, see Continued fractions
  logarithmic derivative 8, 43, 104
Gauss, C.F. 4, 7, 34, 36, 42, 50, 57, 92
Gauss's continued fraction 103, 134
Gauss's theorem 17, 25, 26, 36, 56, 84, 120
Generalized hypergeometric series 8
Gessel, I. 7, 15
Glaeske, H.-J. 276
Glaisher, J.W.L. 192, 262, 315, 327
Glasser, M.L. 6, 226
Goldberg, L. 305
Goldstein, L.J. 276
Gosper, R.W. 7
Gould, H.W. 69
Goulden, I.P. 126, 130
Graham, S. 304

Grosjean, C.C. 294, 329
Grosswald, E. 256, 261, 276
Guinand, A.P. 256, 262, 276

H
Hafner, J.L. 304
Halbritter, U. 276
Halphen, M. 327
Hansen, E.R. 69
Hardy, G.H. 2, 4, 7–9, 11, 15, 18, 20–
    22, 24, 39, 41, 50, 57–61, 63, 86,
    102–103, 145–146, 156, 168, 181,
    185, 190, 192, 214, 225, 240, 261–
    262, 284, 293, 295, 298, 304–305,
    333
Henrici, P. 221, 247–248
Herschfeld, A. 109
Hill, J. 242
Hill, M.J.M. 34
Hurwitz, A. 256, 262
Hypergeometric series 4, 7–102, 186,
    196, 245–248
  asymptotic formulas 12–15, 40–43,
    52, 70–77, 193–205
  balanced 13, 48, 70–77, 80
  confluent 186, 192–193, 202–205
  continued fractions 103, 112, 115–117,
    120, 131, 134–137, 139–140, 142–
    145, 164–165, 180
  differential equations 49, 87–92
  partial sums 11–14, 39, 41–42, 45–47
  products 48, 58–64
  quadratic transformations 48–51, 58,
    64, 92–99

I
Imperfect series 320
Inclusion–exclusion principle 28
Incomplete series 320
Infinite products 230–231, 241
Infinite radicals 108–112
Integrals 31–32, 54–56, 79–80, 166–
    171, 178–182, 185–207, 216–217,
    219–227, 229–237, 263–264, 278–
    279
Iseki, S. 276
Ismail, M.E.H. 6, 321, 323
Iteration of function 335–338

Ivić, A. 304
Iwaniec, H. 304

J
Jackson, D.M. 126, 130, 133
Jacobi, C.G.J. 9, 185
Jacobsen, L. 6, 104, 107, 109, 121–124,
    132, 137, 141, 146, 148, 156–157,
    159, 162, 164–165, 169
Jogdeo, K. 182
Jones, W.B. 105, 116, 121–122, 129,
    134
Jordan, W.B. 145

K
Kampé de Fériet, J. 8
Karlsson, P.W. 7
Katayama, K. 276–277
Khovanskii, A.N. 105, 134, 170
Kirschenhofer, P. 276
Klamkin, M.S. 182
Klein, F. 1, 8
Klusch, D. 299
Knopp, K. 268
Knuth, D.E. 182
Kolesnik, G. 304
Koshliakov, N.S. 262
Krishnamachari, C. 256, 262
Krishnan, K.S. 226
Kummer, E.E. 4, 7, 36–37, 42, 48–50,
    64, 92–93, 96, 117
Kummer's theorem 17, 21, 24

L
Lagrange, J. 253, 256, 261
Lagrange, J.L. 133
Lagrange inversion formula 100–102,
    198, 202
Laguerre, E. 132
Lambert, J.H. 133
Lamm, G. 40
Lamphere, R.L. 5–6, 243
Landau, E. 304
Landau's constant 40
Lavoie, J.L. 19
Lawden, D.F. 182

Lebedev, N.N. 186, 193, 202
Legendre, A.M. 165
Legendre polynomials 65–69
Lehmer, D.H. 330
Lerch, M. 276, 293
Lerch zeta-function 259
LeVeque, W.J. 330
Levitt, J. 329
Lewittes, J. 256
L-function 259, 276–277
Lindelöf, E. 221
Linear series 321
Ling, C.-B. 256, 262, 299
Littlewood, J.E. 305

M
Macmahon, P.A. 28
Madhava, K.B. 150
Malmsten's integral representation for
    Log $\Gamma(z)$ 162
Malurkar, S.L. 256, 261–263, 276–277,
    295
Marsaglia, J.C.W. 181
Masser, D.W. 6, 329
Masson, D. 141, 142, 145
Matala-Aho, T. 38
Mathematical Gazette 182, 295
Matsuoka, Y. 276
McCabe, J.H. 130
McKay, J.H. 108
Mellin, H. 295
Mikolás, M. 276
Mixed series 320
Modified theta-function 5, 314–317
Modular equation 301, 333–335, 338
Modular form 327–329
Modular group 318
Morley, F. 25
Mozzochi, C.J. 304
Muir, T. 129
Multiplier for modular equation 334

N
Nagasaka, C. 276
Nanjundiah, T.S. 261–263, 277, 294–
    295
Narasimhan, R. 262, 276

Newman, D.J. 182
Newton, I. 28
Nielsen, N. 32, 165, 181, 184, 206
Nörlund, N.E. 135, 156

O
Ogg, A. 320
Olivier, M. 162
Olver, F.W.J. 6, 104, 168, 184, 208,
    212–214, 216

P
Paris, R.B. 182
Parseval's theorem 186, 207, 224–225
Partial fraction decompositions 237, 241,
    248–249, 267–275, 277–279, 291–
    293, 314–315
Partition function 305
Perfect series 320
Perron, O. 105–106, 112, 120, 132–135,
    137, 141, 146, 148, 156, 159, 170
Pfaff, J.F. 9, 36, 93
Phillips, E.G. 294–295
Pincherle, S. 142–143
Poisson distribution, 182
Poisson summation formula 5, 225, 233,
    235, 238, 240, 252–253, 264, 316–
    317
Poisson summation formula for sine
    transforms 236, 257, 265, 289
Pollak, H.O. 217
Pollard, H. 226
Polya, G. 109
Ponnuswamy, S. 126, 130
Preece, C.T. 121, 146, 156, 293, 295
Primes 304
Prodinger, H. 276
Pure series 320
Putnam, William Lowell 108

R
Rademacher, H. 305
Ramamani, V. 329
Ramanathan, K.G. 134, 137, 141, 146
Rankin, R.A. 318–320, 326, 328, 332
Rao, M.B. 253, 256, 261–263

Rao, S.N. 276
Razar, M.J. 276
Residue for double pole 324
Residue for triple pole 325
Riemann zeta-function 30–31, 150, 153, 155, 173, 276, 301, 306–314
Riesel, H. 262, 276, 295, 305
Riordan, J. 34
Rogers, L.J. 11, 121, 125, 127, 129, 148, 150, 151, 163
Roy, R. 7, 225

S
Saalschütz, L. 4, 7, 9, 15, 99, 187
St. Paul 1
Samuels, S.M. 182
Sandham, H.F. 256, 262, 293–295
Sayer, F.P. 293–295
Schläfli, L. 59
Schlömilch, O. 253, 256
Schoeneberg, B. 320
Self-repeating series 336
Serre, J.-P. 320
Shepp, L. 217
Sitaramachandrarao, R. 6, 256, 261, 271, 276, 293, 297
Sizer, W.S. 109
Slater, L. 8
Smart, J.R. 276, 293–294
Srivastava, H.M. 61
Stanton, D. 7, 15, 48, 70, 77, 102
Stieltjes, T.J. 141, 149–151, 154, 156, 163
Szabo, A. 40
Szegö, G. 109, 181–182, 184

T
Terras, A. 276
Theory of indices 320
Theta-function 253, 301, 314, 322–323
Theta-function reciprocal 305
Thomae, J. 7, 39
Thron, W.J. 105, 116, 121–122, 129, 134
Titchmarsh, E.C. 189, 191, 223–225, 235, 257, 289, 301, 304, 306–307, 311, 317

Toyoizumi, M. 276
Tricomi, F.G. 44
Trinity College, 5
Tschebyscheff, P. 169
Tschebyscheff polynomials 99

U
University of Illinois 1

V
Väänänen, K. 38
Vandermonde's theorem 36–37
Vaughn, J. 6
Venkatachaliengar, K. 319
Vijayaraghavan, T. 108–109
Vivekananda 1

W
Waadeland, H. 105, 107
Wall, H.S. 105–106, 112, 134
Wallis, J. 106
Wallis's product 241
Wang, T.H. 232
Watson, G.N. 5, 39–41, 43, 47, 51–52, 59, 139, 157, 163–164, 181, 184, 193–195, 214, 256, 262, 293, 295, 297–298, 305
Watson's lemma 203–204, 212–213
Weierstrass P-function 330–333
Whipple, F.J.W. 7, 12, 16, 41
Wigert, S. 311
Williamson, B. 102
Wilson, B.M. 5
Wilson, J.A. 11, 86–87, 219–220, 225
Worpitzky, J. 112
Wrigge, S. 96
Wright, E.M. 304

Z
Zagier, D. 6; 283, 286
Zhang, N. 276
Zhang, S. 276
Zippin, L. 109
Zucker, I.J. 262, 295, 299
Zuckerman, H.S. 305

Printed in the United States
120495LV00002B/16/A

9 780387 967943